KNOWLEDGE SERVICE ENGINEERING HANDBOOK

CRC Press
Taylor & Francis Group
Boca Raton   London   New York

CRC Press is an imprint of the
Taylor & Francis Group, an **informa** business

Ergonomics Design and Management: Theory and Applications

Series Editor
Waldemar Karwowski

Industrial Engineering and Management Systems
University of Central Florida (UCF) – Orlando, Florida

Published Titles

Ergonomics: Foundational Principles, Applications, and Technologies
Pamela McCauley Bush

Aircraft Interior Comfort and Design
Peter Vink and Klaus Brauer

Ergonomics and Psychology: Developments in Theory and Practice
Olexiy Ya Chebykin, Gregory Z. Bedny, and Waldemar Karwowski

Ergonomics in Developing Regions: Needs and Applications
Patricia A. Scott

Handbook of Human Factors in Consumer Product Design, 2 vol. set
Waldemar Karwowski, Marcelo M. Soares, and Neville A. Stanton

> Volume I: Methods and Techniques
> Volume II: Uses and Applications

Human–Computer Interaction and Operators' Performance: Optimizing Work
Design with Activity Theory
Gregory Z. Bedny and Waldemar Karwowski

Knowledge Service Engineering Handbook
Jussi Kantola and Waldemar Karwowski

Trust Management in Virtual Organizations: A Human Factors Perspective
Wiesław M. Grudzewski, Irena K. Hejduk, Anna Sankowska, and Monika Wańtuchowicz

Forthcoming Titles

Manual Lifting: A Guide to the Study of Simple and Complex Lifting Tasks
Daniela Colombiani, Enrico Ochipinti, Enrique Alvarez-Casado, and Thomas R. Waters

Neuroadaptive Systems: Theory and Applications
Magalena Fafrowicz, Tadeusz Marek, Waldemar Karwowski, and Dylan Schmorrow

Organizational Resource Management: Theories, Methodologies, and Applications
Jussi Kantola

KNOWLEDGE SERVICE ENGINEERING HANDBOOK

Edited by
Jussi Kantola • Waldemar Karwowski

CRC Press
Taylor & Francis Group
Boca Raton London New York

CRC Press is an imprint of the
Taylor & Francis Group, an **informa** business

CRC Press
Taylor & Francis Group
6000 Broken Sound Parkway NW, Suite 300
Boca Raton, FL 33487-2742

First issued in paperback 2017

Version Date: 20120409

ISBN 13: 978-1-4398-5294-1 (hbk)
ISBN 13: 978-1-138-07194-0 (pbk)

Library of Congress Cataloging-in-Publication Data

Knowledge service engineering handbook / editors, Jussi Kantola, Waldemar Karwowski.
    p. cm. -- (Ergonomics design and management : theory and applications)
    Includes bibliographical references and index.
    ISBN 978-1-4398-5294-1 (hardback)
    1. Expert systems (Computer science) 2. Information technology. 3. Knowledge management. I. Kantola, Jussi. II. Karwowski, Waldemar, 1953-

    QA76.76.E95K5794 2012
    006.3'3--dc23   2012009448

Visit the Taylor & Francis Web site at
http://www.taylorandfrancis.com

and the CRC Press Web site at
http://www.crcpress.com

Contents

PART I Introduction to Knowledge Service Engineering

PART II Engineering from Data, Information, and Knowledge toward Services

PART III Human Networks in Knowledge Services

PART IV High-Performance Knowledge
Service Systems

PART IV High-Performance Knowledge Service Systems

Preface

Knowledge service engineering is an emerging field in the scientific and application worlds, focusing on the joint systems of data networks, information networks, and human knowledge networks. It aims at acquiring and utilizing data, information, and human knowledge to produce high-performance joint knowledge services to support the knowledge economy of the twenty-first century.

Knowledge service engineering provides practical knowledge as a service to citizens, end users, industrial customers, companies, organizations, and governments. This new subdiscipline aims at developing and maintaining sustainable knowledge services globally. Acquiring and utilizing data, information, and human knowledge networks require different types of engineering methods, which inspire, in many exciting ways, the creation of sustainable knowledge services for the future.

The aim of this handbook is to present the recent advances in knowledge service engineering by accomplished researchers and practitioners from around the world. We hope that it will be helpful to researchers and students in the field, as well as to professionals who develop a variety of innovative knowledge services. We project that many college students will become knowledge service professionals shortly after completing their studies.

We thank all the authors who have contributed to this handbook for the outstanding job they have done in writing their respective chapters. We also thank the staff at CRC Press, who gave us the opportunity to publish this handbook. Cindy Carelli and Jessica Vakili, the senior acquiring editor and the production coordinator, respectively, at CRC Press, and Deepa Kalaichelvan, the project manager at SPi Global, India, have been extremely helpful throughout the editing process. We would also like to express our gratitude to our universities for their support in this project, namely, Korea Advanced Institute of Science and Technology (KAIST), South Korea, and the University of Central Florida (UCF), Florida.

The book is organized into four parts. Part I (Chapters 1 through 5) introduces the main concepts of knowledge services. Part II (Chapters 6 through 13) explores data-, information-, and knowledge-based engineering methods and applications that can be used to develop knowledge services. Part III (Chapters 14 through 17) discusses the importance of human networks in knowledge services. Finally, Part IV (Chapters 18 through 21) provides a description of high-performance knowledge service systems.

We hope that you find the contents of this book interesting and useful in your own work.

Jussi I. Kantola
Korea Advanced Institute of Science and Technology (KAIST)
Daejeon, Republic of Korea

and

Waldemar Karwowski
University of Central Florida
Orlando, Florida

For MATLAB® and Simulink® product information, please contact:

The MathWorks, Inc.
3 Apple Hill Drive
Natick, MA, 01760-2098 USA
Tel: 508-647-7000
Fax: 508-647-7001
E-mail: info@mathworks.com
Web: www.mathworks.com

Editors

Jussi I. Kantola is an associate professor in the Department of Knowledge Service Engineering at Korea Advanced Institute of Science and Technology (KAIST). From 2003 to 2008, he worked in Tampere University of Technology and the University of Turku in various research roles, including research director in the IE department and the IT department. Dr. Kantola received his first PhD in industrial engineering from the University of Louisville, Kentucky, in 1998, and his second PhD in industrial management and engineering from Tampere University of Technology, Finland, in 2006. From 1999 to 2002, he worked as an IT and business and process consultant in the United States and in Finland. His current research interests include ontological engineering and soft-computing applications for industrial and knowledge management.

Waldemar Karwowski, PhD, DSc, PE, currently serves as professor and chairman, Department of Industrial Engineering and Management Systems, and executive director of the Institute for Advanced Systems Engineering at the University of Central Florida. He received his MS (1978) in production engineering and management from the Technical University of Wroclaw, Poland, and his PhD (1982) in industrial engineering from Texas Tech University. He then received his DSc (2004) in management science from the State Institute for Organization and Management in Industry, Poland. He also has received honorary doctorate degrees from three European universities. Dr. Karwowski is the author or coauthor of over 400 scientific publications in the area of human factors, human system integration, soft computing, and industrial engineering and is a certified professional ergonomist (BCPE). He was named the J. B. Speed School of Engineering Alumni Scholar for Research, University of Louisville (2004–2006). He currently serves on the Committee on Human Systems Integration, National Research Council of the National Academies, United States (2007–2012). He also serves as an editor of the *Human Factors and Ergonomics in Manufacturing* journal (John Wiley & Sons, New York) and as an editor in chief of *Theoretical Issues in Ergonomics Science* journal (Taylor & Francis Ltd., London). Dr. Karwowski has served as president of the International Ergonomics Association (2000–2003) and of the Human Factors and Ergonomics Society, United States (2006–2007).

Contributors

Tareq Z. Ahram, PhD, is the lead researcher working at the Institute for Advanced Systems Engineering (IASE) at the University of Central Florida. He received his MSc in human engineering from the University of Jordan (2004), MSc in engineering management (2006) from the University of Central Florida, and PhD in industrial engineering (2008) from the University of Central Florida, with specialization in human systems integration and large-scale information retrieval systems (search algorithms). Tareq has served as an invited speaker at several systems engineering and HSI research meetings and workshops and served as program committee member and special speaker at the Department of Defense Human Systems Integration and Human Factors Engineering Technical Advisory Group—meeting 62, 63, and 64 held at NASA Ames—Human Systems Integration Division.

Melissa Archpru Akaka is a PhD candidate in marketing at the University of Hawaii at Manoa. Her research interests include value cocreation, innovation, and cross-cultural issues. Prior to her graduate and doctoral studies, Akaka worked in the retail and travel industries as well as the not-for-profit sector. She received her MBA and BBA in marketing, both from the University of Hawaii at Manoa.

Xavier Amores-Bravo is an associate lecturer of management and business administration at the University of Barcelona (UB) and the University of Girona (UdG). He also collaborates with the Business School EADA as lecturer in its MBA program. Amores-Bravo is an engineer and graduated in the master's program "Innovation, Information Technologies and Organisations" at the UdG. He has worked for the regional government of Catalonia between 2005 and 2010, promoting innovation and regional economic development. Prior to this, he was a consultant for innovation projects, working for companies and also for Spanish and South American public administrations (promotion of innovation, entrepreneurship, and science park development). His research interests include innovation management and, particularly, innovation in knowledge-intensive business services (KIBS). His PhD project focuses on the management of innovation in Catalan Technology KIBS. Amores-Bravo has been involved in some research projects for the regional government of Catalonia, concerning innovation in SMEs and services. His most recent publication focused on promoting innovation in services, especially in Catalan SMEs (2007).

Abdul Quaiyum Ansari works with the Department of Electrical Engineering at Jamia Millia Islamia (a central university by an Act of Parliament), New Delhi. He has also served as professor and head, Department of Computer Science, and as dean, Faculty of Management Studies and Information Technology, at Jamia Hamdard (Hamdard University), New Delhi. Professor Ansari received his BTech (electrical, low current) from AMU, Aligarh, and his MTech (IEC) and PhD from IIT Delhi and

JMI, New Delhi, respectively. His research interests are in the areas of computer networks, networks-on-chip, image processing, and fuzzy logic. He has published about 80 research papers in international and national journals and proceedings of conferences. Professor Ansari is a fellow of the IETE, IE (I), and the National Telematics Forum (NTF). He is a senior member of the IEEE (United States) and the Computer Society of India, and a life member of the ISTE, ISCA, and the National Association of Computer Educators and Trainers (NACET). He is also presently the chairman of the Delhi chapter of the IEEE Computational Intelligence Society.

Barbro Back is a professor in accounting information systems at Åbo Akademi University in Turku, Finland. Her research interests include accounting information systems, business intelligence, neural networks, data mining, and text mining. She has published her research in the *Journal of Management Information Systems, Accounting Management and Information Technology*, the *European Journal of Operations Research*, the *International Journal in Intelligent Systems in Accounting, Finance and Management, Advances in Accounting*, and other journals. She currently serves on the editorial boards of the *International Journal of Business Information Systems*, the *Journal of Emerging Technologies in Accounting*, and the *International Journal of Digital Accounting Research*.

Kun Bai is a research staff member in IBM T.J. Watson Research Center, Hawthorne, New York. Before joining IBM, he worked with Professor Peng Liu at the Cyber Security Lab, The Pennsylvania State University, University Park, Pennsylvania. His research interests include computer systems and networks, mobile service design, data privacy, database security, data mining, information retrieval, information extraction, and document analysis. Kun has published dozens of scientific papers in high-ranked international conferences and served as TPC member and chair for multiple international conferences and journals.

Meeyoung Cha is an assistant professor at the Graduate School of Culture Technology in KAIST. She received her PhD in computer science from KAIST in 2008. She was a postdoctral researcher at Max Planck Institute for Software Systems (MPI-SWS) in Germany from 2008 to 2010. Meeyoung's research interests include the analysis of large-scale online social networks. She received the best paper award from Usenix/ACM SIGCOMM Internet Measurement Conference 2007 for her work on YouTube. Her work on user influence in Twitter has been featured on *New York Times* websites and on *Harvard Business Review*'s research blog in 2010.

Yoon Seok Chang is a professor of the School of Air Transport, Transportation and Logistics at Korea Aerospace University. He is also the director of the Ubiquitous Technology Application Research Center (UTAC) at Korea Aerospace University. He was a senior research associate at Auto-ID Center, Cambridge University, United Kingdom, and a senior application engineer at i2, United States. He received his PhD from Imperial College London, United Kingdom, in 1997. Chang's research interests include RFID and sensor network, air cargo management, SCM/SCE, and product lifecycle management.

Yuh-Jen Chen is currently an associate professor in the Department of Accounting and Information Systems, National Kaohsiung First University of Science and Technology, Taiwan, Republic of China. He received his PhD and MS from the Institute of Manufacturing Information and Systems at the National Cheng Kung University in 2005 and 2001, respectively, and his BS from the Department of Mathematics of Chung Yuan Christian University, Taiwan, Republic of China, in 1999. His current research interests include enterprise information systems, knowledge engineering and management, and service science.

Ashraf Darwish received his PhD in computer science from Saint Petersburg State University, Russian Federation, in 2006 and is currently an assistant professor at the Faculty of Science, Helwan University, Egypt. Dr. Darwish teaches artificial intelligence, information security, data and web mining, intelligent computing, image processing (in particular, image retrieval and medical imaging), modeling and simulation, intelligent environment, and body sensor networking. He is an editor and member of many computing associations such as IEEE, ACM, EMS, QAAP, IJCIT, IJSIEC, IJIIP, IJITNA, IJCISIM, SMC, and Quality Assurance and Accreditation Authority (Egypt), and a board member of the Russian–Egyptian Association for graduates and the Machine Intelligence Research Lab (MIR Lab), United States. Dr. Darwish is the author of many scientific publications, and his publications include papers, abstracts, and book chapters by Springer and IGI publishers. He keeps in touch with his mathematical background through his research. His research, teaching, and consulting mainly focus on artificial intelligence, information security, data and web mining, intelligent computing, image processing, modeling and simulation, intelligent environment, and body sensor networks. Dr. Darwish has a wealth of academic experience and has authored multiple publications including papers and book chapters on computational intelligence. His publications have appeared in international journals, conferences, and publishing houses such as IEEE Systems, Man, & Cybernetics Society (SMC) on Soft Computing, Springer, and *Advances in Computer Science and Engineering*. Dr. Darwish has been invited as reviewer, invited speaker, organizer, and session chair to many international conferences and journals. He has worked in a wide variety of academic organizations and supervises many master's and PhD theses in different areas of computer science. He speaks four languages on a daily basis. Currently, he is working in the Ministry of Foreign Affairs as cultural and educational attaché of Egypt in Kazakhstan.

Tomas Eklund is a researcher and docent (adjunct professor) at Åbo Akademi University in Turku, Finland. He received his doctoral degree from Åbo Akademi University in 2004. His research interests include data mining, self-organizing maps, financial benchmarking and performance analysis, text mining, and customer relationship management. His publications have appeared in *Information & Management, Information Visualization*, and the *International Journal of Intelligent Systems in Accounting, Finance and Management*. His primary research interests are centered around the combination of data and text mining methods for financial analysis.

Annika H. Holmbom is a PhD student in information systems in the Department of Information Technologies at Åbo Akademi University. Her main research interest is in customer segmentation using data mining methods.

Kari Ingman has been plant manager at Lojer Works Oy since 2009 and is responsible for production, R&D, and service. From 2004 to 2009, he worked for Fastems Oy as robotics export sales manager as well as manager for the sales support team. From 1998 to 2004, he worked for Rocla Robotruck Oy, with responsibility for global sales and project management of automated guided vehicle (AGV) systems. Kari received his MSc from Tampere University of Technology (TUT), Finland for Teco-Engineering Oy, where he continued until 1998 as production manager. Currently, he is pursuing his PhD thesis on sales culture at TUT, Finland.

Bang Chul Jung received his BS in electronics engineering from Ajou University, Suwon, Korea, in 2002, and his MS and PhD in electrical and computer engineering from the Korea Advanced Institute of Science and Technology (KAIST), Daejeon, Korea, in 2004 and 2008, respectively. He was a senior researcher/research professor with KAIST Institute for Information Technology Convergence, Daejeon, Korea, from January 2009 to February 2010. He currently serves as assistant professor in the Department of Information and Communication Engineering, Gyeonsang National University, Korea. Dr. Jung received the Fifth IEEE Communication Society Asia-Pacific Outstanding Young Researcher Award in 2011. His research interests include cognitive radio, cooperative relaying techniques, compressed sensing, physical-layer security, interference alignment, and machine-to-machine (M2M) communication.

Jussi I. Kantola is an associate professor in the Department of Knowledge Service Engineering at Korea Advanced Institute of Science and Technology (KAIST). From 2003 to 2008, he worked in Tampere University of Technology and the University of Turku in various research roles, including research director in the IE and IT departments. He received his first PhD in industrial engineering from the University of Louisville, Kentucky, in 1998, and his second PhD in industrial management and engineering from Tampere University of Technology, Finland, in 2006. From 1999 to 2002, he worked as an IT and business and process consultant in the United States and in Finland. His current research interests include ontological engineering and soft-computing applications for industrial and knowledge management.

Waldemar Karwowski, PhD, DSc, PE, currently serves as professor and chairman, Department of Industrial Engineering and Management Systems, and executive director of the Institute for Advanced Systems Engineering at the University of Central Florida. He received his MS (1978) in production engineering and management from the Technical University of Wroclaw, Poland, and his PhD (1982) in industrial engineering from Texas Tech University. He then received his DSc (2004) in management science from the State Institute for Organization and Management in Industry, Poland. He also has received honorary doctorate degrees from three European universities. Dr. Karwowski is the author or coauthor of over 400 scientific publications in the areas of human factors, human system integration, soft computing, and industrial engineering, and is a certified professional ergonomist (BCPE). He was named the J. B. Speed School of Engineering Alumni Scholar for Research, University of Louisville (2004–2006). He currently serves on the Committee on Human Systems Integration, National Research Council of the National Academies, United States (2007–2011). He also serves as an editor of the *Human Factors and Ergonomics in Manufacturing* journal (John Wiley & Sons, New York) and as an editor in chief of *Theoretical Issues in Ergonomics Science* journal (Taylor & Francis Ltd., London). Dr. Karwowski has served both as president of the International Ergonomics Association (2000–2003) and president of the Human Factors and Ergonomics Society, United States (2006–2007).

M. Ayoub Khan works with the Center for Development of Advanced Computing (Ministry of Communication and IT), Government of India, as a scientist, with interests in CAD VLSI design tools and techniques, hardware–software code signs, VLSI (electronic design automation, circuit optimization, timing analysis), placement and routing in network-on-chip (buffer management, interconnects and port design), etc. He has more than seven years of experience in his research area and has published more than 30 papers in reputed journals and international IEEE/Springer conferences. He also contributes to the research community through various other volunteer activities. Khan has served as conference chair in various reputed international conferences like ICMLC 2010, ICSEM 2010, ICRTBAIP 2010, and ICIII 2010, to name a few. He is member of the professional bodies of IEEE, ISTE, IACSIT, ACEE, and IAENG. He is also a member of the editorial/reviewer boards for *IEEE Communications Letters*, *IEEE Transaction of Industrial Informatics*, Springer *CCSP*, and Elsevier *Computer and Electrical*.

Konstadinos Kutsikos is an assistant professor of information management in the Business School at the University of the Aegean (UotA), Greece. He received his MBA and PhD in computer science from the University of Southern California and his MS in computer science from Stanford University, Stanford, California. He has published numerous business and academic papers and is a frequent speaker at major conferences and industry events. Before joining UotA, Dr. Kutsikos had a 10 year international career balancing an academic, industry, and entrepreneurial background, including a managing consultant position with PA Consulting Group in London. His current research work is focused on business systems design, with particular emphasis on service science and innovation management.

Howon Lee is a professor in the Department of Electrical, Electronic and Control Engineering at Hankyong National University (HKNU). He received his BS, MS, and PhD in electrical engineering from the Korea Advanced Institute of Science and Technology (KAIST) in 2003, 2005, and 2009, respectively. Before joining HKNU, he was a team leader of the knowledge convergence team at KAIST Institute for Information Technology (IT) Convergence until 2012. His current research interests include knowledge communications, IT convergence, cross-layer radio resource management, cognitive radio, Voice over IP (VoIP) service, wireless communications, compressed sensing, and wireless power transfer. He is also the recipient of the JCCI 2006 Best Paper Award and the Intel Student Paper Contest Bronze Prize in 2006.

Jae-Gil Lee received his PhD from the Korea Advanced Institute of Science and Technology (KAIST) in 2005. He has previously served at the IBM Almaden Research Center, working on design and development of a data warehousing product. He currently serves as an assistant professor in the Department of Knowledge Service Engineering at KAIST. Dr. Lee was a postdoctoral fellow in the Department of Computer Science, University of Illinois at Urbana-Champaign. His primary research interests include data mining, data warehousing, and database systems. His trajectory mining papers published at SIGMOD, VLDB, and ICDE have been actively cited in recent years. He received the Best Demonstration Award at ICDE'05 and served as the publicity chair of CIKM'09.

Uichin Lee is an assistant professor in the Department of Knowledge Service Engineering at the Korea Advanced Institute of Science and Technology (KAIST). He received his BS in computer engineering from Chonbuk National University in 2001, his MS in computer science from KAIST in 2003, and his PhD in computer science from the University of California at Los Angeles (UCLA) in 2008. Before joining KAIST, he was a member of the technical staff at Bell Laboratories, Alcatel-Lucent, until 2010. His research interests include distributed systems and mobile/pervasive computing.

Meira Levy is a senior lecturer in the Department of Industrial Engineering and Management, Shenkar College of Engineering and Design, Israel, having completed a postdoctoral fellowship in the Department of Industrial Engineering and Management and at Deutsche Telekom Laboratories, Ben-Gurion University of the Negev, Israel. She received her PhD, MSc, and BSc from the Technion, the Israel Institute of Technology. Dr. Levy has extensive experience in the software engineering industry in technical and management positions. Her research interests include distance learning; knowledge engineering, and management, both from human and technological perspectives, including KM audit and requirements analysis methodologies, modeling and design of knowledge systems; embedding KM frameworks within business processes; and identifying KM culture barriers and knowledge representation. Her research papers have been published in conference proceedings and in *Decision Support Systems*, the *Journal of Knowledge Management*, and the *Journal of Information Systems Education*.

Hongyan Liu is a doctoral student at Åbo Akademi University and a research assistant at Turku Centre for Computer Science (TUCS), Turku, Finland. She received her MSc in economics and business administration from Åbo Akademi University in 2008, and her BSc in industrial safety engineering from Capital University of Economics and Business, Beijing, China, in 1992. She has worked in both joint-venture and state-owned companies in China in various capacities, including business development, human resources, customer service, and financial management. Her research interests include business intelligence, data mining, self-organizing maps, and computational intelligence. Her current research is focused on customer consumption behavior profiling and pricing strategy modeling in the development of the demand response electricity retail market.

Ying Liu is an assistant professor in the Department of Knowledge Service Engineering, Korea Advanced Institute of Science and Technology (KAIST). Prior to this, she worked with Professor C. Lee Giles and Professor Prasenjit Mitra at the Intelligent Information Systems Research Laboratory, The Pennsylvania State University, University Park, Pennsylvania. Her research interests include search engine designing, data mining, structured and semistructured data analysis, information retrieval, knowledge extraction, service designing in different areas such as digital library, mobile communication, healthcare and bioinformatics, etc. Ying has published more than 20 scientific papers in prestigious international conferences and served as a technical committee member and reviewer for multiple international conferences and journals.

Torsti Loikkanen, MSc (economics), senior research scientist, works as a research coordinator for innovation policy studies at VTT Technical Research Centre of Finland. His recent research and consulting activities concern enterprise innovation, evaluation, and assessing the impact of innovation policies and policy organizations, technology foresight, and sustainable development. His recent publications deal with the challenges posed by globalization to national and European innovation policies, to the integration of industrial innovation and sustainable development, to the foresight of energy and environmental technologies, and to the performance indicators and competitiveness of national knowledge economies and innovation systems. At the European level, Loikkanen has been engaged in future-oriented studies for the European Commission and European Parliament through the European Techno-Economic Policy Support (ETEPS) network. He has been a keynote speaker and lecturer at international conferences and events. He has international research and consulting experience in Australia, Africa, and Europe, and is also a member of the steering committee of the international innovation policy network Six Countries Programme. Loikkanen is VTT coordinator in the Joint Institute for Innovation Policy (JIIP) and an invited member of The Finnish Academies of Technology (FACTE).

Robert F. Lusch received his PhD from the University of Wisconsin. He is the executive director of the McGuire Center for Entrepreneurship and the Pamela and James Muzzy Chair in Entrepreneurship in the Eller College of Management at the University of Arizona. He has previously served as dean of the M. J. Neeley School of Business at Texas Christian University and was on the faculty of the University of Oklahoma for 25 years, where he also served as dean. Professor Lusch has served as chairperson of the American Marketing Association and as trustee of the American Marketing Association Foundation. He also has expertise in the areas of service-dominant logic, marketing strategy, marketing theory, and marketing channels. Professor Lusch has served as editor of the *Journal of Marketing* and is the author of 125 scholarly articles in outlets such as the *Journal of Consumer Research*, the *Journal of Marketing*, the *Journal of Marketing Research*, the *Journal of Retailing* and *Behavioral Science*, as well as many other journals. In 1997, the Academy of Marketing Science awarded him the Distinguished Marketing Educator Award, and the American Marketing Association presented him the AMA Harold Maynard Award for contributions to marketing theory in 1997 and 2005. In 2001, he received the Louis W. Stern Award from the American Marketing Association for outstanding contributions to the marketing channels literature. The National Association of Accountants awarded him Lybrand's Bronze Medal for contributions to the accounting literature in 1978. The Marketing Management Association awarded him the Lifetime Contributions to Marketing Award in 2006, and in 2009, he was awarded the AMA IOSIG Lifetime Achievement Award for his research and contributions in the marketing channels literature.

Camilla Magnusson received her PhD in strategic management from Tampere University of Technology and her MA in general linguistics from the University of Helsinki. Her research has appeared in journals such as *Information and Management* and *Marketing Intelligence and Planning*. Dr. Magnusson's research interests include strategic management, design science in information systems, and text visualization.

Gregoris Mentzas is a professor of management information systems at the School of Electrical and Computer Engineering of the National Technical University of Athens (NTUA) and director of the Information Management Unit (IMU), a multidisciplinary research unit at the university. From 2006 to 2009, he served as a member of the board of directors of the Institute of Communication and Computer Systems of NTUA. Mentzas has received numerous contracts, grants, and awards from funding bodies and is or has been the principal investigator in more than 40 research projects in the areas of electronic business and e-government, social computing, knowledge management, and semantic web. He is also the coeditor of the book *Semantic Enterprise Application Integration for Business Processes: Service-Oriented Frameworks* (IGI Global, 2009) and the lead author of the book *Knowledge Asset Management: Beyond the Process-Centred and Product-Centred Approaches* (Springer, 2002). He has published more than 200 papers in international peer-reviewed journals and conferences and received ANBAR citations of excellence for his research.

Nick Milton is the chief knowledge architect at Tacit Connexions, Berkshire, England. He is the author of the books *Knowledge Acquisition in Practice* and *Knowledge Technologies*. He received his BSc in electronics from Southampton University (1981), his BA in psychology from the University of Nottingham (1996), and his PhD in psychology from the University of Nottingham (2003). He worked at the University of Nottingham on the application of artificial intelligence techniques to knowledge management during the 1990s. He also worked as a consultant knowledge engineer on projects that developed expert systems for military applications. In the late 1990s, Nick joined a leading-edge knowledge engineering company as chief knowledge engineer. Since then, he has worked on numerous knowledge acquisition projects and helped establish knowledge management programs at large organizations in the engineering, technology, and legal sectors. He plays an active role in solving real-world problems with knowledge systems and in creating new methods and tools to help capture, model, and use knowledge. Nick maintains strong links with leading-edge researchers in knowledge technologies, such as knowledge-based engineering, ontologies, and semantic webs.

Hye-Jin Min is a PhD candidate in the Department of Computer Science at the Korea Advanced Institute of Science and Technology. Her research interests include natural language processing, computational linguistics, sentiment analysis, emotion recognition form text, and human–robot interaction. She received her BE and MSE in computer science from Kyungpook National University and the Korea Advanced Institute of Science and Technology, respectively.

Anssi Neuvonen received his MA from Helsinki University in 1987. He has been working at VTT in systems design, application development, information architecture, and knowledge management since 1989. He is active in tutoring and lectures on various ECM, EIM, and KM issues. Neuvonen is the leader for a team developing new solutions for enterprise information management at VTT. His current focus is on various in-house development initiatives relating to enterprise search, enterprise taxonomy and ontology, and records and case management.

Vesa A. Niskanen serves as a university lecturer and an adjunct professor (docent) in the Department of Economics and Management at the University of Helsinki, Finland. His research activities include computational intelligence, in particular, fuzzy systems, philosophy of science, and methodology of the human sciences. Dr. Niskanen chairs two special interest groups in the Berkeley Initiative in Soft Computing (BISC) community at the University of California, Berkeley. He has served as a secretary in the International Fuzzy Systems Association (IFSA) for two terms.

Chang Heun Oh is a researcher at Ubiquitous Technology Application Research Center (UTAC) at Korea Aerospace University. His research interests include RFID middleware and product lifecycle management.

Michael P. Papazoglou holds the chair of computer science at Tilburg University, where he is also the executive director of the European Research Institute in Service Science and the scientific director of the European Network of Excellence in Software Services and Systems (S-Cube). His research interests include service-oriented computing, cloud computing, distributed computing, business process management, and large-scale data sharing. He has published over 250 papers and has 22 authored/coedited books to his credit. He has approximately 8500 citations for his work with an H-index of 38. Dr. Papazoglou is the coeditor-in-charge of the new Springer-Verlag series in services science and the MIT Press series on information systems. He is also the adviser to the EU FP-7 program. Dr. Papazoglou is frequently invited to give keynote talks and tutorials on service-oriented computing around the world.

Jong C. Park received his BE and MSE in computer engineering from Seoul National University, Korea, and his PhD in computer and information science from The University of Pennsylvania. His research interests include computational linguistics (syntax–semantics interface) and natural language processing (biomedical text mining and management, language processing for quality of life technologies, and user-customized language interface for the web and human–robot interaction). Dr. Park serves as the founding editor in chief of the *Journal of Computing Science and*

Engineering (JCSE) and the founding general chair of the International Symposium on Languages in Biology and Medicine (LBM). He is a also a member of Sigma Xi, Association for Computational Linguistics (ACL), and KIISE. Dr. Park is an associate professor in the Department of Computer Science and an adjunct professor in the Department of Knowledge Service Engineering, both at Korea Advanced Institute of Science and Technology (KAIST).

Minsu Park is an MS candidate in Graduate School of Culture Technology (GSCT) at Korea Advanced Institute of Science and Technology (KAIST), Daejeon, Republic of Korea. He received his BE in mechanical system design and engineering from Hong-ik University in 2010. His current research interests include the analysis of captured moods of users from large-scale online social networks.

Lior Rokach is a senior lecturer in the Department of Information System Engineering at Ben-Gurion University and in the software engineering program. He is an expert in intelligent information systems and has held several leading positions in this field. His main areas of interest are data mining, pattern recognition, and information retrieval. Dr. Rokach regularly participates in program committees for conferences on data mining and recommender systems. He is the author of over 80 refereed papers in leading journals, conference proceedings, and book chapters. In addition, he has authored six books, including *Pattern Classification Using Ensemble Methods* (World Scientific Publishing, 2009), *Data Mining with Decision Trees* (World Scientific Publishing, 2007), and *Decomposition Methodology for Knowledge Discovery and Data Mining* (World Scientific Publishing, 2005). Dr. Rokach received his BSc, MSc, and PhD in industrial engineering from Tel Aviv University.

Aviv Segev is an assistant professor at the Knowledge Service Engineering Department at Korea Advanced Institute of Science and Technology (KAIST). His research interests include classifying knowledge using the web, mapping context to ontologies, knowledge mapping, and implementations of these areas as expert systems in the fields of web services, medicine, and crisis management. In 2004, he received his PhD in management information systems in the field of context recognition from Tel Aviv University. Aviv was previously a simulation project manager in the Israeli aircraft industry. During the past two years, he has been working on multilanguage context recognition, matching, and prediction algorithms using very large databases containing millions of patents. He has published over 30 papers in scientific journals and conferences.

Bracha Shapira is an associate professor in the Department of Information Systems Engineering at Ben Gurion University of the Negev; she also heads the department. Bracha leads a few research projects regarding the usability and security of mobile applications at the Deutsche Telekom Laboratories at Ben Gurion University. She received her PhD in information systems engineering and her MSc in computer science from the Hebrew University in Jerusalem. Her works on personalization, recommender systems, and privacy have been published in leading journals and conferences.

Peretz Shoval is a professor of information systems in the Department of Information Systems Engineering (ISE) of Ben Gurion University. He received his BA in economics and his MSc in information systems from Tel Aviv University, and his PhD in information systems from the University of Pittsburgh (1981), where he specialized in expert systems for information retrieval. Prior to moving to academia, Professor Shoval held professional and managerial positions in computer and software companies. In 1984, he joined Ben Gurion University, where he started and chaired the Information Systems Program in the Department of Industrial Engineering and Management, which later became the Department of ISE. Professor Shoval's research interests include mainly information systems analysis and design methods, database modeling, and information retrieval and filtering. He has published over 140 papers in journals, has edited books and conference proceedings, and has written several textbooks on systems analysis and design. Among other things, he has developed the ADISSA and FOOM methodologies for systems analysis and design, as well as tools for conceptual data modeling, view integration, and reverse database engineering. In 2009, Professor Shoval was selected by the ACM as distinguished scientist.

Albert J. Simard is the knowledge manager for Defense R&D Canada, Directorate of Knowledge and Information Management. He has coordinated the development of two award-winning, automated national forest fire information systems. Dr. Simard has also developed knowledge management projects, programs, and strategies for the Canadian Forest Service and Natural Resources Canada, including access to knowledge policy and a knowledge services framework for government S&T organizations. Internationally, he has developed strategic plans and frameworks for two global information networks related to disaster management and forestry. He recently led the development of a modeling framework for the Canadian Food Inspection Agency. Dr. Simard has published more than 200 scientific and management articles and has given 300 presentations on forest fires and knowledge management. He currently is developing a knowledge services agenda and architecture that adapt knowledge management to support the business strategy for Defence R&D Canada.

Min Gyu Son is a researcher at Ubiquitous Technology Application Research Center (UTAC) at Korea Aerospace University. His research interests include maintenance management and air cargo logistics.

Junehwa Song is a professor in the Department of Computer Science, Korea Advanced Institute of Science and Technology (KAIST), Daejeon, Korea. He is also a KAIST chair professor. Before joining KAIST, he worked at IBM T.J. Watson Research Center, Yorktown Heights, New York, from 1994 to 2000. He received his PhD in computer science from the University of Maryland at College Park in 1997. His research interests include mobile and pervasive computing systems, Internet systems technologies, cloud computing, multimedia services, etc.

Kirsi Tuominen, MSc (chemical engineering), MBA, has been working since 2003 as the head of Knowledge Solutions at VTT Technical Research Centre of Finland. Knowledge solutions' responsibility is to develop and offer innovative knowledge solutions and services to improve the pertinence, speed, and impact of research and innovation activities. Her previous professional experience includes heading the international technology transfer unit at Tekes, the Finnish Funding Agency for Technology and Innovation, where she was responsible for the Innovation Relay Centre (IRC), Finland. IRC is a European support network for the promotion of research, technology transfer, and innovation. Prior to this, she managed Tekes' overseas network that spanned 16 offices in Europe, North America, and the Far East. Tuominen has also worked in London, Boston, and Brussels in various technology cooperation activities. She was elected as the Knowledge Manager of the Year 2010 by the Association of Finnish Information Specialists. She currently represents VTT at the ICSTI (The International Council for Scientific and Technical Information) executive board.

Sapna Tyagi serves as an assistant professor at the Institute of Management Studies (a reputed B-School in the northern region), Ghaziabad, Uttar Pradesh, India. Her research interests include radio frequency identification system, data warehousing, knowledge discovery, data analysis, decision support, and the automatic extraction of knowledge from RFID data and OLAP. She has more than five years of experience in her research area. Tyagi also contributes to the research community through various volunteer activities (e.g., reviewing and editing). She is a member of professional bodies such as IEEE, IACSIT, and IAENG. She has recently coauthored a book titled *Radio Frequency Fundamental and Applications,* InTechWeb, Education and Publishing KG, Vienna, Austria.

Jaume Valls-Pasola is a professor of management and business administration at the University of Barcelona (UB). He was associate professor at the Universitat Politècnica de Catalunya until 1994, when he was appointed as professor by the Universitat de Girona. In October 2006, he joined the UB, where he is currently the director of the Business Administration and Management Department. Valls-Pasola's main research interest is in the field of innovation management. He has been involved in a number of international research projects (OCDE, FAST Monitor, TSER, etc.). He has also carried out many research projects for the regional government of Catalonia concerning innovation in SMEs and the analysis of the regional system of innovation. His recent research includes the analysis of R&D investments done by the largest companies in Catalonia and a project on success and failure factors of entrepreneurship activities. Since March 2007, Valls-Pasola has been the director of the entrepreneurship chair created at UB with the sponsorship of Banco de Santander. The chair coordinates the XEU network—a joint action of the nine regional public universities in order to promote entrepreneurship.

Hannu Vanharanta began his professional career in 1973 as technical assistant at the Turku office of the Finnish Ministry of Trade and Industry. From 1975 to 1992, he worked for Finnish international engineering companies—Jaakko Pöyry,

Rintekno, and Ekono—as process engineer, section manager, and leading consultant. His doctoral thesis was approved in 1995. In 1995–1996, he served as professor of business economics in the University of Joensuu. In 1996–1998, he served as purchasing and supply management professor in the Lappeenranta University of Technology. Since 1998, Professor Vanharanta has been teaching industrial management and engineering at Tampere University of Technology in Pori. His research interests include human resource management, knowledge management, strategic management, financial analysis, and decision support systems, with special emphasis on knowledge discovery and data mining. Many articles in financial analysis as well as text analysis belong to his research experience. The ultimate goal of his research is to synthesize quantitative and qualitative/linguistic data/information/knowledge for strategic decision making.

Luis M. Vaquero received his BSc (electronics), MSc (electrical engineering), and PhD (information technologies) from Universidad de Valladolid, and MSc (pharmacology) and PhD (medicine) from Universidad Complutense de Madrid. After graduating, he served as research associate in several U.S. centers and as a researcher and IPR manager for Telefonica. He was also a part-time assistant professor at Universidad Rey Juan Carlos. Luis recently joined Hewlett-Packard Labs. His research interests include distributed systems and their scalability as well as lifecycle management.

Stephen L. Vargo, PhD, is a Shidler Distinguished Professor and professor of marketing at the University of Hawai'i at Manoa. His primary research areas are marketing theory and thought and consumers' evaluative reference scales. His publications have appeared in the *Journal of Marketing*, the *Journal of the Academy of Marketing Science*, the *Journal of Service Research*, and other major marketing journals. He also serves on six editorial review boards, including the *Journal of Marketing*, the *Journal of the Academy of Marketing Science*, and the *Journal of Service Research*. Professor Vargo has been awarded the Harold H. Maynard Award by the American Marketing Association for "significant contribution to marketing theory and thought" and the Sheth Foundation Award for "long-term contributions to the field of marketing."

Heiko Wieland is a doctoral student in marketing at the University of Hawai'i at Manoa. His research interests include innovation, social networks, and service-dominant logic. Prior to his doctoral studies, Wieland held various management positions in the technology industry. In 1992, he received his BSc in electrical engineering from Bremen University of Applied Sciences and his MBA in marketing from the University of Pittsburg in 1998.

Meng-Sheng Wu is currently a PhD candidate in the Institute of Manufacturing Information and Systems, National Cheng Kung University, Taiwan, Republic of China. He received his BS from the Department of Mechanical Engineering, National Pingtung University of Science and Technology, Taiwan, Republic of China, in 2002, and his MS from the Institute of Manufacturing Information and Systems, National Cheng Kung University, Taiwan, Republic of China, in 2004. His current research interests include product knowledge management and workflow management.

Part I

Introduction to Knowledge Service Engineering

1 Knowledge Services
A Strategic Framework

Albert J. Simard

CONTENTS

Do what is required;
Harness what is real;
Find what is hidden; and
Engage what is mysterious.

1.1 INTRODUCTION

In the twenty-first century, success will gravitate to knowledge organizations. To function in the information society and the knowledge economy, a knowledge organization must create and manage its knowledge assets to maximize their amount, availability, and usefulness for accomplishing business objectives. To succeed in the twenty-first century, it must become a networked organization that facilitates knowledge sharing to leverage its value through reuse, to enable collaboration, and to mobilize it for addressing issues. To flourish in the twenty-first century, it must use knowledge to fulfill its mandate, extract intelligence from the environment, learn from experience, and adapt to evolving conditions. To lead in the twenty-first century, it must establish and support knowledge networks that create, manage, and use knowledge to achieve sectoral goals and objectives.

In addition to strategic outcomes, such as relevance in the information society and positioning in the knowledge economy, there are many organizational benefits of knowledge management (KM):

- The pace of change in society continues to accelerate, while resources for adapting to change remain fixed. KM facilitates keeping abreast of, interpreting, and responding to emerging trends.
- Knowledge and the capacity to create it are essential to success, relevance, and sustainability. KM focuses on knowledge as a core strategic resource.
- Knowledge is expensive to create. KM leverages the value of knowledge and reduces duplication through increased capturing, availability, sharing, and reuse.
- High employee turnover results in frequent inexperience in new assignments. KM mitigates the loss of productivity through effective transfer of knowledge from outgoing to incoming staff.
- In a dispersed and diverse organization, KM enhances coordination, collaboration, and synergy across large distances, business functions, and work processes.
- KM increases organizational efficiency and effectiveness by enhancing the productivity of knowledge work and facilitating knowledge flow from creation to application.

The breadth and depth of these goals require that KM be embedded at all levels of an organization and across the full range of its interests. It should rise from the depths of the organizational infrastructure to the peak of its business strategy. It should reach across the knowledge ecosystem of interactions with partner, client, and customer networks. When viewed strategically, the range of needs extends beyond traditional KM objectives of managing assets, enabling sharing, and facilitating collaboration. A knowledge services strategy should encompass the full breadth and depth of organizational interests.

There are two different but equally valid views of knowledge services:

1. Enterprise-wide corporate services that enable and support KM
2. Production and delivery of knowledge-based products and services to clients

Simard (2008) explained that these two views represent vertical and horizontal dimensions of running an organization and serving clients, respectively. These two dimensions are equally critical to organizational success. Without an organization, it is not possible to produce and deliver knowledge-based products and services for clients. Conversely, unless an organization serves its clients, it has no purpose or means to sustain itself. A knowledge services strategy must encompass both organizational dimensions.

A strategic framework for knowledge services provides a structural outline of all aspects of KM and the relationships among them. It provides a blueprint for building a knowledge organization. Alternatively, it provides a picture of what the KM puzzle will look like when all the pieces are assembled. It also provides a context that positions and guides the development of specific KM functionality. A knowledge services strategy

- Structures the breadth and depth of KM in the context of an organization's environment and business
- Describes the patterns and relationships that underlie KM
- Provides a sound basis for selecting KM practices, prioritizing work, and planning activities

1.2 MANAGEMENT LEVEL

Knowledge services can be structured with a two-dimensional framework. Figure 1.1 shows the first dimension—management level. It ranges from the depths of an organization's infrastructure (people, governance, processes, technology, and content) through KM (assets, sharing, social networks, and flow), to the heights of the organization's business (knowledge work and transfer).

This approach positions KM between an organization's infrastructure (e.g., finance or human resources) and knowledge work (producing products or delivering services) rather than having it lie on the surface as a poorly understood (expendable) function. The organization provides knowledge services (e.g., repositories, search engines, or collaboration sites) that enable KM to support knowledge work and knowledge transfer. This section begins by defining what we intend to manage.

1.2.1 DEFINING KNOWLEDGE

Knowledge spans all domains of human activity. From a human perspective, knowledge ranges from cognition, through our individual interaction with the environment, to how we choose to accept and interpret the evidence we receive. It also shapes the domains of

FIGURE 1.1 Knowledge management level.

human behavior and group dynamics. From an organizational point of view, knowledge spans all divisions of labor, occupational domains, production processes, and technology. Knowledge resides in both knowledge objects and in the minds of people. It is imperfectly transferred between people, with various degrees of latency and effort, through knowledge objects, systems, and relationships.

Knowledge—the fundamental resource of the knowledge economy—is, by its very nature, context sensitive. This partially explains why there are as many, if not more, definitions of knowledge as there are practitioners in the field. Two-and-a-half millennia of debate among philosophers have not resolved the fundamental nature of knowledge. The philosophical definition: "justified true belied" conceals more than it reveals about this elusive entity. Rather than seeking a perfect definition, a practical one has served well in various organizational contexts.

Knowledge is "understanding arising from conscious or unconscious reasoning about data, information, observations, or experience to reveal cause-and-effect relationships that facilitate the explanation and prediction of physical, natural, or social phenomena" (Simard and Jourdeuil 2011). This definition admits scientific knowledge (validated), individual experience (mental models), organizational knowledge (culture), and skill (learning and practice). It is useful for distinguishing between knowledge (understanding that enables prediction) and information (meaning in context). This definition emphasizes knowledge that is predominantly based on evidence and underlying validity.

The distinction between explicit knowledge and information is important from a knowledge creation perspective. Information cannot be reused to create new information, whereas knowledge can be reused indefinitely to create new knowledge in an endless cycle. For example, an organization's annual report describing outcomes in the previous year says little about what will happen in the coming year. Forecasts based on past experience become irrelevant with the emergence of a revolutionary technology.

There is a broader psychological definition of knowledge: "That which a person accepts as true and encodes in memory for future use" (Pigeau 2004). He outlines five truth-acceptance criteria (perception, authority, consensus, coherence-in, and coherence-out) used by individuals and notes that their weights vary in different situations. This definition admits facts, information, awareness, opinions, and beliefs, as well as evidence-based and validated knowledge. It explains why people are reluctant to abandon deeply held beliefs even in the face of overwhelming evidence, yet they can be fairly flexible in other areas. This definition makes no claims about evidence or the underlying validity of "knowledge." It allows that knowledge is neither complete nor perfect. This definition is useful for developing a comprehensive knowledge services strategy that reflects social interactions that underlie all organizational activity.

1.2.2 KNOWLEDGE MANAGEMENT

KM is defined in terms of managing a knowledge infrastructure. A knowledge infrastructure, in turn, supports embedding, advancing, or extracting organizational value from knowledge. In this view of KM, the corporation provides knowledge services (e.g., repositories, search engines, directories of expertise, or collaboration sites) that support diverse knowledge work performed by subject-matter experts. This approach embeds KM into an organization's infrastructure (like finance or human resources) rather than having it lie on the surface as a poorly understood (expendable) function. A knowledge infrastructure provides the foundation for constructing a knowledge services strategy.

Knowledge management: Develop, implement, and operate a knowledge infrastructure to support creating, managing, and using knowledge to produce and deliver knowledge products and services.

Knowledge infrastructure: People, using work processes and technology, within a governance framework to embed, advance, or extract organizational value from knowledge.

In just two decades, KM has evolved through three generations (Dixon 2010). The first generation focused on creating, acquiring, and preserving explicit knowledge. In fact, elements of managing explicit knowledge objects date back to the Library of Alexandria—2500 years ago. We can generally manage explicit knowledge in a structured way—much like data, information, or scientific publications. Although there are variations in processes, methods, and outcomes of use, all forms of explicit knowledge must be captured, organized, preserved, and made accessible. Differences arise in the methods used to execute these functions.

The second generation of KM emphasized sharing, integrating, and using tacit knowledge. This is clearly different from managing data and information, as it focuses on interactions among people and systems. We can promote and support the exchange of tacit knowledge, for example, with a directory of expertise or communities of practice, but we cannot manage it in the same structured sense as we can for explicit knowledge assets. As tacit knowledge is captured and made explicit; however, systems and processes must be in place to manage it or it will be lost. For those aspects of tacit knowledge that cannot be made explicit, we can manage social and relational opportunities to transfer knowledge among people.

A third generation of KM is evolving—using conversations, collaboration, and synergy within and among communities and networks to create, share, and make emergent knowledge accessible. Social interactions and interfaces can be formal or informal and are largely based on culture. They form and dissolve on an "as-needed" basis. We can encourage and facilitate the creation and use of emergent knowledge. Experience shows that it happens more often in learning organizations with diverse environments that are more likely to capture emerging trends and changes in their environment. Organizations that do not support conversation, collaboration, and communities have difficulty capturing and leveraging emergent knowledge. Management should, therefore, focus on architectures of participation, cultural conventions, and organizational incentives that encourage diversity and learning.

1.2.3 ORGANIZATIONAL BUSINESS

The purpose of managing knowledge is to support business objectives by increasing the value of knowledge and the productivity of knowledge work. Knowledge work is the use of domain, process, and functional knowledge, expertise, or experience to embed, advance, or extract value from content to accomplish organizational goals and objectives. Knowledge work can be divided into three broad categories—inputs, transformation, and outputs. Knowledge inputs acquire and process external content to provide an awareness of the state and trends of drivers of interest to an organization. Knowledge inputs include acquiring competitive intelligence, identifying market demands, and determining organizational priorities. Knowledge transformation is work that changes organizational inputs into outputs. Knowledge transformation includes managing business units, delivering corporate services, creating new knowledge, producing knowledge outputs, or organizational learning. Knowledge outputs are knowledge-based products and services that are transferred to clients, customers, and stakeholders. Knowledge outputs include submitting reports, addressing business issues, disseminating knowledge products, providing knowledge services, and organizational adaptation.

Work is increasingly being done through collaboration or peer production supported by social networking technology. Social networks are individuals, groups, or organizations that are interconnected to each other based on one or more interdependencies, common interests, relationships, or knowledge. Communities of practice connect people who share a common expertise, skill, or profession, such as scientists, policy analysts, IT specialists, or purchasing officers. The business task is to adapt the highly popular Web 2.0 technology to support creating and increasing value in the workplace.

Knowledge transfer is an act, process, or instance of conveying, copying, or causing knowledge to pass from one person, place, or situation to another. Transfer is primarily driven by an individual or organizational responsibility or mandate. It is also primarily one way—from a provider to a user. Transfer involves three processes: business transactions, interactions with recipients, and delivery. A key strategic decision is the positioning and distribution of knowledge products and services among different users and uses. A knowledge transfer strategy classifies different users and attributes of knowledge products and services into six categories (unique, complex, technical, specialized, simplified, and mandatory). It also describes knowledge transfer methods that are best suited to each category.

A knowledge market is the infrastructure that enables, facilitates, or supports the transfer of knowledge products and services among multiple providers, intermediaries, and users. Knowledge markets affect what flows between providers and users, how it flows, and its attributes, such as quality, timeliness, or usefulness. Understanding the challenges and opportunities of such markets allows an organization to best position itself to function, succeed, and lead in the twenty-first century knowledge economy. Knowledge markets function through six types of processes, ranging from simple to complex: communications, exchange, parallel, sequential, cyclic, and network.

1.3   KNOWLEDGE MANAGEABILITY

The horizontal dimension of a knowledge services framework must span the full breadth of organizational interactions, from managing internal knowledge work, through collaborating with partners in joint activities, to interacting with practitioners in open knowledge networks. Each of these environments requires a fundamentally different approach to managing knowledge.

Management regimes range from authoritative hierarchy (control, decisions, unification, and authoritative knowledge), through organizational infrastructure (coordination, structure, replication, and explicit knowledge), to negotiated agreement (collaboration, connectivity, peer production, and tacit knowledge), and finally responsible autonomy (engagement, environment, diversification, and innate knowledge). This section begins by describing concepts from existing frameworks that have been incorporated into knowledge manageability.

1.3.1 EXISTING FRAMEWORKS

Many frameworks have been developed that could be and are used to structure KM. This section outlines five that are relevant here: business operations, business engagement, cognition/learning, tacit/explicit, and knowledge dimensions. These frameworks are discussed in order of increasing generality.

Ross et al. (2006) describe a business framework based on operating models. They emphasize that the role of technology is to simplify capturing and improve the trust and availability of routine and essential knowledge in an organization. This allows an organization to focus on interactions designed to harvest the nuggets in the patterns, trends, and opportunities that they observe. The success of social computing applications shows that people are seeking to find better and cheaper ways to converse.

They developed a four-quadrant framework of business operating models:

1. Replication (independent but similar business units sharing best practices)
2. Unification (single business unit with global process standards and data access)
3. Coordination (business units that need to know each other's transactions)
4. Diversification (independent business units with different clients and expertise)

Weill and Ross (2009) developed a framework for negotiated IT governance. The purpose of this framework is to explicitly recognize the need for negotiated agreements between business and IT leadership, at the enterprise, business unit, and project level on the appropriate balance and priorities for investing in, designing, developing, and implementing an organizational IT infrastructure. This is the realm of negotiations over what is important and deserves investment and what effort should be devoted to transforming tacit knowledge and ways of working into explicit ones.

Senge (2006) developed a framework that emphasizes cognitive disciplines related to learning. He argues that industrial-age concepts of knowledge and management condition us to believe that knowledge is power, and lead us to hoard our stock, and be reluctant to share. He concludes that the crux of organizational success is related to being able to harmonize personal and organizational agendas, identify biases, and create an organizational climate that supports conversations between people in teams. Success also involves management actions that address underlying long-term organizational systems instead of focusing on visible symptoms, which, when addressed, seem to solve apparent problems temporarily but often exacerbate the root problem. He argues that success can only be achieved through five main disciplines (personal mastery, shared vision, mental models, dialogue, and systems thinking). These are essential for developing three core learning capabilities of teams (fostering aspiration, developing reflective conversations, and understanding complexity).

Nonaka and Takeuchi (1995) provide a knowledge creation framework that focuses on the cyclic flow of knowledge from explicit to tacit and back again. This widely-used model represents a humanistic approach to KM. It comprises a four-quadrant cycle emphasizing the transformation of one type of knowledge to another: socialization (tacit–tacit), externalization (tacit–explicit), combination (explicit–explicit), and internalization (explicit–tacit).

The Nonaka–Takeuchi model is well suited to an Eastern philosophy in that what is known is inseparable from the knower and that the individual is inseparably interdependent and embedded with the world and society. Hence, the inherent value of sharing tacit knowledge and the value of collaboration are generally understood and accepted; they need no measurement, business case, or cost-benefit analysis to be endorsed and approved. Pigeau (2004) also posits that what is known is inseparable from the knower, as it is based on what the individual accepts as true and encodes in memory for future action. This also helps to explain why what is known by one may be difficult to transfer to another through the filter of their truth-acceptance criteria. Management must therefore help to resolve the cognitive dissonance that occurs when conflicting inputs collide while maintaining diverse perspectives and avoiding "groupthink."

This model is very different from the western Cartesian philosophy that separates what is known from the knower or that separates observation and the observer. The separation between subject and object has been an underpinning of Western science for four centuries and management practice in the last century. It has also led to an emphasis on measurability and reward (primarily extrinsic) of individual accomplishment. Although the distinction between observation and observer is breaking down at the leading edge of both science and management, it remains well entrenched in Western culture, particularly in bureaucracies. The added value of sharing tacit knowledge and collaboration within a team or organization must be demonstrated to workers that are compensated based on what they know and do and to a management culture that demands measurement. This presents a formidable barrier to establishing KM programs for less tangible forms of knowledge.

Kurtz and Snowden (2003) developed a four-quadrant framework that classifies orders of knowledge in which the application of knowledge varies according to its relative "knowability," predictability, and controllability (Figure 1.2). That is, each quadrant represents an environment that describes the extent to which something is or can be known as well as how the knowledge within the environment can be predicted, used, or controlled. This framework provides the underlying basis for knowledge manageability and will, therefore, be described in greater detail.

The lower right-hand quadrant represents the stable world of common (routine or applied) knowledge. This dimension is characterized by knowledge that is accessible

3. Complex (partially knowable)	2. Complicated (knowable)
• Tacit knowledge • Scientists, experience • Find patterns, understand	• Technical documents • Experts, consultants • Design, develop systems
• Observations • Explorers, innovators • Sense, respond	• Standards, manuals • Bureaucrats, administrators • Categorize, process
4. Chaotic (unknowable)	1. Common (known)

FIGURE 1.2 Orders of knowledge. (Adapted from Kurtz, C.F. and Snowden, D.J., *IBM Syst. J.*, 42, 462, 2003. With permission.)

and agreed to by many people who have little personal investment in the action that results from it. This type of knowledge is often embedded in codified processes, is transactional in nature, and is amenable to automation. It is easily replicated and little skill is required to achieve repeated, reliable, and predictable results. This is the world of bureaucrats and administrators, who use frameworks and guidelines to classify work and then rules and standards to process the work. This is the world where best practices and resource scheduling work well. The most appropriate problem-solving approach is to *sense, categorize*, and *respond*.

Much organizational work is based on common or routine knowledge. Such knowledge primarily requires training in procedures and methods. Work done under this rubric is generally well understood. Common knowledge is primarily explicit and relatively static; it is described in standards, guidelines, manuals, and through standard training dealing with tasks, processes, and procedures. These generally require and permit relatively little interpretation in their application. There may be tacit elements, however, in the form of "tricks of the trade" or "work-arounds." In fact, much of the work regularly involves people to formulate and use "informal" and "ad hoc" means to get work done.

Examples of work that use common or routine knowledge:

- Administration (record keeping, purchasing, and contracts)
- Human resources (hiring, training, evaluation, and discipline)
- Operations (maintenance, scheduling, and reporting)
- Finance (budgets, expenses, and account balances)

The upper right-hand quadrant represents complication—a world that is "knowable" through the professional or technical knowledge of experts and specialists. This dimension is characterized by knowledge that is accessible and agreed to by smaller groups of people who have significant personal investment in the action that results from it. Although codified, this type of knowledge is often transferred only through intensive study or lengthy practical experience. As a result, although it can be replicated by its practitioners, considerable skill and training may be required to achieve repeated, reliable, and predictable results by others. Here, the objective is to develop, implement, and manage more flexible and modular functions, processes, and systems, using existing technical specifications, expertise and explicit knowledge. This is the world where good practice and expert-based research, prototyping, testing, and modeling are the best approaches—representing *sense*, *analyze*, and *respond*.

Examples of work that require complicated knowledge:

- Laboratory analysis (e.g., chemistry, physics, biotechnology, and electronics)
- Production (e.g., operations research, plant layout, and inventory management)
- Developing information systems (e.g., technology, computers, communications, and networks)
- Change management (e.g., planning, human resources, negotiating, and culture)
- Large-scale project management (e.g., logistics, coordinating, budgeting, and evaluating)

The upper left-hand quadrant represents complexity—situations and problems that can be only partially known or understood. This environment is characterized by knowledge that is accessible but not generally agreed by a large enough group of people to be reliably replicated. People who labor in this environment have considerable personal investment in the action that results from the agreement or acceptance of the knowledge that emerges from here. Although knowledge in this dimension can be codified, codification does not guarantee its truth or acceptance. This type of knowledge is often transferred through dialogue, discussion, or debate between sometimes opposing views, which may lead to its acceptance, marginalization, or being discredited. As a result, it is not easily replicated by its practitioners and it is difficult to achieve repeated, reliable, and predictable results.

We can distinguish complex systems and situations from complicated ones by delayed or predictive feedback (causing oscillations from repeatedly overshooting targets), nonlinear responses to stimuli (causing sudden, irreversible jump-shifts), as well as self-organization and coevolution (involving the independent actions of large numbers of "agents") Complexity also includes concepts of self-similarity in patterns and change.

Examples of work that use complex knowledge:

- Innovation (science and technology, research and development, and commercialization)
- Advising (policy development, organizational positioning, and implementation)
- System integration (across organizational functions, domains, and levels)
- Risk mitigation (security threats, public safety, and response to events)

Complex situations and problems may be highly sensitive to small, immeasurable changes in initial conditions. Alternatively, complex behavior can result from interactions among many independent agents or components, coupled with positive feedback, system delays, nonlinear relationships, clustered effects, and irreversibility of processes. Components of complex systems cannot be analyzed independently of the system and their impacts must be considered in the context of the system as a whole. Complex systems may function far from equilibrium. Consequently, small stimuli can lead to unpredictable, irreversible jump-shifts in behavior, nonlinear responses, or emergent states.

In complex situations, hindsight does not lead to foresight. One must *probe*, *sense*, and *respond* in ongoing iterative cycles. This is the domain of scientists and experienced managers. Here, the challenge is to find patterns, enhance positive ones, and dampen negative ones, and to understand to the extent possible, using tacit knowledge, expertise, and experience. In complex situations, science can sometimes explain phenomena in general but only predict specific outcomes statistically.

Finally, the lower left-hand quadrant represents chaos—that which is unknowable. This environment is characterized by "knowledge" that is held by individuals who have significant emotional attachment to and investment in the knowledge itself and any action that results from it. To the extent that such knowledge has been accepted as true and encoded for future use, it is individual and can be held equally by the unskilled as belief, the uninformed as bias, the knowledgeable as opinion, or the master as wisdom. As a result, it is difficult to identify what is valuable and what is not or what is worth the effort to explore, develop, and codify and what should be ignored. In chaotic situations, agents are unconstrained and independent of each other. Although they can be studied through statistics and probability, there are few discernable patterns or cause-and-effect relationships. Here, we are limited *acting*, *sensing*, and *adapting*. That is, taking small steps, sensing the environmental response, and adapting our actions to that response.

Many situations such as turbulence, emergency events, or military operations are chaotic at differing levels. We cannot know precisely how air will flow around a specific obstruction, where or when an emergency will occur, or when a critical part or system will fail. There are no clear or simple cause-and-effect relationships. Some causes can produce different effects due to extreme sensitivity to initial conditions (e.g., weather) or variable development processes (e.g., stem cells). We cannot predict the state of the weather, an economy, or a society beyond a near-term future or very broad trends. Although scenario analysis can consider possible consequences of potential events, precise prediction is not possible. In some cases, even statistics fail us as there is no basis for calculating the probability of occurrence of unique events.

Collectively, these frameworks provide many useful organizing structures: business operations, negotiated governance, systems analysis, learning, explicit and tacit knowledge, and orders of knowledge. However, none of the frameworks is sufficient to structure the full breadth and depth of what KM needs to encompass. Consequently, all of these concepts, and others, have been combined into a knowledge manageability framework developed by Simard and Jourdeuil (2011) that is described in the next section.

1.3.2 Management Regimes

Knowledge manageability considers the extent to which knowledge and knowledge work can actually be managed by an organization. The manageability framework identifies and characterizes four management regimes according to their purpose, motivation, emphasis, interactions, order and type of knowledge, and business model (Table 1.1). The regimes are authoritative hierarchy (decisions, control, authority, and common), organizational infrastructure (objects, coordination, enterprise, and complicated), negotiated agreements (people, collaboration, partnerships, and complex), and responsible autonomy (environment, self-interest, networks, and chaotic). This framework provides a structure for KM both within an organization and across organizational and jurisdictional boundaries because it focuses on the underlying raison d'être of the management regime, the types of interactions it manages, the orders of knowledge managed, and how that knowledge is expressed.

1.3.2.1 Authoritative Hierarchy

Authoritative hierarchy is an industrial-era organizational structure. However, some to substantial aspects of most organizations continue to be managed through this approach. Knowledge creation, management, and use can be *completely* mandated, governed, structured, and evaluated. This approach derives its mandate from legal instruments, such as a constitution, laws, or regulations. Governance is highly structured, through control, restrictions, and penalties. Typical functions include enforcing compliance, responding to events, and maintaining privacy and security. Hierarchies monitor activity, mitigate risks, and measure their outputs. Attributes include mandatory requirements, inflexible structures, and prior approval of actions. This is the realm of common knowledge in Kurtz and Snowden's (2003) orders of knowledge and of Ross et al.'s (2006) unification model.

This regime controls subordinate decisions and actions. Motivation is based on compliance through extrinsic incentives (e.g., compensation, employment,

---

TABLE 1.1
Overview of Four Management Regimes

Management Regime	Authoritative Hierarchy	Organizational Infrastructure	Negotiated Agreement	Responsible Autonomy
Primary purpose	Control	Coordination	Collaboration	Self-organization
Motivation	Extrinsic	Success	Mutual interest	Intrinsic
Emphasis	Decisions, action	Structure, process	Connectivity, interactions	Environment, engagement
Organizational interactions	Authority	Work flow, functions	Partnerships	Network
Order of knowledge	Common	Complicated	Complex	Chaotic
Type of knowledge	Authoritative	Explicit	Tacit	Innate
Business model	Unification	Replication	Coordination	Diversification

performance reviews, and penalties) tied to objects and tasks. Authoritative knowledge is common because it is fully structured and codified, it is agreed to, and it is embedded in the infrastructure. It is also relatively inflexible because once embedded, it requires a comprehensive review, development, and approval cycle to change it and the infrastructure that it is embedded in. Authoritative knowledge may also be externally imposed (e.g., laws, regulation, and standards).

Authoritative knowledge is explicit knowledge that has been formally reviewed and approved for use in producing knowledge-based products or services or that has been institutionalized by being embedded into organizational policies, procedures, or positions. It defines a formal organizational "position." Systems and processes containing embedded knowledge tend to be relatively rigid. Knowledge work is governed by organizational authority that determines how knowledge is produced and used. Examples include reporting relationships, implementing policies and regulations, standard operating procedures, using computer programs, or managing databases.

The knowledge is highly structured and has been embedded into relationships, processes, products, and services. Training and toolsets are the chief means of knowledge transfer, generally minimizing the need or opportunity for interpretation to enable use. Examples include data, records, computer programs, and system architecture. Attributes include formal reporting relationships and approval, inflexibility, static, and mandated use. The stock is increased by various methods, including being embedded, increased training, measurement, and development. Authoritative knowledge is transferred through an organization through means such as dissemination, communications, or direction.

1.3.2.2 Organizational Infrastructure

Organizational infrastructure underlies most contemporary organizations. Knowledge creation, management, and use can be *predominantly* mandated, governed, structured, and evaluated. Infrastructure is implemented by decision making, delegating authority, assigning responsibility, and allocating resources. Governance is through direction, integration, and coordination. Management functions include developing, implementing, and managing projects and programs. Infrastructure is evaluated through efficiency, productivity, and organizational outputs. Attributes include centralization, planning, and running the business. This is the realm of complicated knowledge in Kurtz and Snowden's orders of knowledge, in Ross et al.'s unification business model, and in Nonaka and Takeuchi's (1995) explicit knowledge.

This regime provides the basis for organizing and coordinating functions across the enterprise. Motivation is based on measurable success criteria (e.g., efficiency, productivity, and effectiveness) organized through resources and decisions. It is the primary regime for generating, managing, and using explicit and complicated knowledge. It is also the means for formally expressing and transferring explicit knowledge or capturing and codifying it in reproducible media (e.g., scientific publication and after-action report). Explicit knowledge can represent "received opinion" or an author's "opinion" (notwithstanding that such opinion may have been endorsed by a professional community). This knowledge is somewhat flexible in that it can be superseded by new or improved knowledge and is subject to interpretation by those receiving it. The knowledge is generally complicated because it involves interactions among many elements and interpretation by experts to make action easily reproducible. Organizational infrastructure may apply to a single organization or a group of organizations.

Explicit knowledge is tacit knowledge that has been captured and codified in reproducible media and is interpreted to coordinate organizational processes. Management focuses on the context within which explicit knowledge is used on behalf of the organization. This region may be primarily governed by a domain or profession that determines how knowledge should be interpreted and used. Examples include designing or developing policies, incentives, products, or services.

The focus is on interpreting functional or domain-based explicit knowledge (in the form of semistructured content) that is owned by or available to an organization and that often requires significant education to understand and analysis to usefully serve the organization. Examples include statistics, technical documents, presentations, and libraries. Attributes include either physical or electronic media, intellectual capital, captured, organized, and storable. The stock is increased through collection, tabulation, production, or development, among others. Explicit knowledge is fully transferable through transactions that include giving, licensing, exchanging, or selling.

1.3.2.3 Negotiated Agreements

Negotiated agreements have existed since the origins of trade and commerce, but are becoming increasingly central to the knowledge economy. Knowledge creation, management, or use can be *nominally* mandated, governed, structured, and evaluated. Agreements arise through negotiated contracts, mutual interests, and community participation. They are governed through leadership, support, and community structures. Primary functions include collaboration, facilitation, and sharing. Agreements can be evaluated through connectivity, exchange transactions, and participation. Key attributes include mutual interest, enforceability, and trust. Strategically, this is the realm of negotiations between business units over priorities and levels of investment. Operationally, it is the realm of negotiations between managers and knowledge workers over what work will be done and how resources will be allocated.

This regime emphasizes collaboration among employees and partners or managing tacit knowledge. Motivation is based on qualitative collaboration criteria (e.g., participation, connectivity, and teamwork) supported and facilitated through connectivity and interactions. Partners are persons, groups, or organizations with legal, contractual, or mutual agreements to cooperate in achieving common objectives as principles within an association, organization, or group. Partnership knowledge is generally complex because of uncertainty in the issues under consideration or the difficulty of eliciting it from individuals. Managing tacit knowledge within an organization is similar to managing knowledge in partnerships, although agreements are administered differently in each case.

Tacit knowledge is innate knowledge that is shared among individuals and groups based on agreed rules of conduct. Tacit knowledge and knowledge work are collaborative; they are governed through negotiated agreements. Examples include undertaking science, implementing organizational policies, collaborating to jointly produce a document, or providing advice. This is the world of complex knowledge in Kurtz and Snowden's framework, in Ross et al.'s coordination business model,* and in Nonaka and Takeuchi's tacit knowledge.

The knowledge may be owned or held by more than one organization or individual. It can be brought to bear on joint or collaborative work within the context of an agreement. Examples include patents, processes, resources, skill, and lessons learned. The stock can be increased through various methods, including research, acquisition, experience, or learning. Tacit knowledge is partially transferable through sharing, interactions, conversations, or dialogue.

1.3.2.4 Responsible Autonomy

Responsible autonomy is an emerging management regime founded in network structures that collapse communication and coordination costs. Knowledge creation, management, and use can be *minimally* mandated, governed, structured, and evaluated. Responsible autonomy is based on intrinsic motivation, voluntarism, and accountability. It is self-organized through community-based codes of conduct or rules of behavior. Management functions include recognizing, enhancing, and harnessing intrinsic motivation. Responsible autonomy can be evaluated through active membership and collective benefits. Attributes include peer production, intrinsic motivation, and synergy. This is where emergent knowledge is born, lives, and sometimes dies. This represents the chaotic dimension of knowledge and the diversification business model. This is the area in which the brilliant and the not-so brilliant create new knowledge and concepts and in which they first begin to be socialized, looking to attract resources in the organization to start them on their journey toward consensus, codification, and institutionalization.

This regime is central to engaging individuals and for encouraging sharing of tacit and innate employee knowledge.† Motivation is based on intrinsic criteria (e.g., recognition, respect, and acknowledgement) enhanced by an environment that is conducive to creativity. Innate knowledge is part of the essential character, constitution, and nature of an individual (e.g., intelligence, creativity, and talent). Although it can usually be enhanced, it cannot be transferred. Here, emphasis is on the intrinsic motivation of individuals and a capacity to engage it—through responsible autonomy within a network environment. "Knowledge" is typically chaotic because it is unknowable and it enables unpredictable emergent knowledge and capability and cannot be expressed. Eliciting participation from an external volunteer involves similar techniques to eliciting creativity from an employee.

Innate knowledge is voluntarily used to accomplish organizational objectives. Knowledge work is guided through engagement, including autonomy, mastery, and purpose. In this regime, individuals or groups have autonomy to decide what to do. Responsibility is understood as accountability and is, therefore, not anarchy. This

---

* The mapping between Kurtz and Snowden's orders of knowledge and Ross et al.'s business models is only approximate.

† Innate knowledge is used here as a subset of Polanyi's (1962) implicit knowledge. Implicit knowledge is more general in the sense of knowledge that is part of the essence or nature of something, whereas innate implies presence in an individual from birth; for example, an inborn talent, aptitude, or proclivity.

is consistent with tacit knowledge and knowledge work being within an individual; therefore, they are self-governed through responsible self-interest and voluntarism. Examples include allowing employees to determine what they do with part of their time and enabling self-motivated engagement in creative work.

Emphasis is on intrinsic motivation that encourages individuals to self-organize, or gravitate toward other individuals with similar knowledge, to create new knowledge. Examples include wisdom, judgment, intelligence, or creativity. Attributes include constitutional, natural, or essential characteristics of an individual. The stock of knowledge originates through being embodied, inherited, or inherent talent. Innate knowledge is not transferable, although it can be modeled or copied as it is demonstrated, displayed, or performed. Innate knowledge could also be seen as "wisdom." That is, experience and judgement that enable the correct application of knowledge.

1.3.2.5   Integrating Management Regimes

The most appropriate management regime is based on many factors ranging from an organization's business and culture to particular situations. It also depends on the nature of the work being done and the type of knowledge involved, such as doing research, developing a policy, or managing an operation. Finally, the nature of organizational interactions, such as intervention, partnership, collaboration, or advice plays a significant role. Although most organizations have predominantly one business or culture, their internal interactions normally involve multiple types of relationships. This is what Pigeau (2004) refers to as the "practiced structure" in an organization that differs from the "espoused structure" or the organization shown in its organization chart.

Most organizations manage all or several types of knowledge and knowledge work. For example, developing a policy requires experience to anticipate probable outcomes of different policy alternatives and judgment to compare the desirability and value of those outcomes. Once developed, a policy must be approved and implemented in a way that is consistent with the context in which it will be used. Consequently, for most organizations or groups of organizations, KM requires the application of more than one management regime, if not all four, to achieve organizational objectives.

To accomplish knowledge work, knowledge must flow across the management regimes. It may flow from structured to unstructured regimes. It starts with organizational decisions that determine the work to be done and that provides an infrastructure to support the work. This is followed by first-generation KM practices that capture, preserve, and manage explicit knowledge assets. In negotiated agreements, second-generation KM supports transfer, connectivity, and collaborative use of tacit knowledge to do knowledge work. Finally, third-generation KM enhances the wisdom and creativity of individuals as well as the synergy and emergence of networks.

Alternatively, knowledge often flows from unstructured to structured regimes. It starts with the determination that there is an emergent problem or opportunity that is socialized through groups or communities. Management then seeks advice, builds consensus, and directs that action be taken. Decisions are documented and embedded in the organizational infrastructure if it is important and pervasive enough. The nexus of organizational decision making is often in the negotiated agreement quadrant where competing interests in the organization reach a consensus that the senior executives implement through direction.

Knowledge manageability encompasses a broad spectrum from dynamic, unstructured organizational and knowledge environments to inflexible, highly structured environments. It provides a robust, multidimensional framework for managing all types of knowledge across diverse organizational contexts.

1.4   KNOWLEDGE INFRASTRUCTURE

Just as businesses will still need sound business models to succeed in the twenty-first century knowledge economy, many traditional aspects of management, such as decision making, resource allocation, and supervision remain essential in the knowledge era. A knowledge infrastructure is people using work processes and technology, within a governance framework to embed, advance, or extract value from content (Figure 1.3).

A knowledge infrastructure exists to enable and support the generation, management, and use of knowledge to produce knowledge products and provide knowledge services. All components of the infrastructure are equally necessary to success. Downplay any and productivity will suffer; eliminate any and an organization cannot function.

1.4.1   GOVERNANCE

Governance provides strategic, managerial, and operational authority, accountability, responsibility, and decision making needed to fulfill an organizational mandate. It enables the establishment and operation of projects and programs that produce knowledge products and provide knowledge services. Governance is essential to any collective activity in an organization.

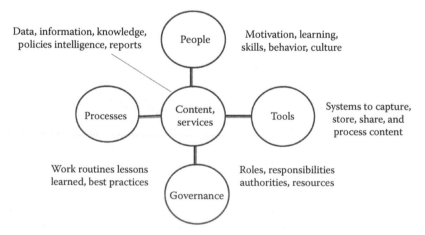

FIGURE 1.3   Knowledge infrastructure.

A number of principles should guide the development of a governance framework:

- A mandate from higher authority is needed for undertaking formal work.
- Authority is essential for directing work in an organizational context.
- Specific individuals should be responsible for accomplishing objectives.
- Collaboration is often needed to accomplish complicated or complex objectives.
- Soliciting client or customer views should be part of the decision process.
- Decisions should be transparent and consensus based.
- Management should emphasize negotiation, facilitation, and coordination.
- Rights and responsibilities of partners should be clearly articulated.
- Assigned responsibility ensures the completion of specified work.
- Transparency is essential to developing trust and credibility.

A governance framework specifies an authority structure of who reports to whom and who is responsible for what. Simard et al. (2007) describe seven roles and responsibilities in any project or program. These may be assigned to one or multiple individuals or groups, depending on the size and complexity of the project or program:

- *Leader*: Person who inspires, provides examples, or sets precedent for a vision or strategy
- *Manager*: Person who organizes people, processes, and technology, to accomplish objectives
- *Advisor*: Person who analyzes expected consequences of alternatives and recommends action
- *Planner*: Person who develops plans and schedules for accomplishing organizational objectives
- *Coordinator*: Person who organizes diverse activities or work to achieve integrated objectives
- *Specialist*: Person who provides subject-matter expertise to support accomplishing objectives
- *Worker*: Person who uses knowledge, processes, or technology to undertake knowledge work

A well-thought-out plan substantially reduces downstream difficulties. Depending on the magnitude and complexity of a project or program, planning can range from a brief document covering all necessary aspects to multiple documents, each covering one aspect:

- *Charter*: Establishes the purpose and mandate, place in the organization, and source of funding
- *Business case*: Describes the context and need, business lines and goals, and level of investment
- *Governance*: Specifies authority, accountability, roles, responsibilities, and decision making

- *Strategy*: Outlines implementation approach, resources needed, and performance evaluation
- *Communication*: Indicates methods for disseminating information and promoting the activity
- *Budget*: Includes salaries, operating expenses, capital investments, and external resources
- *Plans*: Lists milestones and deliverables, provides work schedule, and assigns tasks
- *Reports*: Describe activities, progress, and accomplishments for a specific period
- *Terms of reference*: Provides task group context, mandate, membership, and deliverables

Virtually every act undertaken by an organization's employees involves decision making. Decision making is arriving at a conclusion after considering a matter involving uncertainty, alternatives, flexibility, doubt, or controversy. In routine actions, decisions are normally made only once and the appropriate action is embedded in rules or procedures, in an individual's mental model of how things work, or an internal skill set of how to do things. All decisions involve both quantitative and qualitative aspects. Methods range from observing facts, through calculation, to using experience, and exercising judgment. The more that is understood about an issue, or the more limited the choices, the more that quantitative approaches can be used to support a decision. The less that is known about an issue, or the greater the diversity of choices, the more that opinions, experience, and judgment are necessary.

Decision tools are relatively simple frameworks with fixed processes. They lead decision makers through a categorization procedure, usually comprising a sequence of either/or choices to conclude with a simple classification. They may be used manually or they may be supported with a computer application. Decision tools can be used to classify an object or situation or calculate relative indices, such as low, medium, or high risk. These require interpretation to be used.

Decision-support system (DSS) calculates outputs related to complicated or complex problems. A DSS is normally embedded in an information system that acquires and processes input data and transmits outputs. The DSS manages the database and provides inputs to quantitative models, it manages and integrates the models, and it interacts with the user to execute queries, analyses, and interpret model outputs. The strength of DSS lies in the capacity to integrate multiple databases and models.

Expert systems attempt to reproduce the performance of one or more experts by embedding their knowledge through the use of heuristic reasoning rather than process models. They are a subset of artificial intelligence, positioned between quantitative analysis of DSS and qualitative analysis of human reasoning. Expert systems recommend courses of action or for automated systems, such as network security, take action directly.

1.4.2 PEOPLE

Knowledge is a human construct. Consequently, knowledge work can only be done by people. People use technology and processes to create, manage, and use content to accomplish organizational goals. Consequently, people are central to KM. There are two inseparable aspects of the human dimension of an organizational infrastructure. The social aspect focuses on individual behavior, community dynamics, and organizational culture. Human resource management involves functions such as staffing, supervision, and retention. Some elements of the human dimension are controlled by laws and collective agreements, others are required by organizational policies, and some are guided by frameworks of best practices.

1.4.2.1 Individuals

Knowledge workers think for a living. This is different than the industrial era, where workers primarily used their hands or bureaucrats performed highly structured work. Thus, the nature of work and those who do it has changed in very important ways. Nonaka (1998) observes that "Inventing new knowledge is ... a way of behaving, indeed, a way of being, in which everyone is a knowledge worker." Drucker (1998) explains that "In the information-based organization, the knowledge will be primarily at the bottom, in the minds of the specialists who do different work and direct themselves."

From a human capital perspective, people are seen as a factor of production, along with land, equipment, and capital. Stewart (1997) states that "Human capital grows two ways: when the organization uses more of what people know, and when people know more stuff that is useful to the organization." Drucker (1999) states that "Productivity of the knowledge worker is not—at least primarily—a matter of quantity of output. Quality is at least as important." Similarly, Ruggles and Holtshouse (1999) indicate that "In knowledge work, the means of production is now owned by the knowledge worker. They are volatile and can work anywhere ... Consequently, they must be managed as volunteers, not as employees."

Unlike machines and bureaucratic processes, an organization cannot control what is in people's heads, how it gets there, or how it is used to create and use knowledge. Further, knowledge workers are not readily interchangeable with each other; each has unique knowledge and abilities. Equally important, people are social animals; their place in a social context and the nature of their interactions with others is very important. Handy (1995) observes that "People who think of themselves as members have more of an interest in the future of the business and its growth than those who are only its hired help." Tapscott (1996) extends this idea: "It is only when workers identify with the goals of the organization and trust its managers to act in mutual self-interest that effective knowledge work can be performed."

1.4.2.2 Communities

Wegner et al. (2002) define communities of practice as "groups of people who share a concern, a set of problems, or a passion about a topic and who deepen their knowledge and expertise in this area by interacting on an ongoing basis." They continue with, "It is not communities of practice themselves that are new, but the need for organizations to become more intentional and systematic about managing knowledge, and therefore to give these age-old structures a new, central role in the business."

Cooper (1996) notes that "The real advantage of making it possible for everybody to communicate online isn't cheaper and more efficient communications—the mechanical-age viewpoint. Instead, it is the creation of virtual communities—the information age advantage that was revealed only after it materialized in our grasp." Hoffert (2000) used a meeting place analogy: "Networks have become social halls, and they are occupied by virtual communities that require no physical space, do not encroach on each other, and never go to war because they outgrow their resources."

Communities of practice are different from traditional management processes. Boyett and Boyett (2001) state that "Managers cannot create or direct communities of practice, but they can encourage their growth." Ruggles and Holtshouse (1999) note that "Inside a community, ideas are validated by the shared practice or paradigm of that community. Taking an idea outside the community requires the testing not just of the idea, but of the paradigm itself. Negotiations between communities, therefore, make knowledge more robust and force us to understand the barriers to knowledge sharing."

Communities can have both positive and negative effects on an organization. Positive impacts include increasing effectiveness by supporting the organization, providing advice and encouraging cooperation, creating a feeling of acceptance and belonging, providing a useful communication channel, and fostering openness. Negative impacts include supporting resistance to change, protecting negative values and beliefs, exacerbating conflicts between formal and informal organizations, creating and processing false information, and creating pressure to conform.

1.4.2.3 Culture

Culture is fundamental to all forms of social structure, including organizations. An organization's culture consists of the behavior and norms that stem from the shared attitudes, assumptions, values, and beliefs of its people (Eagen 1994). Perhaps the greatest challenge to implementing KM is the need to change long-standing, deeply entrenched, vertically oriented bureaucracies to dynamic, learning, adaptive, and horizontally oriented structures.

Davenport and Prusak (1998) state that "Values, norms, and behaviors that make up a company's culture are the principle determinants of how successfully important knowledge is transferred." Hamel and Prahalad (1994) put it somewhat more graphically: "Any company that fails to reengineer its genetic coding periodically will be as much at the mercy of environmental upheaval as tyrannosaurus rex." Holmes (2001) presented a similar argument for the public sector: "Cultural resistance is the greatest obstacle to integrated online public services. There will be employees who resent the intrusion of new technology in their daily routine and cling to the old, familiar ways no matter how well processes are improved."

Culture change is not a means to KM, but rather culture change and KM are inextricably intertwined. From an anthropological perspective, we first change a technology, which changes the ideology, which, in turn, changes the culture. On a small scale, culture change emerges through promoting desired behavior which, collectively, leads to increased sharing, collaboration, and engagement.

1.4.2.4 Human Resources

Knowledge workers need different capacities and abilities than industrial-era workers or bureaucrats. Tapscott (1996) observed that "Anyone responsible for managing knowledge workers knows they cannot be 'managed' in the traditional sense. Often they have specialized knowledge and skills that cannot be matched or even understood by management." Whiteside quipped that "Nobody's dumb at IBM ... but it's like herding cats—they just all have their own agenda." von Krogh et al. (2000) state that "Knowledge workers cannot be bullied into creativity or information sharing; and the traditional forms of compensation and organizational hierarchy do not motivate people sufficiently for them to develop the strong relationships required for knowledge creation on a continuing basis." Further, the shifting marketplace for knowledge workers makes it increasingly difficult to retain knowledge and experience. Handy (1995) notes that "If laborers are worthy of their hire, there is no reason to suppose that they won't go where the hire looks better. The assets of the new information-based corporations are, as a result, increasingly fragile." Human resource management will have to evolve beyond managing transactions to engaging creativity.

1.4.3 PROCESSES

Organizational work gets done through processes. In a KM context, work processes are the development, implementation, and operation of methods to acquire, organize, preserve, and provide access to content to enable its creation, management, or use. There are two groups of processes, one for each dimension of an organization: producing products and services, and running the organization. Although the first can be categorized into domains (e.g., manufacturing, research, service, and government) every organization is, to some extent, unique in how it does its work. There is no point in attempting to capture the enormous diversity of production processes here. Conversely, there are categories of organizational processes that are relatively similar across many organizations. Three generic processes are broadly described here as examples: administration, project management, and content management (Figure 1.4).

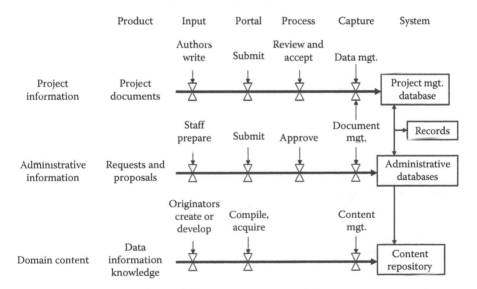

FIGURE 1.4 Generic organizational processes.

Work processes tend to be parallel to each other. They involve similar but separate activities. Their only point of interaction is often just prior to or after content has been captured in one of the databases. Project management and administrative databases interact with each other while both provide input to records management. Because the content repositories are often open to the outside world, they must be isolated from the rest of the organization to preclude providing an unintended pathway into internal databases.

Administration can be classified into four broad categories: acquisition (contracts and purchasing), finance (budget, expenses, and balance), staff activity (training, travel, and leave), and human resources (hiring, retention, and evaluation). Administration emphasizes authoritative hierarchy. Although each process is typically separate, they are conceptually similar. Documents and forms are prepared by responsible individuals, they are submitted and revised by internal staff, they are reviewed and approved by the appropriate authority, they are input into relevant administrative systems, and enterprise-wide activity reports are prepared. Administrative processes typically have little or no interaction with each other or other organizational processes, except that information about most administrative activities is generally managed as official records. Further, anything involving expenses must be entered into the financial system, and staff activity is part of human resource management. Each process is captured by separate enterprise-wide systems, which are used to produce enterprise-level reports for the process.

Projects are normally temporary with fixed start and end dates. Project processes combine authoritative hierarchy (establishment and conclusion) and organizational infrastructure (execution). The role of negotiated agreements may range from minimal to substantial. A project comprises three stages: establishment, execution, and conclusion. Establishment begins with a call for proposals, involving an announcement, submission of proposals, review, and selection. This is followed with project implementation, in which project documents are prepared, submitted through a portal, reviewed, accepted, and input into a project management database. Funds are then allocated to projects to begin work. Periodic reports are prepared and submitted through a similar process to that for project implementation. An annual report provides a basis for annual review. At the end of a project, a final report is submitted, reviewed, accepted, and input into a project management database.

Content may be produced internally or through partnerships with external organizations. Intellectual property (IP) rights for internal content are solely owned by the organization. The content, along with appropriate metadata, is entered into a repository to enable managing it at an enterprise level. There should be a one-way link between internal- and external-facing repositories or a partition between two parts of one repository to limit access to internal content, as appropriate. Rights to content produced through partnerships are jointly owned or held through partnership agreements. Domain content is produced and owned by external organizations. It may be accessed as a library service on behalf of clients and stakeholders. Access to domain content may be free, through subscription, or licensing. Although an organization may provide access and search capability, it has no IP property rights to domain content except as provided for in access agreements

1.4.4 TECHNOLOGY

Technology is essential to producing knowledge products or delivering knowledge services. No organization can succeed in the knowledge economy without information and communication technology. Yet, technology alone is not sufficient. Technology that ignores the need for coevolution with individual behavior, group dynamics, and organizational culture results in systems that simply reinforce the status quo of organizational silos. Tools and systems must reflect the needs of a knowledge organization. Jones (2000) observes that "Technology is expensive and bad choices hurt productivity and often eliminate second chances for deploying a better program."

Technology includes tools or systems that help people to find, create, share, interpret, or use knowledge. Tools include business intelligence software, mind-mapping tools, digital whiteboards, DSSs, and collaboration sites, such as communities of practice or wikis. Applications for finding and accessing knowledge include directories of expertise, inventories of knowledge assets, "lessons learned" systems, web-based search engines, portals, and networks. Traditional tools include telephones, e-mail, data warehouses, document and record management systems, file systems, and yellow pages.

Technology is often the first component of an organization's infrastructure to be considered in developing a KM program for several reasons. Vendors make money by selling technology, so it is promoted as a quick solution to every problem. Any increase in technology increases the span of influence (and control) of IT groups. Equally importantly, technology is real. It provides tangible and visible evidence that something is being done about KM without necessarily having to understand or measure what is actually being accomplished. Notwithstanding these considerations, technology should be the last component of an organization's infrastructure to be developed. One should know what work needs to be done and how it will be done before acquiring the tools to do the work.

Designing, developing, and implementing an evolving technological infrastructure requires balancing between two conflicting needs. On the one hand, increasingly complex tasks and interactivity among tasks, projects, and functions requires fairly rigorous hardware compatibility, application standards, systems interoperability, and communications protocols to enable functionality. On the other hand, knowledge workers will only use technology if it fulfills their diverse needs and compliments their diverse ways of working.

1.4.5 CONTENT

A knowledge organization exists to create, increase the value of, or use content. Content is both input to and output from a knowledge organization. Content is embedded value, in the form of a pattern, message, or signal contained within elements of a content value chain. A content value chain is the flow of a pattern, message, or signal through a sequence of stages in which its form is changed and its value or utility to end users is increased at each stage. Various types of content and related processes are described in the following list:

- *Collections* are groups of objects that facilitate and support the generation of content. Curators acquire, organize, preserve, and provide access to collections. Examples include rocks, minerals, fossils, wood samples, insects, plant materials, or disease tissue.
- *Libraries* are collections of IP. Librarians acquire, organize, preserve, and provide access to their holdings. Examples include books, publications, reports, documents, maps, photographs, images, art work, and audio or video recordings. Libraries generally do not own IP rights to the material that they hold.
- *Data* are recorded, ordered symbols or signals that carry information and patterns. Data managers acquire, organize, preserve, and provide access to data and data-based products. Data are organized into elements, files, data sets, databases, or statistics.
- *Information* is meaning and context, arising from processing, interpreting, or translating data to extract an underlying message or pattern. Information managers acquire, organize, preserve, and provide access to information and information-based outputs. Examples include documents, reports, images, maps, brochures, presentations, or multimedia recordings.
- *Records* are content that is specifically related to organizational management. Records managers acquire, organize, preserve, and provide access to organizational records. Examples include data (e.g., finance, personnel, and operations), information (decisions, meeting minutes, proposals, and reports), and knowledge (e.g., experience, policies, guidelines, and contacts).
- *Knowledge* is understanding arising from reasoning about data or information to reveal cause-and-effect relationships that facilitate the explanation and prediction of physical, natural, or social phenomena. Knowledge managers acquire, organize, preserve, and provide access to knowledge and knowledge outputs. Examples include authoritative, explicit, tacit, or innate.

All content involves some common processes (acquire, organize, preserve, and provide access). This partially explains why there is confusion among them. Yet each process involves specialists who use different practices and vocabulary. Libraries and collections manage physical or digital objects. To use information, one only needs to find, access, read, and understand it. To use data, one must also know its format, codes, and the software that was used to prepare it, as well as be able to extract and manipulate subsets of it. A key aspect of using knowledge is finding and interacting with an expert. Records are not simply a subset of information but include organizational data and knowledge as well. It is the similarity of objectives rather than functionality that unites content processes and requires coordination.

1.5   KNOWLEDGE ASSETS

The view of knowledge as an asset is relatively recent. Stewart (2001) notes that "Because knowledge has become the most important factor of production and knowledge assets the most powerful producers of wealth, the leaders and organizations that take command of their knowledge will occupy competition's high ground." Allee (1997) points out that "Knowledge is an infinite resource that challenges the dominant economic models grounded in scarcity. Viewing and managing knowledge as assets represents a fundamental change of perspective and process for organizations—one which is not yet fully understood." Leonard (1995) observes that "To manage knowledge assets, we need not merely to identify them but to *understand* them—in depth—in all their complexity: where they exist, how they grow or atrophy, how manager's actions affect their visibility."

Knowledge is broadly divided into two categories—explicit and tacit. Explicit knowledge has been expressed, captured, codified, and stored in reproducible media. It is the visible result of knowledge work. Explicit knowledge is what an organization can manage as assets. Tacit knowledge is personal, cognitive knowledge, gained through observation, experience, or practice, stored in the mind of an individual. Tacit knowledge is also embedded in group dynamics and an organization's culture and relationships. It is the invisible accumulation of experiences. Tacit knowledge cannot be "managed" in the same sense as explicit knowledge, but it can be leveraged as an essential capacity for performing knowledge work. Psychological knowledge (e.g., beliefs, perspectives, opinions, and values) is not included as a form of knowledge of interest to most organizations. Although psychological knowledge affects how individuals perceive, interpret, and respond to the world around them and it is essential to social interactions, it cannot be managed as assets.

There are six aspects of managing knowledge assets: acquiring knowledge, organizing knowledge, mapping knowledge, preserving knowledge, integrating knowledge, and IP.

1.5.1 ACQUIRING KNOWLEDGE

An organization's knowledge base (stock of knowledge) is a key determinant of its short-term competitiveness in the knowledge economy. This is what enables the production of knowledge-based products and services. There are three methods of acquiring existing knowledge: capture, inventory, and import.

Capture represents content on reproducible physical, mechanical, analog, or digital media. Capture may begin with explicit or tacit knowledge. Capturing explicit knowledge involves compiling what exists in individual or group files, computers, or in content repositories and requesting that copies of specified content be submitted to central repositories. Capturing tacit knowledge requires personal interviews, conversations, or dialogue designed to elicit what people know but may have difficulty expressing.

Inventory finds, gathers, lists, classifies, and describes existing explicit content that is owned or held by an organization. This includes conducting an inventory of existing content and mapping the content to business or program needs. If participation is voluntary, only a minority of existing content is likely to be captured. A mandatory inventory, with staff assigned to visit individuals and solicit content, will capture a substantially higher percentage of what an organization knows. Results of the inventory should be accessible to and searchable by all staff, partners, clients, or stakeholders, as appropriate. The inventory must be dynamic, with content continuously being updated, or it will start becoming obsolete as soon as it is finished.

Import brings existing external content into the organization. Importing involves three steps: accessing one or more content sources or repositories, searching for and finding appropriate content, and retrieving it by downloading or transferring it to a local repository. Importing external content includes five considerations: accessibility, source, purpose, rights, and security.

1.5.2 ORGANIZING KNOWLEDGE

An unstructured collection of knowledge is of limited use to an organization. We must first develop a systematic, coherent structure for knowledge assets. This is typically done through classification, which is categorizing, assigning, or sorting knowledge into systematic classes, categories, or structure according to classification criteria.

There are three categories of classification schemes: single criteria, multiple criteria, and unstructured. Single criteria schemes, such as subject, alphabetic, or hierarchy, have the advantage of a simple structure coupled with the disadvantage that they are not well suited to complicated subjects with multiple relationships. In the last half century, multiple criteria schemes, such as metadata and automation, have emerged to classify subjects using more than one predetermined criteria. Although they are more complicated than single criteria schemes, they provide flexible structures that enable Boolean searches with multiple terms. Most recently, unstructured schemes, such as full-text indexing or "folksonomies," enable totally unstructured searching. The advantage is infinite flexibility and diversity in organizing and searching content, which leads to the disadvantage that no search can be exhaustive and there are normally large numbers of "hits" that must be filtered.

There are challenges associated with organizing knowledge. Different disciplines use different terminologies. For example, a "tank" means something very different to army, navy, and air branches of the military. The broader the domain, the greater are the difficulties. Linguistic issues impact multilingual schemes as some terms cannot be exactly translated from one language into another. There are also cultural issues in that a term (e.g., "blue") may mean different things in different cultures.

There is no single organizational scheme that can work for all purposes. In particular, a subject-classification system that is appropriate for running an organization (e.g., records of decisions, meeting minutes, budgets, and procedures) is of little use for domain-based knowledge. Conversely, domain-based explicit knowledge (e.g., species names, medical taxonomies, and engineering terminology) is normally structured according to national or international standards and is not subject to the operational needs of an organization. The best solution is to develop or adopt structures that are appropriate to each dimension of an organization and then map each to the other in a way that requires no action on the part of individuals.

1.5.3 MAPPING KNOWLEDGE

Mapping knowledge relates what an organization knows to what it needs to know to accomplish its business objectives. Mapping knowledge is discovering the constraints, assumptions, value, and uses of available knowledge assets, and finding opportunities to leverage existing knowledge. There are three aspects to mapping knowledge: needs, gaps, and priorities.

Knowledge needs describe what needs to be known to accomplish an organization's business and program goals. Although this sounds straightforward, it is not. Evaluation often results in compromise because knowledge is neither perfect nor complete. One is always dealing with uncertainty; the more complex a situation, the greater the uncertainty. Much of what we think we know may be based on assumptions. Or we may understand a similar situation but the differences are significant. Mapping business needs to existing knowledge or existing knowledge to possible uses is often an iterative process that develops successive compromises between what is known and what needs to be known.

Knowledge gaps document the difference between available knowledge and what is needed for supporting organizational goals. Available knowledge is knowledge that has been or can be captured, inventoried, or imported. The gap between knowledge needs and available knowledge is used to establish priorities for generating new knowledge or for evaluating existing capacity to produce or deliver new products or services. Determining knowledge gaps is often an iterative process. We need to know *that*, but only know *this*. Is there another way to approximate *that*? Is *this* good enough? Alternatively, how much can we do with *this*?

Evaluating knowledge: All knowledge is not equal. Knowledge should be evaluated in terms of its relative importance to the organization. Knowledge can be prioritized into six categories:

1. *Critical*: Essential, core knowledge that cannot be replaced
2. *Essential*: Loss or unavailability would risk organizational competitiveness or sustainability
3. *Very important*: Important knowledge that would take substantial cost, time, or effort to replace
4. *Important*: Loss or unavailability would significantly degrade knowledge-based outputs
5. *Useful*: Enhances organizational productivity or the quality of specific products and services
6. *Supplemental*: Useful knowledge that could be replaced with minimal cost, time, and effort

It is worth expending substantial resources to acquire and preserve critical and essential knowledge, while useful and supplemental knowledge should be acquired and preserved only if the cost, effort, and time required are less than its perceived "value" to the organization.

1.5.4 PRESERVING KNOWLEDGE

Preservation is the foundation of KM. Preservation prevents the irretrievable loss of content throughout its life cycle by managing it in permanent physical or electronic media. Preservation links an organization's past, present, and future. Past knowledge that was not preserved is not available today; present knowledge that is not preserved will not be available in the future. There are three elements of preservation: archiving, maintenance, and migration.

Archiving is storing or placing content into physical or electronic repositories for safekeeping and subsequent retrieval. Examples include filing systems, databases, libraries, content repositories, or data warehouses. Although shared drives are often used to store content, they are inefficient, ineffective, and insecure; they are not recommended. Archiving includes eight components:

1. System architecture to integrate the components
2. Prioritization process to select content
3. Classification scheme to organize content
4. User interface to input content
5. Repository to store content
6. Search engine to find content
7. Retrieval system to extract and download content
8. Administrator interface to manage the system

Maintenance is keeping content in its original state, preventing loss or decline, or safeguarding it from damage, vandalism, or theft. Content is no exception to the law of entropy. If left unattended, it will, sooner or later, become lost or unusable for many reasons. Content maintenance includes seven components:

1. Periodically monitor content to identify and restore any degradation.
2. Maintain content integrity to preclude inadvertent or deliberate modification.
3. Implement site security to preclude physical damage to repositories.
4. Provide redundant off-site repositories to ensure recovery of core content.
5. Provide access to content for authorized users.
6. Secure networks to preclude unauthorized access to repositories.
7. Manage content in accordance with a predetermined life cycle.

Migration is transferring content from obsolete storage media to long-term stable media. This is absolutely essential for digital content. Ultimately, despite its relative bulk, weight, and susceptibility to damage, paper remains the best very long-term storage media available. We can still read Leonardo da Vinci's notebooks written five centuries ago but can no longer read some electronic media that are less than three decades old. During the course of the author's career, media have evolved through many stages: paper, punched cards, paper tape, magnetic tape, main frame disks and drums, floppy disks, diskettes, tape cassettes, CD-ROMS, DVDs, flash drives, and USB memory sticks. Each represents a significant increase in storage efficiency and access speed, but a considerable challenge to users who must constantly migrate archived content.

1.5.5 INTELLECTUAL PROPERTY

IP relates knowledge assets to what an organization legally owns. IP rights are the power and authority to possess, use, transfer, or dispose of IP and to receive fees, royalties, or other consideration in exchange for its use. Although easy to state, the concept is somewhat elusive. Lucky (1989) notes that "Even if we subscribe to the notion that a person should own the fruits of his intellectual labor, it is surprisingly difficult to authenticate the provenance of such works." Similarly, Rifkin (2000) observes that "Property is an elusive concept. On the one hand, it seems so easy to identify in all of its various forms. Even the most uneducated of souls recognized property when he sees it and understands viscerally what is meant by the term. On the other hand, few concepts have proven more difficult to pin down."

Intellectual capital refers to all intangible knowledge-based resources that are part of the value and competitiveness of an enterprise. Intellectual capital relates knowledge assets to an organization's balance sheet. Economists have only recently begun to explore the nature of intellectual capital. Stewart (1997) observes that "Nonsubtractive, structurally abundant, front-loaded, and unpredictable: When the most important economic resource has these characteristics, it's no wonder that information-rich businesses such as finance and computer software are notoriously volatile." English and Baker (2006) linked knowledge and intellectual capital: "Knowledge is to intellectual capital what flour is to bread: each is the key ingredient. In fact, the knowledge that underlies the intellectual capital of several renowned companies is so crucial that it accounts for most of the value of their brand."

From a competitive perspective, Tapscott and Williams (2006) note that "Increasingly, and to a degree paradoxically, firms in electronics, biotechnology, and other fields find that maintaining and defending a proprietary system of intellectual property often cripples their ability to create value. Smart firms are treating intellectual property like a mutual fund—they manage a balanced portfolio of IP assets, some protected and some shared." Similarly, English and Baker (2006) state that "Intellectual property is a short-lived advantage that will eventually evaporate. Patents are limited in scope and do expire. Copyrights have loopholes and may be circumvented … Intellectual property content is being file-swapped or pirated. There's hardly any knowledge content that can't be eventually downloaded off the Internet." Sparr (2001) observes that "In cyberspace … the problem of property is complicated by its intangible nature. It is harder to patent a concept than a device and harder to establish ownership over things that cannot be touched, or felt, or seen."

Managing IP entails three considerations: inventory, ownership, and intent. An inventory is simply a compilation of the description, classification, utility, and value of IP. Simard et al. (2007) identified five types of ownership: sole, shared, licensed, held, and borrowed. They also identified three broad intents to managing IP: protection, licensing, and enforcement.

1.6 KNOWLEDGE SHARING

Knowledge in a repository or in the minds of individuals entails a real cost of creation or production, but represents only potential value or utility to an organization. The actual value is only extracted when knowledge is used to accomplish knowledge work. This requires that knowledge "flows" from its source to where it is needed. This is the purpose of knowledge sharing—the second generation of KM—which leverages the value and utility of knowledge through sharing.

Knowledge sharing is disseminating, making available, or providing access to knowledge or knowledge-based products or services without expected reciprocity, fee, or consideration from the recipient. This contrasts with knowledge exchange, which suggests a transaction that includes an expectation of some sort of "quid pro quo." The term "sharing" is preferable, because, in a knowledge organization, people should make their knowledge available to others without specific expectations of a return, knowing who might use the knowledge, or what they might use it for.

The benefits of knowledge sharing are well-recognized. Saint-Onge and Armstrong (2004) state that "An organization's intangible assets are made of capabilities and relationships that are built through the exchange of knowledge ... Knowledge exchange serves as the bases for accelerating learning and systematically developing individual and organizational capabilities." In contrast, Boyett and Boyett (2001) point out that "There are many barriers in traditional organizations to the sharing of tacit knowledge, including reward systems that discourage sharing and the fear on the part of employees that they will be unable to express the inexpressible." Holmes (2001) identifies specific problems: "There are employees who have a natural tendency to hoard information and look suspiciously on and discredit data that others have developed. To many, entering what they know into a computer system and passing it on to someone in another department is not only threatening, it's a pain in the neck."

Sharing bridges the three generations of KM in that it begins with explicit knowledge, continues with tacit knowledge, and reaches into community knowledge. Sharing also encompasses individuals, communities, and organizations. Much of the literature on sharing mixes these processes, resulting in confusion about just what is being shared, how it is shared, and between whom.

1.6.1 MOTIVATING SHARING

Sharing is quintessentially human, but it is also primarily voluntary. Tiwana (2000) observes that "What ... employees need are incentives, not faster computers. Technology provides many enablers except the biggest one of all: an incentive to share knowledge." Although knowledge sharing can be mandated, the extent of sharing generally depends on generosity and altruism or the expectation of reciprocity on the part of those who have it, particularly in the case of tacit knowledge. People who have knowledge must be convinced that they should share it with those who do not. This is an enormous attitude and behavioral shift from the traditional "knowledge is power" culture of the industrial era, with its emphasis on authoritative hierarchy. Because sharing generally benefits organizations more than individuals, incentives are usually needed to motivate such a substantial change in people's behavior.

Garfield (2010) identifies 10 steps that managers could take to encourage people to share their knowledge:

1. Communicate regularly on sharing expectations and rewards.
2. Conduct training, webinars, and knowledge fairs on using sharing tools.
3. Establish and communicate knowledge sharing goals.
4. Convince people in small groups or individually that sharing works.

5. Share sharing success stories.
6. Managers practice good sharing behavior.
7. Implement reward and recognition programs for good sharing behavior.
8. Encourage people to work in communities.
9. Reinforce desired behaviors and discourage undesired behaviors.
10. Align knowledge sharing with organizational processes.

1.6.2 Explicit Knowledge

Dixon (2010) identified a number of assumptions that underlie the value of sharing explicit knowledge:

- Knowledge is an asset to be managed like capital and land.
- It is the responsibility of leadership to manage assets.
- Organizational knowledge can be accurately documented and disseminated through technology.
- Subject-matter experts have the most valuable knowledge to capture.
- Organizational knowledge is relatively stable so it can be stored without loss of value.
- Employees will seek captured knowledge and use it.

Although sharing knowledge is everyone's responsibility, leadership is responsible for enabling the process. Although tacit knowledge is often transferred through conversations, authoritative and explicit knowledge are captured in documents and disseminated through technology. Although everyone has knowledge that is valuable, expert knowledge is often essential. Although some knowledge degenerates over time, some, such as policies and scientific reports, is relatively stable. The weakest assumption is that employees would find existing knowledge and use it. The "If you build it, they will come" syndrome does not work all that well in real life. In general, many of the assumptions are generally valid for explicit knowledge and provide a context for sharing explicit knowledge. Further, explicit knowledge is a prerequisite for authoritative knowledge, which is the basis of decision making and action. Regardless of the maturity of KM, explicit knowledge will remain essential to the success of an organization.

The greatest value of sharing explicit knowledge is realized in relatively stable organizations, where procedures and processes tend to be standardized and do not change rapidly. A service strategy should consider three forms of sharing explicit knowledge: dissemination (one way, push), access (one way, pull), and exchange (two way, transactions).

1.6.2.1 Dissemination

Dissemination involves passive, one-to-many, one-way transfer by pushing content to recipients. It means to spread, disperse, or distribute content to recipients or audiences. This approach is used for increasing awareness or understanding of organizational positions, activities, or accomplishments. Dissemination can be accomplished through various physical and electronic methods, such as communications, publication, distribution, posting, or promulgation. Successful dissemination is rooted in the three elements of the Weaver and Shannon (1949) communications model. This half-century old model can be seen as comparable to the current triumvirate of IT, IM, and KM.

- *Transmission*—How rapidly and accurately can symbols be transmitted?
- *Semantics*—How accurately do the transmitted symbols convey the intended meaning?
- *Effectiveness*—How well does the received meaning influence actions?

Several factors affect dissemination success. Recipients must be interested in the content when it is received or it may be discarded. They must also have the capacity to process and understand it. The content must be compatible with recipients' culture and beliefs and it must come from a trusted source.

The passive nature of the process requires that content be carefully crafted for specific audiences. Who are the intended recipients? What motivates them to take action? To whom do they listen? What is their level of knowledge? Scientific audiences focus on concepts, technical audiences need product specifications, and general audiences need simplified content. Executives are interested in strategic content, managers need mid-level content, and operational staffs want techniques.

Perhaps, the most important drawback to dissemination is its passivity. That is, the process is considered successful when content is delivered to a recipient or audience. Dissemination generally does not consider the effectiveness or impact of the message on the recipient or whether any action follows.

1.6.2.2 Access

Accessing content is an active, many-to-one, one-way transfer involving users pulling content to themselves by physically, remotely, or electronically going to and entering one or more repositories, sites, or locations to search for, find, and retrieve content. From an effectiveness perspective, sharing through accessing has a number of advantages over dissemination:

- A potential user wants the content, as evidenced by active search and retrieval.
- A potential user likely needs the content to make a decision and take action.
- Content is accessed and available when the potential user is receptive to it.
- Presumably, the user can understand the content they are accessing.

Potential users must be aware of the content, know where to find it, and be able to access it. As with dissemination, they must also have the capacity to process and understand it. Accessing content involves five considerations: awareness, permissions, accessibility, search, and retrieval.

1. *Awareness* is having realization or perception of facts, information, events, or conditions. Potential users of content must know where it is and how to find it. Awareness is developed by promoting content repositories, linking them to related sites, and interacting with clients.
2. *Permission* to access content may be granted to individuals or groups. Permissions range from an entire repository to individual documents.

Permissions also range from full administrative access, through management access, to contributor access, to view-only access.

3. *Accessibility* is the extent to which content can be obtained, delivered to, downloaded, or received by users. Accessibility is facilitated by providing content in broadly used standard media, formats, locations, and processing technology.

4. *Searching* is the use of manual methods, search engines, or automated processing to look for or discover content in filing systems, libraries, data warehouses, electronic repositories, or websites. Searching may be limited to specific metadata (e.g., titles, authors, and keywords) or full text.

5. *Retrieval* may be via any or multiple traditional channels (e.g., client visits, site visits, e-mail, telephone, and fax) or online channels (e.g., e-mail, File Transfer Protocol, Internet, intranet, extranet, and World Wide Web).

1.6.2.3 Exchange

Knowledge exchange involves active, one-to-one, two-way transactions. Exchanges typically involve a request from a user coupled with a response by a provider. The exchange may involve a request for and transfer of IP (accessing and downloading a document from a repository), exchange of equal-value IP (among partners), or exchange of IP for other considerations (e.g., agreement, fee, feedback, and visibility). Exchanges may be synchronous—two-way communication with virtually no time delay, allowing rich, real-time interaction. Examples include conversations, presentations, the telephone, and meetings. Exchanges may also be asynchronous — two-way communication with a time delay, allowing a broad reach and interaction at provider and user convenience. Examples include e-mail, a website, frequently asked questions, or notification.

Davenport and Prusak (1998) describes knowledge exchange between individuals within an organization from a "market" perspective. He lists five attributes of knowledge markets.

1. *Price*—Although knowledge is not generally "sold," there is an expectation of reciprocity, the provider's reputation and status may be enhanced, or knowledge exchange is mandated.

2. *Trust*—Trust between provider and user is essential to exchanging knowledge. The exchange process must be equitable, transparent, and ubiquitous.

3. *Signals*—There are signs of knowledge "quality," such as a provider's position in an organization, their level of education, and their reputation.

4. *Inefficiencies*—Users may have incomplete information, there may be asymmetry between market segments, and there may be local peaks and valleys of availability.

5. *Pathologies*—There may be knowledge "monopolies" (only one source) artificial scarcity (unwillingness to participate), and trade barriers (security).

Other attributes of knowledge exchange markets include facilitation and brokering. Facilitation is providing logistical, administrative, and/or advisory support to knowledge providers, users, and/or market transactions. Brokering is acting, arranging, or negotiating transactions, agreements, or contracts as an agent or intermediary on behalf of providers and users. Knowledge facilitators may

- Increase awareness of knowledge availability
- Assist with knowledge dissemination
- Assist with search and retrieval
- Maintain content repositories
- Assist in adapting knowledge to user needs
- Provide a digital infrastructure for exchanges
- Manage the market infrastructure
- Negotiate knowledge exchange agreements

Sharing explicit knowledge can be managed with relatively familiar processes, such as communications, publication, and libraries. There are also substantial bodies of knowledge in related fields that can be adapted to sharing explicit knowledge.

1.6.3 Tacit Knowledge

Tacit knowledge is personal, cognitive knowledge, gained through observation or experience, which is influenced by personal beliefs, perspectives, and values. It resides in the minds of knowledge workers and is often difficult to transform into explicit knowledge. Although it is less manageable and more difficult to share, it is also considered to be the bulk of an organization's knowledge as well as more valuable to the organization. Tacit knowledge may be shared by individuals, groups, or networks.

Most of the benefits of sharing accrue to recipients, groups, and the organization, while most of the costs and risks accrue to providers. However, if providers do not share their tacit knowledge, nothing else follows. Hard-won, real-life, individual experience must be shared to be reused by others in new situations, be converted into group knowledge for validation, or form the basis of organizational adaptation. Consequently, it is important to consider and enhance those benefits that accrue to providers. Simard (2005) listed a number of provider benefits:

- Sharing broadens the *awareness* and reach of the provider's knowledge and increases its visibility, influence, use, and impact.
- Providers will develop a *reputation* as an active, competent, and knowledgeable community participant, thereby promoting personal collaboration and group synergy.
- Feedback from recipients increases awareness about what knowledge is needed and how it is being used, increasing the *relevance* of the provider's work.
- Increasing awareness of the provider's knowledge, reputation, and relevance encourages partnerships and leveraging resources to make more efficient use of existing capacity.
- Provider collaboration in large-scale activities enhances opportunities for synergy and resources beyond what might be available from individual efforts.

Conversely, Dalkir (2005) listed a number of obstacles to knowledge sharing along with ways of counteracting them;

- The notion that knowledge is the "property" of an individual and that "ownership" must be maintained. Reassure individuals that authorship and attribution will be visible.
- A connection between knowledge and those who are knowledgeable is essential. Expert locator systems and links to authors are useful in this regard.
- Individuals are commonly rewarded for what they know rather than what they share. Stop rewarding knowledge hoarding and start providing valued incentives for sharing.
- Lack of trust that the receiver will understand and correctly use the knowledge or that the recipient is unsure about its credibility. This can be resolved through self-regulating communities that validate knowledge and recognize expertise.
- Organizational cultures that encourage discovery and innovation foster sharing, whereas those that nurture individual genius hinder sharing.

1.6.3.1 Methods

There are many methods for sharing tacit knowledge. Most predate the Internet and World Wide Web. Some methods are one-on-one, while others are one-to-many. Some pull tacit knowledge while others push it. Interactions may be among equals, between superiors and subordinates, or between subordinates and superiors. Some focus on individuals while others emphasize groups. The following list describes various methods for sharing tacit knowledge, along with the types of people involved:

- Conversations, discussion, and dialogue (colleagues or peers)
- Questions and answers, problems and solutions (novice/expert)
- After-action review and lessons learned (event or activity/group)
- Capturing, documenting, interviewing, videotaping, and recording (expert/facilitator)
- Brainstorming, spontaneous ideas, and group synergy (group/facilitator)
- Extraction, identify, codify, and organize (expert/knowledge engineer)
- Advising, briefing, and recommending (subordinate/superior)
- Teaching, educating, training, and instructing (teacher/student)
- Storytelling, narratives, and anecdotes (teller, listener)
- Explaining, demonstrating, showing, and describing (technician/user)
- Mentoring, guiding, and leading (leader/apprentice)
- Presenting, lecturing, and speaking (speaker/audience)

1.6.3.2 Fora

A forum is a public space or meeting place for open expression and discussion of ideas. People overwhelmingly prefer sharing tacit knowledge through face-to-face interactions. This is because humans are hard-wired to interpret the many visible and audible clues that accompany and enhance the underlying, intended meaning of spoken words. Face-to-face encounters also greatly facilitate establishing mutual trust, an essential precursor for sharing tacit knowledge. That is, trust on the part of the recipient that they are receiving valid knowledge from someone who knows a subject, and trust on the part of the provider that what they are sharing will only be used in appropriate ways.

Individuals can share tacit knowledge anywhere that two people can meet and talk—in an office, a hallway, the street, via telephone, or e-mail. Although such interactions are not open, these are places where tacit knowledge is traditionally shared. A key advantage of one-on-one conversations is that, once trust is established, the knowledge being shared tends to be more honest and straightforward. The provider is less concerned with appearances, avoiding mistakes, and who might be listening. Many significant breakthroughs begin life as a sketch on the back of a restaurant napkin.

There are many fora that support sharing tacit knowledge among individuals and groups, including conference calls, meetings, workshops, conferences, symposia, knowledge fairs, site visits, classrooms, communities, and networks. Most of these methods can be used in traditional physical space or in cyberspace. The chief advantage of physical interaction is significantly enhanced transfer of meaning and understanding; the chief disadvantage for distributed participants is the high cost of travel. For cyberspace interactions, the advantages and disadvantages are reversed.

1.6.3.3 Technology

For about two decades, technology has been gradually shifting sharing preference from physical space to cyberspace. Early communications provided little or no visual clues to assist in interpreting the meaning of what was being said. What has changed is that a generation of users has grown accustomed to being at both ends of electronic exchange media. Providers have adapted their methods to more clearly transmit meaning and understanding without the context of visual clues ":-)," recipients have increased their skills at interpreting meaning "LOL," and both understand that additional dialogue may be needed. Equally important, global connectivity has substantially increased the average distance between participants at the same time that the cost and bother of travel have also increased substantially.

There are many technologies available for supporting the transfer of tacit knowledge between individuals and groups. Each category has many providers; some are small and local while some are massive and global. Collectively, available services number in the tens of thousands and each service may have thousands to millions of members.

- *Telephony*—Electronic voice communications among two or more individuals
- *Groupware*—Application that facilitates the work of teams, committees, and work groups
- *E-mail*—Asynchronous, electronic, text-based communication among individuals or groups
- *Chat rooms*—Synchronous, electronic, text-based communication among individuals or groups

- *Bulletin boards*—Electronic text-based sites for posting and responding to messages
- *Online forums*—Public websites for posting and discussing threaded topics of interest
- *Web portal*—Single-window access to an information space containing related content
- *Video conferencing*—Meetings among dispersed participants in which they can see each other
- *Expertise locator*—Web-based search engine that finds, lists, and links to subject-matter experts
- *Blogs*—Websites for posting and commenting on personal views or opinions
- *Microblog*—Websites for posting and responding to very brief messages
- *Media communities*—Web-based repositories for media-based content
- *Social book marking*—Websites used to list members' bookmarked web pages
- *Really simple syndication (RSS)*—Web-based alerts that monitor content updates

1.6.4 COMMUNITY KNOWLEDGE

A community is an identifiable group having similar or related culture, interests, or place with interactions, relationships, participation, or interdependency. Communities of practice are described in greater detail in Section 1.6.1. The shift from sharing knowledge between individuals to sharing among members of a community is based more on facilitated accessibility and richness of response rather than a qualitatively different process. Cross and Parker (2004) found that a knowledge worker is five times more likely to turn to another person to find something rather than an impersonal search engine. This also turns out to be much quicker and more successful approach.

Rather than struggling with search terms in the hopes of finding relevant content, a member of a community only has to send a message: "Does anyone know something about ...?" Every member of a community is a facilitator; each member has deep knowledge in their area of expertise—collectively, far more than any individual could possibly possess. Experts do not have to search; they know. Typically, one to three members of a community will respond within a few minutes. If more people know something, they will usually default to the initial responses unless they have something useful to add. The process is generally very efficient for both providers and users.

Dalkir (2005) points out that the response often contains some context as well as content: where it was found, whether the search is on the right track, and if not, where it went wrong. Most importantly, the searcher knows that the information is coming from a known and usually trusted, credible source. In other words, "talking to other people provides a highly valuable learning activity that is primarily a tacit–tacit knowledge transfer, for this type of knowledge is seldom rendered explicit, nor is it captured in any form of document."

Although both explicit and tacit knowledge are shared by communities, much of what is shared is explicit. An expert is more likely to respond to a query if they only have to take a minute or two to reference a link or send a document that already exists. They are less likely to respond if they have to take the time to write one or a few paragraphs, and will probably only do so if the question is "interesting" in some way. Experts are unlikely to respond if they have to create a page or more of text, or they will respond when (if) they have time. Individuals need incentives, such as altruism, enhancing their reputation, or self-interest to distract them from their primary work to share information through a community of practice.

1.7 SOCIAL NETWORKS

Web 1.0 technology enabled anyone, anywhere, to publish and access digital content. As great as that revolution was, it was essentially about information sharing and global accessibility. The full social impact of the World Wide Web is emerging through Web 2.0 technology that enables everyone, everywhere to interact with everyone else, virtually simultaneously. This qualitative shift in connectivity is demonstrating synergy and emergence that is leading to collective creativity of a kind never before experienced in human history.

In a well-connected world, bad product experiences can go "viral" in a matter of days or even hours; waiting until next week's management meeting to address the issue may be too late. Social activists have organized major rallies in a few hours, much to the chagrin of government planners who did not issue a permit. Alternatively, some organizations monitor what people are saying about issues of interest to them. A US presidential campaign used social networks to tip the balance of opinion in its favor.

Social networking is divided into three aspects—communities of interest, communities of practice, and networks. Communities of interest connect people who share a common personal interest or passion, such as friends, hobbies, or neighborhood. Communities of practice connect people who share a common expertise, skill, or profession, such as scientists, IT specialists, or purchasing officers. Although Web 2.0 technology has been primarily developed to support communities of interest, our task here is to adapt the popular technology to support creating and increasing value in the workplace.

The primary difference between communities and networks is scale: size matters. Communities of practice are typically on the order of 30–50 people, most of whom know each other through their subject-matter expertise and experience. Networks are larger—on the order of hundreds to millions of members, in which individual members only know a few members of a network. The large numbers of autonomous participants in networks leads to dynamic and complex behavior, coupled with loss of trust and safety.

1.7.1 COMMUNITIES OF PRACTICE

Communities of practice do not replace organizational infrastructure, functions, or business units. Rather, as stewards of knowledge, they contribute to organizational success through timely and rapid sharing and integration of knowledge to address complex problems or issues. Communities of practice provide many organizational benefits from the perspective of the participants, management, strategy, and outputs:

Participants—Communities can help people to do knowledge work, assist in addressing challenges, increase access to expertise, review and comment on proposed actions, provide a sense of belonging to a stable community, meaningful participation and contribution, a forum for learning and skill development, a source of the latest developments, enhanced professional reputation, and increased visibility.

Management—Communities can connect isolated experts, coordinate related but unconnected activities, rapidly diagnose, analyze, and resolve problems, provide quick answers to questions, reduce development time and costs, improve decision quality, coordinate and standardize across functions, help build common language and methods, enable synergy among participants, strengthen quality assurance, help to develop and retain talent, and aid in retaining knowledge when employees leave or projects end.

Strategy—Communities can develop new strategies, rapidly diffuse a new strategy, embed knowledge and expertise in larger populations, have authority with stakeholders, increase talent retention, develop new knowledge, provide a forum for benchmarking, foresee technological change, quickly react to emerging issues and opportunities, and build core capabilities and competence.

Outputs—Communities can provide tangible outputs, such as reports, standards, manuals, or recommendations. They also provide intangible outputs, such as increased participant skills, a sense of trust among members, diversity of perspectives, cross-fertilization of ideas, capacity to innovate, relationships among people, a sense of belonging, and a spirit of inquiry.

Nickols (2000) identifies six roles that need to be fulfilled for a community to succeed:

1. *Champion*—Ensure support at the highest level, communicate the community's purpose, promote the community, and ensure impact.
2. *Sponsor*—Serve as a bridge between the CoP and the organization, communicate organizational support, and remove barriers, such as time, funding, and resources.
3. *Leader*—Provide thought leadership, validate new ideas, identify emerging trends, approve membership, resolve conflicts, evaluate performance, and prioritize issues.
4. *Facilitator*—Clarify communications, encourage participation, ensure that all views are heard and understood, organize meetings, and administer the community.
5. *Service center*—Interface with all communities, ensure clarity and lack of duplication, inform communities about relevant activities, and inform others of community activities.
6. *Members*—Provide knowledge, expertise, and experience, participate in discussions, raise issues and concerns, alert members to changes, and look for ways to increase community effectiveness.

Saint-Onge and Wallace (2002) classify communities into three structures: informal, supported, and structured. They also describe the characteristics of each structure (Table 1.2). The characteristics are not absolute, but rather indicate shifting patterns across the range of community structures. There tends to be an inverse relationship between creativity and organizational impact. Informal communities tend to be more democratic, dynamic, and creative, whereas structured communities tend to be more committee-like and are more likely to produce outputs that have organizational impact. From a management perspective, the central purpose of communities is to generate creative and innovative outputs that would not likely arise from traditional committees.

Communities are groups of people with individual and collective behavior and attitudes that strongly affect community effectiveness. Given that communities are self-governing, the primary method of affecting community behavior is through carefully choosing community members. This is less important in informal communities, because those who negatively impact community activities or ignore community rules of conduct are simply ignored or removed by the community itself and there is no appeal to higher authority. Structured communities, with invited participants, are most in need of careful orchestration at the outset.

Positive community behaviors

- *Dialogue*—A free flowing exchange of ideas facilitates synergy
- *Trust*—Is essential to enable participants to present honest views
- *Safety*—Controversial opinions will not be externally attributed to participants
- *Meritocracy*—The best ideas rise to the top, based on their merit
- *Equality*—All opinions and points of view are solicited, valued, and equal
- *Outliers*—Are sought after because they likely represent innovative ideas

TABLE 1.2

Organizational Structures for Communities of Practice

Structure/Characteristic	Informal	Supported	Structured
Purpose	Discussion forum	Build knowledge and capability	Platform for joint objectives
Membership	Self-joining	Members invited	Selection criteria
Sponsorship	No sponsor	Business unit	Senior management
Mandate	Defined by members	Defined by members, endorsed by sponsor	Defined by sponsor, endorsed by members
Evolution	Organic development	Purposeful development	Business objectives
Outcomes	Member orientation	Management orientation	Strategic orientation
Accountable	To members	To business unit	To the business
Support	Endorsement	Discretionary resources	Organizational commitment
Visibility	Limited	Business unit	Enterprise-wide

Negative community behaviors

- *Discussion*—A reciprocal pounding on opposing points of view impedes synergy
- *Debating*—In which someone wins and someone loses inhibits creativity
- *Arguing*—Until participants are worn down leads to withdrawal
- *Agenda*—Representing an external constituency impedes progress
- *Positioning*—Authoritative declarations terminate meaningful interaction
- *Assuming*—Superior knowledge patronizes participants and disengages them
- *Majority*—Voting yields plain vanilla averages of current opinions
- *Consensus*—The lowest common denominator acceptable to all participants
- *Groupthink*—Represents collective opinion and rarely rises above the status quo

1.7.2 NETWORKS

A network is an interconnection among large numbers of people, communities, or organizations with common interdependencies, interests, or purpose. The Internet, and its resultant global connectivity among virtually everyone, everywhere has raised the idea of "networking" to heretofore unimaginable levels. Networks have become embedded in every aspect of society—from communications and transportation, to work and recreation, to governing countries and sustaining the planet.

Networks can range from hundreds to millions of participants. Because networks are so large, it is impossible to personally know more than a small percentage of the members. Consequently, there is a loss of trust and safety—two key community attributes. Participants must be cautious about what they post in a network, as some teenagers and employees have learned, to their dismay. Conversely, size greatly enhances the possibility of synergy and emergent knowledge.

Organizational participation in large networks typically involves staff who are plugged into external, national, or global networks related to their profession or areas of interest. As such, they perform a "gatekeeper" function—in the sense of monitoring the environment and bringing emerging external content into the organization, as a form of intelligence gathering. Gatekeepers bring external content to the attention of those who have an interest in it.

The purpose of a network is the underlying driver that causes people to become members, participate, and care about a network. Anklam (2007) identifies six types of networks, although her broad definition—almost any set of relationships—includes both communities of interest and communities of practice.

1. *Personal* networks provide assistance, advice, and fellowship. They include people we know, including family, friends, coworkers, and neighbors. They also include religious, civic, or wellness activities, as well as personal interests, such as clubs or hobbies.
2. *Idea* networks share and develop ideas through creative exchange that allows ideas to build on each other. Generally, the outcome of such conversations cannot be predicted in advance. Idea networks include innovation and advocacy.

3. *Learning* networks augment personal capacity in a particular area involving skill, expertise, vocation, avocation, or knowledge. Learning networks include communities of interest, communities of practice, professional associations, and research networks.
4. *Mission* networks are directed toward social good, in areas such as arts, culture, education, environment, health, human services, religion, and social justice. Mission networks include both nonprofit and nongovernmental groups, at local, regional, and global scales.
5. *Business* networks focus on the production and growth of revenue, profits, and returns to shareholders through increased market reach, product breadth, expertise, and knowledge. Business networks include suppliers, alliances, consulting, and consumers.
6. *Leadership* networks emphasize dialogue among managers. They reshape the frequency, intensity, and honesty of interactions among managers. External networks provide a diversity of ideas while advisory networks involve joint problem solving.

1.7.2.1 Structure

Structure is what holds a network together and how it operates. Anklam (2007) describes four aspect of network structure: governance, management, topology, and texture.

Governance is leading a network in a mutually agreed direction, balancing its operations in the context of its purpose, and maintaining relationships without an authoritative structure. All networks have explicit or implicit leadership in the form of a patriarch or matriarch; a hub, or convener; a director or committee who guide and steer a network. Governance must be flexible, attuned to the network's environment, and adaptable to emerging circumstances. Network governance is not based on an organizational mandate, charter, or business model. Rather, it gradually evolves as a network comes into being.

Management is the use of institutionalized structures for network governance. This may be necessary when the size and scope of a network is large enough that paid staff are needed to run it. It may also be needed when a network needs authority to negotiate with other organizations or institutions. Finally, it may be required when financial investments or intellectual capital are needed from partners, investors, or donors. In such cases, many or all of the trappings of organizational governance may be required, including a mandate, authority, responsibility, rights, roles, resources, and decision-making processes.

Topology is the pattern, configuration, or geometry of the nodes, links, and connectivity in a network. Network theory describes six patterns of connectivity among computers and people: bus, ring, hub and spoke, cluster, hierarchy, and mesh. There are advantages and disadvantages to each network pattern. The first five patterns are simpler to implement and operate but they have two weaknesses: like a chain, disruption of any link disrupts the flow of information to subsequent nodes and influence is disproportionately distributed to those who come above or before a node or are positioned in the center. The mesh, in which everyone is connected to everyone else, is the most robust network structure.

A clustered structure comprises both strong and weak links among participants. Strong links, among family, friends, coworkers, and colleagues provide contextual support, feedback on ideas, and a sense of community. Weak links among casual acquaintances in widely distributed, highly diverse domains, and social groups provide different perspectives on problems that are unlikely to be found among the strong links of a local cluster. A small number of weak links in a network greatly reduces search effort and increases the chances of finding different content and perspectives that lead to creativity and innovation.

Texture describes how a network is "woven" together. Density is the total number of connections among the nodes. The greater the density, the greater is the resilience and effectiveness of a network. Distance measures the number of connections through which content must pass to reach everyone in a network. It also refers to "degrees of separation"—the number of connections needed for a search to find someone. Centrality is the degree to which a network is dependent on a central hub or individual. High centrality places the network at risk should the central hub become disabled. Openness refers to the ratio between external and internal linkages. Low openness minimizes external disturbance and interruptions whereas high openness results in greater influx of new ideas, insights, and emerging trends.

1.7.2.2 Behavior

Networks comprise large numbers of autonomous participants who interact with each other in myriad, often unpredictable ways—a classic example of a complex system. The following list outlines a number of complex network behaviors:

- *Positive feedback*—Unlike diminishing returns, the larger a network the more valuable and productive it becomes. Thus the larger the network, the larger it tends to become.
- *Biological growth*—For a long period, networks require constant attention and they grow relatively slowly. At some unidentifiable point, however, networks cross a threshold, beyond which they grow exponentially and sustain themselves with no further inputs. Once biological growth becomes apparent, it is normally too late for competitors to begin taking action.
- *Synergy and emergence*—Large numbers of participants can result in synergy that yields new knowledge that could not have been created even by the most intelligent members acting individually. Alternatively, the larger the number of people who look at a problem, the greater the chance that someone will solve it.
- *Nonlinear response*—The value of a small initial effort can be leveraged by orders of magnitude. Being first with a popular application can yield a substantial market advantage over those who come later.
- *Volatility*—A network market leader can be displaced by an unforeseen disruptive technology virtually overnight. Missing a Christmas season by just a few weeks can be fatal to a new application because by the next season, the application could be obsolete.
- *Clustering*—Connectivity and activity in a network tend to become unevenly distributed even if all nodes are equal when a network is established. A limited

number of participants will often capture and exploit a disproportionate share of a network's value.

- *Instant response*—The importance of an issue often precedes its recognition by those who are impacted. Political rallies have been organized in a matter of hours; flaws in a new product are widely disseminated on the day of the product launch; or an amateur video receives a million views within a week of posting.
- *Unpredictability*—Viral growth of network postings cannot be managed. Conversely, negative messages about a product that seemingly originate "out of the blue" require immediate action.

1.7.2.3 Strategy

Whether it is called the new economy, the knowledge economy, or the networked economy, fundamentally new ways of creating wealth are emerging. Not only is knowledge becoming the core asset, not only is trade shifting from atoms to bits, but also the creation of value is shifting from organizations to networks. Value is created by active participants in a network rather than by an individual or organization. Consequently, value lies external to member organizations; it is shared by everyone in the network. Capturing value, however, is often uneven; those who own network standards or play key roles in a network typically have an advantage. Kelly (1998) described a set of strategies for success in a network economy that remain relevant a decade after they were published.

Embrace the swarm—Society is increasingly connecting everything to everything else. There is huge power in numbers. Even though individual agents in a network are limited, massive connectivity between large numbers of agents can yield unlimited possibilities for synergy and emergence. *Strategies* include: make technology invisible, embed technology into as much as possible, connect people and things, distribute knowledge, and recognize that more is different.

Increasing returns—The value of a network is proportional to the square of the number of participants. In a network, success is self-reinforcing: the more you have, the more you get. The law of diminishing returns does not apply to networks. Biological growth in technological systems changes everything. Innovations may become significant before they are recognized as such. *Strategies* include: check for external emergence, coordinate smaller webs, create feedback loops, protect long incubations, and recognize that a few home runs pay for everything.

Plenitude—In a network, the more plentiful something is, the more valuable it is. As operating systems, computer applications, and technology become more ubiquitous, their value to everyone increases because more people use them. The value of a technology increases exponentially as the number of systems that it interacts with increases linearly. *Strategies* include: touch as many networks as possible, maximize the opportunities of others, avoid proprietary systems, and do not seek refuge in scarcity.

Focus on the web—Allegiances are shifting from organizations to networks. An organization's focus should shift from maximizing internal value to maximizing network value. Exchanging knowledge-based products and services requires standards that are key to interacting. *Strategies* include: maximize network value, seek the highest common denominator, use existing standards in products, and automate products.

No harmony, all flux—Networks shift organizational dynamics from occasional change to constant flux. Harmony and equilibrium can lead to stagnation and irrelevance. Organizations must negotiate a path between the ossification of order and the destruction of chaos. Sustainable innovation requires persistent disequilibrium without succumbing to or retreating from it. *Strategies* include: exploit flux rather than prohibiting it; you cannot install complexity—it must be grown; preserve the core; and let the rest flux.

Relationship technology—Networks are supported by technology but are developed through relationships. Relationships blur the distinction between employees, clients, and stakeholders as they interact and collaborate. Because relationships involve multiple participants, their combined value exceeds any individual investment. Developing trusting relationships requires respecting partner, client, and stakeholder privacy. *Strategies* include: make customers as smart as you are, connect customers to customers, choose technology that connects people, and think of customers as unpaid employees.

Opportunities before efficiencies—Every opportunity seized launches new opportunities. Pursue new opportunities in preference to optimizing existing products and services. Productivity is for machines; opportunities are for humans. In a network, doing the right thing is much more important than doing things right. *Strategies* include: automate what can be automated, explore unknown territory, the upside potential of one great success outweighs the downside of many failures, and seek cascading opportunities.

1.8 KNOWLEDGE FLOW

To be useful, knowledge must flow both vertically and horizontally across an organization. Stop or impede the flow of knowledge and an organization cannot function. Knowledge not only flows, but also changes both its form and content as it moves from stage to stage of development. Knowledge flow is described from three perspectives: mobilization, across management regimes, and to clients.

1.8.1 KNOWLEDGE MOBILIZATION

Knowledge mobilization is bringing diverse knowledge assets and knowledge creation capacity from multiple sources to bear on a problem or issue. In the context of knowledge manageability, this means integration and interoperability across management regimes, KM functions, multiple domains of knowledge, and diverse content. Three aspects of knowledge mobilization are considered here: transitions, integration, and interoperability.

Transitions: Without transitions across process boundaries, work remains isolated and KM cannot function as a whole. An organization must integrate the four management regimes into a continuous process from creation to application. Transitions from one regime, stage, or domain to another do not occur passively. Organizations have to implement processes that actively advance knowledge across processes. For example, prioritizing existing knowledge requires divergent, forward chaining through knowledge work to desired outcomes. Conversely, identifying knowledge needs requires convergent, backward chaining from desired outcomes through knowledge work to knowledge assets.

Integration is combining or incorporating content and processes into a whole to achieve a common objective. The outcome of integration is a convergence of diverse content and processes to produce a holistic result, in which individual elements cannot be distinguished. Integration requires underlying standards—an intellectual infrastructure that enables and supports the use of diverse content and processes to produce a common output. Integration may require agreements with external sources to allow access to IP. It often requires interfaces between diverse sources of knowledge to allow them to be processed on a common base, such as a GIS map. It may also require interpretation from source measurements to different measures required by different models. Finally, it may require methods to aggregate knowledge from smaller to larger scales.

Interoperability requires compatibility of objectives and architectures from two or more autonomous sources that permits and enables them to interact and jointly operate to achieve a common objective. The outcome of interoperability is an assemblage or aggregation of diverse functionality into a holistic result, in which individual components can be distinguished. Interoperability can be viewed from four perspectives: jurisdiction, hierarchy, function, and system:

1. Most complex issues require interactions across jurisdictional boundaries. Consequently, the extent to which a content owner is ready, willing, and able to share it affects the extent to which it can be integrated into a larger whole.
2. Most organizational frameworks recognize three hierarchical levels: strategy, management, and operations. Knowledge intended for developing strategy is typically too coarse for operational use while that intend to support operations is too detailed for strategic use.
3. Functions are related organizational actions, activities, or work that collectively contributes to a larger objective. Organizational functions, such as human resources, finance, and production often operate as semiautonomous units with incompatible content and management systems.
4. Most content is acquired, managed, and transferred through systems. Analysis of complex issues often requires interaction among more than one system. System interoperability can be achieved through a common architecture for multiple systems, modular design that allows subsystems to plug into larger systems, or interface modules that connect incompatible systems.

1.8.2 FLOW ACROSS MANAGEMENT REGIMES

Four types of organizational knowledge have been defined: authoritative, explicit, tacit, and innate. Organizational success requires an asymmetric two-way flow across the four types. The flow is asymmetric because it must be actively transferred ("pumped uphill") from innate to authoritative whereas flow in the other direction tends to be passive ("flows downhill")

1.8.2.1 Engaging Innate Knowledge

Innate knowledge is part of the essential character, nature, or intellect of an individual. It is often viewed as a natural talent, aptitude, or ability to do certain things. Innate knowledge can be enhanced through training and practice, but it cannot be transferred; it can only be used by the individual who has it. As indicated previously, using innate knowledge and sharing tacit knowledge depend on engaging individuals more than anything else. Engagement emphasizes intrinsic incentives such as sense of ownership, meaningfulness, self-esteem, and enjoyment. Through engagement, people personally commit to and become truly involved in a task or activity. They do so not because it is asked of them, or they expect something, but rather because they want to; because they enjoy doing it.

Tosti and Nickols (2010) identify positive signals that can be sent by managers to promote engagement as well as negative signals that can inadvertently damage it:

- *Value—Positive*: Agree how both parties benefit, commit time to interaction, listen, ask questions about their ideas, ask for help, and advice. *Negative*: Seldom interact, not inviting people to planning meetings, and ignoring suggestions and ideas
- *Honesty—Positive*: Jointly review progress and performance, freely share helpful information. *Negative*: Close monitoring, tight controls, and withholding information
- *Ability—Positive*: Work together, mutual clarity of expectations. *Negative*: Close monitoring and tight controls
- *Judgment—Positive*: Delegate decision-making authority and ask to be kept informed. *Negative*: Require approval for every decision
- *Sincerity—Positive*: Mutual, up-front agreements, including changing assignments. *Negative*: Use overly tight contracts and withhold information

1.8.2.2 Harvesting Community Knowledge

Harvesting knowledge from communities is a higher-level equivalent of capturing tacit knowledge from individuals. A key purpose of harvesting community knowledge is to transform it into authoritative knowledge. Thus, it has already been made explicit by the community. If the knowledge of interest is tacit, methods of capturing it from individuals or groups are discussed in the next section.

Management interest in the output of communities ranges from supported (structured) through encouraged and promoted (semistructured) to valued or ignored (informal). Structured communities typically have processes built into their charters that are designed to transform community outputs into authoritative knowledge (decisions and actions). Semistructured and informal communities require explicit processes to facilitate the transformation.

The external bridging functions of communities described previously—the champion, sponsor, and knowledge center—are keys to harvesting community knowledge. Their primary role is to ensure that community outputs of strategic interest are brought to the attention of senior management, in ways that they relate to, accompanied by recommended actions. Ideally, a predetermined process should be embedded into the decision-making infrastructure to facilitate such transformations.

External communities and networks of interest to the organization almost invariably fall into the informal category from an organizational perspective. One or a few employees provide a gatekeeper (enabling, not limiting) function with respect to external communities. Their first task is to bring external community developments of interest to the organization into the organization. These could take many forms, including best practices, events or situations, recent developments, innovations, or lessons learned. The next task is to validate the input from an organizational perspective. This is an ideal role for communities of practice, with their ability to quickly respond to emerging trends and issues. Once the analysis is complete, recommendations should be moved into the authoritative hierarchy.

1.8.2.3   Capturing Tacit Knowledge

In addition to using tacit knowledge to create and validate new knowledge, it is also important to transform it into explicit form so that it can be preserved as an asset for reuse. Capturing tacit knowledge begins by identifying and prioritizing what should be captured. Attempting to capture everything is both unnecessary and infeasible. Some tacit knowledge is held by many individuals and does not need to be captured. Other knowledge loses its value rapidly and there is little point in capturing it. Critical knowledge held by one individual should be continuously captured and updated. The consequences of unexpectedly losing such knowledge could be substantial. Core expertise held by one or a few individuals should also be captured and updated as significant milestones are accomplished. Important lessons learned and best practices should be documented and disseminated as they occur. Dalkir (2005) summarized a number of methods for capturing tacit knowledge:

- *Structured interviews*—Involve specific goals and questions for a knowledge acquisition session. The interviewer needs good communication skills, an ability to conceptualize, and a good grasp of the subject. Both open and closed questions are used. The former provides context and responses that were not specifically asked for while the latter helps to structure the response.
- *Stories*—Are narratives that are communicated informally. Stories provide a rich context that results in retaining the memory longer than information without context. Stories can increase organizational learning, communicate common values, and serve to capture, code, and disseminate tacit knowledge.
- *Being told*—An interviewer observes a task or process. There is an interaction between the interviewer who expresses their understanding and the interviewee who clarifies and validates the knowledge. The interviewee "thinks aloud" as they undertake the work and the interviewer notes the information used, questions asked, alternatives considered, and actions taken.

- *Observation*—While an expert addresses a problem or undertakes a task, an observer watches, usually accompanied by an audio or video recording. Although the underlying knowledge cannot be observed directly, the actions of the expert can be. The recording forms a permanent record of how a task is accomplished.
- *Road maps*—Are facilitated problem-solving meetings that follow a predetermined agenda. They are used to resolve day-to-day problems in a public forum that can lead to guidelines and standards for continuous process improvement.
- *Lessons learned*—Capture experience in group settings. They document a structured retrospective history of significant events that occurred in an organization's recent past, as described by the participants. Lessons learned involve planning, reflective interviews, distillation, documentation, and dissemination.
- *Action learning*—Is based on learning by doing. Small groups meet regularly, brainstorm alternatives, try out new things, report on progress, and evaluate the results. This is well-suited to narrow domains and specific issues.

1.8.2.4 Authorizing Explicit Knowledge

Transforming explicit knowledge into a funded organizational mandate is the single greatest challenge confronting KM. Yet, this process receives scant attention in the KM literature. The process of transforming explicit knowledge into authoritative knowledge is well understood by bureaucrats who work in enterprise-level staff positions. They simply do not call it by this name; rather, they view it as "getting something approved." Equally important, they are not strongly committed to or involved with any particular initiative. Their job satisfaction comes from having correctly followed the submission process, regardless of whether or not a submission is approved. There are two aspects of authorizing the use of explicit knowledge—establishing and sustaining an initiative.

There are five techniques for establishing a KM initiative: understanding, experience, resource limitations, management concerns, and the submission:

1. *Understanding: Keep it simple.* Managers do not fund things they do not understand. Although knowledge managers must understand the underlying complexity and be able to drill down to it, presentations and briefings should be simple. Each executive will come to understand KM in a way that they can relate to. A single metaphor is unlikely to work for everyone.
2. *Experience: Prepare adequately.* Brief every decision maker prior to the meeting. Solicit and rectify or negotiate reasonable objections to the submission. The knowledge manager should participate in briefings to witness nonverbal as well as verbal reactions. If the preparation is adequate and the compromises are reasonable, one should expect a strong majority of approvals.
3. *Resources: Pick low-hanging fruit.* Start with low-cost, small-effort, small-workload activities. These will demonstrate real benefits of KM and develop

trust that the KM group knows what it is doing. Ultimately, the success of KM will depend on adequate funding at the enterprise level.

4. *Management: Think big but start small.* Divide the initiative into low-risk, short-term project-scale undertakings with concrete, measurable, high-return, early deliverables. Select projects that solve recognized business problems. This should be done in the context of an overall strategy that provides a place for initial project-scale developments.

5. *Submission: Leadership is essential.* If a concern is expressed that was not mentioned during the pre-briefings, or that has been expressed and adequately addressed, or that has little merit, the senior executive must move the initiative forward to avoid an endless cycle of submissions.

The second greatest challenge facing KM is sustaining an initiative over a long term within an ever-changing short-term environment. To be sustainable, KM must reach a "critical mass." It must become institutionalized in every aspect of an organization's infrastructure. There are six techniques for sustaining a long-term initiative in a short-term environment: leadership, governance, reorganization, priorities, support, and culture:

1. *Leadership* change is a key issue. New leaders have a great mass of new material to absorb. Getting their attention, explaining the importance of KM, and securing their support in the context of many organizational priorities takes several months to a year. Short-term deliverables must be scheduled within the executive's expected tenure.

2. *Governance* plays an important role in transitions across leadership changes. Autocratic approaches to KM have resulted in impressive short-term accomplishments that generally revert back to previous, often more deeply entrenched, silos than previously. Representative, federated governance is the only sustainable form of KM governance.

3. *Positioning*—There are many places and levels where KM can be positioned in an organizations. Each position has a particular perspective, which may require a substantial shift from a previous direction of KM. Positioning KM within one or two steps of the executive committee helps to mitigate reorganization issues.

4. *Priorities*—The world is changing at an ever-increasing pace and organizations must adapt accordingly. New leadership often brings new priorities. Priorities on paper can change much faster than underlying initiatives involving commitments and momentum. KM should be closely aligned with an organization's business strategy, which reflects long-term directions rather than short-term priorities.

5. *Support*—Long-term initiatives are typically provided with sufficient resources to begin work, with the expectation that they will return to demonstrate initial success and request resources to continue the effort. However, each return for continued funding represents a risk, given the possible changes that may have occurred since the previous approval. Outputs must be delivered as and when promised. An initiative must also be prepared to adapt to a changed environment.

6. *Culture*—Culture change is so grand in scale and so sweeping in scope that attempting to implement it directly is unlikely to succeed. Rather, KM should address elements of the knowledge infrastructure, such as rewarding desired behaviors, developing favorable policies, and implementing systems that help people do their work. Collectively, each success will move the culture, imperceptibly but assuredly, toward that of a knowledge organization.

1.8.3 KNOWLEDGE PRODUCTS AND SERVICES

Knowledge-based organizations generate, use, and transfer knowledge-based products and services to clients. A knowledge services system provides the underlying infrastructure and processes that enable knowledge services. Yet a knowledge services system does not actually exist as a tangible entity in the real world. It is an artificial construct that combines many components and flows across organizations, sectors, and the society they serve. To understand knowledge services, we must bring together in one place all the processes that collectively transform new knowledge into improved outcomes for clients and stakeholders. We have to be able to "heat water at one end and see steam coming out at the other."

Simard et al. (2007) developed a systems model that traces the flow of knowledge products and services from their original generation to their final use by clients or stakeholders. The system is described as a horizontal value chain that embeds, advances, or extracts value (Figure 1.5). The framework describes the process adequately to enable understanding, permit measurement, and support management. The framework is independent of content, issues, or organizations. It is designed at an organizational level, but is scalable both upward and downward. The primary driver is an organization's mandate or business; a secondary driver is client needs. The framework can function from either a supply or demand approach to knowledge markets. There are two levels of resolution—performance measurement and classifying

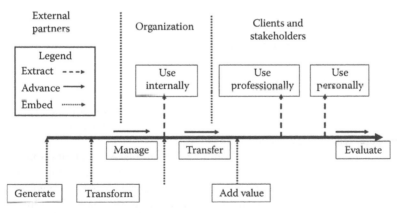

FIGURE 1.5 Knowledge services value chain. (From Simard, A. et al., *Understanding Knowledge Services at Natural Resources Canada*, Department of Natural Resources, Office of the Chief Scientist, Ottawa, ON, 80pp, 2007. With permission.)

service-related activities. There are four types of knowledge products and services: generate content, develop products, provide services, and share solutions.

The knowledge services value chain includes nine stages, which starts with generating content and ends with a final evaluation of the process. Downstream services generally have higher embedded value than upstream services. The boundary between what is internal to an organization and what is external varies. Figure 1.5 shows an S&T organization that uses external partnerships to generate new knowledge and develop knowledge products and transfers products and services to external clients and stakeholders.

The following list outlines the nine stages of the knowledge services value chain. Stages marked with (*) are necessary for the system to function because everything must flow through them, while others may be bypassed without stopping the system. At least one use stage (4, 7, or 8) is needed, however, for knowledge services to yield value to users:

1. *Generate**—Content with intrinsic value and potential utility *must* be generated as the first stage of the knowledge services value chain.
2. *Transform*—Content is transformed into products or services to increase its utility or value to users.
3. *Manage**—The flow of knowledge products or services *must* be enabled to permit internal use or transfer to external users.
4. *Use internally*—Knowledge products and services are used internally to accomplish business goals and objectives.
5. *Transfer**—Knowledge products and services *must* be transferred to external users to enable external use.
6. *Enhance*—Work is done by intermediaries to increase the availability, utility, or value of knowledge products and services.
7. *Use professionally*—Knowledge products and services are used by clients with sector-related knowledge to benefit an identifiable sector.
8. *Use personally*—Knowledge products and services are used by clients for personal benefit.
9. *Evaluate*—The system is evaluated to improve its performance in supplying or fulfilling the demands of knowledge markets.

1.9 KNOWLEDGE TRANSFER

The short-term competitiveness and long-term sustainability of an organization depends on its ability to transfer its knowledge products and services to clients and stakeholders. Knowledge transfer is an act, process, or instance of conveying, copying, or causing knowledge to pass from one person, place, or situation to another. Although knowledge transfer can be said to occur between every stage of internal knowledge work and subsequent stages, internal transfer generally involves many small steps that are integrated under knowledge flow (Section 1.7). This section emphasizes transfer between an organization and its clients and stakeholders, which generally involve a relatively small number of significant actions. Unlike sharing, which is primarily based on voluntarism, transfer is driven by an individual or organizational responsibility or mandate. Unlike exchange, transfer is primarily one way—from a provider or source to a recipient or user. Knowledge transfer is divided into three parts: process, strategy, and markets.

1.9.1 TRANSFER PROCESS

Transferring knowledge to clients and stakeholders involves more than passive delivery of outputs. There are three distinct transfer processes: transactions, interactions, and delivery. Transactions involve conducting or carrying out business to enable the transfer of rights and limits to use, reuse, or redistribute intellectual products or services. This is done through methods such as giving, lending, licensing, exchanging, or selling. Interactions involve enhancing the capacity, readiness, and/or willingness of users to understand and apply outputs to solve their problems. Interaction methods include provide, advertise, explain, promote, support, or intervene. Various methods can be used to deliver products and services, including publishing, proclaiming, disseminating, handing out, sending, or providing. Similarly, various communication channels are used to deliver knowledge, including on- or off-site interactions, kiosks, mail, World Wide Web, e-mail, telephony, or fax.

1.9.2 TRANSFER STRATEGY

A key strategic decision is positioning and distributing knowledge products and services among different users and uses. Different knowledge users and uses require different types of knowledge outputs. Each category of user and use has a set of attributes that determine the most appropriate methods for transferring knowledge products and services (Table 1.3). The attributes in Table 1.3 are not categorical. Rather, they show a progression across a spectrum of knowledge transfer methods, ranging from rich interactions (top) to maximizing reach (bottom).

TABLE 1.3

Knowledge Transfer Attributes and Methods for Different Users and Uses

User	Use	Purpose	Difficulty	Audience	Media, Methods
Internal	Specific problem	Intervention	Unique	One	Customization, conversation
Partner	Joint problem	Collaboration	Complex	Few	Official report, meeting
Body of knowledge	Research	Understanding	Conceptual	Few	Scientific paper, symposia
Intermediaries	Enhancement	Development	Complicated	Some	Specifications, diagrams
Practitioners	Application	Management	Professional	Many	Technical report, demonstration
Individuals	Personal benefit	Awareness	Popular	Large numbers	Description, self-help

Transfer to internal users is closest to the rich end of the spectrum because interactions tend to be unique. It is intended for solving a specific problem, where a decision to use a knowledge output is typically made by one or two persons. Transfer typically takes place through conversations and involves customization of outputs. Transfer to partners is generally complex. It is usually intended for a few individuals who may use outputs for governance or intervention. Transfer takes place through meetings and reports. Transferring concepts to the body of knowledge is at the lower limit of rich transfers. Although broadly published, the intended recipients are the few scientists who are at the "cutting edge" of a (normally narrow) area of research. Transfer takes place through scientific papers and symposia.

Transfer to intermediaries is in the upper-middle region of the transfer spectrum. It involves complicated content that is intended for those intermediaries who can further develop or commercialize outputs. Transfer takes place through technical specifications and diagrams or blueprints. Transfer to practitioners is in the lower-middle region of the transfer richness spectrum. It involves professional knowledge that is transferred to many practitioners through technical reports and demonstrations. Transfer to individuals for personal interests is in the reach region of the transfer spectrum. It involves popularized, descriptive material intended for large numbers of potential users who have sufficient knowledge to help themselves at a personal level.

The most appropriate mechanisms, channels, and media for transferring knowledge outputs at the ends of the spectrum are generally self-evident. Highly interactive conversations, meetings, or scientific fora are typically used for rich content, while the dissemination capacity of the World Wide Web and traditional mail are used when reach is important. As a corollary, the most appropriate methods at either end of the transfer spectrum are generally inappropriate and ineffective at the other end. Selecting appropriate transfer methods is more ambiguous in the middle regions of the spectrum, which blends some aspects of both richness and reach.

Transferring specific knowledge outputs is not limited to any particular channel or media. Rather, multiple channels and media are often used, as long as they are consistent with the nature of the knowledge outputs and needs of users. Media consistency is important, however. For example, providing a web-based form to request a service is negated by also requiring a signature to be mailed in separately or an office visit to obtain materials not available through the web. The latter merely transfers the effort of coding a request to the requester along with the annoyance of not actually providing a full online service.

1.9.3 KNOWLEDGE MARKETS

A knowledge market is the infrastructure that enables, facilitates, or supports the transfer of knowledge products and services among multiple providers, intermediaries, and users. Knowledge markets affect what flows between providers and users, how it flows, and its attributes, such as quality, timeliness, or usefulness.

Understanding the challenges and opportunities of such markets allows development of a knowledge transfer strategy to best position an organization to function, succeed, and lead in the twenty-first century knowledge economy.

Knowledge markets have been described by Stewart (1997), Davenport and Prusak (1998), and Simard (2006), among others. The traditional view of a market is that of sellers and buyers interacting with each other through transactions facilitated by the market infrastructure. In knowledge markets, providers and users replace sellers and buyers, IP is the primary product being exchanged, and money is not essential for a transaction. The term "users" implies more than passive communications or simply delivering knowledge-based products and services to an audience. Rather, it implies active interactions that facilitate and support the uptake and use of knowledge to yield benefits to users.

Knowledge markets normally comprise large numbers of participants. They exist and function irrespective of an organization's conscious participation. An organization can play multiple roles in knowledge markets and it will likely play different roles in different markets. From a content perspective, an organization may be a provider, a user, or an intermediary. As a participant, it may own the market infrastructure, it may manage it, it may develop it, or it may simply participate. With respect to support, it may be a champion, facilitator, provide funding, or develop the infrastructure.

Knowledge markets function through six types of processes, ranging from simple to complex: communications, exchange, parallel, sequential, cyclic, and network.

Communications involves one-to-many, one-way dissemination of information and knowledge products that is managed through authoritative hierarchy. Communications tend to focus on passively delivering a message to an audience. For example, government agencies and businesses want recipients of a message to believe that their position on issues is the most appropriate. Communications products require formal review and approval prior to dissemination and they must adhere to approved formats and branding.

Communications is a marketing process in the sense that many organizations are competing for attention and influence in a marketplace of ideas. Communication is used to increase awareness of organizational brands, positions, activities, or accomplishments. Messages are crafted to balance facts with positioning the organization. The balance ranges from predominantly positioning (political statements) to predominantly factual (peer-reviewed publications). Communication techniques include interviews, press releases, brochures, advertising, presentations, informal documents, formal reports, and publications.

Knowledge exchange markets involve active, one-to-one, two-way transactions that are managed through the organizational infrastructure. The exchange may involve a request for and transfer of IP (accessing and downloading a document from a repository), exchange of equal-value IP (among partners), or exchange of IP for other considerations (e.g., agreement, fee, feedback, and visibility). Exchange markets are similar to traditional markets comprising buyers and sellers, although an exchange of money is not necessary and the process may involve significant provider/user interaction in addition to the transaction.

Knowledge exchange markets have a number of attributes:

* Providers are mandated to create and disseminate content. They also have an interest in continuing relevance in the information society, thus insuring a continuing supply.
* The amount and nature of the content made available is determined by a broad range of provider mandates, roles, and capacities.
* Users need content but have some to considerable difficulty in finding what they need, filtering relevant content, and applying it to their situation.
* Interpretation, adaptation, and application of the content are determined by a broad range of user needs, roles, and capacities.
* Content is technical, highly diverse, and interpreted from many different perspectives and at many scales.
* Content can yield high value to users but because it is relatively abundant, and user budgets are often limited, its value may be less than the cost of production.

Parallel knowledge markets involve transferring knowledge-based products and services from two or more providers and/or to two or more users. Participation is managed through partnership agreements. Parallel markets may be either one way or two way. This approach is generally used when knowledge products or services are jointly developed, owned, or used. Other than needing agreements on roles, rights, authorities, and responsibilities of participants, parallel markets are similar to exchange markets.

Sequential knowledge markets involve multiple organizations that sequentially produce, enhance, and transfer knowledge-based products and services along a content "supply chain" from origin to final use. Output from one organization becomes input to the next organization in the sequence. The role of an organization is typically to receive inputs from the previous organization, transform them into outputs, and transfer them to the next organization. The role may extend up the production chain to monitor and facilitate content flow to ensure a timely supply of inputs. It may also extend down the production chain to monitor and facilitate content flow to ensure adoption and use to achieve outcomes.

Sequential knowledge markets have two significant shortcomings. First, content can be altered or the chain can be broken at any stage, with potentially serious consequences. For example, a local outbreak of a highly contagious disease is reported to the World Health Organization, which, in turn, sends an alert to health officials around the world. However, if a regional health officer fails to pass the information on to local hospitals, there could be a significant delay in diagnosing the symptoms presented by a traveler, which, in turn, could substantially increase the number of cases and mortality.

Second, it is also important for each stage to understand the limitations of previous stages and the needs of subsequent stages. For example, a scientist could develop a genetically modified strain of wheat that, if commercialized and planted, could preclude trade in all wheat from a large region, due to a risk of cross-fertilization coupled with a recipient country's trade barrier for all genetically modified crops.

Despite their drawbacks, however, sequential knowledge markets are relatively common because many organizations often process just one stage of the flow of complex content from its origin to final use.

Cyclic knowledge markets are unique to knowledge. They are best exemplified by the science and technology "knowledge cycle," in which existing knowledge is used to create new knowledge in an endless spiral. A cyclic knowledge market is created when the knowledge services value chain (Figure 1.5) is curved so that it closes on itself to yield a cycle with no beginning or end. The elements or stages of a cyclic market are similar to those for a sequential market except that there is no "final valued product." Rather, many agents embed, advance, or extract value in a continuously flowing knowledge cycle.

In a supply- or capacity-driven market, evaluating performance is the last stage of one iteration of the cycle, as is typical of government programs. In a demand- or need-driven market, market analysis is the first stage of the cycle, which is essential for business survival. In either case, the tendency is to focus on outputs because it is very difficult to trace the flow of knowledge forward from its origin to an end use or backward from a use to its point of origin. Note that this is not a free market. Organizational mandates and resources govern what and how much is created, developed, and transferred.

Network knowledge markets involve interactions among large numbers of participants in a "knowledge ecosystem," in which everyone interacts with everyone else. Each participant may play different roles (creator, provider, intermediary, or user) in different situations. Network knowledge markets are not simply larger, faster, and more complex than traditional markets; they are qualitatively different. Network value results from connectivity, so that open networks are superior to closed networks because the former include more nodes and diverse weak links. Further, because larger networks provide more value than smaller networks, success tends to lead to more success in a network market.

Network knowledge markets are founded on four principles (Sunstein 2006):

1. *Openness*—Collaboration is based on candor, transparency, freedom, and accessibility.
2. *Peering*—A "horizontal voluntary meritocracy" is based on personal values, such as altruism, self-interest, or enjoyment.
3. *Sharing*—The increased value of common products benefits all participants.
4. *Acting globally*—Products are created through knowledge ecosystems rather than by individuals or organizations.

Network value has a number of unique attributes (Kelly 1998):

- Network value is proportional to the number of participants squared. As the size of a network increases, the value of participation increases exponentially.
- Value is created by the collective work of all, not by an individual or organization.
- Network value is external to member organizations. Common value is shared by all members.

- Individual members gain by capturing common value and adding proprietary value that they can market. However, capturing network value is
  often uneven.
- In a network market, IP is more likely to be accessed or licensed
  rather than owned. This enables faster adaptation in rapidly changing
  environments.

1.10   INTEGRATING KNOWLEDGE SERVICES

Most knowledge work involves more than one service. KM lies at the top of an
integrated hierarchy that has information technology at its base and information management in the middle. Broadly speaking, IT provides a technological
infrastructure of hardware and software systems and networks; IM focuses on
an IP infrastructure of terminology, standards, and information flow, while KM
emphasizes a collaborative infrastructure of communities, innovation, and synergy. A business framework should be used to structure and integrate IT, IM, and
KM activities, such as providing repositories, supporting sharing, and facilitating
collaboration, to support knowledge work undertaken by an organization. It is
through their contributions to knowledge work that IT, IM, and KM help to create
value in an organization.

An interaction matrix between the knowledge work performed by Defence
Research and Development Canada (DRDC) and the knowledge services needed to
support the work developed by Simard and Jourdeuil (2011) provides an example of
service integration (Table 1.4). The need for a service to support knowledge work is
indicated by an "x."

The same or similar services, such as repositories, templates, or communities, are
repeated across multiple types of work. This is done for completeness within each
type of work. This, in no way, implies that there should be multiple, independent
services. In some cases, the same service can support multiple types of work with
little or no change. In other cases, related services could be identifiable components
or modules of a larger enterprise-wide system. Strategically, once an overall architecture is developed, activity-based modules, such as repositories for records, intelligence, project management, or lessons learned, can be developed and implemented
one at a time, as resources permit. It also enables each module to be maintained and
updated without shutting down an entire system. Further, it allows each module to
be tailored to the needs of specific activities within an overall context. Finally, a
modular design greatly increases the likelihood of success of a much larger overall
undertaking.

In the DRDC example, 42% of all services are based on IT, 36% are based on
IM, and 22% are based on KM. However, the distinction between IT, IM, and KM
is not all that useful from a work perspective, except in a technical sense of how
they are executed. Of 161 of their detailed work/service descriptions, only 33%
are associated with only one of the three service groups. The majority (56%) are
associated with two, usually a combination of IT with IM or KM. Another 11%
are associated with all three.

TABLE 1.4
Interaction Matrix between Knowledge Work in Defense R&D Canada and Knowledge Services

Work	Access Sites	Search Content	Acquire Content	Work Process	Decision Process	Subject Repository	Record Repository	Templates	Metadata	Expert Locator	Contact Info	Solicit Input	Analysis Tools	Synthesis Tools	Collaboration	Community	Social Network	Capture Experience	Data Management	Specific Tools
Inputs																				
Monitor	×	×	×			×			×						×	×	×			×
Analyze	×	×	×		×	×	×	×		×			×			×				
Needs					×	×	×	×			×	×		×					×	
Prioritize				×	×		×	×							×					
Establish			×	×	×	×	×	×	×				×		×					×
Transformation																				
Programs				×	×	×	×	×	×						×	×		×		
Services				×	×	×	×	×	×							×			×	×
Create	×	×	×			×	×			×				×	×	×			×	
Produce	×	×	×			×	×	×				×			×	×	×		×	×
Mobilize	×	×	×	×	×	×	×			×				×					×	
Learn	×	×	×	×	×	×	×	×	×	×					×	×		×		×
Outputs																				
Reports	×			×		×	×	×							×					
Innovation	×	×	×			×	×			×	×	×	×	×	×	×	×	×		
Integration	×				×	×	×	×	×						×	×	×		×	×
Advice	×	×	×	×		×		×		×	×		×	×		×		×		
Mitigation						×		×			×	×	×	×	×					
Adaptation	×	×	×		×	×	×			×			×	×				×		
Markets																				
Communicate				×	×	×	×	×				×				×				
Exchange	×		×			×		×	×			×								×
Parallel	×	×	×			×		×	×							×	×			
Sequential	×	×	×			×		×	×			×				×	×			
Cyclic	×	×	×			×						×				×				
Network	×	×	×			×						×					×			
Totals	16	13	15	9	11	22	15	16	9	7	4	9	6	7	11	14	7	5	6	7

1.11 CONCLUSIONS

The knowledge services strategic framework provides a road map for becoming a knowledge organization. It spans the full depth and breadth of business functions and interests. It ranges from unstructured approaches for leveraging tacit and collective knowledge to enhance knowledge creation to the structured transformation of explicit knowledge into authoritative knowledge for supporting decision making and taking action.

Undertaking the entire strategy at once is neither feasible nor desirable. It is infeasible because the total investment that would be required greatly exceeds the resources that would likely be available at any one time. Although the overall return on the investment in KM will be substantial, the work will, realistically, have to be undertaken over a period of several years.

It is not desirable because large, enterprise-scale technology projects have a long history of failing to accomplish their goals. In addition, changing an organization's culture is a long, slow process that cannot be rushed. Such changes must be carefully planned and managed, one step at a time. Further, the level of uncertainty increases unacceptably the farther out one projects into the future. This partially results from the complexity of the work, in which early results will reveal unanticipated issues that need to be resolved. It also results from a rapidly changing environment, in which today's long-term plans will be obsolete before they can be fully implemented.

The knowledge strategy most likely to succeed is to "think big but start small." Thinking big provides an overall direction toward which an organization could set a course for the future. Starting small yields initial project-scale deliverables with a high probability of succeeding. Most importantly, it enables learning by doing, and revising long-term plans in the context of early results and environmental changes.

REFERENCES

Allee, V. (1997) *The Knowledge Evolution*. Butterworth-Heinemann, Boston, MA.

Anklam, P. (2007) *Net Work: A Practical Guide to Creating and Sustaining Networks*. Butterworth-Heinemann, New York, 268pp.

Boyett, J. H. and J. T. Boyett (2001) *The Guru Guide to the Knowledge Economy*. John Wiley & Sons, New York.

Cooper, A. (1996) Three models of computer software. *Technical Communications*, p. 234.

Cross, R. and A. Parker (2004) *The Hidden Power of Social Networks*. Harvard Business School Press, Boston, MA.

Dalkir, K. (2005) *Knowledge Management in Theory and Practice*. Elsevier/Butterworth-Heinemann, Oxford, U.K.

Davenport, T. H. and L. Prusak (1998) *Working Knowledge*. Harvard Business School Press, Boston, MA.

Dixon, N. (2010) The three eras of knowledge management. http://www.nancydixonblog.com/ (accessed on October 21, 2010).

Drucker, P. F. (1998) The coming of the new organization. In *Knowledge Management*. Harvard Business School Press, Boston, MA.

Drucker, P. F. (1999) *Management Challenges in the 21st Century*. Harper Collins, Inc., New York.

Eagan, G. (1994) *Working the Shadow Side: A Guide to Positive Behind the Scenes Management*. Jossey-Bass Publishers, San Francisco, CA.

English, M. J. E. and W. H. Baker, Jr. (2006) *Winning the Knowledge Transfer Race*. McGraw Hill, New York, NY, 304pp.

Garfield, S. (2010) 10 Reasons why people don't share their knowledge. http://sites.google. com/site/stangarfield/10reasonswhypeopledontsharetheirknow.pdf (accessed on October 21, 2010).

Hamel, G. and C. K. Prahalad (1994) *Competing for the Future*. Harvard Business School Press, Boston, MA.

Handy, C. B. H. (1995) Trust and the virtual organization. *Harvard Business Review*. May/ June, 7pp.

Hoffert, P. (2000) *Connected Communities*. Stoddart Publishing, Toronto, Ontario, Canada.

Holmes, D. (2001) *E.gov: E-Business Strategy for Government*. Nicholas Brealey Publishing, Naperville, IL.

Jones, P. H. (2000) Knowledge strategy: Aligning knowledge programs to business strategy. Presented to *KMWorld 2000*, Santa Clara, CA, 10pp.

Kelly, K. (1998) *New Rules for the New Economy*. Viking Press, New York, 179pp.

Kurtz, C. F. and D. J. Snowden (2003) The new dynamics of strategy: Sense-making in a complex and complicated world. *IBM Systems Journal* 42(3), 462–483.

Leonard, D. (1995) *Wellsprings of Knowledge*. Harvard Business School Press, Boston, MA.

Lucky, R. W. (1989) *Silicon Dreams: Information, Man, and Machines*. St. Martin's Press, New York.

Nickols, F. (2000) *Community of Practice Start-Up Kit*. Distance Consulting Company, Robbinsville, NJ.

Nonaka, I. (1998) *The Knowledge-Creating Company*. Knowledge Management Harvard Business School Press, Boston, MA.

Nonaka, I. and H. Takeuchi (1995) *The Knowledge Creating Company*. Oxford University Press, Oxford, U.K., 284pp.

Pigeau, R. A. (2004) The human dimensions of cyberspace. Presented to *DRDC Science & Technology Symposium on Computers Everywhere and in Everything*, Ottawa, Ontario, Canada, April 21, 2004.

Polanyi, M. (1962) *Personal Knowledge: Towards a Post-Critical Philosophy*. University of Chicago Press, Chicago, IL.

Rifkin, J. (2000) *The Age of Access*. Ken Tarcher/Putnam, New York, 312pp.

Ross, J. W., P. Weill, and D. C. Robertson (2006) *Enterprise Architecture as a Strategy (Creating a Foundation for Business Execution)*. Harvard Business School Press, Boston, MA.

Ruggles, R. and D. Holtshouse (1999) *The Knowledge Advantage*. Capstone Business Books, Dover, NH.

Saint-Onge, H. and C. Armstrong (2004) *The Conductive Organization*. Elsevier/Butterworth-Heinemann, Burlington, MA.

Saint-Onge, H. and D. Wallace (2002) *Leveraging Communities of Practice for Strategic Advantage*. Butterworth-Heinemann, Boston, MA, 370pp.

Senge, P. M. (2006) *The Fifth Discipline: The Art and Practice of the Learning Organization*. Currency Books, Doubleday a division of Random House, New York.

Simard, A. (2005) Disaster information service: A proposal. Presented to *United Nations World Conference on Disaster Reduction*, Kobe, Japan, January 18–22, 18pp.

Simard, A. (2006) Knowledge markets: More than providers and users. *IPSI BgD Transactions on Advanced Research* 2(2), 3–9.

Simard, A. J. (2008) Knowledge services: The "why" of knowledge management. In *Knowledge Management in Practice*, Srikantaiah and Koeing, eds. American Society for Information Science and Technology Monograph, Information Today, Medford, NJ, pp. 169–198, Chapter 11.

Simard, A., J. Broome, M. Drury, B. Haddon, B. O'Neil, and D. Pasho (2007) *Understanding Knowledge Services at Natural Resources Canada.* Department of Natural Resources, Office of the Chief Scientist, Ottawa, Ontario, Canada, 80pp.

Simard, A. J. and P. Jourdeuil (2011) A knowledge services agenda: Simplifying interactions to make knowledge work more productive. Department of National Defence, Defence Research and Development Canada, Canada. Technical Report DRDC Corporate TR 2011-003, 200pp.

Sparr, D. L. (2001) *Ruling the Waves: From the Compass to the Internet.* Harcourt, Inc. Orlando, FL, 416pp.

Stewart, T. A. (1997) *Intellectual Capital.* Bantam Doubleday Dell Group, New York.

Stewart, T. A. (2001) *The Wealth of Knowledge.* Doubleday—Random House, New York, 400pp.

Sunstein, C. R. (2006) *Infotopia: How Many Minds Produce Knowledge.* Oxford University Press, Oxford, U.K., 273pp.

Tapscott, D. (1996) *The Digital Economy.* McGraw-Hill, New York.

Tapscott, D. and A. D. Williams (2006) *Wikinomics.* Penguin Group, New York, 324pp.

Tiwana, A. (2000) *The Knowledge Management Toolkit.* Pearson Education, Noida, India, 640pp.

Tosti, D. and F. Nickols (2010) *Making Players out of Spectators, Cynics, and Deadwood.* Distance Consulting LLC, Mount Vernon, OH, 8pp. http://www.nickols.us/ MakingPlayers.pdf (accessed on October 21, 2010).

von Krogh, G., I. Kazuo, and I. Nonaka (2000) *Enabling Knowledge Creation.* Oxford University Press, Oxford, U.K.

Weaver, W. and C. Shannon (1949) *The Mathematical Theory of Communications.* University of Illinois Press, Urbana, IL.

Wegner, E., R. McDermott, and W. M. Snyder (2002) *Cultivating Communities of Practice.* Harvard Business School Press, Boston, MA, 284pp.

Weill, P. and J. W. Ross (2009) *IT Savvy (What Top Execs Need to Know to Go from Pain to Gain).* Harvard business School Press, Boston, MA.

2 Managing Innovation in KIBS and Their Growing Importance in Relation to Innovation Systems

Jaume Valls-Pasola and Xavier Amores-Bravo

CONTENTS

2.1 INTRODUCTION: KIBS IN THE KNOWLEDGE SOCIETY

2.1.1 SERVICES AND THE KNOWLEDGE SOCIETY

Miles et al. (1995) define KIBS as

> business service companies, i.e. private service companies which sell their services on markets and direct their service activities to other companies or to the public sector. They are specialised in knowledge-intensive services, which means that the core of their service is contribution to the knowledge processes of their clients, and which is reflected in the exceptionally high proportion of experts from different scientific branches in their personnel.

According to Gallouj (2002), "KIBS firms are organizations that are particularly representative of the knowledge economy, since knowledge constitutes both their main input and output ... the activity of KIBS providers can be said to consist of the production of knowledge from knowledge." Muller and Doloreux (2009) summarize more than a decade of research on the topic of knowledge-intensive business services (KIBS), more specifically using Miles (2005), who defined KIBS as "services that involved economic activities which are intended to result in the creation, accumulation or dissemination of knowledge," and Toivonen (2004), who defined KIBS as "expert companies that provide services to other companies and organizations," and Hertog's (2000) "private companies or organizations that rely heavily on professional knowledge, i.e. knowledge or expertise related to a specific (technical) discipline or (technical) functional-domain and supplying intermediate products and services that are knowledge-based." All these definitions make it clear that knowledge is the core asset of KIBS activities.

In line with the conclusions of previous research (Howells and Tether 2004), recent EU reports such as PRO INNO Europe 2009 highlight the urgent need to provide major support to service innovation, especially as regards KIBS. After many years in which innovation policy and research was focused predominantly on industrial enterprises, technological innovation, and R&D, policy makers have now turned their attention to services.

The main reason for the growing interest in KIBS can be found in the changes that have come about in recent years both in industry and in economic structures in general, changes that have seen increasing attention being paid to the role played by services as regards the competitiveness of modern economies. This interest derives not only from the importance of services linked to industry and what is known as the tertiarization of the industrial sector but also from the specific weight of service companies in relation to enterprise as a whole. For while the overall services sector has shown continuous growth in recent years, this has especially been the case with KIBS. Indeed, as one of the most rapidly growing subsectors they have been regarded as key players in terms of the change in economic structures (Miles 2005). Although the factors responsible for this are too numerous to list in their entirety, mention should be made of increased leisure time, the partial privatization of the welfare state, the incorporation of services into manufacturing industry, the increased need for training, the importance of knowledge mastery as a competitive factor, an aging population, and, in general, the greater number of services linked to health care.

In the shift from an industrial economy to one based on knowledge, it must therefore be remembered that the services sector accounts for approximately two-thirds of employment and GDP in the European Union and other advanced nations. Indeed, data from Eurostat reveal that over the last two decades new jobs have only been created in the services sector. According to the PROINNO report on "challenges for EU support to innovation in services" (European Commission 2009), there is a strong positive correlation between GDP per capita and the proportion of services jobs as a share of total employment. Countries with high income levels—such as Luxembourg, the Netherlands, Sweden, and Denmark—are also characterized by a high proportion of services jobs as a share of employment within their economies. Thus services represent a large part of the economy and are one of the main drivers of growth in advanced industrial economies, this being reason enough to analyze

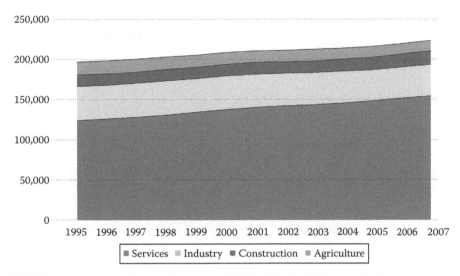

FIGURE 2.1 Total employment by sectors, EU27, 1995–2007 (thousands). (From European Commission, Challenges for EU support to innovation in services—Fostering new markets and jobs through innovation, PROINNO Europe® Paper No. 12, Commission Staff Working Document, 2009. Based on Eurostat.)

them on different levels: as a sector, as enterprises, and as creators of new markets and activities. Furthermore, this takes us beyond the traditional view of industrial policy and competitiveness, which regarded industry as the principal sector of economic growth due to its ability to incorporate new technology and replace jobs with capital, thereby increasing the productivity of its factors. In fact, and as various EU statistics show, services now provide 69.2% of all jobs (Figure 2.1), 71.6% of gross value added, and have been one of the main sectors to have compensated for the loss of jobs in industry. Moreover, business services (NACE codes 71–74), which to a large extent include KIBS, has been one of the most dynamic and rapidly growing sectors over the last decade.

Results from the PROINNO Report (European Commission 2009) highlight that all business services categories (except R&D services) have grown faster than total EU25 growth in employment and value added during the last decade. In particular, computer and related activities (NACE code 72) have been a real driving force for job and wealth creation in the EU25 from 1995 to 2005, representing 6.5% of the annual increase in employment and 7.8% in value added. Agriculture, manufacturing, and public administration were the least dynamic activities in the period under analysis. Communications services registered the second largest increase in value added (7.6% per year) but at the same time one of the lowest employment growth rates within the services sector (see Figure 2.2).

There is debate as to the future evolution of KIBS. Miles (2005) foresees three possible scenarios from a European perspective. In the first, KIBS are still growing due to a demand pull effect and they become the second knowledge infrastructure

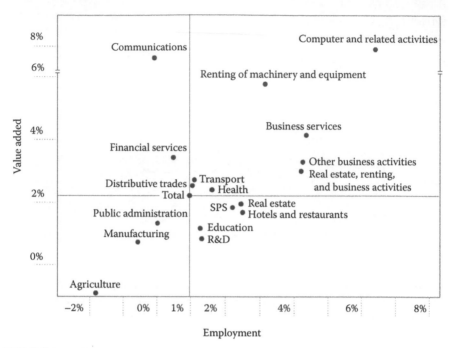

FIGURE 2.2 Employment and value-added annual growth rates in EU25 (1995–2005). (From European Commission, Challenges for EU support to innovation in services—Fostering new markets and jobs through innovation, PROINNO Europe® Paper No. 12, Commission Staff Working Document, 2009. Based on EU KLEMS database, March 2008 release.)

(universities and research centers being the first); the second scenario is a KIBS plateau, where growth is slower, industry is more mature, and some activities that were traditionally carried out by KIBS start to be off-shored to less-developed countries or by companies more oriented toward disseminating innovation rather than actually promoting technological innovation. Finally, there is the two-tier KIBS scenario where there is an increase in KIBS specializing in the integration/coordination of services in coexistence with technologically specialized KIBS. For Miles, these scenarios can take place at the same time and they strongly depend on regional specificities and policy constraints.

2.1.2 ECONOMIC IMPORTANCE OF KIBS

Until very recently, those analyses that considered the services sector in relation to innovation only took into account its role as a consumer or user of the innovation derived from industrial firms, thereby ascribing it a role that is clearly secondary. Nowadays, however, service companies and especially knowledge-intensive companies are regarded as genuine creators of innovation that have a significant impact on the whole economy.

The main authors in this field usually divide KIBS into two subgroups. Miles (2005) proposed a working definition of KIBS that distinguishes between "traditional professional services" (P-KIBS) and "new technology-based services" (T-KIBS). The first comprises accounting, legal, training, and certain consulting services, which are based on administrative knowledge systems and tend to become users of new technology and providers of knowledge and information. The second subgroup, usually referred to as T-KIBS, includes services related to ICT, technological consulting, engineering services, and R&D services. Companies in this subgroup tend to use knowledge to generate intermediate services for their clients' processes (production, commercial, innovation, etc.). Some authors, such as Kox and Rubalcaba (2007), distinguish between knowledge-intensive services (KIS), high-tech KIS (HTKIS), and KIBS. In this context, it is worth noting that both T-KIBS and HTKIS have experienced sharp growth in recent years, in part due to the marked expansion of the telecommunications sector in Europe. The data presented in Table 2.1 illustrate the economic importance of KIBS and their evolution for the period 1995–2005. Figure 2.A.1 presents a classification of these different types of KIS according to NACE classifications, the most usual for identifying KIBS in Europe.

The KIBS industry tops all the rankings relating to employment "creators" and, for their part, HTKIS (or T-KIBS) have significantly contributed to value-added growth (Figure 2.3). The analysis by Kox and Rubalcaba (2007) shows the key role played by KIBS for the European regions: they have growth in 90% of the

TABLE 2.1
Economic Importance of KIBS and HTKIS (1995–2005)

		Value Added			Employment		
	NACE	Thousand Million € (2005)	Relative % 1995	Relative % 2005	Thousand (2005)	Relative % 1995	Relative % 2005
Knowledge-intensive business services (KIBS)	72/73/74	711.4	5.4	6.9	12,259	4.5	6
High-tech knowledge-intensive services (HTKIS)	64/72/73	489.3	3.7	4.8	6,765.4	2.8	3.3

Source: Adapted from Kox, H. and Rubalcaba, L., *Business Services and the Changing Structure of European Economic Growth*, CPB Netherlands Bureau for Economic Policy Analysis, The Hague, the Netherlands, 2007. Based on the EU KLEMS database, March 2008 release.

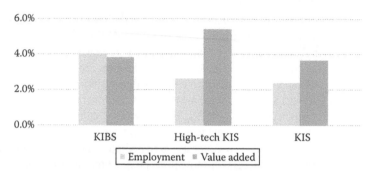

FIGURE 2.3 Annual growth rates of employment and gross value added in KIS sectors, EU25 (1995–2005). *Note*: Value added at constant prices, 1995. (From European Commission, Challenges for EU support to innovation in services—Fostering new markets and jobs through innovation, PROINNO Europe® Paper No. 12, Commission Staff Working Document, 2009. Based on EU KLEMS database, March 2008 release.)

regions (2000–2007 period) although some regional inequalities remain. A comparison with Japan and the United States is also helpful for completing our overview (Figure 2.4).

It is acknowledged that policies on competitiveness as well as studies conducted to understand the innovation process and its impact on the economy (on growth, employment, productivity, and the generation of benefits for society as a whole) have failed to take sufficient account of the services sector and its relationship to innovation.

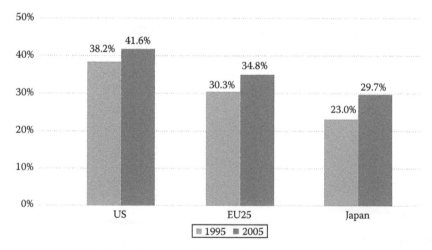

FIGURE 2.4 KIS share of total employment in the United States, European Union, and Japan (1995–2005). *Note*: The KIS classification follows Eurostats' definition and covers the following NACE Rev.1.1 codes 61, 62, 64, J, K, M, N, 092. (From European Commission, Challenges for EU support to innovation in services—Fostering new markets and jobs through innovation, PROINNO Europe® Paper No. 12, Commission Staff Working Document, 2009.)

Indeed, it is only recently that experts in the field have prioritized research into this sector. This has also coincided with an increasingly broader view of the concept of innovation being taken by government agencies and experts, one that is not limited to a technological perspective. Specifically, this broader view includes innovation in organizations and commercial processes and also pays greater attention to services since many of the innovations in this sector are not of a technological nature. As a result the government support available for innovative projects related to services is slowly beginning to contemplate innovation in marketing, in organization, and in business. By moving away from an exclusively technological focus, it may therefore benefit those service companies whose competitiveness policy is far from being technology driven.

2.2 KIBS AS INNOVATION DRIVERS

Consideration of the role of KIBS and their relationship to innovation dates back to studies such as those published by the European Innovation Monitoring System (Miles et al. 1995), which analyzed 15 KIBS in the context of ICT. Prior to this, the definition of KIBS included services to advanced companies (Stanback 1979) and other concepts such as the provision of advanced services (Moulaert and Daniels 1991). In all events, it should be noted that most research on KIBS has focused particularly on their influence as regards the innovation of industrial companies, that is, on KIBS clients (Tuominen 2005), and on studying clusters and local or regional innovation systems in which KIBS have a key role in promoting and supporting innovation.

From a historical point of view, innovation measurement was developed for industrial activities after the appearance of the *Oslo Manual* in the 1990s (the compulsory reference book for innovation measurement promoted and developed by the OECD). As far as innovation in services is concerned, progress in measurement has evolved significantly more recently. The third edition of the *Oslo Manual* (2005) for the first time includes recognition of nontechnological innovations such as marketing and organizational innovations. This new development has to be seen as an important contribution in measuring and analyzing innovation in services. The R&D expenditures of service industries are lower than in industrial activities, but in the case of KIBS and T-KIBS the situation is changing quite rapidly. Results of the Community Innovation Survey (CIS2006) show the differences in innovation behavior between different types of service activities: KIBS enterprises show similar innovation patterns to those of manufacturing enterprises.

Some recent analyses, such as the Challenges for EU support to innovation in services report (European Commission 2009), show that great differences exist between KIS and other services. Innovation in other services tends to be a continuous process consisting of a series of incremental changes. KIBS make significant R&D investments and are much more efficient than other companies when it comes to taking advantage of ideas from their business context.

Recent research results have led to increased interest in the characteristics of KIBS innovation activities and their contribution to the innovation processes of companies linked to KIBS in any type of relationship. In fact, according to authors such as Miles, KIBS constitute a subset of entrepreneurial services that are normally engaged in changing the status of manufacturing products or of products based on information or knowledge (rather than actually producing their own products). The service is sometimes delivered via a system, although the value of the information content is generally greater than the physical asset in itself. According to Miles et al. (1995), KIBS are companies that provide services to other firms and have the following characteristics:

• Knowledge makes an important contribution to service production and provision.
• The service is based on professional skills.
• The services are used as sources of knowledge and skills or as inputs for the development of these skills.
• Customers are by and large other companies.
• There is close interaction between the service and the customer, using part of the customer's existing knowledge while generating new knowledge.

Boden and Miles (2000) note that these are companies that invest heavily in knowledge and training and that may rapidly depreciate as a result of imitation. One could say that they become a source of ongoing innovation on the basis of elements that are intangible, noncontractual, implicit, and tacit, and that they offer human-embodied forms of knowledge and/or technology transfer. The knowledge derived from KIBS normally takes a specialist form, such as in the case of ICT and engineering, knowledge of applications based on previous experience, and knowledge of markets, networks, and interactions with other clients, etc. Toivonen (2004) considers three types of related skills and knowledge that are necessary for KIBS (according to the different service levels defined by Kuusisto and Meyer 2003):

• *At company level*: understanding the economy and business, change, the combination of professional skills, and the value chain
• *At the level of service processes*: methodologies and content of the profession, client know-how and skills for managing processes
• *At the level of personal interaction*: marketing, sales, and cooperation, both social and personal

In relation to these characteristics, Hertog (2000) refers to KIBS as private companies that offer services that are not targeted at the final consumer (excluding from this definition public services such as universities that might offer knowledge-based services, for example, those related to technology transfer). In contrast, other authors such as Green et al. (2001) consider that KIBS include all those service companies that are in possession of highly specialized knowledge. In their analysis of CIS-2, these authors conclude that T-KIBS constitute the most active sector in relation to innovation across the whole economy, and this, in part, is why it is especially interesting to analyze the nature of the innovation process in such companies.

Technology-based KIS are, as their name suggests, focused on the development and management of technological knowledge (Miles et al. 1995; Forseén et al. 2004), and they usually include services related to ICT, technological consulting, engineering, and R&D. As already noted, these companies generally use knowledge to generate intermediate services for their customers' processes (production, commercial, innovation, etc.). Toivonen (2004) concluded that KIBS are, by nature, intermediaries offering knowledge and information that will benefit their customers. The core of these services is the contribution made to knowledge processes, and KIBS will have a large number of experts in a given knowledge area. In short, KIBS are characterized by their ability to gather external knowledge and information and, by combining this with internal knowledge, to transform it into a service output that is often highly customized to the client's needs.

In light of these definitions and given what was said earlier, it is clear that KIBS, and especially T-KIBS, are recognized as being an enormously important source of innovation. Furthermore, the innovation they foster is by no means inferior to that produced in other sectors of the economy. However, the innovation deriving from KIBS does differ in terms of the way it is measured and the type of impact it has on other companies. Indeed, it leads to a different kind of output in the innovation process, and this process may well be managed differently, or at least have certain features that need to be taken into consideration.

The role of KIBS in relation to innovation processes was already beginning to be analyzed in some detail in the pioneering studies by authors such as Howells (2000), with various conclusions being reached as regards their relevance to the study of R&D. These analyses showed that service companies were increasingly investing in R&D and that they were more technology driven, while in the case of T-KIBS they matched industrial firms in terms of R&D but placed greater emphasis on technology. This research also highlighted the central role played by service companies in national innovation systems, as well as their proactive relationship to innovation processes (which was greater than had been reported previously). This aspect was linked to the growth of networking, with service companies operating in partnership with certain industrial firms to promote innovation, while in other cases they acted as clients and users of innovation. In a similar vein, Bilderbeek et al. (1998) argued that in the context of innovation systems KIBS played a key role as facilitators (creating knowledge and facilitating innovation in their clients' processes), as sources of innovation, as vehicles for innovation (distributing innovation among various firms), and as innovative companies in themselves.

Muller and Zenker (2001) also highlighted the fact that KIBS acted as facilitators in the interaction between a multitude of players by disseminating, absorbing, and reprocessing knowledge, thereby impacting upon innovation as a whole, where this is understood in the sense described by authors such as Freeman (1998), Lundvall (1992), and Nelson (1993). In the same vein, Tether and Hipp (2002) state that, although KIBS account for a small proportion of service companies, they have an impact on the whole economy, while Howells and Tether (2004) suggest that KIBS have increasingly become generators of innovation while remaining distributors or users of innovations. This aspect was one of the conclusions reached in the SI4S project (Hauknes 1998), which analyzed various studies conducted at European level.

Adding to the aforementioned text, Kox and Rubalcaba (2007) argued that service companies play a highly important role within innovation systems because they develop nontechnological innovations in fields such as consulting or the reorganization of processes, usher in technological advances through engineering or other areas of knowledge, promote good practices among companies, and are key players in terms of surpassing human capital indivisibilities. From a different analytic perspective, various authors have suggested that the concentration of KIBS at local level fosters both interaction among companies and the dissemination of knowledge and good entrepreneurial practices, which, in turn, facilitate innovation processes. According to the European Commission (2009), "KIBS can mostly be found in advanced European economies, such as Sweden, Denmark or the United Kingdom, but also in urban centres in other countries." There is also a high level of KIBS concentration found in Central Europe (Czech Republic and Slovakia) and Southern Europe (Portugal and Spain), and this seems to confirm the idea that the more knowledge intensive an economic activity is, the more this activity tends to concentrate geographically.

To sum up, the research on KIBS conducted over the last 15 years (Muller and Doloreux 2009) has evolved from a situation where KIBS were identified as support companies for customers' innovation processes or as innovation users to a situation where they are real innovators often working in codevelopment with customers. More and more researchers have started work on looking for the answers to the following questions: Are KIBS innovating companies? If so, what are the KIBS innovation processes like in comparison to those of industrial companies? A huge number of qualitative case studies have been developed, but also quantitative research related to patterns (and varieties) of innovation types, forms, and consequences:

> descriptive statistics to show the nature of innovative activities in KIBS; cross-country comparisons related to innovation in KIBS; multivariate data analyses to compare patterns of innovative processes in KIBS and manufacturing firms; and econometric models to explore the link between innovation and the economic performance of KIBS
>
> Muller and Doloreux (2007)

From the most relevant quantitative studies we can mention that developed by Freel (2006), in which a large comparative study is carried out between industrial companies, KIBS, and T-KIBS using a sample of 1161 companies. Other research surveys include Corrocher et al. (2009) and Amara et al. (2009). Older research projects carried out by Sundbo and Gallouj (1998), Hollenstein (2003), and Tether and Hipp (2002) were also relevant in contributing to a better understanding of KIBS innovation processes. Other topics of interest in this research field include the role of KIBS in relation to the outsourcing activity of industrial firms (Camacho and Rodriguez 2003); the interrelationships and forms of cooperation between SMEs and KIBS in the innovation process (Muller and Zenker 2001);

the importance of KIBS as regards the economic development of the region and change (Bilderbeek et al. 1998; Hertog 2000; Muller and Zenker 2001; Kuusisto and Meyer 2003; Koch and Stahlecker 2006); and the study of developmental processes among new services, which sometimes focuses on KIBS (Cooper and Edgett 1999; Cooper et al. 2001; Vermeulen 2004; Brax 2005).

2.3 MANAGING INNOVATION IN KIBS AND T-KIBS

2.3.1 UNDERSTANDING KIBS INNOVATION ACTIVITIES

Recognition that KIBS, and especially T-KIBS, are not only an important source of innovation but also foster the dissemination of innovation among other firms has led to a growing interest in identifying the processes, good practices, and models that these companies use when implementing and managing the innovation process. In this regard, one sees increasingly systematic approaches to innovation management being taken in the industrial and services sectors, especially in the case of KIBS or T-KIBS. A good example of this is the recent EU initiative known as the *Improve Innovation* project, with its focus on systemic models for improving innovation management and in which numerous service companies have taken part. This growing interest derives not only from recommendations made in regional, national, and European innovation policies, but also from the data obtained in different surveys and case studies (Hauknes 1998; Sundbo and Gallouj 1998; Howells and Tether 2004; Tuominen 2005). Furthermore, analyses of how KIBS and T-KIBS manage innovation have taken into consideration a wide range of aspects, for example, organizational features, those related to company culture, the relationship between strategy and innovation, the link between the business context and its capacity to foster open innovation, and the identification of good practices, which are usually related to innovation management in the company as a whole (such studies normally being focused on industrial firms).

Some authors, such as Sundbo (1997), argue that innovation in service companies cannot be adequately explained by either the paradigm based on technology as an essential part of the innovation process (Dosi 1982) or that founded on entrepreneurship (Kent et al. 1982). Instead, he considers company strategy to be the key factor that determines innovation (Teece 1987; Porter 1990) and he analyzes companies according to the theory of resources and capacities (Grant 1991; Teece 1992). Innovation, Sundbo (2000) concludes, is managed within the framework of the company's formal or informal strategy, and its development is determined by internal resources (Teece and Pisano 1994). The previously mentioned paradigms may, he argues, be suitable for understanding the behavior of new companies (including service ones) and technological innovations, respectively, but they are not applicable to the whole. In subsequent research (Sundbo and Gallouj 1998), he analyzes various kinds of innovation processes in which there is an interaction between management and the organization, between trajectories and the enterprise context, the aim being to characterize a range of service activities and innovation processes. In doing so, he found that the way in which the more technological service companies operated showed several similarities with respect to innovation processes and management used in the industrial sector.

The most comprehensive findings are probably those published by Coombs and Miles (2000), who distinguish between three research approaches to service innovation in relation to industrial firms. The first is what they call an "assimilation approach," in which service companies operate in a similar way to industrial firms as regards the innovation process. The second or "demarcation approach" is where the innovation process in the services sector differs greatly from that found in industry. Finally, they propose what they term a "synthesis approach," whereby certain elements of innovation in service companies acquire increasing importance in the industrial sector and vice versa, that is, the innovation process usually found in industry has an impact on the equivalent process in the services sector. In the case of T-KIBS, the approach taken to the innovation process sometimes resembles that used in the industrial sector. This is because many of these companies are among those which invest the most in R&D, carry out the greatest number of R&D projects, and because many of them have departments or organizational structures with responsibility for innovation.

Different studies have shown the differences between KIBS and industrial firms but also the existence of some possible innovation patterns. For Freel (2006), the general pattern for T-KIBS was suggestive of science-based manufacturing companies as described in Pavitt's well-known original taxonomy. Corrocher et al. (2009) analyzed KIBS from the Lombardy region. Their results also show the difficulty of defining a unique innovation pattern for KIBS and are in line with the results reached by Howells and Tether (2004) and Hollenstein (2003). In a key research paper on patterns of innovation in service industries, Miles (2008) highlighted the difficulties and contradictions in the use of industrial innovation patterns to KIBS:

> only a small segment of service innovation conforms to the typical manufacturing-based model, in which innovation is largely organized and led by formal research and development (R&D) departments and production engineering. Project management and on-the-job innovation are common ways of organizing service innovation. Innovation policy and management have to be much more than R&D policy and R&D management.

But Miles agreed with the idea that T-KIBS are the companies that best fit industrial innovation patterns: "innovation survey data indicates that some service organizations behave very much like high-technology manufacturing. This is especially true of technology-based, knowledge-intensive business services (T-KIBS)."

Miles (2008) sums up the main taxonomy of service innovation proposed by Soete and Miozzo (2001) and the refinements of the model proposed by Hipp and Grupp (2005): again, the Pavitt taxonomy is used because KIBS behave like "science-based and specialized suppliers" in a similar way to how small high-technology companies do. There are many examples, especially in ICT, companies that have strong links with universities and significant innovation and R&D expenditures when compared with other service companies. In addition, the "interactive style" is also emphasized: this is the case for T KIBS and T-KIBS with strong links with customers in development and codevelopment of innovations combining different types of knowledge in order to get innovative solutions.

As a result, it is important to note that many of the analyses carried out of the innovation process in general will also be applicable to T-KIBS. For example, Adler et al. (1992) sought to identify which practices affected the innovative capacity of companies, with the findings being adapted, improved, and systematized by authors such as Chiesa et al. (1996). This led to an audit model of innovation management based on key processes (new concept generation, new product development, process redefinition, and technology acquisition) and other "support" activities (market focus, leadership and culture, resource allocation, and organizational systems). Brown's (1997) model modified this audit model and proposed a view of innovation management that takes into account the need to master a series of subprocesses or key activities. As regards these subprocesses, and especially in relation to services, the literature contains numerous reports on innovation with respect to new service development (see, for example, Cooper and Edgett 1999; Cooper et al. 2001). However, research has also looked specifically at innovation in processes (Davenport 1992), knowledge management (Nonanka and Takeuchi 1995), how to stimulate and manage creativity and the generation of new concepts (Twiss 1986; Cooper 1994; DeGraff and Lawrence 2002; Flynn et al. 2003), the acquisition of technology and its relationship to strategy (Little 1981; Hax and Majluf 1991; Prahalad and Hamel 1991), the creation of a management-led innovative culture and intrapreneurship (Quinn 1991), and the fostering of open innovation (Chesbrough 2003). Although this chapter is not focused specifically on services, it does provide a general view of how these aspects are applied within companies. Chesbrough asserted that "firms can and should use external ideas as well as internal ideas, and internal and external paths to market, as the firms look to advance their technology." The role of open innovation was previously studied in large high-technology firms, but now the role of small and medium-sized firms is also being studied along with the importance of business models for open innovation in service companies (Lichtenthale 2011).

Within the framework of a program designed to improve innovation capabilities, the EU recently turned to a model developed by Kearney known as "innovation house" (Kearney 2006), the aim being to come up with a test or audit that focused on a range of aspects (innovation strategy, the organization of innovation and culture, innovation life cycle management, and enabling factors). This proved to be an interesting compendium of many of the subprocesses that are widely considered when referring to innovation management in companies. In these approaches, innovation is generally focused on specific aspects around which it is possible to standardize certain routines and good practices, develop competencies, or foster the acquisition of key knowledge within the innovation process. As noted earlier, a significant number of KIBS have made use of this audit methodology to analyze innovation management and improve their innovative capacity, and its applicability in this regard makes it a highly interesting approach.

Furthermore, this approach can be complemented by other proposals that focus strictly on services. For example, authors such as Sundbo (1997) propose a structured model of the innovation process, which enables the evolution of each innovation across a series of stages to be monitored and organized. Specifically, the process is divided into four stages: generation of the idea, transformation into an innovation project through the input of an intrapreneur, development by a group, and commercial or organizational implementation by management. Küpper (2001) concludes that all the innovation processes of service companies that are described in the literature usually include the stages of idea generation or selection, conceptual design, piloting, and market launch.

Subsequent to this work, Tidd and Hull (2003) proposed a model of service innovation based on the development and provision of services and adapted to some extent from the approach used in the industrial sector. Other authors such as Mudrak et al. (2005) took as their basis the work of Tidd et al. (2001) and Kemp et al. (2003) and came up with models of innovation management that took into account both the innovation process and the organizational context. Toivonen and Tuominen (2009) look at five innovation patterns in terms of their degree of formality in managing innovation in service industries: internal processes without a specific project, internal innovation projects, innovation projects with a pilot customer, innovation projects tailored for a customer, and externally funded innovation projects. However, it is worth noting that others, such as Tuominen (2005), have stated that most research on KIBS is focused on their impact on innovation within the client company and on the latter's innovation systems, and that very little attention has been paid to the innovation process within KIBS themselves. This is therefore an area that remains open for considerable exploration.

2.3.2 ANALYSIS MODEL OF A REGIONAL COMPETITIVENESS AGENCY

As far as these two fields of knowledge (managing innovation in companies in general and in the industrial sector) are concerned, there appears to be no consensus over which model might best be suited to KIBS when it comes to managing the innovation process.

Consequently, and given the previous efforts made by the regional government of Catalonia (Spain) to promote innovation among SMEs, this field was felt to be one that would benefit from further pilot projects. In fact, the ACC10 (Competitiveness for Catalonia) agency has, since 2001, been carrying out various projects aimed at characterizing and improving innovation capacities and their management among numerous industrial clusters and SMEs. The model used was developed locally and based on the models of Chiesa et al. (1996) and Brown (1997), and it later took into account a wide range of good practices designed to boost innovative capacity in enterprise, for example, those described in the EU report *Innosupport: Supporting Innovation in SMEs* (European Community 2005) and the previously mentioned "innovation house" model (Kearney 2006) within the framework of the European Union's *Improve Innovation* programme.

Based on this experience with models relating to the industrial sector, consideration was then given to aspects of innovation in service companies, for example, the conclusions reached by the European Commission project *Services in Innovation, Innovation in Services* (SI4S), carried out within the framework of the Targeted Socio-Economic Research (TSER) program by Hauknes (1998). These conclusions were subsequently drawn upon by reports produced in other countries, one of the most noteworthy being the analysis conducted by Wong and Annette (2003) of 190 KIBS in Singapore by means of a service innovation survey. Although the heterogeneity of activities carried out by service companies implies different kinds of innovation and related subprocesses, the behavior of T-KIBS can usually be classified according to just a few patterns, such as those described by Miles (2008) (i.e., "professional knowledge" and "large network-based service firms"), and these were also considered in order to select good practices related to the innovation process and its management. In Spain, organizations such as the foundation for technological innovation COTEC (2004) have likewise proposed a set of good practices designed to systematize service innovation and have also carried out more detailed analyses in specific sectors. The final important input concerned the dimensions of innovation proposed by Bilderbeek et al. (1998), who distinguished between new service concepts, new client interfaces, and new service delivery systems. Kuusisto and Meyer (2003) later added to this; their work classifies innovations into six categories: service innovations, technology and product-based innovations, delivery system innovations, customer interface innovations, new network and value chain configurations, and organizational innovations, while others such as Brax (2005) distinguished between innovations related to service concept, service process, and service resources and infrastructure. More recently, Amara et al. (2009) carried out a qualitative study that focused on the identification of KIBS innovation patterns. Six patterns are proposed: product, process, delivery, strategic, managerial, and marketing innovation. Four of them are related to nontechnological innovation.

Many studies have sought to characterize innovation in service companies, KIBS and T-KIBS, this being a necessary step prior to designing interventions that might serve to increase the innovative capacity of companies. A number of key points can be drawn from this analysis of characteristics and activities (see Table 2.2). These analyses are helpful for improving our knowledge of service innovation strategies and management.

Using contributions of the analytical framework for innovation in services referred at the beginning of this section, and on the basis of a review of the literature, the work carried out in recent years with industrial SMEs and case studies with businesses that for various reasons were regarded as being of special interest, the Catalonian Centre for Innovation and Business Development (CIDEM) developed a system for auditing innovation among KIBS, taking into account the economy and the sectoral structure in Catalonia. This led to the publication of a set of guidelines (CIDEM et al. 2007) designed to foster service innovation and targeted at SMEs and the main services sectors in Catalonia.

The emphasis was on T-KIBS. We used a "synthesis approach" based on the Coombs and Miles (2000) taxonomy but adding specific service innovation elements. This initial model led to an audit and a series of good practices for innovation management that took into account the following: (a) the dimensions or types of innovations that could be developed by service companies (new concept

TABLE 2.2

Characterizing Innovation in Service Companies and Policy Tools

The intangibility of service companies: The intangibility of service companies distinguishes them from industrial firms. The innovation process does not lead to the appearance of a new product, prototype, or physical process; instead, the result obtained often takes the form of innovation in organization or in the commercial process, or the emergence of new service concepts. There is a need to consider specific measures and grants for the smallest businesses that might help them to develop these kinds of innovation. At present the emphasis placed by government agencies on innovation in the industrial sector, which is more tangible, has led to service companies being overlooked to a certain extent.

Relevance of the human factor: In service companies the human factor is much more relevant than it is in the industrial sector, and this also applies to their innovation processes, especially those which are knowledge intensive. Initiatives related to better training, specialization in certain fields of knowledge, and the recruitment of highly qualified personnel, such as PhD holders and technicians, are measures that can indirectly improve the innovative capacity of businesses. Some studies comparing KIBS with the manufacturing sector show KIBS firms have higher human capital and training spending intensities than manufacturing firms (Wong and He 2005).

An integrated approach with customers: This is a key element that defines service companies. Measures designed to encourage collaborative innovation projects must (as acknowledged by the majority of large-scale grants or subsidies awarded through national or European projects) take into account the importance of service companies not only as subcontractors in R&D projects but also as drivers or collaborators within the value chain. The management of complex innovation projects is another aspect that must be considered in this context.

Size of the companies: The services sector is primarily composed of SMEs. In other words, the size of the company makes it difficult to systematize innovation in the same way as in other sectors, especially the industrial sector.

R&D and the innovation process: In service companies the organization of the innovation process is not concentrated around R&D departments, as is the case in industry, but more widely spread. A more systematized approach to innovation management has been called for in recent years, with government agencies regarding this as a way of increasing innovation capacity. This has led to the publication of guidelines, handbooks, and standards designed to facilitate innovation management and introduce good practices. However, there have been very few attempts to raise awareness about the specific characteristics of service companies and almost no public funding is set aside to help them become more systematic in terms of innovation, unlike what is available in the industrial sector. This is partly due to the lack of awareness within the services sector itself as to the importance of innovation and innovation management.

Outputs: The types of output or results achieved through innovative activities do not make progress easy: incremental innovations, the difficulty in separating process and product, innovations in client delivery, etc. It is difficult to define common indicators of service innovation that enable the sector to be analyzed and guidelines to be given to the companies involved.

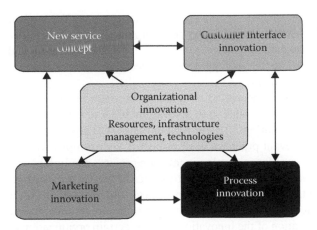

FIGURE 2.5   Five dimensions of service innovation. The CIDEM model. (From CIDEM et al., *Guia Metodològica per la gestió de la innovació en les empreses de serveis*, Generalitat de Catalunya, Barcelona, Spain, 2007.)

services; customer interface innovation or delivery innovation; and process innovation, marketing innovation, and organizational innovation, which includes innovation in management, resources, infrastructure, and technologies used by the firms; see Figure 2.5); (b) the creation of organizational structures to drive internal innovation and foster links with the business context; and (c) the description of a planned innovation process in service companies (identifying opportunities, generating and selecting ideas, managing the portfolio of innovation projects, managing innovation projects, and analyzing innovation results). The model also included a set of recommendations based on the good practices identified through the case analyses and review of the literature.

The model emerges from an adaptation of previous models like the one proposed by Bilderbeek et al. (1998) that was based on a four-dimensional approach: the service concept, the client interface, the service delivery system/organization, and the technological options. We also took into account the work by Kuusisto and Meyer (2003) mentioned earlier and the nontechnological categories proposed by the *Oslo Manual* (2005). Technologies are at the center of the model together with infrastructures and organizational innovation because they play a key role as "facilitators" in relation to the other four dimensions where service innovations take place: concept services, customer interface, marketing, and process innovations.

This model was subsequently applied at regional level to analyze 50 Catalonian T-KIBS with more than 20 employees, examining the relationship between their capacity for innovation management and their degree of business success (based on their financial results in recent years). This study (Sanchez et al. 2010) incorporated a considerable number of the good practices for innovation management that have been identified in the industrial sector, such as those mentioned earlier. However, it also included certain aspects linked to new service development, a field in which numerous studies have been conducted (Cooper and Edgett 1999), the results of which were taken into account when designing the study questionnaire. The study analyzes these cases through face-to-face interviews about organizational aspects, strategy, innovation culture in the company, degree of innovation systematization, and practices for innovation (creativity, technology watch, project management, knowledge management, etc.), among others.

The study tried to determine correlations between business success and good management practices in relation to innovation. It has also been used to validate the model used when auditing KIBS at regional level (CIDEM et al. 2007). A number of particularly relevant practices were identified, for example, working in multidisciplinary teams in order to develop innovation projects, the link between innovation capacity and the degree of investment in internal R&D, the close participation of managers, and the relationship between strategy and innovation. At the same time a set of interrelated variables was identified, for example, a number of practices linked to the systematization of the innovation process, certain organizational practices, the use of tools and techniques designed to support innovative activity, and a range of variables related to the evaluation of the innovation process.

2.4   CONCLUDING REMARKS

A review of the literature reveals that KIBS are increasingly being seen as key players within national innovation systems and their capacity for innovation is arousing interest among analysts. KIBS, and especially T-KIBS, are not only becoming an ever greater source of innovation but are also important drivers of innovation, not least in relation to the innovation processes of their customers. To a certain extent, this means that improving the innovation capacities of KIBS could have repercussions beyond simply the capacities of businesses themselves.

This chapter has summarized the reasons why service companies and KIBS are of growing importance both as objects of study and in the context of innovation policy. Reference has also been made to certain aspects of their own innovation processes and how these are managed. Although many studies, surveys, and articles have focused on national level, research and case studies have also been conducted at regional level, this being the context focused on here. Specifically, this analysis of service companies and T-KIBS began by considering the good practices identified in the industrial sector in relation to the innovation management of SMEs, complementing this with a discussion of research findings about service innovation. As a result it was possible to begin characterizing the innovation processes of T-KIBS and, more specifically, to examine the relationships between these practices and their business success. At regional level, innovation in services has not traditionally been a priority for innovation policies, which makes it more interesting to analyze the results, the environment, and the impact of innovation activities of this kind of company on the innovation system. This analysis may help in designing new policies to foster innovation capacity in a key sector of the knowledge economy, one that has witnessed significant growth in recent years but about which there has been limited research or examination of good practices—at least not at regional level—from the perspective of innovation policy.

APPENDIX 2.A

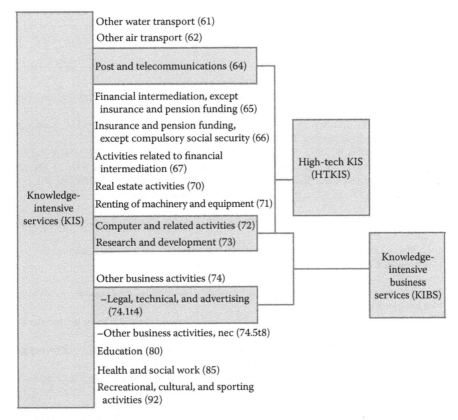

FIGURE 2.A.1 Different types of KIS according to NACE classification. (From European Commission, Challenges for EU support to innovation in services—Fostering new markets and jobs through innovation. PROINNO Europe® Paper No. 12, Commission Staff Working Document, 2009; Based on Kox, H. and Rubalcaba, L., *Business Services and the Changing Structure of European Economic Growth*, CPB Netherlands Bureau for Economic Policy Analysis, The Hague, the Netherlands, 2007.)

REFERENCES

Adler, P.S., McDonald, W.D., and McDonald, F. 1992. Strategic management of technical functions. *Sloan Management Review* 33(2), 19–37.

Amara, N., Landry, R., and Doloreux, D. 2009. Patterns of innovation in knowledge intensive business services. *The Service Industries Journal* 29(4), 407–430.

Bilderbeek, R., Hertog, P., Marklund, O., and Miles, I. 1998. Services in innovation: Knowledge intensive business services (KIBS) as co-producers of innovation. SI4S Synthesis Paper, SI4S-S3-98, August. STEP Group (Studies in Technology, Innovation and Economic Policy).

Boden, M. and Miles, I. 2000. *Services and the Knowledge-Based Economy*. Science, Technology and the International Political Economy Series. London, New York: Continuum.

Brax, S. 2005. Viewing knowledge-intensive business services as systems. *Proceedings of the Conference on Frontiers of e-Business Research Tampere*, September 26–28, 347–359.

Brown, D. 1997. *Innovation Management Tools. A Review of Selected Methodologies.* Luxembourg, Europe: European Commission, Directorate-General XIII.

Camacho, J.A. and Rodriguez, M. 2003. Servicios y globalización. *Comercio Exterior* 53(1), 13–22.

Chesbrough, H.W. 2003. *Open Innovation: The New Imperative for Creating and Profiting from Technology.* Boston, MA: Harvard Business School Press.

Chiesa, V., Coughlan, P., and Voss, C.A. 1996. Development of a technical innovation audit. *Journal Product Innovation Management* 13, 105–136.

CIDEM, Amores, X., and Ayneto, X. 2007. *Guia Metodològica per la gestió de la innovació en les empreses de serveis.* Barcelona, Spain: Generalitat de Catalunya.

Coombs, R. and Miles, I. 2000. Innovation, measurement and services: The new problematique. In: Metcalfe, S.J. and Miles, I. (Eds.), *Innovation Systems in the Service Sectors. Measurement and Case Study Analysis.* Boston, MA: Kluwer, pp. 85–104.

Cooper, R.G. 1994. Third generation new product processes. *Journal of Product Innovation Management* 11: 3–14.

Cooper, R.G. and Edgett, S.J. 1999. *Product Development for the Service Sector.* New York: Basic Books.

Cooper, R.G., Edgett, S.J., and Kleinschmidt, E.J. 2001. *Portfolio Management for New Products.* Cambridge, MA: Perseus Books Group.

Corrocher, N., Cusmano, L., and Morrison, A. 2009. Modes of innovation in knowledge-intensive business services evidence from Lombardy. *Journal of Evolutionary Economics, Springer* 19(2), 173–196.

COTEC. 2004. *Análisis del Proceso de Innovación en las empresas de servicios.* Ed. Fundación para la Innovación Tecnológica, Madrid.

Davenport, T.H. 1992. *Process Innovation: Reengineering Work through Information Technology.* Boston, MA: Harvard Business School Press.

DeGraff, J. and Lawrence, K.A. 2002. *Creativity at Work.* University of Michigan Business School: Management Series. San Francisco, CA: Jossey-Bass.

Dosi, G. 1982. Technological paradigms and technological trajectories: A suggested interpretation of the determinants and directions of technical change. *Research Policy* 2, 147–162.

European Community. 2005. InnoSupport: Supporting innovation in SMEs. Innosupport. European Community.

European Commission. 2009. Challenges for EU support to innovation in services— Fostering new markets and jobs through innovation. PROINNO Europe® Paper No. 12. Commission Staff Working Document.

Flynn, M., Dooley, L., O'Sullivan, D., and Cormican, K. 2003. Idea management for organisational innovation. *International Journal of Innovation Management* 7(4), 417–442.

Forséen, M., Heikkonen, M., Hietala, J., Hänninen, O., and Kontio, J. 2004. Knowledge-intensive service activities facilitating innovation in the software industry. Final Report of the KISA-SWC Finland Project.

Freel, M. 2006. Patterns of technological innovation in knowledge intensive business services. *Industry & Innovation, Taylor and Francis Journals* 13(3), 335–358.

Freeman, C. 1998. *The Economics of Technical Change.* In: Publicado en Archibugi, D.Y. and Michie, J. (Eds.), *Trade, Growth and Technical Change.* Cambridge, MA: Cambridge University Press.

Gallouj, F. 2002. Knowledge-intensive business services: Processing knowledge and producing innovation. In: Gadrey, J. and Gallouj, F. (Eds.), *Productivity, Innovation and Knowledge in Services.* New Economic and Socio-Economic Approaches. Cheltenham and Northampton, U.K.: Edward Elgar.

Grant, R.M. 1991. The resource-based theory of competitive advantage: Implications for strategy formulation. *California Management Review* 33, 114–135.

Green, L., Howells, J., and Miles, I. 2001. Services and innovation: Dynamics of service
innovation in the European Union. Final Report Working paper. PREST and CRIC,
University of Manchester.

Hauknes, J. 1998. Services in innovation—Innovation in services. STEP (Studies in
Technology, Innovation and Economic Policy). SI4S Final report from the project
Services in Innovation, Innovation in services (SI4S).

Hax, A.C. and Majluf, N.S. 1991. *The Strategy Concept and Process: A Pragmatic Approach.*
Englewood Cliffs, NJ: Prentice Hall.

Hertog, P. 2000. Knowledge-intensive business services as co-producers of innovation.
International Journal of Innovation Management 4(4), 491–528.

Hipp, C. and Grupp, H. 2005. Innovation in the service sector: The demand for service-specific
innovation measurement concepts and typologies. *Research Policy* 34(4), 517–535.

Hollenstein, H. 2003. Innovation modes in the Swiss service sector: A cluster analysis based
on firm-level data. *Research Policy* 32, 845–863.

Howells, J. 2000. Innovation & services: New conceptual frameworks. CRIC Discussion
Paper No. 38, August.

Howells, J. and Tether, B. 2004. Innovation in services: Issues at stake and trends. Commission
of the European Communities. Final Report. ESRC, CRIC, Institute of innovation
Research, University of Manchester, Manchester, U.K.

Kearney, A.T. 2006. Innovation management assessment. Improve innovation UE. European
Commission.

Kemp, R.G.M., Folkerings, M., De Jong, J.P.J., and Wubben, E.F.M. 2003. Innovation and
firm performance. SCALES, Zoetermeer. Netherlands Ministry of Economic Affairs.

Kent, C.A., Sexton, D.L., and Vesper, K.H. 1982. *Encyclopedia of Entrepreneurship.*
Englewood Cliffs, NJ: Prentice Hall.

Koch, A. and Stahlecker, T. 2006. Regional innovation systems and the foundation of knowledge intensive business services. A comparative study in Bremen, Munich, and Stuttgart,
Germany. *European Planning Studies* 14, 123–145.

Kox, H. and Rubalcaba, L. 2007. *Business Services and the Changing Structure of European
Economic Growth.* The Hague, the Netherlands: CPB Netherlands Bureau for Economic
Policy Analysis.

Küpper, C. 2001. Service innovation—A review of the state of the art. Working Paper Institute
of Innovation Research and Technology Management.

Kuusisto, J. and Meyer, M. 2003. Insights into services and innovation in the knowledge intensive economy. *Technology Review* 134/2003 TEKES.

Lichtenthale, U. 2011. Open innovation: Past research, current debates, and future directions.
The Academy of Management Perspectives 25(1), 75–93.

Little, A.D. 1981. *The Strategic Management of Technology.* European Management Forum,
Davos, Switzerland.

Lundvall, R.A. 1992. *National Systems of Innovation: Towards a Theory of Innovation and
Interactive Learning.* London, U.K.: Pinter Publishers.

Miles, I. 2005. Knowledge intensive business services: Prospects and policies. *Foresight* 7(6),
39–63.

Miles, I. 2008. Patterns of innovation in service industries. *IBM Systems Journal* 47(1),
115–128.

Miles, I., Kastrinos, N., Bilderbeek, N., and Hertog, P. 1995. Knowledge-intensive business
services—Users, carriers and sources of innovation. EIMS Publication No. 15, EC 1995.

Moulaert, F. and Daniels, P.W. 1991. Advanced producer services: Beyond the microecon-
ics of production. In: Daniels, P.W. and Moulaert, F. (Eds.), *The Changing Geography
of Advanced Producer Services.* London, U.K.: Belhaven Press.

Mudrak, T., Wagenberg, A., and Wubben, E. 2005. Innovation process and innovativeness of facility management organizations. *Emerald Facilities* 23(374), 103–118.

Muller, E. and Doloreux, D. 2007. The key dimensions of knowledge-intensive business services (KIBS) analysis: A decade of evolution. Working Papers "Firms and Region" U1/2007, Fraunhofer Institute for Systems and Innovation Research (ISI).

Muller, E. and Doloreux, D. 2009. What we should know about knowledge-intensive business services. *Technology in Society* 31, 64–72.

Muller, E. and Zenker, A. 2001. Business services as actors of knowledge transformation and diffusion: The role of KIBS in regional and national innovation systems. *Research Policy* 30, 1401–1516.

Nelson, R.R. 1993. *National Innovation Systems—A Comparative Analysis*. New York: Oxford University Press.

Nonanka, I. and Takeuchi, H. 1995. *The Knowledge-Creating Company*. New York: Oxford University Press.

OECD. 2005. Oslo Manual. The measurement of scientific and technological activities. Proposed Guidelines for Collecting and Interpreting Technological Innovation Data. Third Edition. Paris.

Porter, M.E. 1990. *The Competitive Advantage of Nations*. New York: Free Press.

Prahalad, C.K. and Hamel, G. 1991. La organización por unidades estratégicas de negocio ya no sirve. *Harvard-Deusto Business Review* 45, 47–64.

Quinn, J.B. 1991. Managing innovation: Controlled chaos. In: Mintzberg, H. and Quinn, J.B. (Eds.), *The Strategy Process: Concepts, Contexts, Cases*, 2nd edn. Englewood Cliffs, NJ: Prentice-Hall, pp. 746–758.

Sanchez, A., Amores-Bravo, X., Ferras, X., and Ribera, J. 2010. Innovation management in Catalan t-KIBS: Organization, process, environment and behavior patterns. *Proceedings of the 17th International Annual EurOMA Conference*, 2010, Porto, Portugal.

Soete, L. and Miozzo, M. 2001. Internationalization of services: A technological perspective. *Technological Forecasting and Social Change* 67(2), 159–185.

Stanback, T.M. 1979. *Understanding the Service Economy. Employment, Productivity, Location*. Baltimore, MD: The Johns Hopkins University Press.

Sundbo, J. 1997. Management of innovation in services. *The Service Industries Journal* 17(3), 432–455.

Sundbo, J. 2000. Organization and innovation strategy in services. In: Boden, M. and Miles, I. (Eds.), *Services and the Knowledge-Based Economy*. Science, Technology and the International Political Economy Series. London, U.K.: Continuum, Chapter 6.

Sundbo, J. and Gallouj, F. 1998. Innovation in services. SI4S S2 Synthesis Papers. STEP Group. Studies in Technology, Innovation and Economic Policy.

Teece, D. 1987. Profiting from technological innovation: Implications of integration, collaboration, licensing and public policy. In: Teece, D. (Ed.), *The Competitive Challenge: Strategies for Industrial Innovation and Renewal*. New York: Harper & Row, pp. 185–221.

Teece, D. 1992. Strategies for capturing the financial benefits from technological innovation. In: Rosenberg, N., Landau, R., and Mowery, D. (Eds.), *Technology and the Wealth of Nations*. Stanford, CA: Stanford University Press.

Teece, D. and Pisano, G. 1994. The dynamic capabilities firms: An introduction. *Industrial and Corporation Change* 3(3), 537–555.

Tether, B. and Hipp, C. 2002. Knowledge Intensive, Technical and Other Services: Patterns of Competitiveness and Innovation Compared. *Technology Analysis & Strategic Management* 14(2), 163–182.

Tidd, J., Bessant, J., and Pavitt, K. 2001. *Managing Innovation: Integrating Technological, Market and Organizational Change*. Chichester, U.K.: Wiley.

Tidd, J. and Hull, F.M. 2003. *Service Innovation. Organizational Responses to Technological Opportunities & Market Imperatives*. London, U.K.: Imperial College Press.

Toivonen, M. 2004. Expertise as business—Long-term development and future prospects of knowledge-intensive business services (KIBS). Doctoral dissertation series 2004/2. Department of Industrial Engineering and Management, Helsinki University of Technology, Espoo, Finland.

Toivonen, M. and Tuominen, T. 2009. Emergence of innovations in services. *Service Industries Journal* 29(7), 887–902.

Tuominen, T. 2005. Challenges in managing innovation activities in KIBS organisations. *18th Scandinavian Academy of Management Conference*, August 18–20, Denmark. Innovation Management Institute, BIT Research Centre, Helsinki University of Technology, Espoo, Finland.

Twiss, B. 1986. *Managing Technological Innovation*. London, U.K.: Pitman.

Vermeulen, P. 2004. Managing product innovation in financial services firms. *European Management Journal* 22(1), 43–50.

Wong, P. and Annette, S. 2003. The pattern of innovation in the knowledge intensive Business Services Sector of Singapore. *Singapore Management Review* 26(1), 21–44.

Wong, P. and He, Z. 2005. A comparative study of innovation behaviour in Singapore's KIBS and manufacturing firms. *The Service Industries Journal* 25(1), 23–42.

3 Knowledge Services in the Innovation Process*

Torsti Loikkanen, Anssi Neuvonen, and Kirsi Tuominen

CONTENTS

* Loikkanen, T. et al., *Acquisition, Utilization and the Impact of Patent and Market Information on Innovation Activities*, Espoo 2009. VTT Research Notes 2484, 68p, 2009. With permission.

3.1   INTRODUCTION

3.1.1   BACKGROUND

This chapter is about the acquisition, utilization, and impact of information and related knowledge services in different phases of innovation processes. Information and knowledge are strategic assets that provide competitive advantages for enterprises and nations worldwide. Effective, target-oriented, and successful research and development work require qualified and competent scientific staff and research facilities as well as comprehensive information and knowledge of the research and market landscape of the project.

The access to, and utilization of, a versatile pool of information sources is necessary in developing unique and novel ideas or inventions that differ essentially from existing and already invented ones. By efficiently exploiting the research landscape, the innovator shall be aware of whether corresponding work is ongoing or has been carried out elsewhere in order to avoid "reinventing the wheel," for example, to avoid redeveloping already patented ideas. Various market information, related to, for example, industry trends, the future needs of consumers, the size and development of markets, and the profiles of competitors or potential partners, are important in innovation development. Such information is available from different sources, for example, data banks, market research reports, government officials, foresight studies, technology road maps, etc. However, the acquisition, refining, and analyzing of such information can be time consuming and difficult. The costs of versatile information may range from free of charge to extremely expensive.

3.1.2   ABOUT KNOWLEDGE AND INFORMATION

This chapter explores information and knowledge sources and services needed in different phases of the R&D and innovation process. Part of the knowledge that is needed in innovation activities is tacit expert knowledge that cannot be presented, documented, and disseminated in codified form; the rest of the needed knowledge can be documented and transmitted in codified form (e.g., Gibbons et al. 1994; Machlup 1980). Ackoff (1989) makes a distinction between information and knowledge by elaborating that information is data that is processed to be useful and provides answers to who, what, where, and when questions. Knowledge, on the other hand, is the appropriate collection of information, such that its intent is designed to be useful and it answers also to how questions (Ackoff 1989).

In colloquial language, information and knowledge are used often as synonyms. The scholars in the field propose to determine the content of these concepts in more detail according to the context in each question, and the previous research in the field indeed offers a variety of solutions. For example, according to Machlup (1980, p. 8), "one may, with good reasons, insist on distinguishing information from knowledge by having information refer to the act or process by which knowledge (or a signal, or a message) is transmitted." According to Dosi, information "entails well-stated and codified propositions about 'states-of-the-world' (e.g., 'it is raining'), properties of nature (e.g., 'A causes B') or explicit algorithms on how to do things." On the other hand, the definition Dosi (1996) suggests for knowledge includes "(i) cognitive categories; (ii) codes of interpretation of information itself; (iii) tacit skills; and (iv) problem-solving and search heuristics irreducible to well-defined algorithms."* The definition of knowledge of Gibbons et al. (1994) is based on the perceived changes in the ways by which scientific, social, and cultural knowledge are produced. The new mode of knowledge production ("Mode 2") is replacing or reforming established institutions, disciplines, practices, and policies. The traditional knowledge production of "Mode 1" is based primarily on the concept of scientific and academic knowledge. The new knowledge production of "Mode 2" concept is much broader and encompasses, for example, such issues as scientific transdisciplinarity, heterogeneity, social accountability, and reflexivity (Gibbons et al. 1994).

In conclusion, in innovation studies knowledge, information, and related concepts are understood in several ways with different contents and meanings depending on the contributors. The scope and content of knowledge or information concepts, as understood in this chapter, is broadly due to the heterogeneous character of knowledge and information needed in the innovation process. Such knowledge or information may be codified and readily available, usable, and useful for research and the development of enterprises. This chapter discusses services and service organizations that "inform" enterprises by delivering codified information supporting the R&D and innovation work of enterprises. In addition to delivering codified information or knowledge, these service organizations can offer more in-depth consulting services during which also tacit knowledge will be transferred between company and service organization in a mutual and interactive information exchange and learning process. Such services may relate, for example, to foresight exercises in which, for example, the so-called SECI methodology developed by Nonaka can be applied (e.g., Nonaka 1991).

3.1.3 ABSORPTIVE CAPACITY OF BENEFITING OF KNOWLEDGE

This chapter accentuates the importance of access for inventors and innovators to relevant information sources. It is however well recognized that merely providing access to information sources and services will not suffice. In addition, the organization must have sufficient learning and absorptive capacity to utilize and reap

---

* See also many knowledge-related work in Neef et al. (1998).

benefits from the available information. As Cohen and Levinthal (1989, 1990) stress, "... while R&D obviously generates innovations, it also develops the firm's ability to identify, assimilate, and exploit knowledge from the environment—what we call a firm's 'learning' or 'absorptive' capacity." Accordingly, it is also about the knowledge management (or information management) strategy of an organization that forms a context for acquiring, using, and utilizing knowledge and information in research and innovation activities (e.g., Bouthillier and Shearer 2002).

3.1.4 AVOIDING DUPLICATION OF R&D BY ACQUISITION OF KNOWLEDGE

One point of departure for this chapter is a concern about the duplication of research and innovation efforts, and the consequent dissipation of research investments. The duplication of research and innovation is one element of the theory of economic rationale on public intervention in private research and innovation activities. Empirical evidence of duplication of R&D and innovation efforts is available, especially from overlapping patent applications. This evidence refers to relatively insignificant use of patent documents as an information source in R&D work. Inadequate use of information in R&D work applies for other kinds of information as well. The utilization of knowledge services to provide information on industry trends, markets, competitors, potential partners, and other market information is scarce. In particular, small- and medium-sized enterprises (SMEs) have limited capacity to use available services as compared to large enterprises (Kinnunen et al. 1996). Large companies have specialized internal service units and staff (e.g., knowledge professionals, and patent engineers) for this purpose and they also can afford to invest in external knowledge services.

3.1.5 IMPACT OF KNOWLEDGE SERVICES IN INNOVATION ACTIVITIES

An effective access to, and utilization of, relevant information and related knowledge services in different phases of the innovation process can improve innovation activities on the microlevel of inventors in enterprises, especially in SMEs. If information is utilized effectively throughout the innovation system, ultimately this has a positive effect on the performance of the entire innovation system. Accordingly, it is a strategic issue in the national innovation policy.

An additional motivation to explore this topic comes from the scarce research in this area nationally and internationally. Moreover, the topic interestingly relates also to certain topical issues of innovation activities, such as open innovation and impacts of the new phase of globalization on global cooperation in S&T, challenging the traditional IPR approach (see, e.g., Gurry 2007). This chapter aims to contribute to discussion among inventors and innovators in enterprises and research institutes, among policy makers, and also in the innovation policy research community.

3.1.6 STRUCTURE OF CHAPTER

This chapter is structured as follows. Section 3.2 is about the needs of knowledge and necessity of using professional knowledge services in different phases of innovation process. Section 3.2 makes an overview on studies of acquisition, utilization, and impacts of information sources and services in different phases of research and innovation process. Moreover, Section 3.2 gives future perspectives in this area. Section 3.3 considers the acquisition and utilization of knowledge within a context of innovation policy and underlying theoretical and empirical aspects. In the innovation policy, rationale information and knowledge as well as an avoidance of unintended duplication of research and innovation work are among the key elements. Section 3.3 also includes the results of Finnish patent application study indicating duplication of R&D work. Section 3.4 draws conclusions, discusses arising policy implications and recommendations, and makes suggestions for further topics to be studied in this research field.

3.2 NEED OF INFORMATION AND KNOWLEDGE SERVICES IN INNOVATION

This section considers the need, acquisition, and utilization of information sources and related knowledge services in R&D and innovation. Section 3.2.1 identifies different needs of external knowledge and information sources in different phases of innovation process. Section 3.2.2 makes a short overview of studies of the roles and impacts of diverse information sources in research and innovation work of enterprises. Section 3.2.3 discusses the changes of global business and innovation environments, such as the emergence of global alliances, new innovation models, and open innovation model. Section 3.2.4 draws conclusions from Section 3.2.

3.2.1 INFORMATION NEEDS IN INNOVATION PROCESS

Diverse information and a number of information sources are needed in different phases of the innovation process. In fierce global competition, the challenge to acquire and utilize information is constantly growing. As the mobility of information increases, a firm's competitive success critically depends on its ability to monitor and quickly seize external sources of knowledge (Iansiti 1997). A company can leverage basic or generic technologies developed elsewhere, which allows it to focus on developing unique applications that better suit the needs of specific overseas markets (Iansiti and West 1997).

The information needs in different phases of the innovation process can be illustrated by chain-linked model of innovation developed by Kline and Rosenberg (1986) (Figure 3.1).

This complex, nonlinear, and dynamic model of innovation process is developed on the basis of long-lasting consultation of R&D activities of enterprises and related innovation studies. This model is used as an analytical framework in innovation and innovation policy studies.

The chain-linked model of innovation process describes the flow paths of information and cooperation. Symbol C refers to central chain-of-innovation and F/f to feedback loops. K-R refers links through knowledge to research and related feedback loops. Symbol D illustrates influence from research to invention and design and again related feedback loops. Symbol I refers to the support of scientific research and S the support of research in the product area. The model attempts to illustrate the

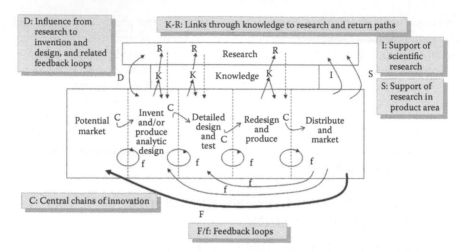

FIGURE 3.1 Chain-linked model of innovation process. (From Kline, S. and Rosenberg, N., An overview of innovation, in *The Positive Sum Strategy: Harnessing Technology for Economic Growth*, R. Landau and N. Rosenberg (eds.), National Academy Press, Washington, DC, 1986.)

complex nature of innovation with feedback loops from markets and research back to invention and design in company.

The actors and institutions playing roles in different phases of the innovation process need diverse and relevant information from respective expert sources and services. Some examples of such information needs are illustrated in Figure 3.2.

In the phases of innovation process, information is needed of the future trends and market potentials, required technologies (road maps), and services (consumer preferences,

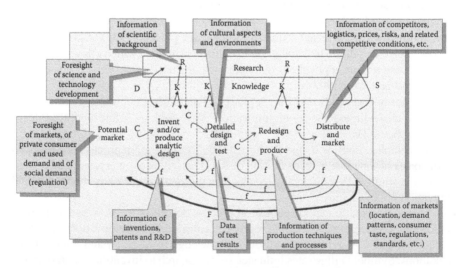

FIGURE 3.2 Examples of knowledge and information needs in different phases of innovation process.

intermediate products, etc.) as well as of new business opportunities. This information may be based on the examinations and surveys of private and social needs and demand in the future, that is, of private consumers and companies as well as of needs for regulation.

In the phases of research and invention, as well as in the later phases of design, piloting, and prototyping, it is important to have information on previous or ongoing studies in other companies or research institutions. In the later stages of the innovation process, in approaching the commercialization or innovation and markets, it is important to have information about distribution, markets, etc. It is essential to have a various sort of information available from the very beginning of the innovation process in order to avoid overlapping or doubling efforts and any ensuing dissipation of resources.

Figure 3.2 is for illustrative purposes only. A more detailed description of information source needs could be provided for each actor, organization, and instruction (inventors, enterprises, R&D organizations, etc.). As Cohen and Levinthal (1989) argue, what is important is also that while R&D obviously generates innovations, it also develops the firm's ability to identify, assimilate, and exploit knowledge from the environment—what we call a firm's learning or absorptive capacity. While encompassing a firm's ability to imitate a new process or product innovations, the absorptive capacity also includes the firm's ability to exploit external knowledge of a more intermediate sort, such as basic research findings that provide the basis for subsequent applied research and development. Also, in light of the dependence of industrial innovation upon external knowledge, absorptive capacity represents an important part of a firm's ability to create new knowledge. In this regard, the exercise of absorptive capacity represents a sort of learning that differs from learning-by-doing, the focus of industrial economists' work on firm learning in recent years (e.g., Lieberman 1984; Spence 1981). Learning-by-doing typically refers to the automatic process by which the firm becomes more practiced and, hence, more efficient at doing what it is already doing. In contrast, with absorptive capacity a firm may acquire external knowledge that will permit it to do something quite different.

Ottum and Moore (1997) summarize that although no single variable holds the key to new product performance, many of the widely recognized success factors share a common thread: the processing of market information. Understanding customer wants and needs ultimately comes down to company's capabilities for gathering and using market information.

As emphasized, for example, by WIPO and IFIA, the successful marketing of inventions and technology means to marry a new invention to a real existing need, and it demands an extensive and very close collaboration and cooperation between three groups of people: those who create inventions and technology, those who explore and create markets, and those who use inventions and technology (WIPO and IFIA 1998). Such cooperation depends to a large extent on their capacity to actively collect, select, analyze, and exchange information (WIPO and IFIA 1998).

3.2.2 Studies of Information Sources in the Innovation Process

This section gives a short overview of studies of the roles and impacts of versatile information sources in different phases of the innovation process. In the literature of

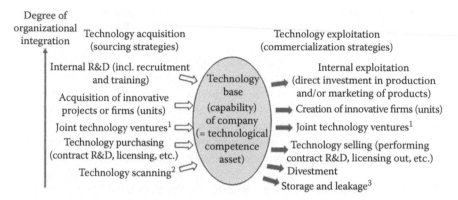

FIGURE 3.3 Generic strategies for acquisition and exploitation of technology. (*Notes*: (1) Joint ventures refer to interfirm R&D cooperation in general, not necessarily formalized, for example, with subcontractors. (2) Scanning includes legal and illegal forms of acquiring technological know-how from outside with no direct purchasing from its original source. (3) This is not a strategy for exploitation but a kind of residue of unappropriated technology, possibly leaking to competitors through their technology scanning efforts.) (From Granstrand, O. et al., Business strategies for new technologies, in *1990 Technology and Investment: Crucial Issues for the 1990s*, E. Deiaco, E. Hörnell, and G. Vickery (eds.), OECD and IVA, Pinter Publishers, London, U.K., pp. 64–92, 1990.)

industrial innovation and innovation policy, diverse roles played by information and knowledge have been discussed from multiple points of view.

One essential issue here is from where and how companies acquire their required new technological knowledge. Granstrand et al. suggest a typology of technology acquisition and exploitation strategies of companies, presented in Figure 3.3 (Granstrand 1999; Granstrand et al. 1990).

The in-house R&D activities of companies can be supported by acquiring R&D from external sources. Typical technology sources are, for example, the acquisition of innovative firms, which have competitive technology, the purchase of capital consisting of new technology, innovative cooperation with other companies, recruitment of technological experts, follow-up or scanning of technological development, and the purchasing of new technology in the form of patents and licensing. These ways of acquisition of new technologies vary in enterprises in different countries (e.g., Granstrand 1999).

In technology acquisition, the importance of external contract research has increased in the United States, Sweden, and Japan (Granstrand 1999). One reason is related to a changed R&D and innovation strategies of corporations. To a certain extent corporations have dispersed their R&D activities to subsidiary level, or increased flexibility by increasing the acquisition of external R&D services and correspondingly decreased permanent in-house R&D personnel. Such trends give opportunities for private and public R&D and consulting firms to enlarge their services. Nevertheless, in many cases corporations still keep on executing their strategic product and process technology R&D in-house.

External information sources for companies include public universities and R&D institutes both of which constitute an important part of the national innovation infrastructure in developed countries.

Cohen et al. (2002) explore the role that public research plays in industrial R&D, and the pathways through which that effect is exercised. The study is based on data from the Carnegie Mellon Survey on industrial R&D. They evaluate the influence of "public" (i.e., university and government R&D lab) research on industrial R&D in the U.S. manufacturing sector. The results indicate that the key channels through which university research impacts industrial R&D include published papers and reports, public conferences and meetings, informal information exchange, and consulting.

Nivala (1994) carried out a study in Finland on external information sources of enterprise innovation. Nivala studied the knowledge acquisition of SMIEs to their development work from technology institutes and other information channels (Nivala 1994). According to the study, the importance of external information and know-how to enterprise development grows with the enterprise size and investment in developing functions. The companies have experienced as most profitable the kind of information channels that are integrated to the enterprise's daily functions and processes. The importance of customers has been great but contributions of the so-called technology-issuing information channels (research institutes, universities, and educational establishments) were significant only for few enterprises.

In a more recent study, the utilization of existing information and information services in SMEs was explored in two empirical case studies (Loikkanen et al. 2009). The first study was about acquisition, utilization, and impacts of patent and market information in the innovation activities of SMEs. The second study charted the services produced by domestic information services to support innovation, from the point of view of SMEs.

The results of these two case studies demonstrate that systematic utilization of external information in SMEs innovation activities is remote. On the one hand, companies are hesitant to use services, or are unaware of the potential gains of using external information in innovation processes. On the other hand, information services targeted to SMEs are scarce (due to lack of demand) and productization of SME-friendly services has not got off the ground among service providers. Therefore, raising the awareness of the gains related to using patent and market information among service providers and buyers is crucial. In addition, creation of value-adding information services and combining different types of data sources is fundamental since technical and scientific documents, as such, tend often to hide important information. The service providers should concentrate on creating services that are value added and offered proactively (in interaction with users) and easily exploitable. The most potential areas (where the acquisition of information is challenging but the benefits are the highest) of information service development include information about market potentials, business trends, competitor analyses and information about legal frameworks, regulations, orders or standards.

The results also indicate that those SMEs that use external information in their innovation activities tend to rely on public organizations, namely, research organizations, industrial associations, and patent offices. Half of the service users considered that using the information services prevented duplication in their product development processes. In general, the concrete benefit of the knowledge services is however hard to identify—this is also reflected on the setting of operational goals by information service providers. The external information that is fed to the company's innovation process functions as one component among many others and it is difficult to separate its unique impact from the final innovation. Regarding the interviewed SMEs, the role of external information is to provide confidence in decision making and to prevent possible patent violations or investment losses. There is a smorgasbord of information services available to fulfill the needs of SMEs, including professional providers and services. However, for creating sufficient demand for such services the communication and marketing of the potential gains of the utilization of quality information in the innovation processes should be increased. Integrating the information utilization into the processes of public R&D financiers should be considered, and productization, as well as the delivery and distribution of such services, should be developed further.

In addition to aforementioned studies, a survey carried out by PricewaterhouseCoopers is of relevance in this context. The survey of PricewaterhouseCoopers (PwC 2009) in Finland to CEOs of the 100 largest companies, of the 100 biggest R&D investors, and of the 200 innovative SMEs were queried about the importance of public and university research in the innovation activities of companies (PwC 2009). According to the results, this type of research collaboration is of minor importance for companies today. The survey concludes that companies are trying to do basic research-related activities by themselves. In SMEs the importance of this particular item was estimated to decrease during the following 2 years, but large firms have more nuanced picture of the situation. The results indicate differences among companies operating in different industries.

As a short overview on selected studies indicate, the roles and impacts of diverse information sources in different phases of the innovation process vary. In the future, more systematic examinations should be executed in order to better understand and make consequent policy conclusions on the versatile roles played by external information sources in industrial innovation.

3.2.3 CHANGING GLOBAL LANDSCAPE OF INNOVATION

Paced by information and communication technologies, the global economy has shifted to a new phase characterized by international capital, production, knowledge, and labor markets. The growth of BRIC countries and particularly of emerging Asian countries such as China and India has a worldwide influence on the economic, scientific, and technological development (e.g., Aubert 2008; Gurry 2008; Thurow 2000).

The globalization of innovation activities and related organizational and local changes, accelerating mobility of labor force and increasing pace of change in general, create new challenges and opportunities for enterprises. They cannot mobilize all resources, competences, and capabilities required by effective knowledge production by themselves internally, but are increasingly dependent on external sources, networks, and cooperation (Chesbrough 2003; EPO 2007; Ernst 2006; R&D Management 2006; Torkkeli et al. 2007; von Hippel 2005). The innovation model according to which the majority of knowledge was created internally will be replaced by a model according to which majority of knowledge will be acquired from external sources. Empirical evidence of increased collaboration in innovation is available, for example, from patent applications. By the OECD, cross-border collaboration on inventions (sharing patents among inventors from two or more countries) almost doubled as regards to global volume, from 4% to over 7% from 1991–1993 to 2001 2003 (OECD 2007). Increasingly effective ICT improves the access to sources of science, patents, and related information, which also decreases the need to own basic or applied R&D but, on the other hand, necessitates additional investments in product development and the acquisition of knowledge and competences worldwide (Chesbrough 2003; Ernst 2006; Gurry 2007).

In addition to the various impacts of globalization of innovation activities, enterprises face other challenges. Competition for highly skilled but affordable labor force increases in narrow niche areas requiring special competencies. Business models of companies must effectively adjust to the new requirements of global markets (e.g., Ernst 2006). In addition to own R&D, the pressure increases to utilize licensing in business, that is, commercialize IPR property and technology acquired from external sources. As Grindley and Teece (1997) emphasize, this should, however, not jeopardize absorptive capacity to develop core competencies and the capability to utilize external knowledge. Different aspects described earlier give new challenges to "IPR regime" and related rationale of innovation policy as well (e.g., Dosi et al. 2006).

The development of knowledge and competences on the basis of external sources, networks, and collaboration change IPR strategies, principles, and practices of enterprises, as well as traditional ways of thinking, private and public interests of patenting, and its incentive effect on innovation (Dosi et al. 2006; EPO 2007; Gurry 2008; von Hippel and von Krogh 2003). von Hippel has explored how and why users, individually and in firms and in communities, develop and freely reveal innovations. He argues that there is a general trend toward an open and distributed innovation process driven by steadily better and cheaper computing and communications. The net result is an ongoing shift toward the democratization of innovation (von Hippel 2005). As von Hippel accentuates, this welfare-enhancing shift is forcing major changes in user and manufacturer innovation practices, and is creating the need for changes in government policies (von Hippel 2005).

According to Chesbrough (2006), rather than restricting innovations to a single path to the market, open innovation inspires companies to find the best business model—whether that model exists within a firm or with an external one. It is increasingly the university systems that will be the locus of fundamental discoveries, and industry will need to work with universities to transfer those discoveries into innovative products, commercialized through appropriate business models (Chesbrough 2006). Boldring and Levine (2008) question the traditional IPR thinking:

> IP is not like ordinary property at all, but constitutes a government grant of a costly and dangerous private monopoly over ideas. We show through theory and example that intellectual monopoly is not necessary for innovation and as a practical matter is damaging to growth, prosperity and liberty.

Although Boldring and Levine represent a kind of extremist critical approach vis-à-vis traditional thinking in IPR, the open source approach in innovation activities seems to be an increasing trend in many innovation development platforms.

3.2.4 CONCLUSIONS

Diverse information and knowledge are needed in different phases of the innovation process. Companies need information about the future trends and market potential, required technologies (road maps) and services (consumer preferences, intermediate products, etc.), and new business opportunities. In the phases of research and invention as well as in later phases of design, piloting, and prototyping, it is important to have information on previous or ongoing studies in other companies or research institutions. In the later stages of the innovation process, in approaching the commercialization or innovation and markets, it is important to have knowledge about distribution, markets, etc. What is important is also a firm's ability to exploit (absorptive capacity) external knowledge (of a more intermediate sort), such as basic research findings that provide the basis for subsequent applied research and development. Compared to so-called learning-by-doing, with absorptive capacity a firm may acquire external information that will permit it to do something quite different. As assumed in Section 3.1, the acquisition of external knowledge is difficult in SMEs as compared to big enterprises, and hence special schemes are needed to support the acquisition of external knowledge in SMEs.

The global economic and innovation environment is also in a dynamic change, which has numerous impacts on innovation, IPR, and the knowledge exchange strategies of companies, as well as on innovation policies. Increasingly new knowledge is created and R&D labs are established in emerging Asian countries, such as China and India. In addition, an increase in the acquisition of information from external sources in product development—whether this information originates from research organizations, company networks, or user communities—challenges the traditional way of thinking, private and public interest of patenting, and its incentive effect on innovation.

3.3 DUPLICATION OF RESEARCH IN THE CONTEXT OF INNOVATION POLICY

Characteristics of knowledge, knowledge production, information, and information asymmetries between economic agents are starting points in the economic theory of innovation policy and the rationale of innovation policy making and public research funding. The traditional analysis of the economic rationale of innovation policy is based on the determination of socially optimal research investment and related market failures, that is, the market invests too much (or too little) into research, as compared to the social optimum. This analysis is controversial and related, for example, to the principles of the equilibrium economics, the determination of the socially optimal level of R&D investments, etc. (see, e.g., Georghiou et al. 2003; Research Policy 2000; STI 1998). Besides market failure, the discussion of the economic rationale of innovation policy consists also of other aspects and elements such as government failure or system failure, and the changes of the science system and community within the past two decades, that has also brought new elements to the analysis of the socioeconomic rationale of innovation policy and public R&D funding (see, e.g., Geuna et al. 2004; Gibbons et al. 1994; Kutinlahti 2005).

In spite of the controversy on the traditional economic rationale of innovation policy and public R&D funding, the market failure and hypotheses as to whether markets invest too much or too little in research as compared to the social optimum are assessed as relevant starting points in this chapter. Section 3.3.1 gives a short overview of the economic rationale of innovation policy based on these two hypotheses. Empirical research gives support to "underinvestment," that is, research investments are less than the social optimum, which legitimates public research funding. "Overinvestment" hypothesis relates to the duplication of R&D work, which is a less examined issue in literature. Section 3.3.2 discusses different theoretical aspects, problems, and evidence of duplication of R&D work, and Section 3.3.3 presents empirical evidence of the duplication of invention in patent applications of the study by the National Board of Patents and Registration (PRH) in Finland.

3.3.1 RATIONALE OF INNOVATION POLICY—THEORETICAL AND EMPIRICAL BACKGROUND

The reasons of government intervention in funding research have been analyzed in the economic literature since seminal contributions of Nelson (1959) and Arrow (1962). The economic rationale, as presented conventionally in the economic literature, is based on two assumptions related to imperfect markets: first on the assumption that markets allocate less-than-optimally to private research ("underinvestment hypothesis") and second on the assumption that markets allocate more-than-optimally to private R&D ("overinvestment hypothesis"). The recent evolutionary innovation system analysis, while questioning assumptions of the mainstream economic analysis, points to various systemic failures or problems (e.g., Chaminade and Edquist 2011; Metcalfe 1995; Woolthuis et al. 2005) as an argument for public intervention in research. The following analysis focuses however solely to the underinvestment and overinvestment hypotheses.*

From society's perspective, the economic inefficiency of innovation markets emerges from suboptimal investments in private R&D. Underinvestment in R&D stems from the peculiar characteristics of the outcome of R&D, new information and knowledge, and associated uncertainties of investing in R&D, due to the problem of the appropriating of returns (see, e.g., Arrow 1962; Nelson 1959). New knowledge and emerging benefits may leak to competitors, which diminish private incentives to invest in R&D. Moreover, uncertainties concerning R&D investment may arise from moral hazards and the adverse selection of researchers in R&D work and organizations (Arrow 1962; Dasgupta 1987). Consequently, private investments in R&D may remain suboptimal at a societal level. The solution in attaining societally efficient knowledge production is public intervention in the form of R&D subsidies, and through establishing a patent institution that grants a temporary monopoly for the original discoverer to exploit the benefits of invention.

---

* The controversy of two hypotheses and related market failures in economic literature is related, for example, to the principles of the equilibrium economics, the related determination of socially optimal level of R&D investments, and challenges of empirical measurement of private and "additional" public returns of R&D investments (see, e.g., STI 1998; Special issue of Research Policy 2000).

According to the "overinvestment hypothesis" (Dasgupta and Stiglitz 1980), the competition in industrial R&D may lead to R&D efforts being duplicated and the consequent dissipation of resources on a societal level. The proposed solution for overinvestment is also government intervention, but this time by taking a coordinating and reconciling role in private R&D investments to avoid R&D being duplicated and thus avoiding also any consequent dissipation of R&D resources. The end result will be more efficient R&D and knowledge production.

The under- and overinvestment hypotheses have been developed within the welfare economic analysis, and a lot of controversy is given by literature of their relevance also from an empirical perspective. In the economic literature as well as among innovation policy documents, the hypothesis of underinvestment gets support more than that of overinvestment. For example, Martin and Scott (2000, p. 438) stress that empirical evidence (e.g., estimated rates of return on investment in R&D) suggests that on balance it is underinvestment that is observed in practice. Griliches concludes that although the theoretical possibility for excessive investment in R&D may be real and may actually occur in isolated cases, the available empirical evidence does not support the conclusion that this is true in general. On the contrary, both the estimates and the likely importance of spillovers suggest that the opposite is true (Griliches 1995).

The "underinvestment" hypothesis is supported by empirical observations in cases of basic research: Griliches (1986), Mansfield (1980), and Link (1981) found a significant "premium" of basic research on the order of a factor of three or higher. This could imply that there may be even more underinvestment in basic research in industry and, by implication, possibly also in science in general (cf. Griliches 1995, pp. 82–83). Moreover, the policy target in many industrialized countries of increasing both public and private R&D expenditures may refer also to the underinvestment in R&D, as understood in comparison with leading R&D investors like Japan, Sweden, and United States. According to Hall and Van Reenen (2000, p. 449), the projects that should be promoted from a social point of view are those with the largest gaps between the social and private returns. Metcalfe (1995, p. 421) interestingly concludes that "private incentive is weakest precisely where the social gain is greatest."

The overinvestment hypothesis seems relevant from practical policy aspects as well because it refers to evident overlapping R&D efforts of enterprises, due to rivalry and confidentiality in competitive market conditions, the result of which may be resource loss. In the absence of "social planner," with perfect information of all R&D and innovation efforts, duplication in research becomes evident. The duplication of R&D and consequent resource loss can be mitigated by coordination of R&D by enterprises themselves, by government, or by any other institution. Economies of joint research and the related sharing of an R&D portfolio among several companies will decrease risks for individual companies and provide lower mutual learning costs. According to the analysis of Beath et al. (1995, p. 134), if product differentiation and product-specific research paths are introduced to the model, the claim of needless duplication of research loses its force.

In conclusion, the theoretical background of the economic rationale of innovation policy is based on characteristics of production of knowledge and information flows and asymmetries between enterprises and other economic actors.* These issues are relevant for the consideration of knowledge and information needs of R&D and innovation in enterprises. The analysis of the economic rationale of innovation policy gives starting points for the consideration of the duplication of enterprise R&D as well, and the next sections provide an overview on the debate of the duplication of research from different perspectives.

3.3.2 VARIOUS ASPECTS OF RESEARCH DUPLICATION

The duplication of R&D and innovation work is discussed in literature mainly in the context of the economic rationale of innovation policy. Generally speaking, the analysis of duplication is relatively scarce in literature and empirical evidence comes mainly from patenting. Because the analysis of duplication of research is not well established, various interpretations of what it means can be presented including, that is, duplications of a problem or how serious problem it is, etc.

The duplication of R&D and innovation work can be considered from various perspectives. It can be considered as a natural part of a market economy where enterprises compete by selling products that fulfill customer needs. Overlapping production activities of enterprises is an inbuilt element of markets. When a strategic objective of R&D and innovation is to support the competitive power of companies in the same markets, related R&D work may also overlap. In the literature of evolutionary economics, the duplication of research is seen as an intrinsic characteristic of technical change. Overlapping is a component of technical change when seen as an evolutionary and collective process. The patent system was created in order to protect the rights of the original discoverer to benefit from invention. The evidence of imitation of patented inventions however indicates that duplication takes place even in spite of patenting.

In the literature of evolutionary economics, the duplication of research is seen as an inherent characteristic of technical change. As Silverberg argued in early 1990s, the acquisition of, and access to, diverse information are both important factors in an innovation process. Technological change and related invention and innovation activities should, though, be seen as being collective and evolutionary processes (Silverberg 1990). As he argues, expediency as such is an example of an externality, and thus poses a challenge to the optimum welfare implications of those styles of general equilibrium analysis, which, at least in theory, fully reconcile individual and social interests. This has led to a sophisticated literature on whether too much or too little R&D will be conducted compared to some posited social optimum, due either to the inadequate private incentive or to the danger of redundancy and duplication of research efforts. Silverberg tries to demonstrate that both externalities and (near) duplication can be very useful, perhaps even necessary components of technical change when seen as being a collective evolutionary process. The key concept in his analysis is learning, which can take place within an individual, the organization and collectively through a network of feedbacks unfolding over time between both cooperative and competitive agents (Silverberg 1990). In conclusion, in the evolutionary economics literature duplication of research is seen as intrinsic and natural phenomenon.

---

* On precise definition of information asymmetry, see, for example, The New Palgrave (1998, Vol. 1).

From the practical perspective, duplication of research and innovation work may be divided into "healthy" and "unhealthy." For example, if research is carried out for already invented and even patented solutions, research work done may be termed "unhealthy" in a sense that inventor cannot reap the benefits of research work and R&D investments made, and thus there has been some dissipation of resources. On the other hand, "near-overlapping" research work may be a positive thing and also needed. It should even be encouraged by public subsidies, for example, if the solution for a final research problem (e.g., developing a new drug) is searched for through several alternative methodological ways, approaches, or schools of thought.

The potential to research duplication grows naturally the bigger geographical areas or the more countries we are considering. Accordingly, potential research duplication is inevitably a challenge not only for big countries but also to supranational organizations such as the European Union. Although also in the European Union the discussion of the union wide duplication of research efforts has been relatively scarce, in the recent Europe 2020 Flagship Initiative Innovation Union document, duplication is listed among the key weaknesses in the European national. It is also a flaw in regional research and innovation systems besides underinvestment in European knowledge foundation and unsatisfactory framework conditions (Box 3.1):

BOX 3.1

"Too much fragmentation and costly duplication. We must spend our resources more efficiently and achieve critical mass." ... "Avoid fragmentation of effort: national and regional research and innovation systems are still working along separate tracks with only a marginal European dimension. This leads to costly duplication and overlap which is unacceptable at a time of tight finances. By better pooling our efforts and focusing on excellence, and by creating a true European Research Area, the EU can enhance the quality of research and Europe's potential for major breakthroughs and increase the effectiveness of the investments needed to get ideas to market" (EC 2010).

The following short analysis attempts to organize the analysis of the duplication of research from many different perspectives and vis-à-vis scarce material available from the literature of this issue and empirical evidence available from patenting and some related studies.

3.3.2.1 Proximity and Duplication of R&D

Proximity, whether related to geographical or cultural proximity or proximity between research and innovation communities, is an important factor related to the duplication of research. Intuitively it is imaginable that the wider, for example, the geographical proximity is (e.g., between continents), the more duplication of research exists. Duplication may also be specific and differ between scientific and technological fields or between industrial sectors.

Giuri et al. (2007) explored geographical proximity and the exchange of knowl edge among inventors in their study on European innovation. The organizational proximity proved to be the most important category. Interactions in the same organization are on average more important than interactions with people in other organizations, especially when they are geographically close.

The relevance of the role of proximity is enhanced also by the econometric study of patenting. One way to capture knowledge flows among firms and industries is to classify firms into different technological clusters according to the technological classifications of their patents (Fung 2004). Jaffe (1988), for instance, relies on the patent offices' classification system to identify the proximity of firms in technological space. Proximity between two firms measures the degree of overlap or duplication in their research interests. Hence, a relevant spillover pool pertinent to a firm can be constructed by summing up the R&D efforts of all the other firms weighted by their proximity. Although overlapping in research areas gives rise to a greater opportunity for knowledge spillovers to occur, Fung reminds us that cross-citations do not necessarily reflect research overlap because a firm conducting certain research may cite a patent originated from a totally different research area. For example, an automobile manufacturer may cite a patent owned by a computer chip maker (Jaffe 1988).

As this brief glance at the literature glance shows, proximity certainly plays an important role in the duplication of R&D and innovation work.

3.3.2.2 Duplication of Research and Patenting

Patenting is used as a proxy indicating R&D and innovation performance in statistical and economic analyses. Patenting is a policy instrument aimed at enhancing the value of innovation. Patenting is also important with respect to the topic of this chapter, that is, the acquisition and utilization and impact of knowledge and information sources in the innovation process. Patents are an important part of information flows between competing enterprises, as shown by Cohen et al. (2002). Patents are seen to play a more central role in diffusing information across rivals in Japan than in the United States and according to Cohen et al. this appears to be a key reason for greater intra-industry R&D spillovers there, suggesting that patent policy can importantly affect information flows (Cohen et al. 2002).

One of the basic rationales of the patent system is to provide an incentive for creation of new technology and inventions. The patent system does this by offering to inventors exclusive rights to commercially exploit patented inventions for a limited time, in return for the disclosure of the inventions to the public (WIPO and IFIA 1998). Patent information is believed to "provide a unique planning resource for managing a firm's technology or product development and for systematically evaluating its competitive position relative to other companies in a market area" (Ashton and Sen 1988).

The patent system has two interrelated functions: the protection function and the information function (WIPO and IFIA 1998). According to WIPO and IFIA,

> The fact that a patent gives an inventor exclusive rights on a special field and by doing so limits the possibilities of access to this special technology for other enterprises is compensated by the information about the newly developed technology which is to be laid open by the inventor.

This information function of the patent is not only the main impetus for the continuous development of the technology, but it is also of increasing importance for industrial property offices. They well recognize that providing information to the public is of equal importance to the granting or registration of patents, trademarks, and designs and have created Internet and other ICT-based information systems to offer access to patent databases (WIPO and IFIA 1998).

According to the White Paper from May 2003, prepared by Butler Group, the benefits from patenting are related to avoidance of unnecessary R&D costs, avoidance of litigation, maximization of license management opportunities (both outgoing and incoming), reduction in product time-to-market, market intelligence, and targeting and evaluation of mergers and acquisitions.

Patenting naturally plays a different role in IPR and innovation strategies of different companies and industries. For example, according to Cohen et al., semiconductor firms, as driven by a rapid pace of technological change and short product life cycles, tend to rely more heavily on lead time, secrecy, and manufacturing or design capabilities than patents to recoup investments in R&D (see Cohen et al. 2002).

3.3.2.3 Duplication of R&D and Imitation in Patenting

Industrial property offices have developed information systems in order to avoid duplication of R&D work. Awareness of the state-of-the-art technology in a particular technical field can avoid duplication in research work by indicating that the desired technology already exists (WIPO and IFIA 1998). It can also provide ideas for further improvements, or can give an insight into the technological activities of competitors and, by reference to the countries in which patents have been applied for, the marketing strategies of competitors. A search for state-of-the-art technologies can also identify newly developing areas of technology in which future R&D activity should be monitored (WIPO and IFIA 1998).

However, WIPO and IFIA (1998) express their concern about the unexpectedly low use of patent documents as a source of technological information. WIPO and IFIA refer to a relatively old survey from 1985, which dealt with the problem of technology and innovation in Austria. The survey found that only 4% of the enterprises used patent literature as an innovative instrument.*

As WIPO and IFIA stress, the low utilization of patent information is regrettable, because it is a fact that in the EC countries billions per year—the U.K. Patent Office has mentioned a figure of 20 billion pounds—are wasted to develop things that are already developed and documented in the description of patent specifications (WIPO and IFIA 1998). WIPO and IFIA also refers to another study that confirmed that a large amount of redundant research takes place, since it was found that 30% of all R&D in Europe duplicates work already done (WIPO and IFIA 1998). WIPO and

---

* According to the Austrian study, the influence and use of patent information increases in relation to the size of the research and development institution or the enterprise; 18.5% of companies with >100 staff reported to actively use patent literature. Only 2%–3% of enterprises with <100 staff use patent literature in the first stage of an R&D project. This result correlates with a much more intensive patent activity in larger enterprises. Only 5% of the enterprises in this study had 500 and more employees, but 55% of the patent applications originated from this group.

IFIA make proposals to industrial property offices in order to further develop patent information services in the future (WIPO and IFIA 1998).

Imitation is, strictly speaking, not duplication of R&D but is, however, a related issue. First of all, there are a lot of studies of imitation and consequent successful science and technology policies, for example, from Japan and South Korea in which a systematic imitation from western countries has played an important role (e.g., Freeman 1987). The aim of patenting is to guarantee that the discoverer company can be the only exploiter of the invention, at least during the period the patent is in force, and in this way to encourage innovation. There are, however, problems in protecting proprietary rights based on patenting due to imitation. For example, a study of nearly 50 product innovations in the electronics, chemicals, pharmaceuticals, and machine tools industries shows that <4 years after they were marketed, some 60% of these innovations patented by their investors had been imitated, and, moreover, at a cost about one-third lower than that which the innovating enterprise had spent on developing them (Mansfield et al. 1981). A report of the U.S. International Trade Commission has estimated that American industry lost almost U.S.$ 24 billion as a result of imitations in 1986 alone (U.S. International Trade Commission).

The White Paper gives evidence from the survey of 261 companies by IRN Services. Over 70% of companies admit to investing in research that led to (or was leading to) a previously patented solution; the estimated cost of this averaged out at 30% of the total R&D spend. When one applies these findings to specific vertical market sectors, the true scale of "waste" is obtainable. The top three vertical market sectors in terms of R&D spend are information technology with U.S.$ 73 billion, automotive with U.S.$ 46 billion, and pharmaceuticals with U.S.$ 44 billion. Some simple mathematics allows us to gauge the wastage in hard cash (as opposed to percentages): information technology wastes U.S.$ 15.3 billion, automotive wastes U.S.$ 9.7 billion, and pharmaceuticals wastes U.S.$ 9.2 billion. In total, across the top eight vertical market sectors (in terms of R&D spend), the estimated total wastage is U.S.$ 48 billion.

There are several factors worth noting here: this loss has a direct impact on the bottom line for every company, it does not include contributory losses due to increased time-to-market, it does not include contributory losses due to litigation, and a significantly high percentage of this waste in R&D could be avoided with access to patent publications and an understanding of their true content (Mansfield et al. 1981). Moreover, it is worth giving some additional thought to the last point mentioned earlier. Simply having access to patents is, by itself, of little benefit. Access to the meaning of the patent, what it exactly details, and the ability to discover that information easily and quickly are the key points. Effectively tapping this key information by advanced knowledge services is clearly a competitive advantage for a firm operating in global markets.

As the short overview of duplication or R&D and imitation of patenting in this section indicates, these are very complex issues that must be taken into account in R&D and IPR strategies of enterprises, as well as in IPR strategies of innovation policies and of related information service providers. Although services delivering market and patent information to customers should be developed to become more effective, the problem of the misuse of these services, for example, in the form of illegal imitation, and principles and measures to respond to this problem should be recognized as well.

3.3.3   DUPLICATION OF INVENTIONS IN PATENT
APPLICATIONS—FINNISH EVIDENCE*

As much as 80% of the value creation of companies is based on intellectual capital. However, the instruments concerning the management of this capital are poorly utilized. This especially concerns the protection of intellectual property and the usage of existing technical knowledge and patent information. Above all, this applies to the SMEs. The poor usage of the IP system creates problems in R&D as well as in marketing and other areas of business. Besides protection, the intellectual property system provides also a unique source of technical and competitor information for the use of marketing and product development.

According to Koen et al. (2001), the innovation process may be divided into three areas: the fuzzy front end, the new product development process, and commercialization. The first part, fuzzy front end, is generally regarded as the one where greatest opportunities for improvement of the overall innovation process appear. Attention is increasingly being focused on the front-end activities in order to increase the value, amount, and success probability of high-profit concepts entering product development and commercialization. Even 80% of the costs for the whole innovation process are committed at the very early phase of the project. Making a wrong choice in this phase means that all later decisions are based on misconception and mistaken premises. Using relevant information—including available front-end knowledge services—in the beginning of the innovation process may significantly reduce the risks of the whole project. When the project proceeds, the possibility of influencing the total costs of the product diminishes rapidly. So it is important that all possible essential information is both available and in a readily absorbable format at the earliest possible phase of the innovation process.

3.3.3.1   On the Role of Patent System

The patent system is an information dissemination system, and it exists for the purpose of promoting the technological progress, the prosperity of the nation. The essential thing in the pursuit of this goal is the stimulation of product development and the new technical information disclosed in patent publications.

The first prerequisite for an invention to be patentable is that it has to be novel. Exclusive right is not granted to old, previously known techniques. Second, besides being novel, the invention must also involve an inventive step. The third precondition for patentability is industrial applicability. Besides conventional industries, patents include solutions, methods, and devices needed in commerce, building industry, farming, forestry, gardening, fishing, handicrafts, etc. A patentable invention is a concrete embodiment of an idea: a technical solution, a device, a product, a process for making a product, for instance, or a new use for a previously existing product.

There are many prejudices concerning patents. One of the most common is the belief that patents are applied for and granted to complete new products or very exquisite inventions. The reality is quite the opposite: >80% of all patents relate to existing

---

* Section 3.3.3 is based on the contribution of the team of the National Board of Patents and Registration (PRH), led by Mr. Mika Waris, to the study of Loikkanen et al. (2009) on which this chapter is mainly based.

technical solutions. Complete patented products are rather rarities. Instead of being a completely novel solution, a product very often contains several technical solutions protected by an existing patent. For example, a typical mobile phone contains easily >1,000 patented solutions, and a modern car can include even 10,000 patents.

For every applied patent, there is a very detailed patent publication. Every patent publication describes the latest cutting edge developments in the sector, the existing problem, and the discovered solution based on the invention. Thus, to sum up, the information contained on patent documents is of a very detailed, applied technical type. The content is also presented in a structured and formal way. Today, more than 60 million patent documents have been published worldwide. The information is increasingly available on the Internet and also free of charge.

Patent documents are one of the largest public data sources in the world. They are also a unique source: the European Patent Office (EPO) estimates that >80% of the technical information contained in a patent document is never published anywhere else.

According to the survey conducted by EPO in 2003, >80% of companies already using patent information considered the information in patents as very important or important. Patent information was considered to be most useful in early stages of product development, during predevelopment, and invention stage—that is during the fuzzy-front-end phase.

3.3.3.2 Evidence of the Duplication of Invention in Patent Applications in Finland

The study by PRH in Finland provides empirical evidence of the duplication of invention in patent applications, and perspectives to avoid duplication by utilizing patent information services. PRH carried out a study of domestic patent applications and the findings based on the novelty examination made by PRH's examiner. The key data and results of the study are presented in Table 3.1. The empirical data used is the novelty examination reports made by PRH of 11,775 patent applications in years 2000–2005. The amount of granted patents during the same time period was 5144, which is 43.7% of the respective applications.

According to results, patents were not granted to 33.2% of patent applications because of the references (obstacles) for novelty and/or inventiveness, that is, the patent applications were filed for an already published invention meaning that the inventive work done overlapped with former invention efforts. In these cases, there is, *first*, a risk of developing solutions to which someone else has exclusive rights, and, *second*,

TABLE 3.1
Domestic Patent Applications in Finland 2000–2005

Applications	Examinations	Obstacle for Novelty/ Inventiveness	Granted	No. of References/ Publications	References Describing Technological Level
14,398	11,775	3,906 (33, 2%)	5,144 (43, 7%)	537 (10, 4%)	4,607 (89, 6%)

a risk of dissipation of research investments. Consequently, by the use of patent information before or during the R&D phase, these pitfalls could have been avoided.

The study also confirmed the previously mentioned fact that only a small part of patents relate to totally novel solutions. In this sample, only 10.4%, or 537, of all granted patents were such solutions where there was no reference publications found, in other words, they were completely new inventions. The radical inventions can be found among these patents. Still, using patent information in the early phases of the process gives you the certainty of the uniqueness of the invention.

Respectively, in 89.6%, or 4607, of the granted patents there were publications describing technological level and thus the use of patent information would have been useful in these cases. In these cases, the advantages of using patent information are

- Saving time to market
- Better quality of the invention
- Possibility to license technology (MOB)
- Stronger protection of the invention
- Useful knowledge of the competition in the field
- Useful knowledge of state-of-the-art technological solutions

In conclusion, duplication of research and innovation efforts is taking place for various reasons, as discussed earlier. Duplication is considered in the literature mostly as a negative phenomenon but not only as that. In evolutionary economics, it is even considered as a necessary component of technical change when seen as an evolutionary and collective process. Empirical evidence of the extent of duplication of R&D and invention efforts is available from patenting, and from studies of imitating.

Unintended duplication of research can be mitigated by efficient coordination of research and innovation efforts, stricter control of IPRs (e.g., in cases of imitation of patents), and also by more effective and efficient use of services delivering patent and market information. Coordination of R&D can be executed either by private or by public activities. Coordination of national research efforts is an inbuilt responsibility of public agencies granting public funding to enterprises, R&D institutes, and universities. Coordination and reconciling of research efforts is however limited to publicly funded and promoted research and innovation activities. Evidence from overlapping patent applications indicates that this coordination is not sufficient for patenting enterprises and inventors. Although there is no information available of the entire overlapping R&D and innovation efforts of national system of innovation, or from a European or global level, overlapping patent applications give a rough indication that duplication of research and innovation efforts is a real phenomenon, and that it should be put on the agenda of nationally and intranationally coordinated innovation policies.

Mitigation of the duplication of R&D and innovation efforts, whether explicitly or implicitly, has recently been taken into account in innovation policy practice. This has been achieved mainly by encouraging networking and collaboration between key players in the field of national R&D and innovation. One of the key strategic objectives of national R&D programs in Finland has been to urge enterprises, universities, R&D institutes, and other relevant organizations to collaborate more closely in their R&D efforts. There are also examples of innovation policy approaches aimed at avoiding duplication of R&D and innovation efforts, for example, from Japan and the European Union. According to Porter et al. (2000), the Japanese government has played a significant role in organizing and financing cooperative R&D projects. The rationale has been to spread the fixed costs of R&D among many participants and to avoid the wasteful duplication of effort by allocating research tasks among the participants. A celebrated example of government-sponsored cooperative R&D was the VSLI project, designed to help Japan catch-up in semiconductor technology. European Research Area (ERA) is another example in trying to avoid unintended duplication of R&D and other innovation efforts.

To sum up, the results of the PRH study showed that in Finland 33.2% of the domestic patent applications were filed for already published invention, meaning that the inventive work done was overlapping with former invention efforts. In 89.6% of the granted patents, there were existing publications describing technological level and thus the use of patent information would have been useful in these cases. Only 10% of all granted patents were such solutions where there was no reference publications found. Still, using patent information in the early phases of the process gives you the certainty of the uniqueness of the invention.

3.3.3.3 Toward More Effective Patent Information Analysis

Using and analyzing patent information during an early phase of R&D project and a product development process is beneficial in multiple ways. The R&D team can

- Avoid duplication of R&D work
- Identify specific new ideas and technical solutions, products, or processes
- Identify the latest developments in a specific technological field in order to be aware of the latest development
- Assess and evaluate specific technology and identify possible licensors (make-or-buy decisions)
- Improve an existing product or process
- Develop new technical solutions, products, or processes
- Identify existing industrial property rights to avoid infringement actions
- Assess novelty and patentability of one's own research and development activities

If we look at the patent information source, we notice that

- It is the largest, unique source of technical information in the world
- The information is very up-to-date, it is updated by the NPO authorities throughout the world
- It is very structured information, the structure has been defined similar in all countries using the patent system, so you can find certain information independent of the country of origin
- It is well classified
- The information is very detailed and reliable
- About 80% of all technical information is available only in this information source

It is known that large companies employ patent information engineers who can tap to this information pool by, for example, conducting patent searches and analyses. They can then share the results with the product development personnel in an organized manner. In SMEs, the situation is much more complex. Regarding SMEs, more studies are needed in order to determine how the use of patent information could be integrated and utilized in their innovation processes.

The main difficulties in using patent information at SMEs are the lack of information about methods and tools to retrieve and analyze the patent information, the lack of internal resources that could be used for this analysis, and the lacking of adequate patent information services.

3.4   CONCLUSIONS

The utilization of external information in R&D and other innovation activities in SMEs is relatively scarce. There is a lot of room to improve information utilization within SMEs and services in organizations delivering information. The more effective utilization of information can decrease the duplication of R&D, invention, and innovation work and save resources on the company level as well as in the whole society. Because the acquisition and utilization of external information in innovation relates closely to issues of intellectual property, the awareness of IPR should accordingly be enhanced.

Regarding the delivery of information to companies, aiming to support their R&D and innovation, it is important to note that the question is not about mechanistic transmission of information from a service provider to customer. The challenge is to create continuous platforms for mutual learning (on individual, organizational, and collective levels) by promoting networking and feedback. This will not happen immediately: the common information platform will unfold only over time between both cooperative and competitive agents. The simple access to information sources and services is then clearly not enough. To verifiably benefit from the information, the organization must have sufficient absorptive and learning capacities to utilize and reap benefits from it. Accordingly, it is the knowledge management strategy of an organization that forms the context for acquiring and utilizing external information in research and innovation activities.

3.4.1   POLICY IMPLICATIONS

Minimizing observed unintended duplication of invention and innovation work can be achieved by a systematic development of knowledge services delivering customer, competitor, market, and patent information and analyses to innovation processes of especially SMEs but also other organizations performing R&D. Another related area is the improvement of coordination of R&D and innovation activities among all relevant organizations promoting, funding, and carrying out R&D. This can mitigate the duplication in R&D, invention, and innovation and improve the performance of the whole innovation system. The importance of raising the awareness of the latent value of patent information for all relevant actors within innovation ecosystem—that is, not only the conventional protective or restrictive information aspect of patents—cannot be stressed too much.

The following concrete proposals are aimed to improve the acquisition and utilization of information in innovation activities. Although the study presented in this chapter was carried out in Finland, these recommendations have relevance also in other knowledge-intensive countries. The proposals are targeted at research organizations, R&D financiers, policy makers and public services, SMEs, and information service providers.

3.4.1.1 Education and Research Organizations

The awareness of intellectual property issues should be raised by increasing related education and training in universities and polytechnics, especially as a part of the studies in business and engineering education tracks. In addition, teaching staff should be trained to tackle IP-related issues and to realize the potential gains of using patent and market information in innovation-related activities.

The utilization of patent and market information should be integrated in joint research projects with SMEs. Using relevant information during the project planning and implementation reduces the risk for investment losses and refines the project.

3.4.1.2 R&D Funding Organizations

The role and training of IP issues should be increased at the organizations funding and coordinating R&D efforts. Comprehensive information of ongoing R&D, invention, and innovation work is fundamental for effective coordination of public R&D funding. Moreover, research funding applications supplied by private and public R&D organizations should consist of sufficient surveys on past and ongoing research work in order to avoid the duplication of R&D, invention, and innovation work.

Knowledge services adding value to R&D, invention, and innovation work shall be integrated into decision-making processes of organizations funding and coordinating R&D. Surveys of IP and market information (e.g., patent surveys, overviews on technologies, analyses of competitors and competition environment, and novelty examinations) shall be provided to and used in decision making when applicable.

3.4.1.3 Policy Makers and Public Services

The awareness of the overall importance of the IP system as well as its use in avoiding of duplication of R&D, invention, and innovation work shall be increased among public policy makers (ministries, agencies, intermediary organizations, etc.). In addition, international collaboration on the development of IP services and IP system should be enhanced, for example, by exploring good or best practices of IP management and services globally.

Public innovation service providers could encourage utilization of patent and market information services by SMEs. For example, knowledge service vouchers or similar instruments could be introduced to SMEs in order to facilitate their "first purchase." This could then pave the way for wider use of such services.

Exposing business service experts and advisors to benefits of patent and market information can improve their ability to serve SMEs in innovation-related information needs. The utilization of patent and market information services shall be integrated in the processes of the services that are especially targeted to internationalization efforts of SMEs.

3.4.1.4   Small- and Medium-Sized Enterprises

SMEs should develop their awareness, abilities, and tools of patent and market information in order to improve the utilization of related services in their innovation processes. This would significantly reduce the risk of harmful duplication and consequently investment losses and patent violations. This would also make the use of SMEs' limited R&D resources more effective and enhance the productivity of their research and development work.

The understanding of the IP-related issues and the use of patent and market information should be integrated within the innovation process of the company. Utilization of such information in the early phase of the development project, that is, in the "fuzzy front end," enables the improvement of the overall innovation process. Integrating the information utilization also into the later stages (e.g., testing, prototyping, production, marketing, etc.) gives opportunities to redirect, accelerate, or refine innovation process.

3.4.1.5   Information Service Providers

Information service providers should create and productize value-adding knowledge services and service packages targeted especially to the needs of SMEs' innovation activities.

Information service providers and business management consultants should join forces in order to develop services and consulting activities targeted at SMEs. In addition, efficient ICT solutions, for example, rapid collaboration platforms should be adopted in the future when delivering and processing the services.

3.4.2   FUTURE RESEARCH NEEDS AND TRENDS

3.4.2.1   Future Research Needs

Various further research options in the field of knowledge services and innovation process are evident. Innovation activities of enterprises—including those of patenting, IPR strategies, and operations—are in transition due to changes in innovation dynamics. These again are in a state of flux due to the globalization of innovation activities, opening up of innovation and increasing pressure to cooperate and network due to increasing R&D costs, etc. As a consequence of all these changes, traditional IPR policy is expanding toward a comprehensive knowledge policy (Gurry 2007) and strategy. It is important to be aware of this, and to examine these changes on the enterprise level. Of particular interest is R&D, innovation, and IPR strategies and operations, as well as—on the level of innovation policy making—policies of patenting, IPRs, and the entire knowledge system.

The examination of overlapping patent applications and more broadly the duplication of R&D should be deepened in further research efforts. The specific research topics include more in-depth exploration of the characteristics of the duplication of research in different industrial sectors, in different technology areas, profiles of inventors, enterprises having and not having public R&D subsidies, etc. As the literature indicates, proximity (both in geography and in other ways) plays an important role in the duplication of R&D and innovation work.

Accordingly, the relationship between proximity and the duplication of R&D should be explored in detail in the future. As the overview of duplication of R&D and imitation of patenting in this chapter indicates, these are very complex issues that must be taken into account when developing R&D and IPR strategies at the enterprises. The same applies to innovation policies and to related information and knowledge services as well. Although services delivering patent and market information to customers should be enhanced further, the challenges arising from misuse of these services (e.g., illegal imitation) should be studied as well. Strategies and measures to counter this increased illegal imitation should probably also be taken in the research agenda.

In addition, the external innovation sources of enterprises should be explored in more detail, that is, the concrete impact of private and public information sources on enterprise innovation. The relevant questions in this respect are, for example, What is the impact of university and government R&D laboratory research on industrial R&D or what is the impact of private knowledge-intensive service activities (KISAs)? These issues imply a further question, namely, What are the key channels through which university research impacts industrial R&D? Are these, for example, published papers and reports, public conferences and meetings, informal information exchange, consulting, contract research, etc.? All these relevant research issues need to be discussed in follow-up studies.

3.4.2.2 Future Trends

In the current challenging innovation ecosystem, it is not enough to tackle issues in a piecemeal fashion, that is, by delivering services to customers simply in ad-hoc situations and contexts. Instead of offering one-off services for a specific information need, it is more rewarding and ultimately more cost-effective to take a more forward looking approach and offer broader solutions to customer problems, that is, knowledge landscapes (Figure 3.4). The major feature of all knowledge landscapes is their interactivity and adaptivity to changing customer needs. In practice, this means shifting part of the analysis workload to the customer. Customers are able to tinker and adjust the landscape to customize it more closely to their needs. This trend is welcome not only because the customers are usually the best experts in their subject fields but also because closer cooperation with clients boosts the development of more innovative solutions and services.

Exploring and analyzing the vast volumes of information is becoming increasingly difficult. Information visualization and visual data mining can help to deal with the flood of information. Information and data visualization is of prime importance for both internal and external knowledge services. There are a large number of information visualization techniques that have been developed over the last decade to support the exploration of large information and datasets. Current text mining tools offer variety of options to pinpoint technology and market trends, track main players in a particular field, and visualize the often complex and multilayered interrelationships and dependencies between various stakeholders and domains. Mapping the ongoing changes in innovation landscape by teasing out hidden patterns and dynamically following their development over a period of time are the main benefits for information professionals serving knowledge-intensive organization.

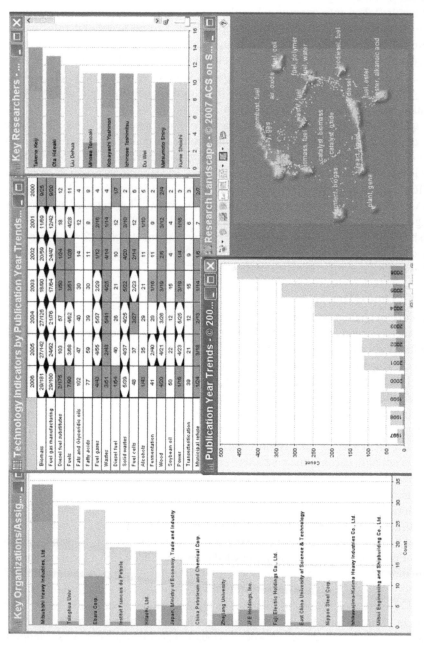

FIGURE 3.4   Patent landscape, made by STN AnaVist registered trademark of American Chemical Society.

Processes and practices for external information and internal knowledge management are converging. Many of the emerging tools and trends in scientific, technical, and market knowledge services are nowadays adapted from practices and solutions developed originally for managing more effectively the internal information and in-house knowledge creation at companies. One of the knowledge advantages of today is to integrate wisely external information and in-house knowledge. Enterprise search concept can be extended to external information (Box 3.2).

BOX 3.2

Enterprise search concept, its major benefits for organization:

- Intensifying the utilization and *reuse* of internal information at the organization
- Strengthening *access*: one search reaches out to all critical in-house information silos ⇒ multi-repository pan-enterprise search
- Intuitive grouping and *clustering* of search results
- Various options for *drilling down* into results
- Discovering unforeseeable *connections* between people and things

---

REFERENCES

Ackoff, R.L. (1989). From data to wisdom. *Journal of Applies Systems Analysis*, 16, 3–9.

Arrow, K.J. (1962). Economic welfare and the allocation of resources for invention. In: *The Rate and Direction of Inventive Activity*. Princeton University Press, Princeton, NJ, pp. 609–625 (republished in: *The Economics of Technological Change*, N. Rosenberg (ed.). Edward Elgar Pub, London, U.K., 1971, pp. 164–181; and in: *The Economics of Information and Knowledge*, J. Lamberton (ed.). Penguin Books, London, U.K., 1971, pp. 141–159).

Ashton, W.B. and Sen, R.K. (1988). Using patent information in technology business planning. *Research Technology Management*, 32(6), 42–46.

Aubert J.-E. (2008). Towards knowledge economies: Global trends and issues. In: *Going Global—The Challenges for Knowledge-Based Economies*, M. Squicciarini and T. Loikkanen (eds.), Publications of the Ministry of Employment and the Economy, Finland, Innovation 20/2008, pp. 62–70.

Beath, J., Katsoulacos, Y., and Ulph, D. (1995). Game-theoretic approach to the modeling of technological change. In: *Handbook of The Economic of Innovation and Technological Change*, P. Stoneman (ed.). Blackwell Handbooks in Economics, Blackwell, pp. 132–181.

Boldring, M. and Levine D.K. (2008). *Economic and Game Theory against Intellectual Monopoly* (to be published by Cambridge University Press, available in July 2008).

Bouthillier, F. and Shearer, K. (2002). *Understanding Knowledge Management and Information Management: The Need for an Empirical Perspective, Information Research*, Vol. 8, No. 1, October. University of Borås, Sweden.

Chaminade, C. and Edquist, C. (2011). Rationales of public policy intervention in the innovation process: Systems of innovation approach. In: *The Theory and Practice of Innovation Policy, An International Research Handbook*, R.E. Smits, S. Kuhlmann, and P. Shapira (eds.). Edward Elgar, Cheltenham, U.K., pp. 95–114.

Chesbrough, H. (2006). *Open Innovation, The New Imperative for Creating and Profiting from Technology*. Harvard Business School Press, Boston, MA.

Cohen, W.M. and Levinthal, D.A. (1989). Innovation and learning: The two faces of R&D. *Economic Journal, Royal Economic Society*, 99(397), 569–596.

Cohen, W.M., Nelson, R.R., and Walsh, J.P. (2002). Links and impacts: The influence of public research on industrial R&D. *Management Science*, 48(1), 1–23.

Dasgupta, P. (1987). The economic theory of technology policy. In: *Economic Policy and Technological Performance*, P. Dasgupta and P. Stoneman (eds.). Cambridge University Press, pp. 7–23.

Dasgupta, P. and Stiglitz, J. (1980). Uncertainty, industrial structure and the speed of R&D. *Bell Journal of Economics*, 11, 1–28.

Dosi, G., Marengo, L., and Fagiolo, G. (1996). Learning in evolutionary environments. IIASA Working Paper No. 96–124.

Dosi, G., Marengo, L., and Pasquali, C. (2006). How much should society fuel the greed of innovators? On the relations between appropriability, opportunities and rates of innovation. *Research Policy* 35, 1110–1121.

EC (2010). Europe 2020 Flagship Initiative, Innovation Union, SEC(2010) 1161, COM(2010) 546 final, Brussels, 6.10.2010, European Commission.

Ernst, H. (2006). Using patent information for strategic business planning, WHU, Otto Beisheim School of Management, Vallendar, Germany.

EPO (2007). Scenarios for the future, how might IP regimes evolve by 2025? What global legitimacy might such regimes have? European Patent Office, 2007.

Freeman, C. (1987). *Technology Policy and Economic Performance, Lessons from Japan*, Pinter Publishers, London, U.K.

Fung, M.K. (2004). Technological opportunity and productivity of activities. *Journal of Productivity Analysis*, 21, 167–181.

Georghiou, L., Smith, K., Toivanen, O., and Ylä-Anttila, P. (2003). Evaluation of Finnish innovation support system. Ministry of Trade and Industry, Publications 5/2003.

Geuna, A., Salter, A.J., and Steinmueller, W.E. (2004). *Science and Innovation: Rethinking the Rationales for Funding and Governance*, New Horizons, The Economics of Innovation Series, Edward Elgar Publishing Ltd., Cheltenham, U.K.

Gibbons, M., Limoges, C., Nowotny, H., Schwartzman, S., Scott, P., and Trow, M. (1994). *The New Production of Knowledge. The Dynamics of Science and Research in Contemporary Societies*. SAGE Publications, London, U.K.

Giuri, P., Mariani, M., Brusoni, S., Crespi, G., Francoz, D., Gambardella, A., Garcia-Fontes, W., Geuna, A., Gonzales, R., Harhoff, D., Hoisl, K., Le Bas, C., Luzzi, A., Magazzini, L., Nesta, L., Nomaler, Ö., Palomeras, N., Patel, P., Romanelli, M., and Verspagen, B. (2007). Inventors and invention processes in Europe: Results from the PatVal-EU survey. *Research Policy*, 36(8), 1007–1127.

Granstrand, O. (1999). *The Economics and Management of Intellectual Property, Towards Intellectual Capitalism*. Edward Elgar, Cheltenham, U.K.

Granstrand, O., Oskarsson, C., Sjöberg, N., and Sjölander, S. (1990). Business strategies for new technologies. In: *Technology and Investment: Crucial Issues for the 1990s*, E. Deiaco, E. Hörnell, and G. Vickery (eds.), OECD and IVA, Pinter Publishers, London, U.K., pp. 64–92.

Griliches, Z. (1986). Productivity, R&D, and basic research at the firm level in the 1970s. *American Economic Review*, 76, 141–154.

Griliches, Z. (1995). R&D and productivity: Econometric results and measurement issues. In: *Handbook of the Economics of Innovation and Technical Change*, P. Stoneman (ed.). Blackwell, Oxford, U.K.

Grindley, P. and Teece, D. (1997). Managing intellectual capital: Licensing and cross-licensing in semiconductors and electronics. *California Management Review*, 39.2, 1–34.

Gurry, F. (2007). Intellectual property, knowledge policy and globalization. In: *Going Global The Challenges for Knowledge-Based Economies*, M. Squicciarini and T. Loikkanen (eds.). Publications of the Ministry of Employment and the Economy, pp. 44–53.

Gurry, F. (2008). Intellectual property, knowledge policy and globalization. In: *Going Global— The Challenges for Knowledge-Based Economies*, M. Squicciarini and T. Loikkanen (eds.), Publications of the Ministry of Employment and the Economy, Finland, Innovation 20/2008, pp. 44–53.

Hall, B. and Van Reenen, J. (2000). How effective are fiscal incentives for R&D? A review of the evidence. *Research Policy*, 29, 449–469.

Iansiti, M. (1997). From technological potential to product performance: An empirical analysis. *Research Policy*, 26(3), 263–390.

Iansiti, M. and West, J. (1997). Technology integration: Turning great research into great products. *Harvard Business Review*, 75(3), 69–78.

Jaffe, A. (1988). Demand and supply influences in R&D intensity and productivity growth. *Review of Economics and Statistics*, 70(3), 431–437.

Kinnunen, E. and Holappa, J. (1996). Patentti-informaation käyttö tuotekehityksen tukena suomalaisissa teollisuusyrityksissä, Patentit–teollisuus–tekniikka, Teknillinen korkeakoulu, koulutuskeskus Dipoli. (The use of patent information as support to product development in Finnish industrial companies, Patents–Industry–Technology, Helsinki University of Technology, Education Centre Dipoli.)

Kline, S. and Rosenberg, N. (1986). An overview of innovation. In: *The Positive Sum Strategy: Harnessing Technology for Economic Growth*, R. Landau and N. Rosenberg (eds.). National Academy Press, Washington, DC.

Koen, P., Ajamian, G., Burkart, R., Clamen, A., Davidson, J., D'Amore, R., Elkons, C., Herald, K., Incorvia, M., Johnson, A., Karol, R., Seibert, R., Slavejkov, A., and Wagner, K. (2001). Providing clarity and a common language to the "fuzzy front end." *Research Technology Management*, 44(2), 46–55.

Kutinlahti, P. (2005). *Universities Approaching Market. Intertwined Scientific and Entre-Preneurial Goals*, VTT Publications 589, VTT, Espoo, Finland.

Link, A.N. (1981). Basic research and productivity increase in manufacturing: Additional evidence. *The American Economic Review*, 71(5), 1111–1112.

Loikkanen, T., Konttinen, J., Hyvönen, J., Ruotsalainen, L., Tuominen, K., Waris, M., Hyttinen, V.-P., and Ilmarinen, O. (2009). *Acquisition, Utilization and the Impact of Patent and Market Information on Innovation Activities*, Espoo 2009. VTT Research Notes 2484, 68 p.

Machlup, F. (1980). *Knowledge: Its Creation, Distribution, and Economic Significance, Volume I, Knowledge and Knowledge Production*. Princeton University Press, Princeton, NJ.

Mansfield, E. (1980). Basic research and productivity increase in manufacturing. *The American Economic Review*, 70(5), 863–873.

Mansfield, E., Schwartz, K., and Wagner, S. (1981). Imitation costs and patents: An empirical study. *Economic Journal*, 91, 907–918.

Martin, S. and Scott, J.T. (2000). The nature of innovation market failure and the design of public support for private innovation. *Research Policy*, 29, 437–447.

Metcalfe, S. (1995). The economic foundations of technology policy: Equilibrium and evolutionary perspectives. In: *Handbook of the Economics of Innovation and Technological Change*, P. Stoneman (ed.). Blackwell, Oxford, U.K.

Neef, D., Siesfeld, G.A., and Cefola, J. (eds.) (1998). *The Economic Impact of Knowledge*. Butterworth-Heinemann, Boston, MA.

Nelson, R.R. (1959). The simple economics of basic scientific research. *Journal of Political Economy*, 67(3), 297–306.

Nivala, K. (1994). Tietokanavien merkitys PKT-yritysten innovaatiotoiminnassa, tarkastelunäkökulmana teknillisten oppilaitosten teknologiapalvelun kehittäminen. Ylivieskan teknillinen oppilaitos, YTOL-julkaisu No. 16. (The importance of knowledge sources in innovation activities of SMEs, from the Angle to develop technology services of Polytechnic Institutions, Polytechnical Institute of Ylivieska, YTOL Publication No. 16. In Finnish.)

Nonaka, I. (1991). The knowledge creating company. *Harvard Business Review*, November–December, 96–104.

OECD (2007). Innovation and growth: Rationale for an innovation strategy, OECD, Paris.

Ottum, B.D. and Moore, W.C. (1997). The role of market information in new product success/failure. *Journal of Product Innovation Management*, 14, 258–273.

Porter, M., Takeuchi, H., and Sakakibara, M. (2000). *Can Japan Compete?* Perseus Publishing, Cambridge, MA.

PwC. (2009). Innovating through the downturn. PricewaterhouseCoopers, Innovation Performance Advisory.

R&D Management (June 2006). Special Issue on Opening up the innovation process, Vol. 36, No. 3, 223–366.

Research Policy (2000). The special issue of research policy. *The Economics of Technology Policy Review*, 29(4–5), April.

Silverberg, G. (1990). Adoption and diffusion of technology as a collective evolutionary process. In: *New Explorations in the Economics of Technological Change*, C. Freeman and L. Soete (eds.), Pinter Publishers, London, U.K., 1991.

Spence, A.M. (1981). The learning curve and competition. *The Bell Journal of Economics*, 12(1), 49–70, Spring.

Squicciarini, M. and Loikkanen, T. (eds.) (2008). *Going Global—The Challenges for Knowledge-Based Economies*. Publications of the Ministry of Employment and the Economy, Innovation 20/2008 (sponsored by the DG Enterprise of the EU and the Ministry of Employment and the Economy of Finland).

STI 1998, STI Review No. 22 (1998). Special Issue on "New rationale and approaches in technology and innovation policy, review science, technology, industry," OECD.

The New Palgrave. (1998). *A Dictionary of Economics*, Vols. 1–4. Palgrave Macmillan, South Yarra Victoria, Australia.

Thurow, L. (2000). Globalization: The product of a knowledge-based economy. *The Annals of the American Academy (ANNALS)*, AAPSS, 570, 19–31.

Torkkeli, M., Hilmola, O.-P., Salmi, P., Viskari, S., Käki, H., Ahonen, M., and Inkinen, S. (2007). Avoin innovaatio: Liiketoiminnan seitinohuet yhteistyörakenteet, Lappeenranta (Open innovation: Gautzy cooperation structures of business), Lappeenranta University of Technology (LUT), Lappeenranta, Finland, Publication of Industrial Engineering and Management Department, Research report 190.

von Hippel, E. (2005). *Democratizing Innovation*. The MIT Press, Cambridge, U.K.

von Hippel, E. and von Krogh, G. (2003). Open source software and the "private-collective" innovation model: Issues for organization science. *Organization Science*, 14(2), 208–223.

WIPO and IFIA (1998). WIPO-IFIA International Symposium on Inventors and Information Technology. Document prepared by the International Bureau of WIPO, WIPO/IFIA/BUD/98/2.

Woolthuis, R.K., Lankhuizen, M., and Gilsing, V. (2005). A system failure framework for innovation policy design. *Technovation*, 25, 609–619.

4 Managing Value Creation in Knowledge-Intensive Business Service Systems

Konstadinos Kutsikos and Gregoris Mentzas

CONTENTS

4.1 INTRODUCTION

Knowledge and innovation are the fuel of growth in modern economies. At the same time, there is a tremendous need for service innovations—that is, new ways of creating value with intangible and dynamic resources to raise the quality and effectiveness of services.

In order to address these issues, scholars are taking an increasing interest in two relevant fields of study, namely, knowledge-intensive business services (KIBSs) and service systems. Their common challenges are twofold: understand how, when, and where service value is created, especially when multiple organizations are involved in developing a service or the knowledge assets that are accessed through a service (service value cocreation), and manage the development, provision, and operation of services, so that service value is realized.

To address these challenges, we developed an innovative framework for managing service value cocreation in KIBS/service systems. Cocreated services play the role of the bridge that enables external service systems to participate in knowledge asset coproduction (knowledge objects and knowledge services) within a KIBS. In return, cocreators may own claim rights and/or financial options for future value realization, stemming from the commercial exploitation of the cocreated assets by the KIBS provider.

In this framework, value management is based on a service portfolio model (SPM) for classifying services in knowledge-intensive systems. This model captures the quotient of participation from external organizations, the type of knowledge assets that are created, and maps the relevant services that provide access to the cocreated assets.

The SPM model drives a service value model (SVM) that captures an array of potential business models. SVM moves beyond the classic focus on service pricing; instead, it aims to address issues related to time-based value (i.e., future value realization) and derivative value. SVM, in turn, drives a balanced scorecard–like performance management model for measuring the performance of cocreated services. Particular emphasis is given on defining a proper set of stakeholders and then customizing performance indicators per each stakeholder.

This chapter is dedicated to the description of our research on this approach to value management in KIBS. Section 4.2 presents a brief overview of value cocreation in service systems. Section 4.3 provides a general overview of our framework and Section 4.4 discusses in detail one of its two layers. Examples of our framework in action are shown in Section 4.5 and future directions of our research are discussed in Section 4.6.

4.2 VALUE AND VALUE COCREATION

The traditional view on value is referred to as goods-dominant (G-D) logic and is based on the value-in-exchange meaning of value (Vargo and Lusch 2004; Vargo et al. 2005). In G-D logic, value is created by the firm and distributed in the market, usually through exchange of goods and money. From this perspective, the roles of "producers" and "consumers" are distinct, and value creation is often thought of as a series of activities performed by the firm.

The alternative view, service-dominant (S-D) logic, is tied to the value-in-use meaning of value (Vargo and Lusch 2008; Vargo et al. 2008). In S-D logic, the roles of producers and consumers are not distinct, meaning that value is always *cocreated*, jointly and reciprocally, in interactions among providers and beneficiaries through the integration of resources and application of competences.

Service value cocreation has received increased attention in service science, an emerging discipline that promotes service cocreation in service systems (Spohrer et al. 2007). The study of service systems (Spohrer et al. 2008) emphasizes collaboration and adaptation in value cocreation and establishes a balanced and interdependent framework for systems of reciprocal service provision. These systems can be business entities that survive, adapt, and evolve through exchange and application of resources (particularly knowledge-based services and knowledge objects) with other systems. Simply put, service systems engage in (mostly knowledge) exchanges with other service systems to enhance adaptability and survivability—thus, cocreating value—for themselves and others (see Figure 4.1).

The increasing emphasis on knowledge exchanges as a modus operandi brings service systems very close to KIBSs. Bettencourt et al. (2002) defined KIBSs as enterprises whose primary value-added activities consist of the accumulation, creation, or dissemination of knowledge for the purpose of developing a customized

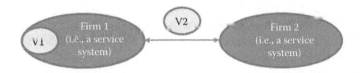

	V1: Value-in-use (derived value)	V2: Value-in-exchange
Value creation process	• Value is based on integration of own and acquired resources (either private or public) • Value is "proposed" to other service systems through value propositions (service offerings)	• Value is embedded in goods or services
Value measures	• Survivability of the beneficiary service system • Financial options—value may be realized at a different point in time than its creation	• Price received • Value creation and realization coincide

FIGURE 4.1   Value cocreation among service systems.

service or product solution to satisfy their clients' needs. Wong and He (2005) further stated that the definition of KIBS provides a platform to study groups of services that are very actively integrated into innovation systems by joint knowledge development with their clients and that consequently create considerable positive network externalities and possibly accelerate knowledge intensification across economy.

Hence, we will treat service systems and KIBS as interchangeable terms that describe entities engaged in the coproduction and provision of knowledge assets (knowledge services and knowledge objects) for service value cocreation. Focusing on the latter, Chen et al. (2008) identify and analyze three factors for sustainable service excellence in service systems: the instilling of the value cocreation concept in the service system, the balancing of innovation and commoditization dynamics, and the configuration of core resources in the service system, that is, people, technology, organization, and shared information.

The first factor (value cocreation) was described in detail earlier. The second factor, balancing of innovation and commoditization activities, defines the activity dynamics required to realize the value cocreation goals (or the actions that must be taken) for the service system. In general, a service provider must be able to innovate and provide differentiating value to clients that others cannot provide. Spohrer indicates that "often the key to successful interactions is that the provider has special knowledge, technology infrastructure, authority, relationships, history, or other competences that allow the provider to do for the customer what they cannot do for themselves, and no other competitor can do for the customer. Successful interactions are based on unique value propositions that ensure the provider can exist, persist, and possibly thrive."

On the other hand, to allow the service system to grow and scale up, appropriate level of commoditization activities must also be implemented. This is because highly innovative activities may also imply high costs—an impediment to the growth of the overall system. Hence, the innovation and commoditization dynamics must be carefully tuned throughout service life cycles in order to ensure the overall and long-term service excellence.

The third factor is the right configuration of core resources that can implement the right actions. Maglio and Spohrer (2007) define four types of resources in service systems, that is, people, organization, technology, and information. People and organizations are critical to carry out required tasks. Shared information and technologies are key ingredients to facilitate people and organizations in execution. To that extent, Kohlborn et al. (2009) defined a generic business service management framework for the development of enterprise-wide service-related capabilities. It consists of four clusters:

Service life cycle management covers all stages from service initiation to service retirement. It addresses the need for a coordinated and systematic management of services in the different stages of their life cycle: development (analysis, design, implementation, and publishing), operation, and retirement.

Service value management ensures the creation of business value by services and the integration of service-centered activities into a service system's overall business strategy.

Service relationship management covers the interactions with customers and suppliers of services. It is in most cases not a pure internal management discipline but relies on effective relationships with external partners, mainly customers and suppliers, but also R&D institutes, universities, etc. For service systems, it may also refer to the management of the relations with internal customers and suppliers.

Finally, all these activities are supported by *service enablement*, which consists of management functions addressing quality, data, and technology requirements.

4.3   SERVICE MANAGEMENT FRAMEWORK FOR KIBS SERVICE SYSTEMS

As described earlier, one of the key elements of value cocreation in any service system is the exchange of knowledge flows (knowledge-based services and knowledge objects). Based on this fact, our research is focused on exploring value cocreation in a service system that owns knowledge-intensive assets (as a key internal resource) and provides access to them through service offerings—that is, a KIBS provider. These assets may seem to be relevant only to organizations (service systems) that base their business models on information and/or knowledge-centric activities, such as publishing, software, education, research, or consulting activities. However, even manufacturing firms have a wealth of knowledge assets to expose to other organizations.

As a result, we developed a service management framework that captures both the generic characteristics of a service system and the particular characteristics of KIBSs (see Figure 4.2).

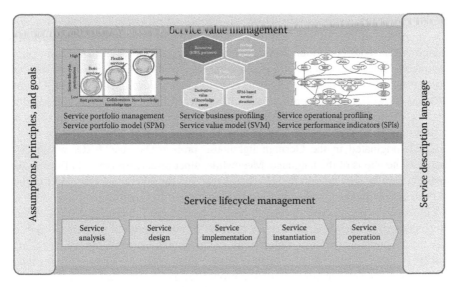

FIGURE 4.2   Framework for managing service value cocreation in KIBS.

The upper layer, the *Service Value Management* layer, is responsible for

- *Service portfolio management*: Its key component is the SPM, a demand-driven model for categorizing knowledge-intensive services. The goal of the service portfolio management component is to explicitly consider the requirements of managing a set of services. A service portfolio may include services that are not developed yet but only exist as service ideas. The feasibility of the latter will be further analyzed in order to make an informed decision about the allocation of resources to realize these ideas in alignment with the overall service portfolio. This, in turn, may lead to service bundling.
- *Service business profiling*: Its key component is the SVM, for defining the value proposition of each service category of the SPM. The goal of the service business profiling component is to define the value profile of each service, which in turn may provide feedback to the Service Portfolio Management module. A strategic assessment for each offered service highlights the viability and strategic alignment of a service based on the underlying business model.
- *Service operational profiling*: Its key component is a set of service performance indicators (SPIs), for quantifying key aspects of value (as defined in the SVM) that need to be accounted for at design time and monitored during service execution.

The lower layer, the *service life cycle management* layer, defines a service life cycle model, depicting the key phases of service development—from service analysis to service operation.

Linking the two layers and ensuring their smooth coordination is a *service descrip-tion language*, which provides a uniform way of defining the business aspects of a service system/KIBS provider. To that extent, we adopted unified service description language (USDL) as the vehicle to explore the intricacies of KIBS descriptions. Our choice is based on the fact that USDL is currently being developed into a W3C stan-dard and our work complements certain of its aspects. Indeed, our work is leading to potential extensions of certain USDL's modules and we are currently participating as Invited Experts in W3C's USDL Incubator.

USDL originated in the German lighthouse project Theseus/TEXO, which pro-duced two iterations of the language. Meanwhile, other research projects in Europe and Australia have heavily contributed to what is now the third iteration, bringing together the business, operational, and technical perspectives of a service (Cardoso et al. 2009).

The business perspective describes properties that are fundamental for the char-acterization of a service. USDL relies on a set of nonfunctional properties such as availability, payment, pricing, obligations, rights, penalties, bundling, security, and quality. In order to provide a suitable language that can be understood by business stakeholders and service consumers, the properties have been clustered into groups: roles (providers and consumers), service level, marketing, legal, interaction, bun-dling, and an extension mechanism.

The operational perspective describes the operations executed by services. Important aspects modeled include operations, functionality, classifications, mile-stones, and phases. The USDL approach to the functional description of services is multifaceted since it allows using natural language, keywords (i.e., tagging), and ontologies as fundamental structures to express the functionality of a service. This perspective includes concepts borrowed from the area of project management. For example, phases allow creating groupings to provide a high-level description of the business process associated with a service.

The technical perspective allows specifying technical information of services exposed by an organization. This perspective acts as a central point for referencing existing web standards in order to describe technical aspects of services such as interfaces (e.g., WSDL) and transaction protocols.

In the following sections, we focus on the service value management layer and provide details on certain key aspects of the three models that comprise this layer.

4.4   SERVICE VALUE MANAGEMENT LAYER

In a service system (i.e., KIBS provider), integration of internal and external resources is the essence of value cocreation, as expressed by e-services (value propositions) offered to other service systems. Value cocreation is then a function of (a) the par-ticipation of other organizations (other KIBS/service systems) in the enhancement of developed services, or the development of new ones. This participation may range from no participation to full-scale involvement and (b) cocreated knowledge assets, which may range from current best practices already validated by the service system to new knowledge acquired from (or coproduced with) other organizations.

Synthesizing the earlier concepts led to the development of a service value cocreation model, called SPM, shown in Figure 4.3.

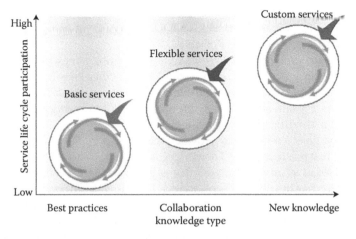

FIGURE 4.3   Service Portfolio Model (SPM).

This model allows us to create rich value propositions by defining the following classes of offered services:

- *Basic services*—These are standardized services that encapsulate best practices (generic or industry-specific) owned by the KIBS provider. Participation of external organizations in the service development life cycle is limited. Knowledge assets handled by this service class are expanded through updates generated internally, by the KIBS provider's own value creation process.
- *Flexible services*—These are configurable services that are based on a wide menu of options offered by the KIBS provider. External organizations participate in the service development life cycle to codevelop new service configurations or new service configuration options, assisted by the KIBS provider's resources (human, technical, etc.). Services of this class expand stored knowledge by providing new syntheses.
- *Custom services*—These are highly customizable and user-driven services. The involvement of external organizations in the service development life cycle is high and should require an equally significant investment of the provider's resources (human, technical, financial, etc.). Services of this class may significantly expand stored knowledge—for example, with industry-specific practices.

These classes may then drive a first-level strategy development by the KIBS provider, in terms of balancing innovation vs. commoditization: the KIBS provider will be able to make initial commercial assessments on the desired mix of innovative (but costly) and commoditized (but easily copied) e-services that it will offer.

Furthermore, it is important to note that the aforementioned service class definitions are not static categorizations of services into rigid blocks. In fact, there are certain underlying characteristics that make SPM service classes a dynamic set of constantly evolving groups of services, such as the lifetime of knowledge assets and their consumption as a portfolio of objects.

In terms of the *lifetime of knowledge assets*, over time, knowledge once classified as "new" slowly becomes "best practice," commodity knowledge or even gets obsolete. In our SPM model, this is equivalent to moving on the horizontal axis, from right to left. As a result, a service may migrate to a "lower" service class, with rippling effects on its business and operational profiles.

At the same time, a service may provide access to a *portfolio of knowledge assets* (ranging from best practices to new knowledge). As a result, such a service may be developed and offered as a "service package," that is, a service may be comprised of subservices, each belonging to a different service class. The business and operational profiles of a service package correspond to a synthesis of the profiles of the individual subservices.

The concept of service package can be further generalized: a service offered by a KIBS provider may be comprised of multiple independent KIBS services, thus leading to the development and offering of "service bundles." Hence, a KIBS service bundle is a network of services: it pulls together KIBS service packages and independent KIBS services into a single new service entity.

Hence, an SPM-based service can be structured as

* A single SPM service
* A service package of SPM services
* A service bundle of SPM single services and SPM service packages

4.4.1 SERVICE VALUE MODEL

The significance of the SPM model lies with the fact that the SPM service classes highlight the co-ownership of knowledge assets (knowledge objects and knowledge services) that are cocreated. However, realizing and sharing value from such co-owned assets may not be a straightforward task, as a number of issues must be addressed:

* *Time value of knowledge assets*: There may be a delay between knowledge assets' cocreation and their value realization. This delay can be attributed to a number of factors, such as development of new services (or refinement of existing services) that will provide commercial access to these assets, or market promotion and adoption of such services.
* *Derivative value of knowledge assets*: The newly cocreated knowledge assets may be used as building blocks for the development of other knowledge assets by the KIBS provider.
* *Complexity of service structures*: Given that an SPM-based service can be structured, at one extreme, as a bundle of independent single services that operate on different types of knowledge objects, sharing revenues among service cocreators may require complex legal and operational arrangements.

These issues, in turn, complicate the development of a KIBS provider's business model and its components, such as revenue streams and value propositions. To that

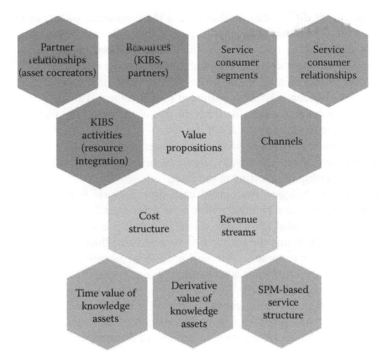

FIGURE 4.4   Service Value Model (SVM).

extent, and in order to explore the effect of these issues on classic business modeling approaches, we expanded the business model canvas, proposed by Osterwalder and Pigneur (2010), and developed the SVM, as shown in Figure 4.4.

The *cost structure* component accounts for future value realization and sharing by encompassing

- Claim rights, that is, descriptions of all responsibilities, duties, or obligations that the KIBS provider has toward service value cocreators
- Financial call options, that is, descriptions of the rights that service value cocreators have in terms of acquiring the cocreated knowledge assets

The process of calculating the value of these rights and options is directly linked to intellectual property rights (IPRs) and can be broken down into four main methods: capitalization of historic profits, gross profit differential methods, excess profits methods, and the relief from royalty method (King 2002).

The *capitalization of historic profits* arrives at the value of IPRs by multiplying the maintainable historic profitability of the asset by a multiple that is assessed after scoring the relative strength of the IPR. For example, a multiple is arrived at after assessing a cocreated knowledge asset in the light of factors such as trend of profitability and marketing. While this capitalization process recognizes some of the factors that should be considered, it has major shortcomings, mostly associated with historic earning capability.

The *gross profit differential methods* are often associated with trademark and brand valuation. These methods adopt the differences in sale prices, adjusted for differences in marketing costs. That is, the difference between the margin of the branded and/or patented service and an unbranded or generic service.

The *excess profits method* looks at the current value of the net tangible assets employed as the benchmark for an estimated rate of return. Although theoretically relying on future economic benefits from the use of the asset, the method has difficulty in adjusting to alternative uses of the asset.

Relief from royalty considers what the purchaser could afford, or would be willing to pay, for the license. The royalty stream is then capitalized, reflecting the risk and return relationship of investing in the asset.

Beyond these techniques, discounted cash flow analysis is probably the most comprehensive appraisal technique. After a rigorous examination of the earnings capability of the cocreated knowledge assets, potential profits and cash flows need to be assessed carefully and then restated to present value through use of a discount rate. The potential for revenue growth will need to be assessed by reference to the enduring nature of the knowledge asset by quantifying its useful lifetime (functional, technological, business, or legal). Then, all expenses must be included, along with estimates on the residual or terminal value (if any) of the knowledge asset.

Beyond the cost structure component, the *partner relationships* component captures realized SPM-based value through a number of different commercial vehicles:

- Shared service revenues arrangements, among knowledge asset cocreators.
- Equity, for example, through the development of new joint venture to better exploit cocreated assets (or provision of KIBS provider's equity to knowledge asset cocreators).
- Marketplaces, for cocreated services call options. Such marketplaces do not exist but we expect that it will be a natural development once cocreated services become widely adopted and used.

In summary, Figure 4.5 depicts how the SPM model drives the development of a service's SVM profile.

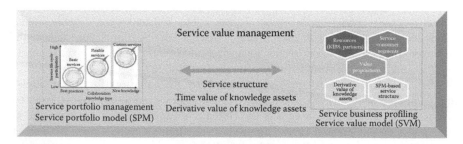

FIGURE 4.5 From service portfolios to specific service profiles.

4.4.2   Service Performance Indicators

The service operational profiling module supports the articulation, communication, monitoring, and updating of the strategic service goals defined by the SVM module. Its design and development is driven by the SPM-based service structure and the service's SVM profile; it is further based on a modified version of the third generation of balance scorecard framework (BSF). We grouped the proposed metrics into three perspectives: value, stakeholder, and internal.

The value perspective encompasses a set of strategic objectives, as identified by the SPM-based service structure and the SVM-based service profile. These are (a) increase participation of external partners in cocreation activities, (b) expand knowledge assets (knowledge objects and knowledge services), and (c) increase value of cocreated services' call options.

The stakeholder perspective identifies the performance expectations of the entities that will work together toward achieving the objectives of the value perspective. These entities are the KIBS provider and external partners (knowledge asset cocreators). The KIBS provider will use relevant metrics to assess the operational and commercial viability of the offered services, as well as to enable strategic decisions about the mix of services in the SPM model. External partners will use the proposed metrics to evaluate the costs and benefits of their involvement in the cocreation process. Combined with own performance indicators, they may develop comprehensive metrics for assessing risks and benefits of being involved in cocreation activities.

The internal perspective identifies the activities of the KIBS provider that enable implementation of the stakeholder perspectives.

Tables 4.1 through 4.4 provide a set of generic SPIs that we defined for a KIBS provider involved in service coproduction. These indicators are based on the perspectives described earlier. (*Note*: those indicators that are common in multiple perspectives are highlighted in italics.)

In summary, Figure 4.6 depicts how the SVM model drives the development of SPIs.

TABLE 4.1
Indicators for the Value Perspective

Objective	Indicators
Increase participation of external partners in cocreation activities	Number of regular participants
	Annual changes in participants
Expand knowledge assets (knowledge objects and knowledge services)	Annual number of new knowledge objects
	Annual number of new service packages (existing services)
	Annual number of new service packages (new services)
	Annual number of new services
Increase value of cocreated services' call options	Number of outstanding call options
	Total present value of outstanding call options

TABLE 4.2

Indicators for the Stakeholder Perspective (KIBS Provider)

Objective	Indicators
Service commercial viability	Annual revenue per service (pay-per-use, fixed membership fees, shared revenues with cocreators, etc.)
	Annual costs per service
	Customer satisfaction evaluation
Service innovation	*Annual number of new services*
	Annual number of patents registered
	Annual investment in service R&D

TABLE 4.3

Indicators for the Stakeholder Perspective (External Partners— Cocreators of Knowledge Assets)

Objective	Indicators
Increase value of KIBS provider's knowledge assets	Annual number of knowledge assets that partner has cocreated
	Annual number of cocreated service packages
	Annual shared revenue per service (shared with the KIBS provider)
	Total value of owned call options
	Total value realized from sold call options
Improve collaboration	Annual usage of KIBS provider's cocreated services

TABLE 4.4

Indicators for the Internal Perspective

Objective	Indicators
ICT management	% of coverage of services demanded by third parties (service consumers, service cocreators)
Branding and marketing	Annual investment in service marketing and branding
	Annual investment in service R&D
	Customer satisfaction evaluation

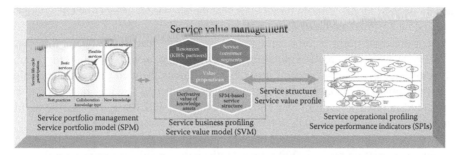

FIGURE 4.6 From service profiles to specific Service Performance Indicators.

4.5 FRAMEWORK IN ACTION

As part of our research within the SYNERGY project (an EU, FP7 project) (Mentzas and Popplewell 2009; Popplewell et al. 2008), we applied our service management framework in the realm of knowledge-intensive virtual organizations (VOs).

The aim of the project is to support networked enterprises in their successful participation in VOs. A SYNERGY service system is developed, comprised of knowledge assets relevant to collaboration creation and operation, along with services to discover, capture, deliver, and apply such knowledge (Popplewell et al. 2008).

A key part of SYNERGY service system's knowledge infrastructure is collaboration patterns (CPats), defined as a prescription that addresses a collaborative problem that may occur repeatedly in a business setting. It describes the forms of collaboration and the proven solutions to a collaboration problem and appears as a recurring group of actions that enable efficiency in both the communication and the implementation of a successful solution. The CPat can be used as is in its application domain or it can be abstracted and used as a primitive building block beyond its original domain (Papageorgiou et al. 2010; Verginadis et al. 2009).

CPat services offered by a SYNERGY service system enable other service systems to access the CPats repository. These services provide recommendations for actions and tools to be used, awareness of the state of collaborators and the state of the collaboration work, as well as statistical analysis based on previous and ongoing VO collaborations.

As a scenario of our framework in action, consider the case of a SYNERGY KIBS provider ("the provider") that participates in a VO comprised of pharmaceuticals (i.e., a group of service systems) that want to develop and test a series of new drugs (Verginadis et al. 2009). One of the collaboration activities within this VO is the design of a joint laboratory experiment, which includes planning and scheduling of pre-experiment tasks (e.g., defining and agreeing on the experiment's objectives).

An existing CPat (i.e., a knowledge object) called "OrganizeExperiment" already encapsulates best practices for similar tasks, accumulated over time through this CPat's uses in other VOs. At run time, two types of CPat services can be invoked: (a) CPat Recommender service, which recommends actions to continue collaboration, as well as tools related to the CPat's collaboration tasks and (b) CPat Awareness service, which provides awareness of the state of collaboration tasks. Both these services are *basic services*, as per our SPM model. This, in turn means, that the knowledge object (the CPat) will be updated by the provider, if new best practices arise.

At a different point in time, an external event may lead to changes in existing CPats or even dictate creation of new ones. For example, new rules imposed by public health authorities may require new lab experiments through the participation of new partners in the VO. New CPats and/or new knowledge services may need to be cocreated in order to capture new collaboration tasks, which may be specific to this case or may be generic enough to become available to future service users.

A CPat Design service will enable VO participants and the provider to engage in such a cocreation process. VO representatives may need to consult with the provider's human experts and collaborate offline, for example, on defining a commercial exploitation plan for a new service offering related to the new CPats. As this is cocreation of new knowledge assets, all participants should agree on SVM parameters for the new service, for example, for revenue sharing or joint equity. These will then be translated into relevant SPIs (e.g., share revenue %) that will be recorded in the new service's USDL description and taken into account during its life cycle. It is important to note that although the CPat Design service is a *custom service*, as per our SPM model, it may depend on other SYNERGY services, such as the SYNERGY basic services described earlier.

These services can be then used during the life cycle of a VO. For example, the partner search and selection (PSS) process is one of the key processes in the operation of a VO. At the VO creation stage, the PSS process is devoted to the identification, assessment, and selection of VO participants. At VO operation, it can be invoked when a new partner needs to be added to the VO (e.g., because new capabilities are needed, or an existing partner needs to be replaced).

In either case, the execution of a PSS process depends on a number of key success factors, including selection criteria, for filtering the "right" candidates. Such criteria may be technical, financial, strategic, operational, etc.; matching algorithms, for multi-criteria partner filtering; and collaboration scorecard/history of eligible candidates.

From the SYNERGY KIBS provider viewpoint, the aforementioned PSS elements can be mapped to corresponding knowledge objects, stored within the SYNERGY KIBS provider infrastructure. The latter may then offer basic services (as per our SPM model) for accessing best-of-breed selection criteria. A number of VO management entities (from different VOs in the same industry sector) may work with the SYNERGY provider to cocreate one or more specialized, industry-specific matching algorithms.

In order to access the latter, new custom services (as per our SPM model) will have to be developed, so that other industry players may be able to use the new matching algorithms. The SVM model for the new service may base its revenue streams on sharing the success of any VOs that will use the new service and the associated knowledge objects: success can be measured in terms of our proposed SPIs. Hence, service cocreators have every interest to make the custom service of the highest quality possible; by essentially claiming the equivalent of an equity stake in VOs that will use the cocreated service/knowledge objects, the PSS process becomes intertwined with the SYNERGY service system.

4.6 CONCLUSIONS

There is no doubt that there is a tremendous need for service innovations and new ways of creating value with intangible and dynamic resources, to fuel economic growth and to raise the quality and effectiveness of services, especially for knowledge-intensive industries.

To that extent, our research work aims to provide a framework for profiling and managing value cocreation in KIBS/service systems. We are currently developing this framework around three basic pillars: our service value management layer; established service life cycle management processes that are enhanced to account for SPM-driven value cocreation parameters; and the USDL v3.0 service description language, as a way of instilling SPM-driven parameters in the end-to-end development of a service (value proposition).

Initial results from applying the aforementioned in business settings of collaborating SMEs are encouraging (SYNERGY project). Looking beyond its current state, our framework could be further exploited in two key directions:

The SPM and SVM models could be used as the basis for defining new approaches of strategy development for KIBS providers. For example, our options-based approach to service value may play a key role in defining new business models, such as marketplaces for service value options.

In addition, the SPM model could be enriched, in order to capture additional aspects of service value cocreation (thus becoming a multi-axes model). This, in turn, could lead to a more detailed understanding of service classes and thus of business models that can be used (or created).

Overall, there is still much to discover about service value cocreation in KIBS and the complex process of resource integration and service-for-service exchange. It is our firm belief that the synthesis of both similar and competing ideas, from various social, economic, and technical disciplines, will be required, thus propelling the generation of new directions in KIBS research and practice.

ACKNOWLEDGMENT

Work presented in this chapter has been partially supported by the European Commission under the 7th Framework Programme on ICT in support of the networked enterprise through contract No. 216089.

REFERENCES

Bettencourt, L. A., Ostrom, A. L., Brown, S. W., and Roundtree, R. I. (2002), Client co-production in knowledge-intensive business services. *California Management Review*, 44, 100–128.

Cardoso, J., Winkler, M., and Voigt, K. (2009), A service description language for the Internet of services. In *First International Symposium on Services Science* (ISSS'09), (eds.) Alt, R., Fähnrich, K.-P., and Franczyk, B. Logos-Verlag, Berlin, Germany.

Chen, Y., Lelescu, A., and Spohrer, J. (2008), Three factors to sustainable service system excellence: A case study of service systems. In *2008 IEEE International Conference on Services Computing*, Honolulu, HI, vol. 2, pp. 119–126.

King, K. (2002), The value of intellectual property, intangible assets and goodwill. *Journal of Intellectual Property Rights*, 7, 245–248.

Kohlborn, T., Korthaus, A., and Rosemann, M. (2009), Business and software lifecycle management. In *2009 Enterprise Distributed Object Computing Conference* (EDOC'09), Auckland, New Zealand, pp. 87–96.

Maglio, P. and Spohrer, J. (2007), Fundamentals of service science. *Journal of the Academy of Marketing Science*, 36(1), 18–20.

Mentzas, G. and Popplewell, K. (2009), *Knowledge-Based Collaboration Patterns in Future Internet Enterprise Systems. ERCIM News*, 77pp.

Osterwalder, A. and Pigneur, Y. (2010), *Business Model Generation*. John Wiley & Sons, Hoboken, NJ.

Papageorgiou, N., Verginadis, Y., Apostolou, D., and Mentzas, G. (2010), Semantic interoperability of e-services in collaborative networked organizations. In *The International Conference on e-Business* (ICE-B 2010), Athens, Greece.

Popplewell, K., Stojanovic, N., Abecker, A., Apostolou, D., and Mentzas, G. (2008), Supporting adaptive enterprise collaboration through semantic knowledge services. In *4th International Conference Interoperability for Enterprise Software and Applications* (*I-ESA*), 2008, Berlin, Germany.

Spohrer, J., Anderson, L. C., Pass, N. J., Ager, T., and Gruhl, D. (2008), Service science. *Journal of Grid Computing*, 6(3), 313–324.

Spohrer, J., Maglio, P. P., Bailey, J., and Gruhl, D. (2007), Towards a science of service systems. *Computer*, 40(1), 71–77.

Vargo, S. L. and Lusch, R. F. (2004), Evolving to a new dominant logic for marketing. *Journal of Marketing*, 68, 1–17.

Vargo, S. L. and Lusch, R. F. (2008), Service-dominant logic: Continuing the evolution. *Journal of the Academy of Marketing Science*, 36(1), 1–10.

Vargo, S. L., Lusch, R. F., and Morgan, F. W. (2005), Historical perspectives on service-dominant logic. In *The Service-Dominant Logic of Marketing: Dialog, Debate and Directions*, (eds.) R. F. Lusch and S. L. Vargo, pp. 29–42. M.E. Sharpe Inc., Armonk, NY.

Vargo, S. L., Maglio, P. P., and Akaka, M. A. (2008), On value and value co-creation: A service systems and service logic perspective. *European Management Journal*, 26, 145–152.

Verginadis, Y., Apostolou, A., Papageorgiou, N., and Mentzas, G. (2009), Collaboration patterns in event-driven environments for virtual organizations. In *AAAI 2009 Spring Symposium on Intelligent Event Processing*, Palo Alto, CA.

Wong, P. K. and He, Z. L. (2005), A comparative study of innovation behaviour in Singapore's KIBS and manufacturing firms. *The Service Industries Journal*, 25, 23–42.

5 Knowledge Service Engineering

A Service-Dominant Logic Perspective

Stephen L. Vargo, Robert F. Lusch,
Heiko Wieland, and Melissa Archpru Akaka

CONTENTS

5.1 INTRODUCTION

This handbook seeks to further advance the emerging field of knowledge service engineering (KSE) by providing a theoretical foundation to guide future research in this area. The development of KSE focuses on understanding the ecology of information technology and the dynamics of a "knowledge-based" economy, or society, in which the survival and success of an enterprise is based on the acquisition of knowledge (Adler 2001). This chapter contributes to the development of KSE by providing a theoretical framework that is centered on *service as the basis of social and economic exchange and knowledge as a primary resource for value creation.* We present alternative views for conceptualizing service, knowledge, and value creation and argue that a service-dominant (S-D) logic (Vargo and Lusch 2004, 2008) framework will provide KSE with a stronger and more robust theoretical foundation for exploring practical issues than the dominant, alternative orientation.

Current literature distinguishes between two main perspectives on service. The traditional view, which originates from neoclassical economics research tradition, is based on the underlying assumption that goods—units of output—are the bases for economic exchange. Often, the term "product" is used to indicate that these units of output can be either tangible ("goods") or intangible ("services"). Vargo and Lusch (2004, 2008) have referred to this underlying logic as goods-dominant (G-D) logic. As the world, especially the developed countries, is increasingly and mistakenly perceived as moving toward a service economy (Vargo and Akaka 2009), the G-D logic view somewhat intuitively suggests that firms need to modify the strategies that they have adopted for manufactured goods by adding a service component or by adapting them to the unique characteristics of services. We argue that a G-D logic view is restrictive because it points to a narrow, dyadic, and unidirectional conceptualization of value creation in which the firm produces value and the customer destroys (consumes) it.

An alternative view of service(s) reconsiders the idea of a "service revolution" and argues, more foundationally, that *service provision is and always has been central to economic exchange and value creation* but has been relatively ignored in the manufacturing-based perspective (G-D logic) (Vargo et al. 2010). This alternative view, S-D logic (Vargo and Lusch 2004, 2008), defines service as the *application of competences for the benefit of another party*.

The first objective of this chapter is to propose S-D logic as a more appropriate perspective for understanding economic exchange in general and for KSE in particular. An S-D logic lens focuses on service as the basis of all exchange and knowledge as one of the central resources in value creation. We show that the appropriateness of an S-D logic orientation is not limited to certain "types" of firms or economies. This is because, in S-D logic, all firms and economies are service-based and thus, as we explain, knowledge-based as well.

The second objective of this chapter is to further explicate and explore the role of knowledge in service and value creation. This is obviously central for KSE research. The S-D logic perspective emphasizes the primacy of knowledge and other intangible and dynamic (operant) resources, over static and tangible (operand) resources (e.g., raw materials and goods). This shift from operand to operant resources is achieved by refocusing the view of exchange, from units of output that one party provides to another to the process of using competences for the benefit of another party. The other party then reciprocates with its own applied competences—that is, *service is exchanged for service*. This is a major distinction between S-D logic and G-D logic since G-D logic is centered on operand resources. We show that from an S-D logic view, all economies are considered knowledge economies and that the importance of knowledge is not unique to certain industries or historical or contemporary economic eras. Arguably, this broadens the scope and importance of the development of KSE and emphasizes the need to advance the understanding of knowledge- and service-driven exchange.

The chapter's third objective is to reframe the firm's activities within a broader value-creation context, which encompasses the interaction among suppliers, customers, and other exchange partners in customer and supplier networks and establishes the market as a system of service exchange. This broader view suggests that resources beyond those acquired by the firm, such as knowledge (and other operant resources) of customers, suppliers, and other stakeholders, must be integrated in order for value to be created. In particular, the knowledge of the customer is necessary for value to be derived (Ballantyne and Varey 2006). Interactions of the parties involved in and external to the exchange continually provide operant resources—knowledge and skills. The use of, and thus understanding of, these extended networks of interaction among actors (network-with-network models) is important for the field of KSE. This is because in most, if not all, value-creation processes neither the providers nor the beneficiaries of service possess sufficient resources to create value alone. The S-D logic network perspective therefore sees the "customer" and the customer's resources as not only integral to but *primary* in the value cocreation process. Additionally, we show that S-D logic conceptualizes these networks as dynamic systems.

Essentially, based on the three objectives mentioned earlier, this chapter argues that S-D logic can provide a solid theoretical foundation for the field of KSE. S-D logic offers a framework for the study of service and clarifies the roles of knowledge and relational networks in service provision. This foundational framework provides a deeper and broader view of exchange than the alternative G-D logic paradigm and, we argue, establishes a cornerstone for studying and understanding information-technology ecology and the dynamics of a "service-centered," "knowledge-based" society.

5.2   GOODS-DOMINANT LOGIC

The roots of the traditional perspective, G-D logic, can be traced back at least to Smith (1776) and his seminal work in *The Wealth of Nations*. In the context of the eighteenth century and the Industrial Revolution, with its limitations on personal travel and lack of electronic communication, exports of tangible goods were the primary source of national wealth creation. Based on this context and his intent on advancing the wealth of England, Smith distinguished the activities related to the creation of tangible goods as "productive" from those activities that did not result in tangible output, which he labeled as "unproductive." This foundational work was subsequently adopted by economic philosophers who developed it into economic science. The dominant logic of science at that time was Newtonian mechanics, a model of matter embedded with properties. Following this tradition provided a certain level of scientific respectability and, thus, economics became the study of products (goods) as matter embedded with properties (e.g., "utility"). This goods-centered economic paradigm took hold and formed the foundation of many subsequent business disciplines (Vargo and Morgan 2005; Vargo et al. 2006).

Based on this tradition, G-D logic views goods, or more generally products—indicating either tangible (goods) or intangible (services)—as units of output. These units of output are viewed as the bases for economic exchange. G-D logic can be summarized as follows (see Vargo and Lusch 2004):

1. Economic exchange is fundamentally concerned with units of output (products).
2. These products are embedded with value during the manufacturing (or agricultural, or extraction) process.
3. For efficiency, this production ideally (a) is standardized, (b) takes place in isolation from the customer, and (c) can be inventoried to even out production cycles in the face of irregular demand.
4. These products can be sold in the market by capturing and stimulating demand in order to maximize profits.

In short, G-D logic describes the purpose of the firm as the production and distribution of units of output. These units of output are embedded with value during the manufacturing (or agricultural or extraction) process and are thereby made attractive to the consumer. These value-laden units of output then become the object of exchange with the customer being the recipient of the embedded value. More specifically, from this view, the customer's primary role is the consumption (destruction) of the value created by the firm (Normann 2001).

G-D logic conceptualizes "services" (usually plural) as either an add-on to goods (e.g., after-sales service) or a special type of product. According to this traditional view, services can be conceptualized as having the following four attributes: intangibility, heterogeneity, inseparability, and perishability (Zeithaml et al. 1985), which the literature refers to as "IHIP" characteristics (Lovelock and Gummesson 2004). In other words, G-D logic characterizes services by their *shortcomings* (as compared to their desired characteristics listed in point 3 earlier) and views them as somewhat inferior "goods" with undesirable qualities that pose challenges for the firm's operational activities. As mentioned, a G-D logic view somewhat intuitively suggests that firms need to modify the strategies that they have adopted for manufactured goods. On the academic side, this trend led to the establishment of subdisciplines such as service marketing and service operations (management).

5.3   S-D LOGIC—ALL ECONOMIES ARE SERVICE ECONOMIES

Although G-D logic was, arguably, reasonably adequate when economics was primarily concerned with the manufacturing of goods, its restrictions have, also arguably, become increasingly apparent as attention has expanded to include service(s) and, more generally, value creation as a relational process. Over the last several decades, many scholars have called for a more encompassing and solid paradigmatic foundation (e.g., Schlesinger and Heskett 1991; Shostack 1977). S-D logic represents the convergence of these general calls as well as more directed calls for reformulating thought in specific areas of academic interest in business, such as service and relationship marketing (e.g., Grönroos 1994; Gummesson 1995), resource-advantage theory (e.g., Hunt 2000), core competency theory (e.g., Prahalad and Hamel 1990), network theory (e.g., Hakansson and Snehota 1995), consumer culture theory (e.g., Arnould and Thompson 2005), and others, which collectively point toward an alternative logic of the market (for detailed evolution, see Vargo and Lusch 2004, 2008; Vargo and Morgan 2005).

The conceptualization of "service" is the foundational distinction between S-D logic and G-D logic. S-D logic defines "service" (singular) as the application of competences (knowledge and skills) for the benefit of another party. The use of the singular *"service"* as opposed to the plural *"services,"* as traditionally employed in G-D logic, is intentional and significant. It signals a shift from thinking about value creation in terms of *operand resources*—usually tangible, static resources that require some action to make them valuable—to *operant resources*—usually intangible, dynamic resources that are capable of acting on operand and even other operant resources to create value (Constantin and Lusch 1994). S-D logic conceptualizes service as the process of doing something for and with another party, and thus always as a collaborative *process*. Whereas G-D logic conceptualizes *services* (plural) as special types of *units of output*, which are somewhat inferior to goods, S-D logic views *service* (singular) neither as a special type of product nor as an add-on to goods. While goods maintain an important role within S-D logic, they are seen as intermediaries or vehicles in the provision of service, rather than primary to exchange and value creation. Service provisions can therefore be applied directly or indirectly, through goods. Thus, S-D logic's central tenet that "service is exchanged for service" implies that service is superordinate to goods. It also implies that all economies are service economies and all businesses, including those traditionally considered "manufacturing," are service businesses.

Stated somewhat differently, value creation occurs at the intersection of providers and beneficiaries and is always determined by the latter. According to S-D logic, value is always created through use, rather than through manufacturing and delivery. This notion that the beneficiary is always an active participant of the value-creation process—that is, a cocreator of value—is another tenet of S-D logic and is discussed in more detail in the following.

It is important to highlight that S-D logic represents a shift in the *logic of exchange*, rather than a shift in the *type of product* that is under investigation. It is a shift that Vargo and Lusch (2004) insist is already taking place. They point out that evidence of this "new logic" can be found in somewhat diverse academic fields such as information technology (e.g., service-oriented architecture, SaaS), human resources (e.g., organizations as learning systems), marketing (e.g., service and relationship marketing and network theory), the theory of the firm (e.g., resource-based theories), etc., as well as in practice. This evidence highlights the shift toward service-centered thinking and the importance of advancing the understanding of KSE.

From an S-D logic view, service provision is not just now becoming abundant nor has it only recently gained importance. What is often referred to as the "service revolution" (e.g., Rust and Espinoza 2006) is actually an aberration of a G-D logic perspective in which manufacturing (and agriculture and extraction) has been considered primary. The emergence of new fields, such as KSE, shows that the service nature of exchange is increasingly becoming apparent. However, it is important to stress, within S-D logic, that all economies are and always have been service economies. Thus, S-D logic is not limited to certain "types" of enterprises or industries. This focus on service as the underlying basis of exchange is of particular significance for KSE research. It suggests that the study of service, as well as knowledge, is important for understanding all social and economic exchange. In this way, the advancement of KSE, centered on service- and knowledge-based exchange, has the potential to provide a deeper understanding of exchange in general. We posit that S-D logic offers a strong theoretical framework from which KSE research can be developed.

5.4   ALL ECONOMIES ARE KNOWLEDGE ECONOMIES

As discussed, one of the key differences between S-D logic and G-D logic is the critical distinction between operant and operand resources. Operant resources, as described, are capable of acting on operand and even other operant resources to create value. Operant resources are usually intangible and dynamic whereas operand resources are usually tangible and static. Almost by definition, G-D logic is centered on operand resources. S-D logic, on the other hand, redirects the focus of exchange to operant resources by shifting from units of output to the process of using individual and collective competences for the benefit of another party.

Thus, the ability to compete in markets is both a function of individual and collective knowledge. This does not mean, however, that S-D logic rejects the value of operand resources. Even through an S-D logic lens, operand resources are still viewed as important since they become valuable in the context of and in combination with operant resources—for example, through modification, combination, and use (see Zimmermann 1951). However, S-D logic implies that value is created through activity and points toward the primacy of human resources within firms (Lusch et al. 2007) and underscores the necessity of seeing the customer as endogenous to value creation (this will be discussed in more detail in the next section). Even a tangible good, such as a car, is just an intermediary in the provision of service. In this example, a car can be conceptualized as the integration of large amounts of mostly operant resources. These operant resources include the ability to design, assemble, and distribute a car (operant resources of provider and its suppliers), the ability to drive a car (operant resources of customer), the ability to build and maintain roads (operant resources of public entities), and the ability to produce and distribute gasoline (operant resources of other market entities).

In this way, all economies are knowledge economies because the centrality of operant resources, knowledge in particular, is not unique to certain industries or historical or contemporary economic eras. That is, through an S-D logic lens, all enterprises are driven by resources that are capable of acting on other resources (i.e., operant resources) and thus, just as service has always been the basis for exchange, operant resources, such as knowledge and skills, have always been the drivers of service provision. What is now drawing attention toward service and knowledge and other intangible, dynamic phenomena is the ability to exchange information in a relatively pure form—that is without being transported by people and/or matter ("liquification" in Normann's [2001] terms)—through digitization.

In this discussion, it is important to distinguish between information and knowledge. Harris (1996) defines data as known fact, information as analyzed data, and knowledge as combinations of information and context. This contextual conceptualization of knowledge is also recognized by Shin et al. (2001) who describe knowledge as a combination of a process element and information and who define it as a form of information that is ready to be applied to decisions and actions (this school of thought coincides with Davenport et al. 1998; Kogut and Zander 1992). It is also important to point out that knowledge resources can, under most circumstances, become obsolete. The following paragraph shows that the contextual element of knowledge and value creation requires enterprises to adapt their capabilities. Knowledge can only be truly dynamic if it is enabled by continuous learning.

For most of human civilization, information was embedded in physical matter and, in most cases, could not be separated. Artifacts that humans developed were essentially frozen ideas or knowledge or what Vargo and Lusch (2004) refer to as "informed matter"; wheels, gears, pulleys, clocks were all matter impregnated with human ingenuity. However, due to the recent proliferation of digital communication and computation, most informational resources now have the potential of being liquefied and physical control or ownership of these resources is often unnecessary (Lusch et al. 2006). Although S-D logic suggests that there is no service revolution and that knowledge has always been the driver for service provision, the liquification trend and emerging fields such as KSE highlight the fact that we are experiencing an "information revolution" (Rust and Thompson 2006). Increasing specialization, the exponential growth in service-centered thinking, and the increasing ability to exchange information in a relatively pure form provide evidence of this information revolution, and, at the same time, highlight the inadequacy of G-D logic.

As Rust (2004) points out, IT is a key driver of the need for and acceptance of S-D logic. The access to increasing amounts of information, combined with the contextual nature of the exchange in general and knowledge in particular, highlights the importance of understanding a broader value-creation context, which encompasses the interaction among suppliers, customers, and other exchange partners. The next section will establish the market as a system of service exchange and explore the implications for KSE research.

5.5 VALUE CREATION AND SERVICE ECOSYSTEMS

As mentioned, the concept of value cocreation is another tenet of S-D logic that mandates that value is always cocreated through the combined resource integration of service providers and beneficiaries. In general, value cocreation implies that all actors involved in economic exchange are both providers and beneficiaries as they provide mutual service for each other's benefit (Vargo 2009). Vargo and Lusch (2011) use of the term "actor" emphasizes the idea that all entities (e.g., firms, customers, and other stakeholders) are integrators of (mostly operant) resources and cocreators of value. Therefore, except in a relative sense, when providers and beneficiaries of service are both considered as resource integrators, the producer–consumer distinction vanishes (Vargo 2009; Vargo and Lusch 2011). This is because, within S-D logic, all entities involved in an economic exchange are considered resource integrators, capable of contributing to value cocreation.

The focus on resource integration redirects the focal point of value creation, from "value-in-exchange" to "value-in-use." While S-D logic does not reject the importance of "value-in-exchange" (nominal value, market price) as an important feedback to the firm and an intermediary of service provision, it points toward a more phenomenological and experiential conceptualization of value. However, even "value-in-use" (value derived through the application of a resource) does not fully reflect S-D logic thought—that is, it is transitional—and the more fitting term of "value-in-context" is recognized. "Value-in-context suggests that value is not only always cocreated; it is contingent on the integration of other resources and is contextually specific" (Vargo et al. 2010, p. 141). This suggests that operant resources, such as knowledge, are only valuable in use and that the use of knowledge is also contextual.

The heterogeneous perception of value that is phenomenologically determined by each actor is based on his or her existing competences and a network of resources, including market (e.g., other firms) and nonmarket (e.g., public and private) resources. This contextual perspective of value perception therefore highlights the importance of time and space as key variables in the determination of value. Additionally, the value-in-context perception can be based on utilitarian measures (e.g., functionality) or higher-order measures (e.g., happiness, confidence, and security) of value and can be determined by an individual actor or groups of actors (Chandler and Vargo 2011). Table 5.1 provides an overview of the major differences between S-D logic and G-D logic from the viewpoint of value creation.

TABLE 5.1
G-D Logic vs. S-D Logic on Value Creation

Value Driver	G-D Logic Value-in-Exchange	S-D Logic Value-in-Context
Creator of value	Firm, often with input from firms in a supply chain	Firm, network partners, and customers
Process of value creation	Firms embed value within "goods" or "services," value is "added" by increasing attributes	Firms propose value through market offerings, customers continue value-creation process through use and in context
Purpose of value	Increase wealth for the firm	Benefit from the service (knowledge and skills) of others
Measurement of value	The amount of nominal value, price received	The utilitarian or social benefit accessed through use
Resources used	Operant resources are embedded in operand resources and delivered	Operant resources are transferred, sometimes via operand resources, and new operant resources emerge
Role of firm	Conducts value-added activities to create value for customers to consume	Places an offering in the market and proposes its value
Role of goods	Units of output, operand resources that distribute are embedded with value	Vehicle for operant resources, enables access to benefits of firm's competences
Role of customers	To "use up" or "destroy" value created by the firm	Cocreate value through continued production, marketing, distribution, and use of a firm's offering

Source: Adapted from *Eur. Manag. J.*, 26(3), Vargo, S.L., Maglio, P.P., and Akaka, M., On value and value co-creation: A service systems and service logic perspective, p. 148, Copyright 2008, with permission from Elsevier.

5.5.1 FROM VALUE NETWORKS TO SERVICE ECOSYSTEMS

As mentioned in the previous section, S-D logic advocates and relies on a network model for conceptualizing the value-creation process. According to Gummesson (2006, p. 342 emphasis in original), "When relationships embrace two or more people or organizations, complex patterns and contextual dimensions emerge—*networks*. We therefore also talk about networks of relationships. What happens between parties in the relationships is called *interaction*." Human interactions continually provide operant resources—knowledge and skills—from actors involved in and external to the exchange. Mutual service provisions require networks of interactions since no actor alone possesses adequate resources for value creation. In particular, the knowledge of the customer is a key resource, without which value cannot be cocreated and determined (Ballantyne and Varey 2006).

Thus, S-D logic uses a network-with-network (see Figure 5.1) conceptualization that includes provider and customer networks and is similar to Gummesson's (2006) "many-to-many" marketing concept. This integrative, network-with-network model of value creation is not limited to individuals and firms or to economic exchange. It applies equally to all actors and institutions (e.g., families, firms, cities, and nations) in the creation of value for themselves and for others through the integration of resources acquired through both economic and social exchange.

The central role of interactions among networked actors is not new to the study of social and economic exchange (e.g., Hakansson and Snehota 1995). However, most network conceptualizations are provider centric. That is, they focus primarily on networks that provide firms with resources, which are used to create value for customers. S-D logic, on the other hand, sees the "customer" and the customer's resources, including its knowledge and network, as not only integral to but also primary in the value cocreation process. Thus, S-D logic suggests that this more comprehensive value configuration (Normann 2001)—interactions among networks of economic and social actors—needs to be considered as the venue of value creation.

In an attempt to more fully develop this complex higher-level system framework, Vargo and Lusch (e.g., Vargo and Lusch 2011; Vargo 2009) have extended the notion of a business ecosystem (e.g., Iansiti and Levien 2004) to a "service ecosystem," in which

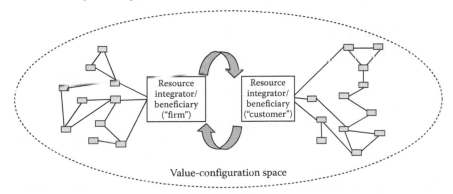

FIGURE 5.1 Value-creation networks. (Adapted from Vargo, S.L., *J. Serv. Res. Vol.*, 11(2), 211, 2008.)

firms operate as part of larger networked structures. The first extension, as mentioned, recognizes the customer (and the customer's network) as involved in the value-creation process. The second extension incorporates the idea that every resource integration event changes the nature of the whole structure to some degree and hence the context for the next event of resource integration. To reflect this dynamic, self-adapting, and relational model of value creation, Vargo and Lusch (2011) define a service ecosystem as a "spontaneously sensing and responding spatial and temporal structure of largely loosely coupled value proposing social and economic actors interacting through institutions, technology, and language to (1) co-produce service offerings, (2) exchange service offerings, and (3) co-create value."

Based on this service ecosystem's view, markets are composed of loosely coupled actors, linked through competences, service relationships, and information. These links of operant resources enable the formation of value propositions for service provisions. S-D logic thereby adopts and extends Haeckel's (1999) notion of sensing and responding organizations to a system's view. Spontaneously sensing and responding service ecosystems are conceptualized as structures that are constantly adapting to changing contextual requirements and are simultaneously creating these changing contexts in the process (e.g., Giddens 1979).

S-D logic's dynamic view of systems and markets differs from the more static, conventional view of networks. Vargo and Lusch (2011, p. 5) argue that "as much as the idea of resource networks contribute to the understanding of value creation and context, its consideration sometimes lacks a critical characteristic of systems, which are dynamic and potentially self-adjusting and thus simultaneously functioning and reconfiguring themselves." Thus, a system's view differs from a network perspective because, from an ecosystem's view, each instance of resource integration, service provision, and value creation changes the nature of the system to some degree and thus the context for the next iteration and determination of value creation. S-D logic therefore conceptualizes networks not just as networks (aggregations of relationships) but as dynamic systems. Understanding these system dynamics has significant practical implications for the information technology ecology. In the next section, we explain why a service-ecosystem perspective offers such an important venue for studying value creation.

5.6   ADVANTAGES OF AN S-D LOGIC, SERVICE-ECOSYSTEM PERSPECTIVE

Information technology research (e.g., service-oriented architecture) has already started to indicate that the traditional make-sell-destroy mentality of G-D logic cannot offer the rich and fertile foundation that is needed to develop the emerging field of KSE. To address the limitations of the G-D logic view, we argue that the S-D logic conceptualization of a service ecosystem provides KSE with a platform for understanding how actors integrate available resources and how they interact with other actors as they cocreate value for themselves and for others.

Two examples might make this concept and its distinctions more salient. The traditional G-D logic view would consider the *Journal of International Taxation* to be a logical way to provide large amounts of information to subscribing tax attorneys.

The publisher produces the magazine and embeds it with information (value) that is dyadically and unidirectionally delivered to the customer (attorney). An S-D logic view, on the other hand, can be best elucidated with a more complex value proposition. An iPad, for example, viewed through an S-D logic lens, can provide attorneys with access to extensive (and growing) resources available through the firm, other market-facing, public, and private sources that are infinitely integratable, allowing flexible, context-specific value creation. It enables the attorney, for example, to store relevant tax code information (public resource) on the device, to schedule and track appointments, to visit law publications online (market-facing resources), to purchase law-related books via iTunes (resources of a firm), and to listen to daily podcasts (private or market-facing resources). Additionally, the iPad platform also provides connection to other actors and their operant resources via the email and phone functionalities. It is important to highlight that both the journal and the iPad can be viewed through an S-D logic as well as a G-D logic lens and that S-D logic is not limited to certain "types" of firms or products. The journal and iPad examples were chosen to contrast the two lenses and to show that only S-D logic can offer a full vision of complex value-creation ecosystems and thus a vision of the more extensive innovation possibilities. We therefore suggest that only a strong, truly service-centric foundation of the nature of economic exchange enables KSE practitioners to shift their focus from goods to service.

The iPad example shows that S-D logic offers a revised perspective on the roles of the actors involved in economic exchange, changing the purpose of the enterprise, from producing and selling goods (or services, or goods-services bundles) that have value, to supporting customers in their own value-creation processes, through service. S-D logic therefore recasts the role of customers from destroyers of value to (primary) value cocreators. It offers a perspective of economic exchange and value creation that is profoundly different from the one offered by G-D logic. S-D logic provides a holistic perspective of markets and economic exchange, in which value creation is a collaborative process that is driven by mutual service provision.

The S-D logic framework encourages the enterprise to focus on the customer's context and resources (in addition to its own). Value can only be created by finding solutions to the customer's problems as defined by the customer's own, unique context, including other available resources. This, in turn, recasts the role of the firm into supporting and assisting the customer in gaining access to resources in ways that allow value creation. We argue that the potential rewards for these service-driven insights are enormous because they suggest that the competitive advantage of a firm can be recast from a logic that focuses on making better products to increase market share to one that is driven by defining and dominating new markets. With regard to KSE, this ecosystems framework suggests that rather than focusing on firm-centered resources and *developing new or better products* firms can draw on vast amounts of knowledge from multiple sources, through service-for-service exchanges, to *(re)design or (re)engineer new markets*. As discussed, the iPad (especially in its present state) represents a wonderful example. As many publishing companies wrestled with decreasing subscription and advertisement revenue, Apple found a way to assist customers in idiosyncratically integrating and organizing their increasingly complex world of available resources. Apple did not enter a market; it created one.

The difference between the logic of making better products and creating new markets can be best elucidated by another example. Consider the difference in potential between the typical car company's approach of incrementally improving features, quality, and supporting service for cars and Ford's new Sync initiative, which connects information supplied by customers and others, made available through the "cloud," and accessed/displayed in its cars through voice activation, as contextually required. Sync, for example, provides phone, text message, and music functionalities; incorporates turn-by-turn directions, Google maps, and internet search listings; and automatically connects to a 911 operator after an accident with airbag deployment. Viewed from G-D logic, this represents making a car with better technology but, from an S-D logic perspective, this represents a value-proposition platform that gives customers access to technology that allows them to (uniquely) integrate information and communication resources as they relocate, while, almost incidentally, providing reliable transportation. Like the iPad, the Sync technology represents an open-source system, with yet-to-be-defined possibilities based on customer-generated applications. It has the potential of (re)defining a market (Hochman 2010).

The long-term success of Ford's Sync initiative and Apple's iPad has yet to be revealed. However, in both cases it is clear that an S-D logic perspective of resource integration and value cocreation provides a richer, more robust view of the value-creation process and thus is a perspective that offers KSE increased possibilities for studying innovation and market creation.

5.7 CONCLUSION

A deep understanding of service is central to the emerging field of KSE. In order to propose a strong and robust theoretical foundation, we described and contrasted two distinct lenses through which social and economic exchange can be viewed. The first lens, G-D logic, characterizes services by their shortcomings and views them as somewhat inferior "goods." In this tradition, management principles developed for the production and distribution of goods are adopted and modified for the "production" and "distribution" of services. The second lens, S-D logic, provides an alternative framework for economic exchange that is based on service-driven principles. It considers service, the process of using one's resources to create value with and for the benefit of another actor, as the fundamental basis of exchange and thus requires a drastic departure from G-D logic.

We argue that service provision has always been central to economic exchange and value creation and that a G-D logic view is restrictive because it points to a narrow, dyadic, and unidirectional conceptualization of value creation in which one party creates (produces) value and the other party destroys (consumes) it. We show that the appropriateness of an S-D logic orientation is not limited to certain "types" of enterprises or economies, since, in S-D logic, all enterprises and economies are service based. To further provide evidence that S-D logic offers a strong foundation for KSE research, we highlight the role of knowledge in service and value creation. S-D logic redirects the focus of exchange to operant resources by shifting from units of output to the process of using competences for the benefit of another party. In these service provisions, the other party then reciprocates with its own applied competences. Thus, S-D logic emphasizes the primacy of knowledge and other intangible and dynamic (operant) resources, rather than static and tangible (operand) resources. In this way, the knowledge of the customer is a key resource, without which value cannot be cocreated and determined. We therefore posit that all economies are knowledge economies and that large amounts of knowledge are not unique to certain types of economic exchanges. However, we acknowledge that the field of KSE is reflective of a rapid increase in the ability to exchange information in a relatively pure form, through digitization.

Finally, we show that S-D logic reframes the enterprise's activities within a broader value-creation context, which encompasses the interaction among suppliers, customers, and other exchange partners in customer and supplier networks, and establishes the market as a system of service exchange. Successful knowledge systems cannot be limited to dyads and unidirectional exchange. Interactions with actors external to the exchange continually provide operant resources—knowledge and skills. The understanding of these extended service ecosystems (network-with-network models) is therefore critically important for the field of KSE.

In summary, S-D logic offers a framework for the study of KSE that clarifies the roles of service, knowledge, and relational networks in exchange. It points toward more insightful practical implications and directions, by reframing value creation as a collaborative, relational, and dynamic process based on mutual service provision. We therefore propose that S-D logic can provide a theoretical foundation that will offer KSE a rich and robust view of the value-creation process and, thus, a perspective from which the possibilities for innovation and successful market creation are more likely.

REFERENCES

Adler, P. S. (2001), Market, hierarchy, and trust: The knowledge economy and the future of capitalism, *Organization Science*, 12(2), 215–234.

Arnould, E. J. and C. J. Thompson (2005), Consumer culture theory (CCT): Twenty years of research, *Journal of Consumer Research*, 31(4), 868–882.

Ballantyne, D. and R. J. Varey (2006), Creating value-in-use through marketing interaction: The exchange logic of relating, communicating and knowing, *Marketing Theory*, 6(3), 335–348.

Chandler, J. D. and S. L. Vargo (2011), Contexualization and value-in-context: How context frames exchange, *Marketing Theory*, 11(1), 35–49.

Constantin, J. and R. F. Lusch (1994), *Understanding Resource Management*, Oxford, OH and Burr Ridge, IL: Planning Forum.

Davenport, T. H., D. W. Long, and M. C. Beers (1998), Successful knowledge management projects, *Sloan Management Review*, 39(2), 43–57.

Giddens, A. (1979), *Central Problems in Social Theory: Action, Structure, and Contradictions in Social Analysis*. Berkeley and Los Angeles, CA: University of California Press.

Grönroos, C. (1994), From marketing mix to relationship marketing: Towards a paradigm shift in marketing, *Asia–Australia Marketing Journal*, 2(August), 9–29.

Gummesson, E. (Ed.) (1995), *Relationship Marketing: Its Role in the Service Economy*. New York: John Wiley & Sons.

Gummesson, E. (Ed.) (2006), *Many-to-Many Marketing as Grand Theory*. Armonk, NY: M.E. Sharpe.

Haeckel, S. H. (1999), *Adaptive Enterprise: Creating and Leading Sense-and-Respond Organizations*. Boston, MA: Harvard Business School Press.

Hakansson, H. and I. Snehota (1995), *Developing Relationships in Business Networks*. Boston, MA: International Thomson Press.

Harris, D. B. (1996), Creating a knowledge centric information technology environment, http://www.htca.comickc.htm

Hochman, P. (2010), Ford's big revival, *Fast Company*, April (105), 90–95.

Hunt, S. (2000), *A General Theory of Competition: Resources, Competences, Productivity, Economic Growth*. Thousand Oaks, CA: Sage Publications.

Iansiti, M. and R. Levien (2004), *The Keystone Advantage: What the New Dynamics of Business Ecosystems Mean for Strategy, Innovation, and Sustainability*. Boston, MA: Harvard Business School Press.

Kogut, B. and U. Zander (1992), Knowledge of the firm, combinative capabilities, and the replication of technology, *Organization Science*, 3(3), 3838–3397.

Lovelock, C. and E. Gummesson (2004), Whither services marketing? *Journal of Service Research*, 7(1), 20–41.

Lusch, R. F., S. L. Vargo, and M. O'Brien (2007), Competing through service: Insights from service-dominant logic, *Journal of Retailing*, 83(1), 5–18.

Lusch, R. F., S. L. Vargo, and M. Tanniru (2010), Service, value networks and learning, *Journal of the Academy of Marketing Science*, 38(1), 19–31.

Normann, R. (2001), *Reframing Business: When the Map Changes the Landscape*. New York: John Wiley & Sons.

Prahalad, C. K. and G. Hamel (1990), The core competence of the corporation, *Harvard Business Review*, 68(3), 79–91.

Rust, R. (2004), If everything is service, why is this happening now, and what difference does it make? *Journal of Marketing*, 68, 23–24.

Rust, R. T. and F. Espinoza (2006), How technology advances influence business research and marketing strategy, *Journal of Business Research*, 59(10–11), 1072–1078.

Rust, R. and D. V. Thompson (2006), How does marketing strategy change in a service-based world?: Implications and directions for research, in *The Service-Dominant Logic of Marketing: Dialog, Debate, and Directions*, R. F. Lusch and S. L. Vargo (Eds.). Armonk, NY: M.E. Sharpe.

Schlesinger, L. A. and J. L. Heskett (1991), The service-driven company, *Harvard Business Review*, 69(September–October), 71–81.

Shin, M., T. Holden, and R. A. Schmidt (2001), From knowledge theory to management practice: Towards an integrated approach, *Information Processing & Management*, 37(2), 335–355.

Shostack, G. L. (1977), Breaking free from product marketing, *Journal of Marketing*, 41, 73–80.

Smith, A. (1776), *The Wealth of Nations* (1976 ed.). Oxford, U.K.: Clarendon.

Vargo, S.L. (2008), Customer integration and value creation paradigmatic traps and perspectives, *Journal of Service Research Volume*, 11(2), 211–215.

Vargo, S. L. (2009), Toward a transcending conceptualization of relationship: A service-dominant logic perspective, *Journal of Business & Industrial Marketing*, 24(5/6), 373–379.

Vargo, S. L. and M. A. Akaka (2009), Service-dominant logic as a foundation for service science: Clarifications, *Service Science*, 1(1), 34–41.

Vargo, S. L. and R. F. Lusch (2004), Evolving to a new dominant logic for marketing, *Journal of Marketing*, 68(January), 1–17.

Vargo, S. L. and R. F. Lusch (2008), Service-dominant logic; further evolution, *Journal of the Academy of Marketing*, 36(1), 1–10.

Vargo, S. L. and R. F. Lusch (2011), It's all B2B ... and beyond: Toward a systems perspective of the market, *Industrial Marketing Management*, 40(2), 181–187.

Vargo, S. L., R. F. Lusch, M. A. Akaka, and Y. He (2010), Service-dominant logic: A review and assessment, *Review of Marketing Research*, 6, 125–167.

Vargo, S. L., R. F. Lusch, and F. W. Morgan (2006), Historical perspective on service-dominant logic, in *The Service-Dominant Logic of Marketing*, R. F. Lusch and S. L. Vargo (Eds.). Armonk, NY: M.E. Sharpe.

Vargo, S. L. and F. W. Morgan (2005), Services in society and academic thought: An historical analysis, *Journal of Macromarketing*, 25(1), 42–53.

Vargo, S. L., P. P. Maglio, and A. Melissa (2008), *On Value and Value Co-Creation: A Service Systems and Service Logic Perspective*. Elsevier, pp. 145–152.

Zeithaml, V. A., A. Parasuraman, and L. L. Berry (1985), Problems and strategies in services marketing, *The Journal of Marketing*, 49(2), 33–46.

Zimmermann, E. W. (1951), *World Resources and Industries: A Functional Appraisal of the Availability of Agricultural and Industrial Materials*. New York: Harper.

Part II

Engineering from Data, Information, and Knowledge toward Services

6 Knowledge-Based Systems Engineering

Ying Liu and Kun Bai

CONTENTS

6.1 FROM DIKW TO SERVICE

The birth of the Internet era generates extensive speculation about the potential consequences of the development for data storage, information management, knowledge identification, understanding extraction, result utilization, better service designing, implementation, etc. Despite the growing literatures and the rapid development of new techniques, most previous works focus on the theoretical foundation or modeling designing. Numerous algorithms and methods are proposed to improve the processing performance from data to information, from information to knowledge, and from knowledge to intelligence; however, there are still gaps between these methodologies and application services. In this chapter, we not only introduce the representative techniques but also analyze why they are so important for service designing. This chapter is a tentative analysis on providing a bridge to associate those theoretical methodologies with real-life service engineering.

6.1.1 TRADITIONAL KNOWLEDGE-BASED ENGINEERING VS. SERVICE ENGINEERING

Knowledge engineering (KE) was defined in 1983 by Edward Feigenbaum and Pamela McCorduck as follows: "KE is an engineering discipline that involves integrating knowledge into computer systems in order to solve complex problems normally requiring a high level of human expertise" (Feigenbaum and McCorduck 1983). KE is closely related to many computer science domains such as artificial intelligence (AI), database, data mining, expert systems, decision support, and geographic information systems (Feigenbaum and McCorduck 1983).

More specifically, knowledge-based engineering (KBE) is a "discipline with roots in computer-aided design (CAD) and knowledge-based systems but has several definitions and roles depending upon the context" (Milton 2008). KBE has a great deal in common with software engineering and it has a wide scope that covers design, analysis (computer-aided engineering [CAE]), manufacturing, support, etc. Although most of the early-stage researches on KBE focused on activities related to product lifecycle management and multidisciplinary design optimization, more applications, especially service-oriented applications in diverse fields, with larger scopes are involved.

Service engineering is the overall discipline of producing, consuming, and managing services and service-oriented architectures (SOA) (Allen 2006). As the progress of software engineering, Service engineering research is continuously stimulated and leveraged, especially in the design and implementation of service-oriented systems. Researchers and engineers are keeping a lookout for new opportunities to innovate, design, and manage the service operations and processes of the new service-based economy. Typical service engineering methods across multiple areas include service architecture and engineering (SAE), service-oriented analysis and design (SOAD), service-oriented development of applications (SODA), service-oriented transformation (SOT by IBM), etc.

6.1.2 DATA → INFORMATION → KNOWLEDGE → SERVICE

Because both KE and KBE are in knowledge level, understanding the nature of knowledge, the generation of knowledge, as well as the relationship among data, information, knowledge, and service is crucial not only for topic understanding but also for crossing the bridge between knowledge and service engineering.

6.1.2.1 DIKW Hierarchy

The "DIKW hierarchy," also known variously as the "wisdom hierarchy," the "knowledge hierarchy," or "information hierarchy" refers loosely to a class of models for representing purported structural and/or functional relationships between data, information, knowledge, and wisdom (Zins 2009).

Russell Ackoff, a system theorist, gave definitions to each term as follows (Ackoff 1989) (Rowley and Richard 2006). Figure 6.1 shows the hierarchical relationship among DIKW, which was proposed in Rowley (2007). In addition, Figure 6.1 also demonstrates the relationship between DIKW and service.

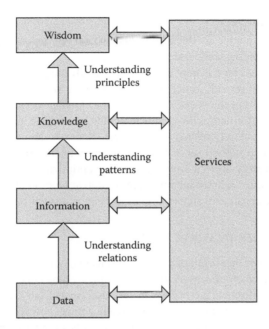

FIGURE 6.1 Relationships among DIKW hierarchy and service.

1. *Data*: Data is the unfiltered fact of the world. Human beings can perceive the data and process with our brain. It is simply existent without any meaning beyond its existence. Different people can have different interpretations from their various perspectives.
2. *Information*: We are in an age often referred as the information age. Information is data with meaning by ways of relational connection and patterns. Such meaning does not have to be useful. Usually, information provides answers to "who," "what," "where," and "when" questions. Relational database is an example of information to store and interpret data.
3. *Knowledge*: Knowledge is an appropriate collection of information. Different from information, knowledge is a deterministic process, which is to be useful. It is often formulated as an implication. It is an application of data and information; answers "how" questions.
4. *Understanding*: Understanding is an interpolative and probabilistic process, which can analyze existing knowledge and synthesize new knowledge. The main difference between understanding and knowledge is the difference between "learning" and "memorizing." Usually understanding is used to answer "why" questions.
5. *Wisdom*: Wisdom is an extrapolative, nondeterministic, and nonprobabilistic process, which can facilitate human beings to understand extremely challenging problems. In addition, it is the essence of philosophical probing instead of the pure understanding.

6.1.2.2 DIKW and Service

Although there are numerous definitions and interpretation on data, information, knowledge, understanding, and wisdom, most of them are still in the theoretical levels. What we are interested here is how to take the best advantage of these concepts, and how to apply them to provide better services in various areas to satisfy diverse user demands. Different from the sequential relationship among the aforementioned concepts (from data to information, from information to knowledge, etc.), service is existing beyond them but closely related to each level. Not only data and information are valuable foundation to support service designing and implementation, but also knowledge and wisdom can be applied directly to service optimization and maintenance.

In the next section, we introduce the representative techniques in each level and analyze the potential usage to support the modeling of service systems.

6.2 TECHNIQUES IN DIKW LEVELS

We are living in an age that is full of sophisticated technologies such as computer, satellites, Internet, etc. Numerous sensors and devices are keeping collection of tremendous amount of data and information every day in various formats from diverse sources. With the advent of computers and means for mass digital storage, how to understand and use the massive collection of data accurately and efficiently is an overwhelming problem.

6.2.1 KNOWLEDGE EXTRACTION TECHNIQUES

In order to process diverse data efficiently and accurately, we should understand the nature of data more thoroughly than in the past. According to the data storage and structure level, data can be divided into several categories: structured data, unstructured data, and semistructured data.

By comparison, unstructured data has no identifiable structures. Unstructured information is typically text intensive, which may contain specific data such as dates, numbers, and facts as well as text and other data types that are not part of a database.

With the development of Internet and web data, structured data became one of the top five web trends in 2009 (MacManus 2009). Relational databases and spreadsheets are examples of structured data. They usually are stored in fixed fields within a record or file.

Semistructured data is located between structured data and unstructured data. It contains a form of structured data, which does not conform to the formal structure of database tables. The data model of semistructured data also associates with relational databases but nonetheless contains tags or other markers to separate semantic elements and enforces hierarchies of records and fields within the data. Therefore, it is also known as schemeless or self-describing structure (Buneman 1997). Representative semistructured data includes XML, other markup languages, e-mail, and EDI, which are all forms of semistructured data.

Because of the variation of data structures and natures, accurately and efficiently knowledge extraction from diverse data types plays a vital role in all the further service applications. Selecting the best knowledge extraction techniques is the essential task. In the following section, we introduce all the representative techniques for each type of data.

6.2.2 Techniques on Structured Data

6.2.2.1 Database and Database Management System

To deal with the chaos of nature data, structured database is designed to store data neatly. In addition, effectively and efficiently retrieve particular information from a large data corpus is an urgent and challenging problem. Structured database as well as the database management system (DBMS) is proliferated.

Database is a collection of information in a structured way. We can say that it is a collection of a group of facts. The telephone directories in yellow page are database examples of individuals or organizations. To save data into a database, we have to understand the nature of data and the structure of tables in databases.

The first task is creation of a database and its management. A common and powerful method for organizing data for computerization is the relational data model. Computer-based databases are usually organized into one or more tables. A table stores data in a format similar to a published table and consists of a series of rows and columns. Each column in a database table will have a name, often called a field name. The term field is often used instead of column. Each row in a table will represent one example of the type of object about which data has been collected. There are three types of relationship between entities in database: one-to-one, one-to-many, and many-to-many. Figure 6.1 displays real examples to these three relationships. A relational database uses the concept of linked two-dimensional tables that comprise of rows and columns. A user can draw relationships between multiple tables and present the output as a table again. A user of a relational database need not understand the representation of data in order to retrieve it. Relational programming is nonprocedural.

Every row in a table in a relational database must be unique. There must not be two identical rows. One or more columns are therefore designated the primary key (or called the unique identifier) for the item contained within. Foreign keys are columns in a table that provide a link to another table.

A DBMS (Maier et al. 1986) is a software program that enables the creation and management of databases. It is designed to use one of five database structures to provide simplistic access to information stored in databases. The five database structures are the hierarchical model, the network model, the relational model, the multidimensional model, and the object model.

Generally, these databases will be more complex than the text file/spreadsheet example. Data management involves creating, modifying, deleting and adding data in files, and using this data to generate reports or answer queries. Using a DBMS files can be retrieved easily and effectively.

There are many DBMS packages available in the market. Some of the more popular relational DBMS include

- Microsoft Access
- Filemaker
- Microsoft SQL Server
- MySQL

In 1971, IBM researchers created a simple non-procedural language called structured English query language or SEQUEL. This was based on Dr. Edgar F. (Ted) Codd's design of a relational model for data storage where he described a universal programming language for accessing databases. In 1979, Relational Software released the world's first relational database called Oracle V.2. In the late 1980s, ANSI and ISO (these are two organizations dealing with standards for a wide variety of things) came out with a standardized version called structured query language or SQL. There have been several versions of SQL and the latest one is SQL-99. SQL-92 is the current universally adopted standard. MySQL and mSQL are DBMS. These software packages are used to manipulate a database. All DBMSs use their own implementation of SQL. It may be a subset or a superset of the instructions provided by SQL 92.

In order to better organize and manage data, remove redundancies and increase the clarity in organizing data in a database; this is a normalization step. Normalization of a database helps in modifying the design at later times and helps in being prepared if a change is required in the database design. Normalization raises the efficiency of the database in terms of management, data storage, and scalability.

As web applications gain popularity in today's world, surviving DBMS from an attack is becoming even more crucial than before because of the increasingly critical role that DBMS is playing in business/life/mission-critical applications. Although significant progress has been achieved to protect the DBMS, such as the existing database security techniques (e.g., access control, integrity constraint, and failure recovery), the business/life/mission-critical applications still can be hit due to some new threats toward the back-end DBMS. For example, in addition to the vulnerabilities exploited by attacks (e.g., the SQL injections attack), databases can be damaged in several ways such as the fraudulent transactions (e.g., identity theft) launched by malicious outsiders and erroneous transactions issued by the insiders by mistake. When the database is under such a circumstance (attack), rolling back and re-executing the damaged transactions are the most used mechanisms during the system recovery.

This kind of mechanism either stops (or greatly restricts) the database service during repair, which causes unacceptable data availability loss or denial-of-service for mission-critical applications, or may cause serious damage spreading during on-the-fly recovery where many clean data items are accidentally corrupted by legitimate new transactions. Database damage management (DBDM) is a very important problem faced today by a large number of mission/life/business-critical applications and information systems that must manage risk, business continuity, and assurance in the presence of severe cyber attacks.

Database should be equipped with many functions, such as DBDM. The self-healing capability of a database is an important problem faced today by a great number of mission/life/business critical applications. As the Internet applications gain popularity and are embraced in industry to support today's e-business world, more and more threats toward the back-end database systems are identified. Surviving the back-end DBMS from e-crime is becoming even more crucial than before because of the increasingly critical role that the DBMS is playing and due to the more critical and valuable information stored in databases, which is now processed through the web and is worldwide accessible. A database system with the self-healing capability aims to assure these applications, such as banking, online stock trading, air traffic control, etc., with high data integrity and service availability because these applications are the cornerstones of a variety of crucial information systems and infrastructures that must manage risk, business continuity, and data assurance in the presence of severe cyber attacks.

In general, database security concerns the confidentiality, integrity, and availability of data stored in a database. A broad span of research from authorization (Griffiths and Wade 1976; Jajodia et al. 1997; Rabitti et al. 1994) to inference control (Adam 1989), to multilevel secure databases (Sandhu and Chen 1998; Winslett et al. 1994), and to multilevel secure transaction processing (Atluri et al. 1997) addresses primarily how to protect the security of a database, especially its confidentiality.

Security-control methods suggested in the literature can be classified into four general approaches: conceptual, query restriction, data perturbation, and output perturbation. Criteria for evaluating the performance of the various security-control methods are identified. Security-control methods that are based on each of the four approaches are discussed in Adam (1989), together with their performance with respect to the identified evaluation criteria.

The addition of stringent security specifications to the list of requirements for an application poses many new problems in DBMS design and implementation, as well as database design, use, and maintenance. Tight security requirements, such as those that result in silent masking of withholding of true information from a user or the introduction of false information into query answers, also raise fundamental questions about the meaning of the database and the semantics of accompanying query languages.

6.2.2.2 KDD/Data Mining

Comparing with the huge amount of data, only a small part of data can be parsed and stored in database. Numerous amounts of data are stored in files, databases, and other repositories in diverse formats. It is increasingly important to develop powerful means for analysis and perhaps interpretation of such data and for the extraction of interesting knowledge that could help in decision making.

Data mining, also popularly known as knowledge discovery in databases (KDD) (Han et al. 2011), refers to the broad process of finding knowledge in data and emphasizes the "high-level" application of particular data mining methods. The key techniques fulfill the nontrivial extraction of implicit, previously unknown, and potentially useful information from data in database. While data mining and KDD are frequently treated as synonyms, data mining is actually part of the knowledge discovery process. It is highly related to many research areas, for example, database, machine learning, pattern recognition, statistics, knowledge acquisition for expert systems, AI, knowledge representation, data visualization, and so on.

Data mining is a relatively young and interdisciplinary field of computer science. It is the process of discovering new patterns from large datasets involving methods from statistics and AI but also database management. In contrast to for example machine learning, the emphasis lies on the discovery of *previously unknown* patterns as opposed to generalizing *known* patterns to new data.

6.2.2.2.1 Data Mining Processes

The KDD process contains several main steps to identify new knowledge from the raw data corpora. Please find the information of the whole system process in Figure 6.2

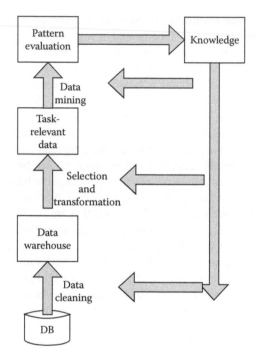

FIGURE 6.2   Data mining is the core of the knowledge discovery process.

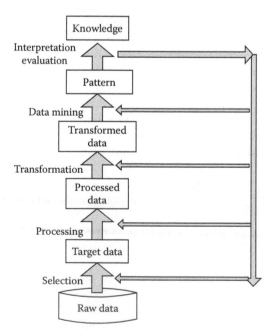

FIGURE 6.3   Outline of the steps of the KDD.

(modified based on process; Han et al. 2011) and main components in Figure 6.3 (modified based on process; Fayyad and Piatetsky Shapiro 1996). Data mining process is an iterative process instead of a single-round work. Before we start the data mining sequence, we have to finish the following tasks: (1) developing an understanding of the application domain, (2) acquiring the relevant prior knowledge, (3) understanding the goal of the end users, and (4) creating a dataset by selecting a set of representative data or variables, and discovering knowledge from it.

Based on the target dataset, we have to finish the following steps:

1. Data cleaning and preprocessing
   a. This is a fundamental but crucial step to find and remove all the noisy, irrelevant, or outlier data. Moreover, missing data fields should be filled based on wise strategies.
2. Data integration
   a. This step includes combining different data sources with heterogeneous structure into a common data source.
3. Data selection
   a. Instead of processing the entire dataset, we only analyze the data that is relevant to the research problem.
4. Data transformation/reduction
   a. This is an important data preprocessing step to prepare the datasets by transforming the selected data into appropriate forms, finding useful features to represent the data depending on the goal of the task, and using dimensionality reduction or transformation methods to reduce the effective number of variables.
5. Data mining task selection
   a. Selecting the appropriate data mining task is a challenging problem as well as deciding whether the goal of the KDD process is classification, regression, clustering, or other problems.
6. Choosing the data mining algorithms
   a. This is the core step of KDD. In this step, we apply a set of powerful techniques as well as models and parameters on the processed dataset, to extract hidden knowledge based on given measures.
7. Pattern evaluation
   a. Although the previous step usually generates a set of results, evaluating whether the mining work is efficient and accurate is a challenging problem. In this step, strictly interesting patterns representing knowledge are identified based on given measures.
8. Pattern interpreting and knowledge representation
   a. The last but not the least step is to visually represent the discovered results to the final users. The guideline is to make sure users can understand and interpret the results.

6.2.2.3   Data Mining Techniques

There are diverse KDD algorithms. The unifying goal of them is to extract knowledge from data in the context of large databases. Methods include classification, clustering, associate rules, etc. In this chapter, we introduce two representative techniques: classification and clustering (Tan et al. 2006).

6.2.2.3.1 Classification Techniques

Classification is a supervised data mining method. In a simple word, it can assign unlabeled objects to one of several predefined categories, based on the knowledge learnt from a set of trained data. It is a pervasive research because it encompasses many diverse applications, such as disease analysis, spam message detection, potential user identification, etc.

As shown in Figure 6.4, the input is a set of unlabeled records X. The purpose is to decide the class label for each $x \in X$. Classification method is the task of learning a target function f: $y = f(x)$ that map x to one of the predefined class labels. x is the attribute set of an input, and y is a specific class that will label x. The function f is also called as a classification model. Classification models can be divided into two types: descriptive modeling and predictive modeling.

Classification technique (or classifier) includes different approaches to building classification models. There are a number of standard classification methods in use. Typical classification models include rule-based classifier, decision tree classifier, neural network, naïve Bayes classifier, support vector machine, etc. Each model has its own suitable application field.

Figure 6.6 shows a general approach for building the classification model. Each classification technique employs a learning algorithm to set up a model that best fits the relationship between the attribute set and categories based on the training set. The model should not only fit for the training set, but it should also correctly predict the label for new records in the test set with good generalization capability. Any classification method uses a set of *features* or *parameters* to characterize each object, where these features should be relevant to the task at hand.

We consider here methods for *supervised* classification, meaning that a human expert both has determined into what classes an object may be categorized and

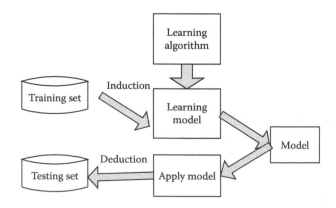

FIGURE 6.4 General approach for building a classification model.

also has provided a set of sample objects with known classes. This set of known objects is called the *training set* because it is used by the classification programs to learn how to classify objects. There are two phases to constructing a classifier. In the training phase, the training set is used to decide how the parameters ought to be weighted and combined in order to separate the various classes of objects. In the application phase, the weights determined in the training set are applied to a set of objects that do *not* have known classes in order to determine what their classes are likely to be.

Neural network methods probably are most widely known. The biggest advantage of neural network methods is that they are general: they can handle problems with very many parameters, and they are able to classify objects well even when the distribution of objects in the N-dimensional parameter space is very complex. The disadvantage of neural networks is that they are notoriously slow, especially in the training phase but also in the application phase. Another significant disadvantage of neural networks is that it is very difficult to determine *how* the net is making its decision. One of the famous applications of neural network is the classification problem on digitized photographic plates made by Odewahn et al. (1992).

Decision tree method is also used on diverse fields. The tree is constructed in which at each node a single parameter is compared to some constant. If the feature value is greater than the threshold, the right branch of the tree is taken; if the value is smaller, the left branch is followed. After a series of these tests, one reaches a leaf node of the tree where all the objects are labeled as belonging to a particular class. Decision trees are usually much faster in the construction (training) phase than neural network methods, and they also tend to be faster during the application phase. Their disadvantage is that they are not as flexible at modeling parameter space distributions having complex distributions as either neural networks or nearest neighbor methods.

6.2.2.3.2 Cluster Techniques

Unlike the classification techniques, cluster analysis is an unsupervised process. In other words, there will be no training data, and human beings do not know the category information in advance. Cluster analysis divides data into meaningful and useful groups (clusters). To do it, clusters should understand the nature data structure and property.

There are many applications based on cluster analysis, such as data summarization or information retrieval. With the help of clustering, people can automatically find classes and divide objects into groups. Especially for those areas with huge data scalability, such as biology or web data, even human beings are unable to understand it thoroughly.

Cluster analysis provides an abstraction from individual data objects to the clusters and these cluster prototypes can be used as the basis for a number of data analyses or data processing techniques: summarization, compression, efficiently finding nearest neighbors, etc.

Clusterings include various types: hierarchical (nested), hierarchical vs. partitional (unnested), exclusive vs. overlapping vs. fuzzy, and complete vs. partial. We introduce two simple but important clustering techniques:

1. K-means clustering (Lloyd and Lloyd 1982)
    a. This is a prototype-based, partitional clustering technique that attempts to find a user-specified number of clusters, which are represented by their centroids.
2. Agglomerative hierarchical clustering
    a. This clustering method works on a collection of closely related clustering techniques that produce a hierarchical clustering by starting with each point as a singleton cluster. These singleton clusters are repeatedly merging the two closest clusters until a single, all-encompassing cluster remains.

6.2.3 KNOWLEDGE EXTRACTION FROM UNSTRUCTURED DATA

6.2.3.1 Information Extraction Techniques

Confronted with huge data collection, information retrieval is simply not enough for decision making. The demands from users to find target information among the huge collection of data keep growing. As the development of Internet and the accumulation of web data, searching becomes one of the most used actions on the Internet. Search engines as an instrument of searching are very popular and frequently used sites. This is the reason why webmasters and every ordinary user on the Internet must have good knowledge about search engines and searching.

A search engine is an information retrieval system designed to help finding information (Vaughan and Thelwall 2004). Most commonly search engines are web search engines, which search for information on the public web, for example, Google, Yahoo! search, Microsoft MSN Search, ASK.com, and Bing.com. Other kinds of search engines are enterprise search engines, which search on intranets, personal search engines, and mobile search engines. Different selection and relevance criteria may apply in different environments or for different uses. More recently, more and more light is shed on specialty search engines. Some of them support search on various kinds of documents, for example, map search (http://maps.google.com/maps), video search (http://www.youtube.com/), image search (http://images.google.com/, http://photo.net/gallery/caption-search/), and patent search (http://www.google.com/patents), as well as on document components, for example, citation search (http://citeseerx.ist.psu.edu/), acknowledgement search (http://citeseer.ist.psu.edu/), and caption search (http://biosearch.berkeley.edu/).

6.2.3.2 General Search Engines vs. Specialty Search Engines

The role of every search engine is to give users search results that will show the URLs that contain the theme users are looking for. From the topic theme and user purpose perspectives, search engines can be divided into two types: general search engines and specialty search engines. Although these two types of search engines have different

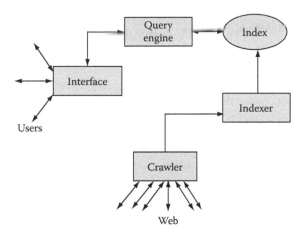

FIGURE 6.5 Typical web search engine.

motivations and work on different datasets, they follow the same general search engine framework with the same main components as shown in Figure 6.5: crawler, indexer, ranking component, and interface.

Crawler is also called *spider, ants, automatic indexers, bots, web robots*, etc. It is a computer program that browses the World Wide Web to collect up-to-date data. In general, it starts with a list of URLs to visit, called the *seeds*. As the crawler visits these URLs, it identifies all the hyperlinks in the page and adds them to the list of URLs to visit, called the *crawl frontier*. URLs from the frontier are recursively visited according to a set of policies.

Indexer is a crucial component of each search engine. It can facilitate fast and accurate information retrieval. The purpose of storing an index is to optimize speed and performance in finding relevant documents for a search query. Without an index, the search engine would scan every document in the corpus, which would require considerable time and computing power.

Each search engine usually will collect and index millions or billions of pages. Given a search query, there are still thousands or millions of matched results. How to present users with the best results that match with the search query is a challenging problem. The matches will be ranked by the ranking component, so that the most relevant ones come first. Of course, the search engines do not always get it right. Non-relevant pages make it through, and sometimes it may take a little more digging to find what you are looking for. Vector Space Model (Salton et al. 1975) based on term frequency (TF-IDF) (Spärck 1972) and PageRank (Page et al. 1999) are two representative ranking schemes for textual document search and web search.

General search engines are useful when users are looking for general and light information, but if users are looking for specialty, specialty search engine is more useful. There are a lot of specialty search engines for music, pictures, marketing, e-zines, nature, science, sport, cars, etc. In their search results, users will get lower number of URLs, and every URL will contain certain theme. In this chapter, we use a specialty search engine on scientific tables as an example to show the detailed work of a typical search engine.

Most prior work on information extraction has focused on extracting information from texts in digital documents. However, often, most important information being reported in an article is presented in tabular form in a digital document. Recently, more and more researchers realized the importance of extracting specific information from documents and supporting the free data integration and sharing among structured, unstructured, and semistructured data. Table extraction and search is a successful service example.

Tables are ubiquitous in scientific publications, web pages, financial reports, news papers, magazine articles, etc. More and more researchers in chemistry, biology, and finance fields adopt tables to display the latest experimental results as well as the statistical information in a concise layout. In the traditional media, tables have a history that predates that of sentential text (Hurst 2006). Tables present structural data and relational information in a two-dimensional format and in a condensed fashion. Researchers always use tables to concisely display their latest experimental results, to list the statistical data, and to show the financial activities of a business, etc. Other researchers, who are conducting the empirical studies in the same topic, can quickly obtain valuable insights by examining these tables. For example, a biochemist may want to search tables containing experimental results about "mutant genes" or an economist may look for the tables about "the GDP growth of USA in 2000–2007."

Figure 6.6 shows the architecture of TableSeer, which consists of a number of important components: (1) a table crawler, (2) a table metadata extractor, (3) a table metadata indexer, (4) a table ranking algorithm, and (5) a table searching query interface. In summary, TableSeer crawls scientific documents from the digital libraries, identifies the documents with tables, detects each table using a novel document page box-cutting method, extracts the metadata for each identified table, ranks the matched tables against the end user's query with the TableRank algorithm, and displays the ordered results in a user-friendly interface.

Tables present a unique challenge to an information extraction system. Extracting sharable table metadata from documents no matter what document formats they are in is a challenging problem for several reasons: (1) tables are shown in documents with diverse medium types, like HTML, PDF, etc.; different media use different schemes to save table information and present tables in different layouts. (2) Different press categories (e.g., journals or conference proceedings, scientific papers as well as financial reports) may have their own requirements for the table layout. For example, different from computer science fields (e.g., JCDL proceedings), most chemistry/biology journals only require vertical lines to separate table with paper context. Within the table body, some journals display rows one by one without any separators while some others alternatively change the background color for a better visual effect. (3) Table cells are so diverse that different content elements can be stored in one cell; (4) some tables have affiliated elements; (5) no formal rules/standards on table designing; and (6) different applications and purposes have different requirements on the table metadata extraction.

TableSeer harvests online scientific documents by crawling open-access digital libraries and scientists' web pages. The crawler supports a number of document media, such as PDF, HTML, WORD, PowerPoint, etc. TableSeer is designed to be able to handle all document media listed earlier. In this chapter, the table crawlers pay attention to the scientific documents in the PDF format because they gain more and more popularity in digital libraries. In addition, comparing to other document

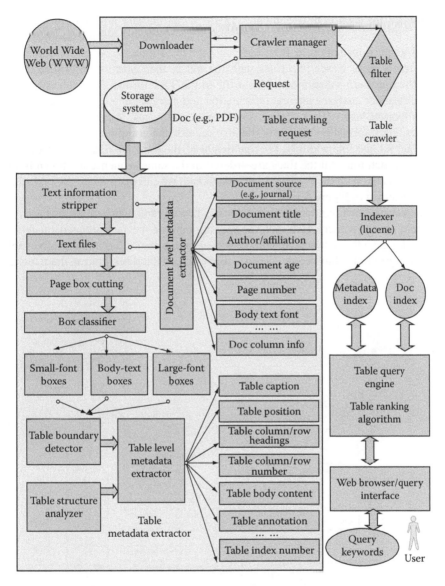

FIGURE 6.6 Architecture of table search engine. (From Liu, Y. et al., TableSeer: Automatic table metadata extraction and searching in digital libraries, in *Joint Conference on Digital Libraries*, June 18–23, Vancouver, British Columbia, Canada, 2007, pp. 91–100.)

media that have been extensively studied in the table detection field, PDF documents have been overlooked. Our table crawlers use a depth-first crawling policy with a maximum depth of five (from the seed URLs).

The table metadata extractor comprises of three key parts: (1) a text information stripper (TIS), (2) a table boundary detector with the page box-cutting method, and (3) a table metadata extractor from both the document level and the table level.

Initially, for each PDF document, TIS strips out the text information from the original PDF source file word by word through analyzing the text operators and the related glyph information in PDF Reference Fifth Edition, Version 1.6. TableSeer reconstructs these words into lines with the aid of their position information and saves the lines into a document content file in the TXT format. For each document page, TableSeer analyzes the text information and merges them into different physical component levels (lines, paragraphs, boxes, and pages) according to their font and position information. The text file also records the coordinates of the leftmost word (X_0, Y_0), the line width W, the line height H, the font size F, as well as the text content of each line. All the lines are ordered in the same sequence as shown in the PDF document. With the extracted table metadata, we can manipulate the table data freely as shown in Figure 6.7.

One key question in information retrieval is how to rank matched results based on their relevance to a query. However, the existing ranking schemes are inadequate and are not designed for table search. Our TableSeer search engine has an innovative table ranking algorithm—TableRank. Given a user query, TableRank returns the matched

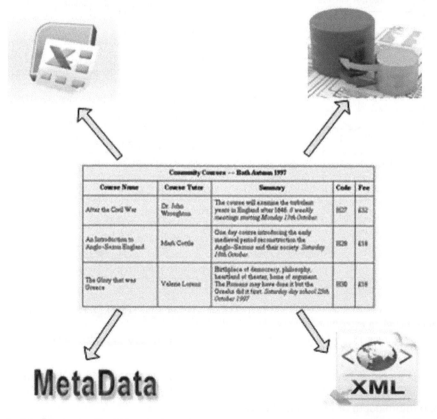

FIGURE 6.7　The table data storage and utility. (From Liu, Y. et al., TableSeer: Automatic table metadata extraction and searching in digital libraries, in *Joint Conference on Digital Libraries*, June 18–23, Vancouver, British Columbia, Canada, 2007, pp. 91–100.)

	$m_1(\text{MW}_1)$				$m_l(\text{MW}_l)$			TLB	DLB
	$t_{1,1}$	...	$t_{x,1}$	...	$t_{1,k}$	...	$t_{z,k}$	...	...
$tb1$	$w_{1,1,1}$	...	$w_{x,1,1}$	...	$w_{1,1,k}$	...	$w_{z,1,k}$	...	...
$tb2$	$w_{1,2,1}$	...	$w_{x,2,1}$	...	$w_{1,2,k}$	...	$w_{z,2,k}$	...	...
$\vdots$	...	$\vdots$	...	$\vdots$	...	$\vdots$	...	...	...
tb_b	$w_{1,b,1}$	...	$w_{x,b,1}$	...	$w_{1,b,k}$	...	$w_{z,b,k}$	...	...
Q	$w_{1,q,1}$			...	$w_{1,q,k}$			...	...
ITTF	...	...	...	...	...	...	...	...	...

FIGURE 6.8 The modified table vector space as well as the query vector. (From Liu, Y. et al., TableSeer: Automatic table metadata extraction and searching in digital libraries, in *Joint Conference on Digital Libraries*, June 18–23, Vancouver, British Columbia, Canada, 2007, pp. 91–100.)

tables in a descendant order according to their relevance scores. Different from the popular web search engines, our TableRank rates the <query, table> pairs instead of the <query, document> pairs. TableRank tailors the traditional vector space model to rate the <query, table> pair by replacing the document vectors with the table vectors. As shown in Figure 6.8, each row is a query or a table vector. To determine the weight for each term in the vector space, we design an innovative term weighting scheme table term frequency–inverse table term frequency (TTF-ITTF), a tailored TF-IDF (Salton and Buckley 1988) weighting scheme. Compared with TF-IDF, TTF-ITTF demonstrates two major advantages. First, it calculates the TF in the table metadata file instead of the whole document, which prevents the false positive results. Second, it calculates the weight of a term with a comprehensive view. TableRank constructs the vector matrix and measures the similarity between the query and each table by computing the cosine of the angle between these two vectors.

Equation 6.1 shows that the final weight of a term $w_{i,j,k}$ comes from three levels: the term-level weight, $w_{i,j,k}$, at term level, the table-level weight boost, $\text{TLB}_{i,j}$, and the document-level weight boost, DLB_j, which are described in Liu et al. (2007), respectively:

$$sim\left(tb_j, Q\right) = \cos\left(tb_j, Q\right) = \frac{\sum_{i=1}^{8} w_{i,j,k}w_{i,q,k}}{|tb_j||Q|} \qquad (6.1)$$

Given a search query, which is similar to other regular web search queries, TableSeer search engine can return a set of matched tables based on the ranking calculation between each <query, table> pair in the crawled and indexed table repository. The matched tables can be shown in different ways: by table, by document, by metadata, or even in new formats.

6.2.3.3 Natural Language Processing

Natural language processing (NLP) (Manning and Schütze 1999) is a field of computer science and linguistics concerned with the interactions between computers and human (natural) languages. It began as a branch of AI. According to the definition in Manning and Schütze (1999), NLP is a theoretically motivated range of computational techniques for analyzing and representing naturally occurring texts at one or more levels of linguistic analysis for the purpose of achieving human-like language processing for a range of tasks or applications.

There are more practical goals for NLP, many related to the particular application for which it is being utilized. Modern NLP algorithms are grounded in machine learning, especially statistical machine learning. Research into modern statistical NLP algorithms requires an understanding of a number of disparate fields, including linguistics, computer science, and statistics.

6.3 CONCLUSION: FROM KE TO SERVICE ENGINEERING

Previous researchers in multiple areas, such as data management, information theory, or AI, spent a large effort on the development of formalisms, models, inference mechanisms, and methods. When these theoretical works are tested and applied in real problems, we call them as knowledge-based systems (KBSs). Typically, the development efforts were restricted to the realization of small KBSs in order to study the feasibility of the different approaches. Though these studies offered rather promising results, it is a challenging task to transfer these technologies into commercial uses, in order to build large-scaled systems. Small working experiments do not guarantee a successful application to large-scaled real systems. So the goal of the new discipline KE is turning the process of constructing KBSs from an art into an engineering discipline. This requires the analysis of the building and maintenance process itself and the development of appropriate methods, languages, and tools specialized for developing KBSs.

Today, we have far more information than we can handle: from business transactions and scientific documents, to satellite pictures, text reports, and military intelligence. All the techniques we introduced earlier are just a tiny partial of the techniques we can use directly. They can be widely used in many applications: from biology to medicine, from business to web information retrieval, from environmental science to telecommunication.

Although there are tons of powerful methods that are available to use directly, there are many problems we still have to consider. Besides the model and method selection, we also have to know how people understand, analyze, and interpret the world, what is the problem to solve, where to collect the data and how to clean it up, how to understand the results and improve the performance, etc.

Using database as an example, database and DBMS are widely used in people's ordinary life. Online database software is delivered as software as a service (SaaS). Online database software combines fully customizable web application solution and an easily accessible web database. Using clustering as another example, it can not only improve the quality of the biological taxonomy defining, but also find the earth's climate patterns in the atmosphere and ocean. Similarly, cluster analysis can be used to identify the difference among illness and symptoms. Huge business information collection is also a valuable field for clusters to grouping customers and marketing activities. The latest application of NLP can be found in an IBM-designed NLP computer code named Watson, which competed in the game show Jeopardy in 2011. Moreover, representative services based on NLP include machine translator, question answering, speech understanding, audio recognition, spoken dialogue, etc. Numerous online search engines are keeping provide the information retrieval services to billions of users. Cloud computing represents the trends not only in research but also in service: provides computation, software, data access, and storage services. It is a way to increase capacity or add capabilities on the fly without investing in new infrastructure, training new personnel, or licensing new software. Because of the space limitation, this chapter is a starting point on the long way to fully convert techniques into services and bridge the gap between two of them.

REFERENCES

Ackoff, R. L., From data to wisdom. *Journal of Applies Systems Analysis*, 16, 3–9, 1989.

Adam, M. R., Security-control methods for statistical database: A comparative study. *ACM Computing Surveys*, 21(4), 515–556, 1989.

Allen, P., *Service Orientation: Winning Strategies and Best Practices*, Cambridge University Press, Cambridge, MA, May 29, 2006.

Atluri, V., Jajodia, S., and Bertino, E., Multilevel secure databases with kernalized architecture: Challenges and solutions. *IEEE Transactions on Knowledge and Data Engineering*, 9(5), 697–708, 1997.

Buneman, P., Semistructured data. In *Proceedings of ACM Symposium on Principles of Database Systems*, May 11–15, Tucson, AZ, 1997, pp. 117–121.

Fayyad, U. and Piatetsky-Shapiro, G., From data mining to knowledge discovery: An overview. In Fayyad, U., Piatetsky-Shapiro, G., Smyth, P., and Uthurusamy, R., eds., *Advances in Knowledge Discovery and Data Mining*, AAAI Press, Menlo Park, CA, 1996, pp. 1–34.

Feigenbaum, E. A. and McCorduck, P., *The Fifth Generation*, 1st edn., Addison-Wesley, Reading, MA, 1983. ISBN 9780201115192, OCLC 9324691.

Griffiths, P. P. and Wade, B. W., An authorization mechanism for a relational database system. *ACM Transactions on Database Systems (TODS)*, 1(3), 242–255, 1976.

Han, J., Kamber, M., and Pei, J., *Data Mining: Concepts and Techniques*, 3rd edn., Morgan Kaufmann, Burlington, MA, 2011.

Hurst, M., Towards a theory of tables. Special issue on detection and understanding of tables and forms for document processing applications. *International Journal on Document Analysis and Recognition*, 8(2–3), 123–131, 2006. DOI: 10.1007/s10032-006-0016-y.

Jajodia, S., Samarati, P., Subrahmanian, V. S., and Bertino, E., A unified framework for enforcing multiple access control policies. In *ACM SIGMOD*, May 13–15, Tuscon, AZ, 1997, pp. 474–485.

Liu, Y., Bai, K., Mitra, P., and Giles, C. L., TableSeer: Automatic table metadata extraction and searching in digital libraries. In *Joint Conference on Digital Libraries*, June 18–23, Vancouver, British Columbia, Canada, 2007, pp. 91–100.

Lloyd, S. P., Least squares quantization in PCM. *IEEE Transactions on Information Theory*, 28(2), 129–137, 1982.

MacManus, R., Top 5 Web Trends of 2009: Structured Data, 2009. http://www.readwriteweb. com/archives/top_5_web_trends_of_2009_structured_data.php (accessed on September 2009).

Maier, D., Stein, J., Otis, A., and Purdy, A., *Development of an Object-Oriented DBMS*, Portland, OR, 1986, pp. 472–482. ISBN 0-89791-204-7.

Manning, C. D. and Schütze, H., *Foundations of Statistical Natural Language Processing*, MIT Press, Cambridge, MA, 1999. ISBN 978-0-262-13360-9.

Milton, N., *Knowledge Technologies*, Polimetrica, Monza, Italy, 2008.

Odewahn, S. C., Stockwell, E. B., Pennington, R. L., Humphreys, R. M., and Zumach W. A., Automated star/galaxy discrimination with neural networks. *Astronomical Journal*, 103, 318–331, 1992.

Page, L., Brin, S., Motwani, R., and Winograd, T., The PageRank citation ranking: Bringing order to the web, Technical Report. Stanford InfoLab, 1999.

Rabitti, F., Bertino, E., Kim, W., and Woelk, D., A model of authorization for next-generation database systems. *ACM Transactions on Database Systems (TODS)*, 16(1), 88–131, 1994.

Rowley, J., The wisdom hierarchy: Representations of the DIKW hierarchy. *Journal of Information Science* 33(2), 163–180, 2007. doi:10.1177/0165551506070706.

Rowley, J. and Richard, H., *Organizing Knowledge: An Introduction to Managing Access to Information*. Ashgate Publishing Ltd., Aldershot, U.K., 2006, pp. 5–6. ISBN 9780754644316.

Salton, G. and Buckley, C., Term-weighting approaches in automatic text retrieval. *Information Processing & Management*, 24(5), 513–523, 1988. doi:10.1016/0306-4573(88)90021-0.

Salton, G., Wong, A., and Yang, C. S., A vector space model for automatic indexing. *Communications of the ACM*, 18(11), 613–620, 1975.

Sandhu, R. and Chen, F., *The Multilevel Relational (MLR) Data Model*, Vol. 1. ACM Press, New York, 1998, pp. 93–132.

Spärck, J. K., A statistical interpretation of term specificity and its application in retrieval. *Journal of Documentation*, 28(1), 11–21, 1972. doi:10.1108/eb026526.

Tan, P., Steinbach, M., and Kumar, V., *Introduction to Data Mining*, Pearson Addison Wesley, Boston, MA, 2006. ISBN 0321321367.

Vaughan, L. and Thelwall, M., Search engine coverage bias: Evidence and possible causes. *Information Processing & Management*, 40(4), 693–707, 2004.

Winslett, M., Smith, K., and Qian, X., Formal query languages for secure relational databases. *ACM Transactions on Database Systems (TODS)*, 19(4), 626–662, 1994.

Zins, C., Conceptual approaches for defining data, information, and knowledge (PDF). *Journal of the American Society for Information Science and Technology* (Wiley Periodicals, Inc.), 58(4), 479–493, 2009. doi:10.1002/asi.20508.

7 Smart Metering and Customer Consumption Behavior Profiling
Exploring Potential Business Opportunities for DSOs and Electricity Retailers[*]

Hongyan Liu, Tomas Eklund, and Barbro Back

CONTENTS

7.1 INTRODUCTION

With the large-scale smart meter rollout in Finland, all Finnish commercial customers and residential consumers will have remotely-read electricity meters by the end of 2013. The distribution system operators (DSOs) and the electricity retailers are imminently facing issues including (1) how to fully utilize this investment and technological advance to enhance their day-to-day operations, (2) how to continuously improve cost efficiency, and (3) how to create new growth opportunities sustainably.

---

[*] Liu, H., Eklund, T., and Back, B., Smart metering and costumer consumption behavior profiling: Exploring potential business opportunities for DSOs and electricity retailers, *CIRED 2011*, June 6–9, 2011, Frankfurt, Germany, Paper 0887, 2011b. With permission.

Here, we propose a business intelligence approach, namely, customer consumption behavior profiling, which we believe is the next step in the effort to develop dynamic price–based demand response applications, new energy and price mixes, and a more active electricity retail market.

The prevailing pricing models (Räsänen et al. 1995, 1997) are mainly based on customer categories (i.e., industry, service and trade, housing, etc.) or housing type in the case of residential consumers. It is highly likely that even in the same customer category/housing type, the consumption patterns may vary considerably due to customers' business nature/lifestyle diversity (Keppo and Räsänen 1999). Before the implementation of smart metering, it was impossible to accurately identify real-time customer consumption patterns on a large scale.

The aim of this study is to apply a data-mining technique in the form of the self-organizing map (SOM) to group customers according to their actual consumption. We investigate 14,000 Finnish customers' longitudinally measured electricity consumption data during 2007–2009. The data are provided by one regional DSO— Ålands Elandelslag (ÅEA, a nonprofit ownership cooperative), whose distribution area has distinct geographical features and customer structure. We profile the customers without regard to their conventional classification (i.e., customer categories and housing type). Then, we compare their contractual electricity rates in light of their actual consumption patterns.

The questions of interest are the following: (1) Can the SOM-based approach provide added value to the DSOs and electricity retailers? (2) How can a business intelligence approach built upon smart metering contribute to obtaining and maintaining an efficient and well-functioning electricity retail market in the long run?

In the next section, we will briefly introduce the Åland area and the SOM method. Thereafter, the experiment and the results will be presented. Conclusions will be drawn in the final part of this chapter, together with proposals for further research.

7.2 BACKGROUND

7.2.1 ÅLAND AREA

Åland is a Finnish archipelago region with nearly 300 habitable islands, which is situated in between mainland Finland in the east and Sweden in the west as shown in Figure 7.1. It consists of 16 municipalities with Mariehamn as the regional capital. ÅEA is responsible for the electricity distribution to 15 municipalities (excluding Mariehamn). Its distribution area covers 14,097 customers, of which Jomala is the largest (2,290) and Sottunga is the smallest (184 customers). Åland's geographical features determine that its economy is heavily dominated by shipping, trade, and tourism. The majority of the housing is in the form of summer cottages, detached houses, or town houses, while multistoreyed buildings only account for a very small portion.

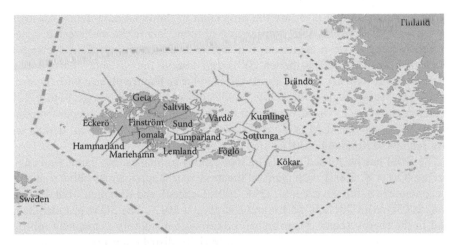

FIGURE 7.1 Åland's geographic location.

According to Statistics Åland, in 2009, Åland's electricity consumption by sector is as follows: households (45.04%), agriculture (7.01%), industry (11.77%), services (21.22%), and the public sector (14.97%), respectively. It shows that households, services, and the public sector constitute the majority in terms of electricity consumption in Åland. This differs from the electricity consumption breakdown on mainland Finland, where industry's electricity consumption amounts to 46%, whereas housing and agriculture, and services and construction, consume 29% and 22%, respectively (*source*: Energiateollisuus).

7.2.2 SOM Method

The SOM is a data-mining approach based upon artificial neural networks (ANNs). ANNs are designed to mimic the basic learning and association patterns of the human nervous system, and consist of a number of neurons (simple processors) connected by weighted connection. ANNs learn by adjusting the weight of each connection, increasing or decreasing the importance of the input (information) being transferred, until a desired output is achieved. Essentially, they are nonlinear, multivariate regression techniques, better able to handle erroneous and noisy data than parametric statistical tools (Bishop 1995).

The SOM is a two-layer ANN that uses the unsupervised learning approach, that is, the SOM does not require target output values for training (Kohonen 2001). The essence of the SOM is to map high dimensional data onto a spatial map (usually in the form of a two-dimensional lattice of hexagonal nodes). The SOM uses the competitive learning algorithm, meaning that the nodes on the output layer compete with each other to be the best matching node (i.e., the winner) whose connection weights to the input pattern are the closest in terms of the Euclidian distance. At the same time, the SOM algorithm allows the output nodes in the neighborhood of the winner to adjust their weights accordingly. Theoretically, all the nodes on the output layer are the projection of the input data items. As such, the intrinsic relationships (e.g., similarities) of input data in the multivariate space are reflected on a two-dimensional topological map, that is, visual clustering is performed (Haykin 1999; Kohonen 2001). In addition, the variables that are used for training the map are usually displayed in color as feature planes, with "warm" colors representing high values while "cold" colors for low values (see Figure 7.2). Therefore, it is easy to visually interpret the characteristics of each cluster from the feature maps. As the SOM is a well-known method, for mathematical details we refer readers to Kohonen (2001).

The SOM has been used widely in applications in finance, medicine, and engineering (Oja et al. 2002), with many of the more recent applications being financial or economic in nature (Back et al. 1998, 2001; Deboeck and Kohonen 1998; Eklund et al. 2003; Sarlin 2011a,b; Sarlin and Marghescu 2011). The SOM has also been used in the energy sector, for example, power system stability assessment, online provision control, load forecasting, and electricity distribution regulation assessment (Nababhushana et al. 1998; Rehtanz 1999; Riqueline et al. 2000; Lendasse et al. 2002; Liu et al. 2011a). Like other ANNs, the SOM is acknowledged for its robustness in handling nonlinear and multivariate data, especially with regards to dimension reduction and large datasets. In particular, the SOM is recognized for its visualization capabilities. Compared to other clustering techniques, the SOM does

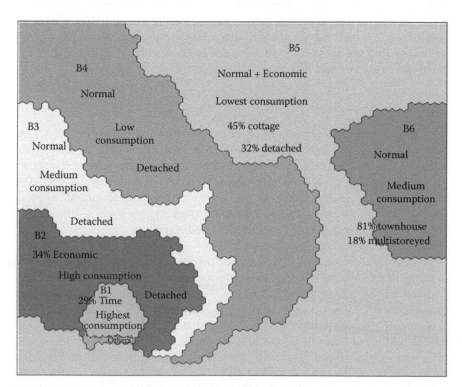

FIGURE 7.2 Brändö consumption clusters.

not require predefining a desirable number of clusters, which means that little a priori knowledge of the data is required. For these reasons, this study seeks to apply the SOM in the field of electricity customer consumption behavior profiling in the Åland context. To our knowledge, the SOM has not previously been applied by others in this domain.

In this study, Viscovery SOMine v.5.0 (http://www.eudaptics.com/) is used. SOMine uses an expanding map size and the batch training algorithm, allowing for very efficient training of maps (Kohonen 2001). The SOM-Ward clustering method is also used to identify clusters based on actual consumption behavior, which eliminates the need for subjective identification of clusters (Vesanto and Alhoniemi 2000).

7.3 EXPERIMENT AND RESULTS

7.3.1 EXPERIMENT

The data used in this study are from ÅEA meter readings for the period of 2007–2009. For each meter, the readings are registered with 27 h 20 min time intervals, due to the communication technology adopted (Turtle automated meter reading system). The variables included in the analysis are as follows:

Consumption (kWh)—Is derived from consecutive readings that are measured per 27 h 20 ± 8 min

Peak load value (kW)—Is the highest load aggregated from three consecutive 20 min intervals during each 27 h 20 min period

Tariff code (TaCo)—Is the contractual electricity rate the customer has chosen from five categories: normal rate, economic rate, time rate, irrigation rate, and temporary working rate, which are provided by ÅEA (available at http://www.el.ax/files/tariffhafte_20110101.pdf, in Swedish)

Housing type (HoTy)—Is based on historical statistics, provided by ÅEA as a reference variable, including five categorical attributes: summer cottage, detached house, townhouse, multistoreyed building, and others

7.3.2 RESULTS

We will present two municipalities' (Brändö and Geta) results in the following. The reason for selecting these two municipalities is because they have a similar customer base size (Brändö: 600, Geta: 667), but their geographical features and customer composition differ significantly. For instance, Brändö is in the northeast of Åland's territory, where out of 1648.51 km², 1540.44 km² is water (93%). Geta, on the other hand, is connected to the mainland of Åland, and its size is 606.56 km², of which about 86% is water. Table 7.1 illustrates the differences in customer composition between Brändö and Geta.

Figures 7.2 and 7.3 present the clustering results, where Brändö results in six consumption groups (B1–B6) while Geta has five clusters (G1–G5). The feature planes

TABLE 7.1

Number of Enterprises in Brändö and Geta in 2007

	Brändö	Geta
Agriculture, forestry, and fishing	10	1
Industry	7	3
Construction	13	12
Trade, hotel	19	10
Transport	10	2
Finance and insurance	6	5
Public services	7	0
Total	72	33

Source: Statistics Åland.

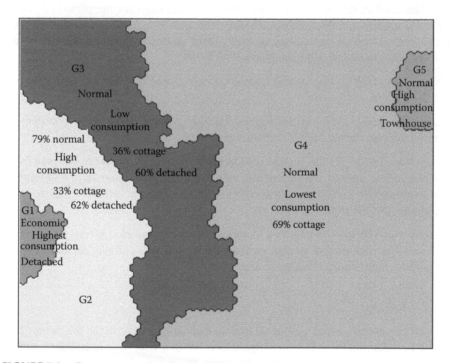

FIGURE 7.3   Geta consumption clusters.

(Figures 7.4 and 7.5*) and cluster characteristics breakdown (Tables 7.2 and 7.3) reveal some interesting facts: in Brändö, cluster B1 has the highest average values in terms of consumption and peak load (235 kWh, 19.88 kW), and 29.8% customers in B1 chose time rate while 70.1% of them adopted normal rate. On the other hand, cluster B2 has the second highest value in consumption and peak load (94 kWh, 7.67 kW),

---

* The figures are only fully interpretable in color.

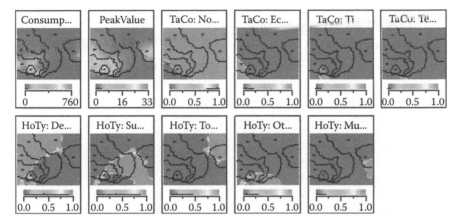

FIGURE 7.4   Feature maps of Brändö clusters.

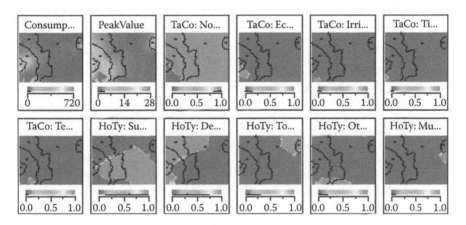

FIGURE 7.5   Feature maps of Geta clusters.

and the customers' choice of contractual electricity rate differs—34.5% chose economic rate, 64.4% went for normal rate, and 1.1% had time rate. The majority of the customers (65.66%) are in cluster B5, which has the lowest average consumption and peak load values (7 kWh, 0.47 kW). There are 3.3% and 1.3% customers who adopted economic rate and time rate, respectively, in B5. Customers in B3, B4, and B6 identically preferred the normal rate, but their consumption profiles can be identified based upon their housing types. In Geta, similar features as for B3, B4, and B6 in Brändö can also be seen in G3, G4, and G5. Even though the majority of the customers in Geta have chosen the normal rate (e.g., G3, G4, and G5), the customers in G1 and G2 differ from the others. The average consumption and peak load values in G1 are the highest among the five clusters (205 kWh, 15.22 kW). 61.9% of the customers in G1 chose the economic rate, while the remaining 38.1% still chose the normal rate. No customers use the time rate. On the other hand, in G2 the customers' average consumption is ranked the second highest (68 kWh), and the average peak

TABLE 7.2
Characteristics of Clusters in Brändö

	B1	B2	B3	B4	B5	B6
Cluster size %	0.42	4.33	6.37	18.84	65.66	4.38
Consumption (kWh)	235	94	47	27	7	42
Peak load (kW)	19.88	7.67	4.66	2.22	0.47	4.22
Normal rate (%)	70.1	64.4	100	100	95.1	100
Economic rate (%)	0.1	34.5	0	0	3.3	0
Time rate (%)	29.8	1.1	0	0	1.3	0
Temporary rate (%)	0	0	0	0	0.3	0
Detached (%)	1.8	64.9	75.7	66.4	32.1	0
Summer cottage (%)	30.1	22.4	19.9	30.0	45.6	0
Townhouse (%)	0.2	0	0	0	13.0	81.7
Others (%)	67.8	12.6	4.4	3.6	6.8	0
Multistoreyed (%)	0	0	0	0	2.6	18.3

TABLE 7.3
Characteristics of Clusters in Geta

	G1	G2	G3	G4	G5
Cluster size (%)	1.04	8.41	15.42	73.95	1.18
Consumption (kWh)	205	68	34	7	66
Peak load (kW)	15.22	5.46	2.95	0.58	5.81
Normal rate (%)	38.1	79.5	100	100	100
Economic rate (%)	61.9	8.3	0	0	0
Irrigation rate (%)	0	5.4	0	0	0
Time rate (%)	0	3.8	0	0	0
Temporary rate (%)	0	3.1	0	0	0
Summer cottage (%)	12.0	33.3	36.9	69.6	0
Detached (%)	87.9	62.4	60.7	18.7	0
Townhouse (%)	0.1	0	0	7.6	96.5
Others (%)	0	4.3	2.4	3.0	0
Multistoreyed (%)	0	0	0	1.1	3.5

load value (5.46 kW) is slightly lower than the second highest value in G5 (5.81 kW). However, 8.3% and 3.8% of the customers in G2 have chosen the economic rate and time rate, respectively.

7.4   CONCLUSION

In this study, we use the SOM to profile customers in the Åland area, based on their measured electricity consumption data. Our purpose is to examine what kind of benefits a business intelligence approach can offer to DSOs or electricity retailers.

The results indicate that the majority of customers in cluster B1 (70.1%), B2 (64.4%), and G2 (79.5%)—which have high consumption profiles in their respective municipalities—chose the normal rate, instead of the economic rate or time rate, which should favor customers with high consumption. The reason behind might lie in the fixed components of the tariffs, but illustrates well how consumption profiling could be beneficial. To this end, we perceive that a SOM-based customer consumption behavior profiling method can provide added value to DSOs or retailers, though it requires further examination to evaluate the profiling results. Additionally, it implies that if analyzing customers according to their consumption similarity and/or deviation, it will assist DSOs and retailers to develop a better understanding of their customers, which in turn could aid them to design electricity rates that can facilitate demand response applications. For example, the attributes of customers in G1 (with high consumption profile) and especially B5 (with low consumption profile), who favor of economic rate or time rate, could be good indicators to gauge other customers who share similar attributes for time-of-use (TOU) rate promotion. Meanwhile, through encouraging more customers to adopt TOU rate, DSOs can better mitigate the peak load formation in line with the supply capability, in order to secure the quality of supply. Hence, we can argue that a data mining–based business intelligence approach is a promising starting point in the effort to obtain and maintain an efficient and well-functioning electricity retail market in the long run.

ACKNOWLEDGMENTS

The authors thank ÅEA for providing the data for this study. The authors also gratefully acknowledge the financial support of the Academy of Finland (grants no. 127656 and 127592).

REFERENCES

Back, B., Sere, K., and Vanharanta, H. 1998. Managing complexity in large data bases using self-organizing maps. *Accounting Management and Information Technologies* 8(4): 191–210.
Back, B., Toivonen, J., Vanharanta, H., and Visa, A. 2001. Comparing numerical data and text information from annual reports using self-organizing maps. *International Journal of Accounting Information Systems* 2(4): 249–269.
Bishop, C. M. 1995. *Neural Networks for Pattern Recognition*. Oxford, U.K.: Oxford University Press.
Deboeck, G. and Kohonen, T. 1998. *Visual Explorations in Finance Using Self-Organizing Maps*. Berlin, Germany: Springer-Verlag.
Eklund, T., Back, B., Vanharanta, H., and Visa, A. 2003. Using the self-organizing map as a visualization tool in financial benchmarking. *Information Visualization* 2: 171–181.
Haykin, S. 1999. *Neural Networks: A Comprehensive Foundation*, 2nd edn., Upper Saddle River, NJ: Prentice Hall.
Keppo, J. and Räsänen, M. 1999. Pricing of electricity tariffs in competitive markets. *Energy Economics* 21: 213–223.
Kohonen, T. 2001. *Self-Organizing Maps*, 3rd edn., Berlin, Germany: Springer.

Lendasse, A., Lee, J., Wertz, V., and Verleysen, M. 2002. Forecasting electricity consumption using nonlinear projection and self-organizing maps. *Neurocomputing* 48: 299–311.

Liu, H., Eklund, T., and Back, B. 2011a. Visual data mining: Using self-organizing maps for electricity distribution regulation. *Proceedings of The International Conference on Digital Enterprise and Information Systems (DEIS 2011)*, July 20–22, 2011, London, U.K., Communications in Computer and Information Science (CCIS), Vol. 194. Heidelberg, Germany: Springer-Verlag, pp. 631–646.

Liu, H., Eklund, T., and Back, B. 2011b. Smart metering and costumer consumption behaviour profiling: Exploring potential business opportunities for DSOs and electricity retailers. *CIRED 2011*, June 6–9, 2011, Frankfurt, Germany, Paper 0887.

Nababhushana, T. N., Veeramanju, K. T., and Shivanna. 1998. Coherency identification using growing self organizing feature maps (power system stability). *IEEE Proceedings of EMPD'98. International Conference on Energy Management and Power Delivery*, March 3–5, 1998, Singapore, Vol. 1, pp. 113–116.

Oja, M., Kaski, S., and Kohonen, T. 2002. Bibliography of self-organizing map (SOM) papers: 1998–2001 addendum. *Neural Computing Surveys* 3: 1–156.

Räsänen, M., Ruusunen, J., and Hämäläinen, R. P. 1995. Customer level analysis of dynamic pricing experiments using consumption-pattern models. *Energy* 20(9): 897–906.

Räsänen, M., Ruusunen, J., and Hämäläinen, R. P. 1997. Optimal tariff design under consumer self-selection. *Energy Economics* 19: 151–167.

Rehtanz, C. 1999. Visualisation of voltage stability in large electric power systems. *IEEE Proceedings Generation, Transmission and Distribution* 146: 573–576.

Riqueline, J., Martinez, J. L., Gomez, A., and Goma, D. C. 2000. Possibilities of artificial neural networks in short-term load forecasting. *Proceedings of the IASTED International Conference Power and Energy Systems*, July 3–6, 2001. Anaheim, CA: IASTED/ACTA Press, pp. 165–170.

Sarlin, P. 2011a. Clustering the changing nature of currency crises in emerging markets: An exploration with self-organising maps. *International Journal of Computational Economics and Econometrics* 2(1): 24–46.

Sarlin, P. 2011b. Evaluating a self-organizing map for clustering and visualizing optimum currency area criteria. *Economics Bulletin* 31: 1483–1495.

Sarlin, P. and Marghescu, D. 2011. Visual predictions of currency crises using self-organizing maps. *Intelligent Systems in Accounting, Finance and Management* 18(1): 15–38.

Vesanto, J. and Alhoniemi, E. 2000. Clustering of the self-organizing map. *IEEE Transactions on Neural Networks* 11(3): 586–600.

8 Smart Perishable Cargo Management with RFID and Sensor Network

Min Gyu Son, Yoon Seok Chang, and Chang Heun Oh

CONTENTS

8.1 PERISHABLE CARGO

Perishable cargo can be defined as goods that will deteriorate over a given period of time or some cargo that will deteriorate if exposed to adverse temperature, humidity, or other environmental conditions. Perishable cargo includes goods from the health care and food sectors, including pharmaceutical products and nonhazardous biological materials, and any time- and temperature-sensitive goods that need to be properly prepared, packaged, and handled. International Air Transport Association (IATA) Perishable Cargo Regulation suggests methods for packing and keeping perishable cargo to reduce the impact of temperature during transportation. Table 8.1 shows the types of perishables defined by IATA Perishable Cargo Regulation (IATA 2009).

IATA Perishable Cargo Regulation suggests management factor considering the characteristics of perishable cargo such as temperature, humidity control, regulation of carbon dioxide (CO_2) levels, and air circulation for the appropriate transport of perishable cargoes.

TABLE 8.1

Types of Perishable Cargo

Class	Types of Perishable Cargo
1	Fruit and vegetables
2	Fresh-cuts and prepared salads
3	Seafood and fish
4	Meat and meat product
5	Dairy product
6	Bakery
7	Frozen food
8	Ornamentals
9	Pharmaceuticals

8.2   TRACKING AND TRACEABILITY FOR PERISHABLE CARGO

The main fact that differentiates food supply chains from other chains is that there is a continuous change in the quality from the time the raw materials leave the grower to the time the product reaches the consumer (Apaiah et al. 2005). Perishables such as meat, fish, milk, and more can change hands many times before reaching to the consumer. Keeping perishables in safe and good quality is a significant challenge as it moves through the supply chain. The quality of perishables is dependent on how the products are handled at every touch point throughout the food chain.

The efficiency of a traceability system depends on the ability to track and trace each individual product and distribution (logistics) unit, in a way that enables continuous monitoring from primary production (e.g., harvesting, catch, and retirement) until final disposal to consumer. Traceability schemes can be distinguished into two types: logistics traceability that follows only the physical movement of the product and treats food as commodity and qualitative traceability that associates additional information relating to product quality and consumers safety, such as preharvest and postharvest techniques, storage and distribution conditions, etc. (Folinas et al. 2006).

IATA requires traceability for cargo from start point to the destination because perishable cargos may impact on the health of customers directly. In the past, barcode technology was used for tracking and recently radio frequency identification (RFID) is considered as a potential candidate. However, RFID can only track the time stamp, location, and product information not the freshness or status of cargo. Because of such limited capability, there are some researches that integrate RFID with sensor technology.

Asimakopoulos et al. (2007) suggested RFID-based perishable cargo tracking using METATRO architecture. Jedermanna et al. (2008) presented a method of managing perishable product quality by collecting data log using sensor-integrated RFID tag during transportation. Brusey and Thorne (2006) compared two methods of RFID and sensor integration such as physical integration and logical integration (Figure 8.1).

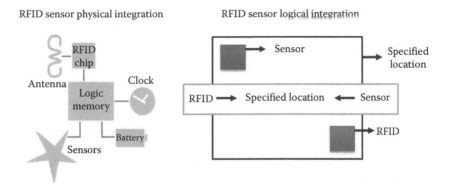

FIGURE 8.1   Two types of RFID and sensor integration.

Currently, system requirements suggested by IATA perishable cargo regulation are as follows:

- Identification of specific object information
- Procedure for information gathering and process efficiency and automation
- Identification of each point in the supply chain

The management of perishable items demands both logistics and qualitative traceability and needs tracking of environmental conditions such as temperature, humidity, light, etc. The ability to collect this information and the use of such information to ensure product quality in "real time" provides tangible benefits to the industry. It provides a greater assurance of product quality and enables quick identification of problems (Wilson and Clarke 1998). Transparency of a supply chain network is important as all the stakeholders of network have a shared understanding of access to product- and process-related information they requested without loss, noise, delay, and distortion. Transparency enables to achieve efficient recalls on the chain level when necessary and support early warnings in case of a possible emerging problem through pro-active quality monitoring system to optimize the supply chain (Beulens et al. 2005).

8.3   MANAGEMENT OF PERISHABLE CARGO

In this section, we explain a process flow of a perishable cargo and issues in management. Table 8.2 shows general processes and activities of each process in perishable cargo terminal.

We have identified key data for tracking perishable cargo as in Table 8.3. Since goods are moving by unit load device (ULD), ULD number could be a key for tracking cargo. Other data such as master airway bill (MAWB) number, house airway bill (HAWB) number, etc. can be used after mapping the number with ULD number.

TABLE 8.2
Perishable Cargo Process

Category	Process Flow	Activity
1	Receiving	Receipt of perishable cargo at the truck dock
2	Inspection	Inspection considering the status of packaging, existence of label, weight, etc.
3	Storage	Keeping perishable cargo at the special storage (freeze/cold/heat storage)
4	Buildup	Building up perishable cargo to the ULD
5	Elevator Transfer Vehicle (ETV)	Keeping ULD cargo in the ETV rack
6	Airside	Loading of cargo to the aircraft

TABLE 8.3
Data Requirements for Perishable Cargo

Data	Description
Special code	Code to identify perishable cargo
Pieces weight	Weight for individual cargo to calculate cargo cost (SKID, loose)
Security	Process of checking security before receipt of cargo
MAWB No.	Master airway bill number
HAWB No.	House airway bill number
ULD No.	Unit load device number. It could be either pallet or container
Comment for handle with care	Information given by shipper or internal instruction of airline
Packing status and packing material	This is standard information to judge the state of perishable cargo
Location information	Information about the location of perishable cargo
Label information	Label information for the shape/types of perishable cargo
Temperature information	Temperature information for special cargo

After interviewing several airlines in Korea, Hong Kong, Singapore, and Germany, we have identified problems of current practices as follows:

1. Location of perishable cargo
   a. Perishable cargo is damaged due to mislocation (e.g., temperature-sensitive cargo could be located outside chill room).
   b. Currently identification is done manually and location information is not collected.
2. Packing container status management
   a. Improper temperature of cold container can make damages to the cargo and could be the reason for claims.

3. Environmental monitoring
   a. There is no monitoring system for special cargo warehouse (e.g., perishable environmental factors such as ethylene, temperature, and relative humidity [RH]).
4. Manual task
   a. Inspection is processed manually.
5. Log data on the environmental data
   a. Environmental data (e.g., temperature and humidity) is not collectable.
   b. Environmental factors (e.g., temperature and humidity) impact on quality of cargo.

8.4   AIDC-BASED SMART PERISHABLE CARGO MANAGEMENT

In this section, we introduce an automatic identification and data collection (AIDC) system–based smart cargo management system. The system can monitor the location of perishable cargo in real time and also collect environmental information using sensor networks. Since RFID and sensor can only collect raw data, the steps for information and knowledge creation are essential. In our research, we defined three levels of knowledge creation as in Figure 8.2. As in the figure, data represent unorganized and unprocessed facts (e.g., raw data from RFID and sensor such as tag ID and temperature data). Information means an aggregation of data (processed data) that makes decision making easier (e.g., finding location of perishable goods using the relative location of tag and reader). Knowledge is an understanding of a subject matter that has been acquired through proper study and experience. Knowledge is usually based on learning, thinking, and proper understanding of the problem area (e.g., some action or decision for perishable goods).

In order to monitor the location of perishable cargo, one should logically infer the location of cargo considering physical location information of reader/interrogator (e.g., we assume that the reader is in fixed location and tag is moving with perishable cargo).

Since the installation of AIDC does not provide the location information of perishable cargo directly, we did data mapping to create location information as in Table 8.4.

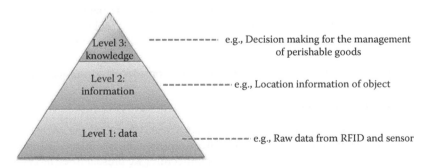

FIGURE 8.2   Layer for data, information, and knowledge.

TABLE 8.4
Tag to Cargo Data Mapping

Process	Tag Data–Cargo Data	Reader Data–Location Data
Receiving	Tag ID–MAWB No.	Reader Logical ID–physical location ID
Storage	Tag ID–MAWB No.	Reader Logical ID–physical location ID
Buildup	Tag ID–MAWB No.	Reader Logical ID–physical location ID
ETV	Tag ID–MAWB No.	Logical ID–physical location ID
Loading	Tag ID–MAWB No.	Logical ID–physical location ID

TABLE 8.5
Classification Temperature

Categories	Insulation (for Warm Retention)	Insulation (for Cold Retention)	Chill	Freeze
Temperature range	15°C–20°C	2°C–15°C	−9°C to −2°C	Below −10°C

In the table, "–" means data mapping between tag and cargo data (e.g., one can track MAWB number using tag ID). Tag ID stands for ID for tag that is attached to perishable cargo (by mapping MAWB number with Tag ID, it is possible to track information). By using the concept of reader logical ID, one can group readers to a logical reader and map logical reader location with physical location information.

The status of perishable cargo is influenced by environmental factors (e.g., temperature, humidity) at the specific location (where the specific cargo is). In this research, we surveyed temperature ranges of special cargo (perishable cargo in this case) and installed sensors to monitor temperature in real time.

By installing sensors at the important locations of all the perishable cargo process, we can collect data log to identify exceptions in environmental factors. Such information can be used as a proof against claim.

There are 380 types of perishable cargos and each has different storing temperature. In this research, we classified temperature ranges of perishable cargo into four categories based on IATA perishable cargo regulations as in Table 8.5.

By collecting log data for perishable cargo considering location and time, one can react efficiently against unexpected event and collect ground information for claims. Figure 8.3 shows the concept of log data collection considering location and time.

8.4.1 ARCHITECTURE FOR PERISHABLE CARGO MANAGEMENT

In this section, we suggest system architecture for AIDC-based perishable cargo management system. As a hardware structure, we present an integrated architecture of RFID and sensor networks. As software architecture, we present three layers such as data layer, control layer, and presentation layer.

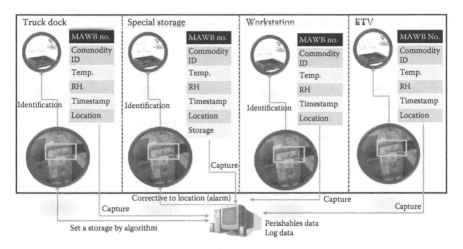

FIGURE 8.3   Concept of log data collection for perishable cargo considering location and time.

8.4.1.1   Hardware Architecture

We employed 900 MHz RFID (passive), 433 MHz RFID (active), and 2.45 GHz-based sensor networks for raw data collection. RFID is used for the identification of cargo and served as a tool for collecting data for specific cargo (Figure 8.4). Tracking of perishable cargo can be performed in two ways: airway bill level (AWB) and ULD level. In our study, we applied 900 MHz for tracking airway bill level and 433 MHz for tracking ULD level. In case of AWB, we applied disposable tag since AWB level cargo is not returning to the shipper, while applying reusable tag to ULD level. Sensor network is applied for sensing environmental information for storage area.

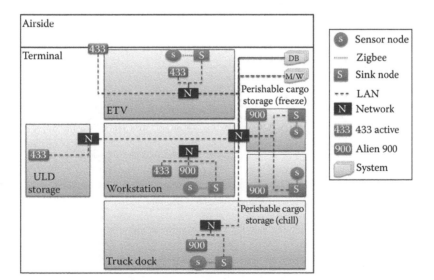

FIGURE 8.4   Hardware architecture for perishable cargo management at the test bed.

8.4.1.2   Software Architecture

There are three layers for software as in Figure 8.5: data layer, control layer, and presentation layer. Data layer is used to manage hardware (e.g., RFID and sensor network) and collecting data from hardware. Control layer manages data transferred from data layer, filters those data, transfers filtered data to presentation layer, and manages overall system performance. Presentation layer is an application layer that converts raw data to information and knowledge and provides decision support to the user.

Figure 8.6 shows example database structure for control layer.

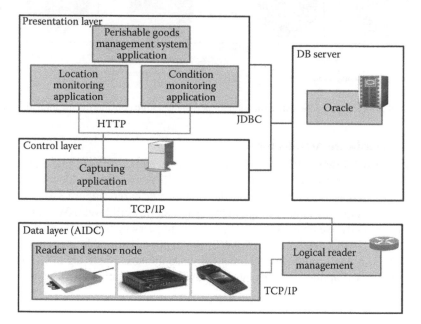

FIGURE 8.5   Software architecture.

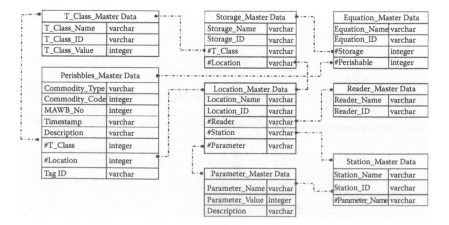

FIGURE 8.6   Data classification and mapping structure for control layer.

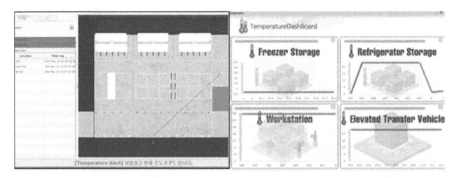

FIGURE 8.7   Location monitoring and temperature monitoring application.

8.4.2   REAL-TIME LOCATION/TEMPERATURE MONITORING AND ADVANCED WARNING

Figure 8.7 shows real-time location and temperature monitoring applications. As in the right figure, "dot" is cargo identified and "straight line" is trajectory of cargo. Instead of text-based information, we provide visible information for the convenience of user. Such location information of perishable goods is generated from data mapping from AIDC data and location information in real time. With the aid of this system, user can identify the location of perishable cargo in real time.

For storage monitoring, using sensor network, one can collect temperature data in real time. The right figure of Figure 8.7 shows monitoring data of temperatures. Using such data, one can identify environmental information on the storage. By mapping location data and storage information, we generated warning capability (we considered this level as knowledge level) as in the left figure of Figure 8.7. As in the figure, "dark gray" storage area has temperature problem (in this case, temperature was 7°C, which was outside the regulation, −9°C to −2°C). By generating management knowledge from temperature data, location data, and information for storage, the system makes decision on the management of perishable goods (e.g., actions to take). The system is also equipped with short messaging service (SMS) capability and e-mail to give alarming action when there is unexpected event.

8.5   CONCLUSIONS

Currently, there has been only procedural guideline for perishable cargo in aviation industry, even though huge benefit can be realized by adopting AIDC-based tracking system. In this research, we proposed AIDC-based perishable cargo management system. Recently, lots of RFID-based real-time data collection systems have been introduced. However, not many research has explained the relation between raw data, information, and knowledge in practice. The purpose of our research was to present a system that utilizes raw data, processes information from raw data, and creates knowledge from various data and information. It is considered that the proposed system could reduce damages of perishable cargo and prepare countermeasures against customer claims by keeping log data and by providing various actions such as SMS. Currently, the system was tested at the cargo terminal in Korea, but in order to maximize the benefit, it should be applied to full scope of supply chain, from the manufactures to the customer.

ACKNOWLEDGMENT

This research was supported by a grant (36-2007-C-Airport) from Development for the Intelligent Airport System Program funded by Ministry of Land, Transport, and Maritime Affairs of Korean government.

REFERENCES

Apaiah, R., Hendrix, E., Meerdink, G., and Linnemann, A. (2005). Qualitative methodology for efficient food chain design. *Trends in Food Science & Technology*, 16(5), 204–214.
Asimakopoulos, G., Louvros, S., and Triantafillou, V. (2007). Technol. Educ. Inst. of Mesolongi, Nafpaktos, METATRO: A real time RFID enabled haulage monitoring system for perishable comestibles. *IEEE Conference on Emerging Technologies and Factory Automation*, September 25–28, 2007, Patras, Greece, pp. 456–459.
Beulens, A. J. M., Broens, D. F., Folstar, P., and Hofstede, G. J. (2005). Food safety and transparency in food chains and networks. *Food Control*, 16(6), 481–486.
Brusey, J. and Thorne, A. (2006). Aero-ID sensor integration: Scope of work, AEROID-CAM-003.
Folinas, D., Manikas, I., and Manos, B. (2006). Traceability data management for food chains. *British Food Journal*, 108(8), 622–633.
IATA. (2009). *IATA Perishable Cargo Regulations*, 9th edition (English). Montreal, Quebec, Canada: IATA.
Jedermanna, R., Ruiz-Garciab, L., and Langa, W. (2008). Spatial temperature profiling by semi-passive RFID loggers for perishable food transportation. *Computers and Electronics in Agriculture*, 65(2), 145–154.
Wilson, T. P. and Clarke, W. R. (1998). Insights from industry food safety and traceability in the agricultural supply chain: Using the Internet to deliver traceability. *Supply Chain Management*, 3(3), 127–133.

9 Quality Analysis of User-Generated Content on the Web

Jong C. Park and Hye-Jin Min

CONTENTS

9.1   INTRODUCTION

With the development of Web 2.0 technologies, public organizations regularly provide news, literature, or articles but at the same time common users also actively post and share their opinions or daily episodes through blogs, wikis, or other social media on the web. Thanks to the continuous growth of the volume of such user-generated content, users can easily access a wider variety of information nearly on any topic. For example, if we want to learn how to cook a particular dish, such as "Korean BBQ," we can check out information about its numerous recipes on the web generated by other users. Such information is usually provided in diverse formats chosen by its authors. For instance, the authors may post recipes with photos they have taken. They may also post video so that the viewers may grasp the whole cooking process better because the video shows every step of it both in further detail and in a seamless way. User-generated content of this kind has the following benefits. First, since the authors usually generate the contents from the reader's (or viewer's) perspective as much as possible, the contents can fully address the points that the readers may have most difficulty in understanding. Second, the publishing time is comparatively quite short, so that the readers may get the most recent and relevant information. Third, the readers may receive feedback most readily from the authors via feedback sharing tools such as RSS or push notification services.

However, user-generated content of this kind has its limitations, too. First, due to the vast amount of data produced every day, it is extremely hard for the readers to keep track of all the available data. Second, due to the lack of an editorial control, the quality of the contents may become easily unreliable. One approach to addressing these limitations would be to customize retrieved information to the needs of specific users, taking full advantage of the knowledge acquired from the analysis of the user's navigational behavior (usage data) in relation to other pieces of information in the web context, which is called "web personalization." Another approach would be to ensure quality assessment of the contents, again taking full advantage of the data acquired from the analysis of the contents and their metadata.

The quality of the content can be assessed by directly analyzing the content. The result of such assessment would be more trustworthy than that of the metadata, though the modeling of the content is harder than that of metadata, due, among others, to the diverse formats of the content, ranging from short or long text to multimedia data such as photo or video. On the other hand, the analysis of the metadata of the content would be easier because most types of data are represented as fixed structures such as numerical values or predictable textual data. One of the common examples of metadata is a user's profile, which is provided in most of the websites these days. For example, the world's largest user-generated content video system, "YouTube," provides the user's profile for a given video, including the total number of views, the number of subscribers, and the user's personal information, such as the name, joined date, or age. The quality can be detected to a degree by analyzing such numerical or fixed text data with an underlying assumption that information of this kind indirectly indicates the user's popularity, trustworthiness, or similarity to certain other groups of users.

In spite of the usefulness of the content and metadata for quality assessment, some issues must still be addressed and resolved. First, the notion of quality is dependent on the characteristics of the content, so that we need to define the quality considering the characteristics of the content and users and the measure for it. Second, the measured quality can be quite varied, depending in particular on the measurers or the method of measurement, so that the gold standard about the quality should be determined before the assessment is made.

In this chapter, we review recent work on analyzing and providing user-generated content in an effective way, focusing on "quality detection." First, we explain the types of content whose quality detection is important and introduce the related proposals to measure such quality in Section 9.2. We then describe existing approaches to quality detection with respect to the types of feature as extracted from the contents and their metadata in Section 9.3. We also present several applications using the contents with their detected qualities in Section 9.4. Finally, we discuss the remaining issues in Section 9.5 and conclude this chapter in Section 9.6.

9.2 QUALITY ANALYSIS OF USER-GENERATED CONTENT

Various content sharing applications such as Flickr for images, del.icio.us for bookmarks, YouTube for video, and a lot of software for blogs have contributed to the rapid growth of the popularity of Web 2.0 content. Most of the content sharing applications provide several social tools for users to interact with one another by sharing their opinions or feelings on the content. For example, YouTube provides an interface for the readers/viewers to leave comments on the published videos. In addition, it provides a tool for the users to give a rating about the comments. These ratings may in turn serve as a measure for assessing and filtering relevant opinions more effectively or for skipping offensive or inappropriate comments. However, such ratings furnished by the general public are mostly effective on contents or comments that have been published over a certain period, so that newly published contents or comments may be easily overlooked, despite the potentially useful information or helpful comments that the latter may have.

For instance, consider the topic "how to cook bulgogi (Korean marinated beef)." We can retrieve about 68 search results as of March 2011 for the user-generated video contents about the topic on "YouTube." By ordering such results on popularity, it is easy to find contents that are considered useful by the public. Figure 9.1 shows one of the most popular contents. One of the noticeable strengths of this content is that its author is a native Korean, so that her recipe comes quite close to the traditional Korean dish. However, most of the popular contents tend to be 2 or 3 years old, and are rather dated by the present standard. Ordering reviews by popularity would thus make it fairly hard for naïve readers to locate newly uploaded videos such as the one shown in Figure 9.2. While this newer video is not yet widely viewed, it has the following attractive points. First, the authors use both Korean and English, making the video more accessible to the general public in the world. Second, the structure of the content in this video is similar to that of a talk show, making it more fun to view. As such, we can anticipate that there will be rapidly growing viewers for this type of

FIGURE 9.1   One of the most popular contents (uploaded 2 years ago): This screen shot was captured from the video named "Bulgogi Grilled Marinated Beef."

FIGURE 9.2   Comparatively not so popular content (uploaded 1 month ago): This screen shot was captured from the video named "Bulgogi with Rola, Khloe, and Justin."

video that has not yet become so popular. These two types of content, popular but old and new but attractive, have both their own trade-offs, and thus may not be ordered easily. However, the current ordering by measures such as "popularity" alone would surely provide biased results and may not satisfy the needs of diverse viewers.

To overcome these biased results, studies on assessing and predicting the quality of the contents have been conducted extensively of late. Table 9.1 shows an analysis of some of the diverse types of user-generated content.

TABLE 9.1
Types of User-Generated Content for Quality Analysis

Type of Content	Sources of the Content	Underlying Measure for the Quality
Image data	Image sharing websites (e.g., Flickr)	Attractiveness
Podcasting	Podcast providers (e.g., CNN and BBC radio)	Popularity
Comments on video data	Video sharing websites (e.g., YouTube)	Acceptability
Online reviews	Online shopping websites (e.g., Amazon.com, eBay, and CNET)	Helpfulness/usefulness
Forum data	Online web forum sites (e.g., Nable.com and Google groups)	Informativeness
Blog posts	Aggregated blogs on the web	Credibility

These types of data are frequently uploaded and updated, so that the readers/viewers may have a hard time navigating over a large amount of content repositories and finding the content with high quality. The underlying measures for the quality are varied among conducted studies, especially with respect to the types of content and the main purpose for quality assessment. We examine in the next section the methods to define the quality of the content and the types of measure that are utilized.

9.2.1 Gold Standard for the Quality of the Content

The quality of the content may be formulated as a collection of some important properties of the content, whose types may in turn work to discriminate high-quality contents from low-quality ones. The most representative property may be dependent on the types and utilities of the content as shown in Table 9.1. For example, the image data on the web is usually utilized for decorating the websites or enhancing the understandability of certain concepts, so that the more attractive the image data is, the more attention people will pay to the target website or concept. Through reviews about items in online shopping sites, potential customers acquire useful pieces of advice or guidelines for the purchase and/or use of the target item, making helpfulness or usefulness as the most important quality of such reviews. As for blog posts, readers may prefer credible posts when they search for information about a particular topic, making credibility of the bloggers as the most important measure for quality.

In order to assess the quality of the content automatically, we would thus need a gold standard for the quality for the given content. Such gold standards as proposed in the literature can be classified into two types, based on how they are constructed: (1) crowd vote–based metric and (2) gold standard model of the content. They are examined further in the following.

9.2.1.1 Crowd Vote–Based Metric

Most online user-generated content sharing applications or websites provide a social interface for the users to comment on or rate the content. The summary of ratings by users is given as the number of counts with graphical image buttons (e.g., 10 with "Like" or "Thumbs up" button on YouTube) or a short description (e.g., "4 of 4 people found the following review helpful" on Amazon.com). Since such ratings are normally given by users who share the rationale for the service, it may be regarded as a reliable and practical method of assessing the quality of reviews, though the outcome may still depend on the nature of the contents.

Due to the difference in the perceived nature of the contents, there are several proposals for estimating functions that incorporate the votes by users. For instance, Kim et al. (2006) defined a simple helpfulness function f for a given review r based on the number of helpful votes (*No. of Helpful Rating*) and unhelpful votes (*No. of UnHelpful Rating*) from Amazon.com in order to identify the gold standard for the quality of online reviews, as shown in Equation 9.1:

$$f\left(r \in R\right) = \frac{\text{No. of Helpful Rating}\,(r)}{\text{No. of Helpful Rating}\,(r) + \text{No. of UnHelpful Rating}\,(r)} \quad (9.1)$$

O'Mahony and Smyth (2009) defined another helpfulness function for online reviews based on the amount of positive feedback that they received from others, where a review is considered helpful if and only if >75% of the feedback is positive. Pedro and Siersdorfer (2009) defined an attractiveness function for image data in Flickr, taking into account the number of times a given photo was assigned to the favorite lists of the viewed users. They explained the appropriateness of such a function by a correlation analysis with information in other community feedback such as the number of views/comments. Ghose and Ipeirotis (2007) considered other characteristics to estimate the helpfulness of product reviews, such as *Elapsed Date*, the difference between the collection date of the reviews and the release date of the product, *Readability*, the readability of the reviews, and *AvgProb* and *DevProb* for the probabilities of the subjectivity of reviews. They also included a variable called MODERATE to encode the intuition that consumers are more likely to post reviews with extremely positive or extremely negative ratings than those with moderate ratings. This intuition comes in turn from the observation that extremely positive or negative experiences on a product are more likely to encourage the interpersonal communication behavior (Dellarocas et al. 2007).

9.2.1.2   Gold Standard Model of the Content

Estimating functions based on the number of votes by the crowd may be regarded as reflecting the opinions of the larger general public and thus reliable and practical, but some researchers came up with gold standard models of the content by looking more closely into the nature of the content and the purpose of the quality estimation. The motivation is that crowd vote–based estimating functions suffer from various biases, some of them are shown in the following (Liu et al. 2007):

- *Imbalance vote bias*: It is observed that users are more likely to evaluate opinions of others positively than negatively. For some reviews, the number of helpful votes does not necessarily reflect the quality of the review.

For instance, one of the reviews received 40 *helpful* votes out of 40 votes though it contains only a rough description of the product features, thus with questionable quality.

- *Winner circle bias*: The idea is that the more votes a review receives, the higher credibility it earns from the readers or potential customers. In addition, reviews ranked high by the number of votes that are taken until the moment of browsing them may gain more votes. To substantiate this idea, Liu and colleagues collected top 50 reviews about digital cameras ranked by the number of "helpful" votes at the Amazon.com site. The reviews are then sorted by a descending order of the number of "helpful" votes. The reviews with top first and second ranks received 250 and 140 votes, respectively, while most of the rest have no more than 50 votes. The numbers of votes decreased exponentially.
- *Early bird bias*: The idea is that the publication date may affect the outcome of the votes. According to their observation, the earlier a review is posted about a given product, the more votes the review will receive.

To prevent biases of these kinds, some guidelines for the gold standard model of the content with high quality have been proposed. Liu et al. (2007) defined four categories of review quality: best, good, fair, and bad. The key properties of such classification are the coverage of (interesting and important) aspects of products, an appropriate amount of evidence for the opinions, and a good format for the content. Similarly, Hoang et al. (2008) considered three levels of content quality for both reviews and question-answering data: good, fair, and bad. In addition, Lu et al. (2010) proposed a gold standard for the review quality with a real value rating between 0 and 5 in order to deal with a wider range of the quality values. Tsur and Rappoport (2009) created "virtual core reviews" as a gold standard for the quality of book reviews. They defined a "virtual core review" as one that contains the most prominent words related to the given book's story or comments on important aspects (e.g., genre and author). Such words appear infrequently but are nevertheless significant.

After establishing the guidelines, human evaluators are asked to manually give an appropriate rating or assign a proper category to each content following the guidelines. Since such guidelines may be interpreted subjectively by each evaluator, the annotation is usually done by several human evaluators. After the individual annotation, we need to see if the annotation is made consistently among the annotators and to utilize only agreed results for further observation or experiment. There are several standard methods to combine the annotated results. One is simply to take some portion of the content to which all the annotators assign the same rating or category. Another method is to measure the inter-annotator agreement. The Kappa statistics (Cohen 1960) is widely used for measuring such an agreement. The Kappa value ranges from 0 to 1, where the classes "almost perfect agreement" (0.81–1.0) and "substantial agreement" (0.61–0.80) are considered acceptable, though such judgment depends also on the characteristics of the domain for the annotation data. For example, the Kappa value of the agreement in Liu et al. (2007) was 0.8142, showing high consistency among the annotation results.

9.2.2 EVALUATION STRATEGY

The evaluation strategies for the quality assessment with respect to the gold standard are divided into two types: (1) comparison of the results with the gold standard and (2) binary classification of the content either as "high quality" or as "low quality" regarding the evaluation as a classification problem.

9.2.2.1 Comparison of the Results with the Gold Standard

Strategies to compare the results of the assessment with the gold standard can be further divided into two types. One is a ranking-based correlation analysis. The proposed methods of this type assign a rank to each content based on the regression model and compare the ranks with that of the gold standard by a correlation analysis, for which Spearman's rank order correlation coefficients are widely utilized (Kim et al. 2006; Zhang and Varadarajan 2006; Huang et al. 2009). Another correlation measure, called Kendall's Tau-b, is also utilized when identifying the correlation between two ordinal variables, where one is an independent variable and the other a dependent variable. For example, Pedro and Siersdorfer (2009) used it in order to compare the automatically generated order by their method with the order of the ground truth.

In order to assign ranks to the content, some researches approximate the existing ranking algorithm based on the quality measure (e.g., helpfulness). For example, Tsur and Rappoport (2009) approximated the helpful vote-based ranking employed by Amazon.com. One of their approximated ranking rules is to rank the reviews that receive more than 15 votes, 0.5 higher than the others, so that their helpful/total ratio (the same as Equation 9.1) comes out slightly better.

Methods of the other type measure the performance of the proposed method by average evaluation scores by comparing the resulting quality to that of the gold standard. One of the commonly used scores is mean squared error (MSE), which measures how much the predicted quality deviates from the annotated quality of as gold standard, where a smaller value indicates a more accurate prediction. Some researchers cross-compared the predicted quality by making several balance sets consisting of selected data from different measures. For example, Tsur and Rappoport (2009) compared the ranking result of their algorithm with the ranking results based both on user votes and on randomly chosen publication dates.

9.2.2.2 Binary Classification

Some researches transformed the evaluation process into a binary classification problem, by reclassifying the content into two categories (e.g., "helpful" and "unhelpful" or "high quality" and "low quality"). If the number of originally classified categories is bigger than two, they regarded intermediate categories (e.g., "good" or "fair" quality) as positive (e.g., high quality). Table 9.2 shows several classifiers that are utilized in various domains. In particular, the most extensively utilized among the state-of-the-art classifiers is based on Support Vector Machine (SVM) (Liu et al. 2007; Weimer et al. 2007; Hoang et al. 2008; Pedro and Siersdorfer 2009; Siersdorfer et al. 2010). Note that unlike other work, Hoang et al. (2008) chose Maximum Entropy (MaxEnt) for training a classifier in order

TABLE 9.2
Utilized Classifiers[a]

Work	Domain	# of Categories	# of Feature Categories/# of Features	Classifier	Kernel
Liu et al. (2007)	Product reviews from Amazon.com	2 (High quality/low quality)	3/19	SVM	Linear
Weimer et al. (2007)	Forum discussions about "software"	2 (Good/bad)	5/12	SVM	Gaussian RBF kernel
Agichtein et al. (2008)	Question-answering data from Yahoo Answers	2 (High/ normal + low)	4/20 (questions); 3/20 (answers)	C4.5 decision tree classifier	No kernel
Hoang et al. (2008)	Product reviews and Q/A data	2 (Good + fair/bad)	4/14	Maximum Entropy	No kernel
Tsagkias et al. (2009)	Podcast feeds	2 (Popular/ nonpopular)	4/20	SVM, naïve Bayes, J48	
Wanas et al. (2008)	Online discussion posts	3 (High/ medium/low)	5/17	SVM	RBF kernel
O'Mahony and Smyth (2009)	Hotel reviews from TripAdvisor	2 (Helpful/not helpful)	4/26	JRip, J48, naïve Bayes	No kernel
Pedro and Siersdorfer (2009)	Photos in Flickr	2 (Attractive/ unattractive)	3/10	SVM	Linear
Siersdorfer et al. (2010)	YouTube Comments	2 (Accepted/ not accepted)	1/N terms	SVM	Linear

[a] The SVM classifiers are implemented by several toolkits such as SVM*light* (Joachims 1999) or LIBSVM (Chang and Lin 2011) with various kernel functions such as linear or RBF kernels. Decision tree-based classifiers such as C4.5 or J48 and propositional rule learner, Repeated Incremental Pruning (Rip), such as JRip, are mostly implemented by the Weka data mining software (Hall et al. 2009). In most cases, the SVM classifiers or tree-based classifiers outperform the baseline classifiers such as naïve Bayes classifiers.

to integrate various features expressed as functions, as MaxEnt can be used to estimate the expected value for each feature from training data and treat the value as a constraint on the model distribution.

9.3   APPROACHES TO ASSESSING THE QUALITY OF THE CONTENT

The work can be divided into three categories, depending on the data utilized for estimating the quality.

9.3.1 Metadata Analysis–Based Approaches

Most of the popular online websites these days provide metadata of user-generated content. One of the common examples of such metadata is a user's profile. For example, the world's largest user-generated content video system "YouTube" provides the user's profile for the given video as shown in Table 9.3. It includes the total number of views of the given content, the total number of views of the total upload content, the number of subscribers, the user's personal information such as name, joined date, age, last visit date, and friends, and comments from other users.

Such metadata are also accessible via the open application program interfaces (APIs) provided by the websites. For example, Amazon.com allows users to access its database through Amazon Web Services (AWS) API.* For customer reviews, they provide the average rating and total number of reviews, and for each customer review, they provide the individual rating, helpful vote, total number of votes, reviewer information (e.g., customer ID, location, interesting people, and friends), and publication date. Table 9.4 shows available open APIs provided in many websites.†

While some types of data directly related to such measures as the numbers of votes by other users (e.g., *helpful* vote or *like*) are utilized for estimating the true quality as the gold standard, other types of data indirectly related to the contents are also utilized for incorporating useful features that capture the quality of the content more reliably. Among them, reviewer information is utilized for incorporating personal reputation and credibility of the author. Huang et al. (2009) utilized three features to assess author's behaviors with respect to credibility and expertise: personal reputation, seller degree, and expertise degree on eBay site. For personal reputation, they used feedback information from other readers. They also utilized the author's roles such as seller or buyer in an e-commerce transaction since users take part in transactions both as a seller and a buyer on eBay site. They computed the seller degree of each author that indicates how much each author takes part in all

---

TABLE 9.3
User's Profile of "YouTube"

Subcategories of User's Profile	Description
Name	User name
Channel views	Number of views of all the contents in the particular channel
Total upload views	Total number of views of all the uploaded contents
Joined	Joined date for YouTube
Website	User's website
Subscribers	Other users who subscribe to the user's content
Friends	Other users who register the user as a friend

---

* http://docs.amazonwebservices.com
† For more APIs, the reader is referred to an open APIs portal website, such as http://www.programmableweb.com/apis/

TABLE 9.1

Open APIs

Category	Description	Example APIs
Answers	Community-driven reference service	Yahoo Answers
	SMS question and answer services	ChaCha
Blog Search	Blog search services	Technorati, PubSub, Blogwise
	Blog promotion tracking service	FeedBurner
Mapping	Mapping service	Google Maps, Yahoo Maps
Photos	Photo sharing service	Flickr, Panoramio, Google Picasa
Social	Microblogging	Twitter
	Social networking service	Facebook, LinkedIn, MySpace
	Location aware social network	Foursquare, Gowalla
Shopping	Product marketplace service	Amazon eCommerce, eBay, Shopping.com
	Review service	Allogarage, The Chicken, QuarkRank
	Social e-commerce service	Tinypay.me
	Online shopping cart service	3dCart, Big Cartel, Google Checkout
	Comparison shopping service	DataFeedFile.com
News	Community-driven news links and ratings	Digg
Video	Video sharing and search	YouTube, AOL Video, Yahoo Video Search

transactions as a seller. They defined $t–r$ score in order to calculate the degree of each role. The score indicates relevance between a transaction and a review based on their similarity by the cosine function with TF–IDF score of the titles of the reviews and the transactions. The computing method works to calculate the ratio of the $t–r$ score as the seller role to the $t–r$ score as both the seller and the buyer roles of the author in all the transactions resulting from the reviews of the author. The degree ranges from 0.0 (100% buyer) to 1.0 (100% seller). As for the degree of expertise, they approximated it by the $t–r$ score of all transactions and reviews of the author. Their assumption is that users with more experiences are likely to post reviews with more professional backgrounds. They found that personal reputation with feedback score and degree of expertise correlates positively with the percentage of "helpful" votes while the seller degree correlates negatively with it. Similarly, Hoang et al. (2008) adopted the notion of authority feature, which indicates whether or not the content is written by a credible author. They represented it by using the number of contents previously written by the same author, and the number of votes or scores received by other readers. O'Mahony and Smyth (2009) also considered the reputation of the author by estimating it from the helpfulness of all the (hotel) reviews by the author. They also included the mean and standard deviations of the number of reviews written by all authors, and the mean and standard deviations of the number of reviews about all hotels.

In addition to authors' reputation, Lu et al. (2010) observed that the quality of an author depends also on the quality of his or her peers in the social network and utilized authors' identities and social context for improving the review quality prediction. For a social context, they considered both in-degree and out-degree of the author and PageRank score of the author employed from Ciao UK, a community review website. The website provides a social network tool so that users can add trusted members to their network called "Circle of Trust," if they think that their reviews are consistently helpful. They also incorporated some constraints based on four hypotheses about how authors behave individually or within the social network. These include author consistency hypothesis, trust consistency hypothesis, co-citation consistency hypothesis, and link consistency hypothesis. They argued that such social context information improves the accuracy of quality prediction when the training data is sparse.

While metadata for text-based contents are represented as numerical values, metadata for non-textual contents are often represented as text or terms. For example, tags provided for images are regarded as metadata. Pedro and Siersorfer (2009) ranked the tags for image data by utilizing mutual information (MI) based on information theory (Yang and Pedersen 1997) as a measure of independence of tags (terms) and categories ("attractiveness" and "unattractiveness"). Their results show that tags related to nature such as sunset or animals, tags related to photography such as Canon or Nikon, artistic, or color-related terms, and positive terms are all connected to attractiveness. On the other hand, terms related to family or private event such as birthday, party, or camping are found connected to unattractiveness.

Quality assessment based on these metadata is more robust than content analysis–based assessment, because content analysis depends heavily on explicitly mentioned information in the content and the accuracy of extracting features is comparatively lower than that of metadata. On the other hand, metadata are accumulated continuously and mostly structured, so that the assessment results become more consistent. However, there is no corresponding information for newly posted content, which leads again to the data sparseness problem. The hybrid approaches that take advantage of extracted information from both the metadata and the content itself are proposed to overcome these limitations. The performance of these hybrid approaches is also higher than that of utilizing one type of data alone. We discuss further details in Section 9.3.3.

9.3.2 CONTENT ANALYSIS–BASED APPROACHES

Researches on detecting the quality of the user-generated content based on the information from the content itself have also been conducted with various criteria. In this section, we examine two types of user-generated content: non-textual data including images, podcasts, and video data and textual data.

9.3.2.1 Non-Textual Data: Images, Podcasts, and Video Data

Pedro and Siersdorfer (2009) utilized visual features from the image data in order to predict the attractiveness of image data. They examined the collected images in Flickr and found that, even though two images have the same semantic concepts, their attractiveness can be different mainly because of their different artistic values. However, such artistic values are hard to quantify. Instead, they assumed that higher colorfulness and increased contrast and sharpness make images more attractive, based on a widely accepted thesis on human's perception of images. With the assumption, they extracted two major features along with their attributes as follows:

- Color: brightness, saturation, colorfulness, naturalness, contrast
- Coarseness (the degree of detail contained in an image): sharpness

They determined the value of each subfeature by using the well-known computing method for each feature, as shown in Table 9.5.

Tsagkias et al. (2009) examined a list of indicators of user-preferred podcasts in the PodCred analysis framework, a framework for assessing the credibility and quality of podcasts released over the Internet in order to predict the podcast preference. The list of indicators consists of four categories such as *podcast content*, *podcaster*, *podcast context*, and *technical execution* with the following subcategories:

- *Podcast content*: topic podcasts, topic guests, opinions, cite sources, etc.
- *Podcaster*: fluency, presence of hesitations, fast speech speed, credentials, personal experiences, etc.
- *Podcast context*: "podcast addresses listeners," "episodes receive many comments," etc.
- *Technical execution*: opening jingle, background music, background noise, simple domain name, feed-level metadata, episode-level metadata, high-quality audio, "feed has a logo," associated images, "logo links to podcast portal," etc.

For classification and ranking, they focused on the category *technical execution* and selected four indicators, feed-level metadata, "feed has a logo," and "logo links to

---

TABLE 9.5
Description and the Computing Method of Color and Coarseness Features

Subfeatures	Description	Computing Method
Brightness	Measure of the amplitude of the light wave	Computing the average of the luminance values
Saturation	Measure of the vividness	Computing the difference of the intensities of different light wavelengths
Colorfulness	Measure of its difference against gray	Computing the colorfulness index using the distribution of chroma values
Naturalness	Measure of the degree of correspondence between images and human perception of reality	Computing the naturalness indexes (Huang et al. 2006)
Contrast	Relative variation of luminance	Utilizing the commonly used RMS contrast
Sharpness	Clarity and level of detail of an image	Utilizing the Laplacian function and normalizing it by the local average luminance in the surroundings of each pixel

podcast portal," because they are easy to extract and well represent surface characteristics of podcasts. They also included *regularity* in order to incorporate a temporal pattern for the release of a podcast, and *podcast episode length* in order to identify the correlation with preferred podcasts.

Siersdofer et al. (2010) analyzed comments and comment-rating behaviors from YouTube. They utilized the sentiment of terms in each comment as key features in order to predict "acceptability" for comments. For this purpose, they selected the top 2000 terms based on the MI measure from information theory and computed their sentiment values with the triples (positive, negative, and objective) from the sentiment lexicon, SentiWordNet (Esuli and Sebastiani 2006).

9.3.2.2   Textual Data

Most of the text data that have been analyzed so far contain opinions, experiences, or emotions of authors, such as online reviews or forum data for a specific topic, because such text data are subjective and the quality among the content may vary. In addition, spams and advertisements are also included since filtering such data is also regarded as an important task.

As for one approach, feature-based classification, the most frequently utilized features extracted from text data are as follows.

9.3.2.2.1   Domain-Independent Features

- *Structural/surface feature*: This feature category refers to the format or the writing style in the target text. Widely used features are the number of sentences in the text, the average length of sentences and the whole text, the number of words in the text, the percentage of sentences ending with particular marks such as "?" or "!", the HTML formatting tags such as bold tags, and the percentage of words in capital letters, which are often related to shouting.
- *Syntactic feature*: This feature refers to the linguistic properties of the text. Commonly used features are part-of-speech tags of each word; the percentage of parsed tokens for particular classes such as nouns, verbs, adjectives, or adverbs; and verbs conjugated in the first person.
- *Lexical feature*: This feature captures the words observed in the text. Some examples are frequency of spelling errors or swear words and *n*-gram-based word statistics (e.g., the TF–IDF statistic of each word occurring in the text).

9.3.2.2.2   Domain-Specific Features

- *Lexical/surface feature*: The utility of some lexical features is dependent on the domain of the text. For example, Liu et al. (2007) utilized the number of products and brand names in the title and the body of the review for review quality detection. Weimer et al. (2007) utilized forum-specific word-level features that are only present in forum postings. For instance, they considered the part of the text that are inside quotes of other posts (*Quote Fraction*) or the number of URLs and the paths of the file system, assuming that the quality of post in a software domain may be affected by the amount of practically usable information such as demos, or source codes. Wanas et al. (2008) also considered forum-specific features such as referencing features including quotation and replies, surface features such as timeliness or format of punctuation, and posting-component features such as web links and their quality and question-answering threads.

- *Readability feature*: This feature captures clues in the text with respect to readability, which is determined by the inherent characteristics of the domain text. For example, product reviews contain pros/cons, and some guidelines, so cue words relevant to this kind of information contribute to the readability. In addition, cue words (e.g., "The first characteristic is ...") or HTML tags (e.g., <br>) for separating paragraphs may also be useful.
- *Relevance feature*: Wanas et al. (2008) considered relevance as one of the most important guidelines in recognizing a given post for the purpose of scoring online discussion posts. In order to use relevance, they extracted topic words from the forum content and leading post in a thread.
- *Subjectivity feature*: Reviews or forum data also contain positive or negative opinions on the target product or topic. Compared to objective facts, such information can be more practically helpful for potential customers or experienced customers. Some of the example features are as shown in Table 9.6.

Another approach to content analysis of the text data is to identify dominant terms that capture the gold standard model of the content. Tsur and Rappoport (2009) proposed an algorithm called RevRank for automatically ranking user-generated book reviews. The algorithm calculates the distance between a feature vector for the given review and an optimal vector for the gold standard review based on the dominance of the terms with an external balanced corpus. The method is fully unsupervised, so neither labor-intensive nor error-prone training is necessary.

Some researchers performed feature selection in order to examine effective features among feature candidates for quality assessment. O'Mahony and Smyth (2009) examined how important each type of feature is relative to others by ranking

---

TABLE 9.6
Examples of Subjectivity Features

Type	Example Features
Word level	Positive and negative sentiment words in the text acquired from the sentiment dictionaries (e.g., SentiWordNet[a] and General Inquirer[b])
Product feature level	The percentage of positive, negative, and subjective product features
Sentence level	The percentage of positive, negative, and subjective sentences, the percentage of comparative sentences

[a]   http://sentiwordnet.isti.cnr.it/ (Esuli and Sebastiani 2006).
[b]   http://www.wjh.harvard.edu/~inquirer/homecat.htm (Stone et al. 1966).

the features based on information gain (Reza 1961). They reported that the reputation features were the most significant and that the next significant one was the sentiment features. Content features reflecting the completeness of text and social features reflecting the degree of the user-hotel review graph did not have a high impact on determining the helpfulness of reviews.

9.3.3 Hybrid Approaches

Hybrid approaches that utilize extracted information from both metadata and the content overcome the limitations resulting from the characteristics of each type of source data. In fact, many researchers have extracted and utilized both metadata-based features and content-based features in their work, though they weighed the relative importance of each type of data against the other type differently. Some researches regarded one type of feature as the baseline of their experiment. Hoang et al. (2008) considered "authority" as a set of features for the baseline method and added additional features from the analysis of the review text or question-answering data such as formality, readability, and subjectivity. They reported that the average precision score with only "authority" features is 0.76 and 0.91 for review text and question-answering data, respectively, but the performance is enhanced up to 0.96 and 0.97, respectively, with the help of other textual features. The most helpful feature is subjectivity, while readability made no contribution.

On the other hand, Lu et al. (2010) regarded textual features such as structural, syntactic, lexical, and sentiment features as key features for the baseline method. They reported that the MSE is reduced as they considered additional social context and constraint features. They found that the feature set consisting of textual features, social context features, and reviewer consistency features is the most effective, based on the results of the experiments with different product categories (cell phone, beauty products, and digital cameras). Interestingly, they also reported that simple metadata feature sets consisting of only social context features performed poorly and did not outperform the text-only feature-based method. They argue that both metadata and context-based features should be incorporated to enhance the quality estimation.

Some work has other interesting results. Ghose and Ipeirotis (2007) proposed a ranking mechanism based on the subjectivity of review contents, sales rank of the product, and the ratio of helpful votes by utilizing both an existing text mining technique and economic methods. Pedro and Siersdorfer (2009) utilized both visual features as content features and text features as metadata features in order to classify "attractive" photos and ranking photos by "attractiveness." They found that the combined features (visual + text) are the most effective. However, if only one kind of features is allowed, visual features would produce better results than text features. Hence visual features can still be useful when textual features are not available yet.

Table 9.7 summarizes several hybrid approaches we have discussed in this section.

TABLE 9.7
Summary of Hybrid Approaches

Work	Domain	Metadata Feature	Content Feature	Performance
Kim et al. (2006)	Product reviews from Amazon.com	Stars (STR)	Structural (e.g., HTML tags, punctuation, and review length), lexical (e.g., n-grams), syntactic (e.g., percentage of verbs and nouns), semantic (e.g., product feature mentions), and metadata (e.g., star rating)	The most informative features are the length of the review, unigrams, and product ratings
Ghose and Ipeirotis (2007)	Product reviews from Amazon.com	Sales data (e.g., price, rating, number of reviews, and the readability)	Subjectivity	Subjectivity analysis can give useful clues to the helpfulness of a review and to its impact on sales
Hoang et al. (2008)	Product reviews from Amazon.com and Answers in community-driven Q&A from Naver's knowledge search in Korea (kin.naver.com)	Authority (indicating the author's trustworthiness)	Formality (the writing style of target document), readability (format of the document), and subjectivity (opinions of authors)	Formality is the most effective feature category; subjectivity features improve the performance on product review data
O'Mahony and Smyth (2009)	Hotel reviews	Reputation features, social features, and sentiment scores assigned by users	Terms, ratio of uppercase and lowercase characters in review text, and review completeness	Best performance with all four types of features; among single features, reputation features are the best
Pedro and Siersdorfer (2009)	Images in Flickr	Terms from textual annotation (a ranked list of tags from a set of 12,000 photos)	Visual features of image (brightness, colorfulness, naturalness, contrast, and sharpness)	The combination vectors obtained from textual and visual features (text + visual) provide the best ranking performance
Lu et al. (2010)	Online reviews from Ciao, United Kingdom	Number of reviews by the author, average rating for the author, in-degree of the author, out-degree of the author, and PageRank score of the author	Text-statistics features, syntactic features, conformity features, and sentiment features	Performance is better when incorporating social context

9.4 APPLICATIONS OF QUALITY ASSESSMENT OF THE CONTENT

In this section, we introduce several applications that use the contents with assessed quality.

9.4.1 FILTERING THE CONTENTS WITH LOW QUALITY: ONLINE DISCUSSION POSTS

Wanas et al. (2008) proposed several potential applications by utilizing automatic scoring of online discussion posts. First, the scoring method could provide high-quality discussion posts as an alternative way when manual ratings by humans are not available. Second, a set of key words related to forum-specific features could be used for browsing the posts in a subforum. Third, it can also serve as a means for encouraging users to participate and engage in discussions.

9.4.2 SUMMARIZATION OF THE CONTENT: PRODUCT REVIEWS (OPINION SUMMARIZATION OF REVIEWS)

Liu et al. (2007) proposed a review summarization system that embeds a process of low-quality review detection. The system filtered the reviews with low quality and used those with high quality for the review summaries. They visually compared the results without a filtering process (called a "one-stage" approach) with those embedding the filtering process (called a "two-stage" approach). They reported that the two-stage approach contains fewer text segments and argued that such fewer segments result from filtering out low-quality reviews. In addition, they compared the results from the two approaches with a professional "editor's review" from CNET focusing on one of the product features or "image quality." The results show that the approach of filtering out low-quality reviews produces results that are closer to the gold standard result from CNET.

9.4.3 CONTENT RECOMMENDATION I: POPULAR PODCASTS

The PodCred is a framework for assessing the credibility and quality of podcasts released over the Internet (Tsagkias et al. 2008). They classified and ranked a group of podcasts listed by iTunes with respect to "popularity." They reported that the method separated iTunes popular podcasts from nonpopular ones. In addition, the podcasts that belong to the leading popular podcast group were ranked at the top of the list by the method. With such method of predicting podcast preference, they have developed applications such as podcast recommendation or collection browsing support in order to provide listeners with a variety of preferred podcasts.

9.4.4 CONTENT RECOMMENDATION II: HELPFUL HOTEL REVIEWS

O'Mahony and Smyth (2009) presented a method to recommend hotel reviews based on the helpfulness estimation of the reviews. They evaluated their method with balanced test sets consisting of highly and poorly recommended reviews by their definition of "helpfulness" (a review text is defined helpful if and only if >75% of feedback for the review is positive), the most recent highly scored and poorly scored reviews ranked by date, and

randomly selected highly scored and poorly scored reviews. According to the performance results, all three ranking schemes achieved good performance. Among the ranking schemes, 60% of the recommended reviews by their definition were considered as
helpful, compared to 28% of those ranked by date and 25% of those randomly selected,
respectively. From the results, the highly scored reviews by their classification method
can be optimistically considered as the basis for high-quality review recommendation.

9.4.5 CONTENT RECOMMENDATION III: ATTRACTIVE PHOTOS

Pedro and Siersdorfer (2009) suggested a ranking method of photos based on their
attractiveness. For evaluation, they compared their ranking results with the ground
truth, which is based on the number of times each photo was assigned to a favorite
list by Flickr users. Based on the correlation results, they showed that their proposed
method focusing on "attractiveness" would enhance the result of image retrieval or
image recommendation along with other criteria.

9.4.6 PREDICTING RATINGS OF COMMENTS AND VARIANCE
OF COMMENT RATINGS: VIDEO DATA

Siersdorfer et al. (2010) studied the relationship between the variance of comment ratings and polarizing videos in order to see the possibility of utilizing such variance as
an indicator for polarizing videos. They asked users to evaluate the polarity of each
video in their data set (100 videos consisting of 50 top and 50 bottom videos in the
order of their variance) on a three-point Likert scale. Their statistical test result shows
that polarizing videos tend to produce more diverse comment ratings. In addition, they
also studied the relationship between the variance of comments ratings and polarizing
topics, which are derived from video tags. For each tag, they computed the average
variance of comment ratings over all videos with the tag. Tags related to hot issues such
as presidential, Islam, Iraq are highly ranked, and some tags related to preferences such
as "xBox" are also highly ranked. The statistical significance also showed that more
diverse comment ratings tend to be connected to polarized topics.

9.5   REMAINING ISSUES IN QUALITY ANALYSIS
OF USER-GENERATED CONTENT

9.5.1   GOLD STANDARD AND STANDARD EVALUATION

As examined, each type of gold standard for the quality of the content has its own
pros and cons. The crowd vote–based metric follows the wisdom of the crowd, one of
the big ideas behind Web 2.0. Such idea has been applied to publishing diverse types
of content democratically, where a group of nonexperts determine which contents are
important, and people outside the group may view them with these rankings. However,
members of the crowd might become too conscious of the opinions of others, so that
they tend to follow and conform to one's preceding idea rather than coming up with
different ideas. Three different types of bias, namely, "imbalance vote bias," "winner
circle bias," and "early bird bias" in Section 9.2.1.2 showed this tendency.

The gold standard model from a limited number of evaluators is free from such biases, because the annotation is made at the same time without considering other evaluators' annotation. However, the agreement is not always unanimous or reliable, especially when the quality of the content is highly dependent on the evaluator's subjective assessment. Such subjectivity comes from various aspects of the contents. For example, it is related to the characteristics of the contents itself. The attractiveness of photos or image data is inherently based on artistic characteristics, so the subjective matter in measurement is unavoidable. In other cases, it is also related to the evaluator's background knowledge. For example, as for informativeness, one evaluator may believe that an episode of installing the viewer program for a digital camera review is quite useful, but other evaluators, who already know the intention behind such instruction, may believe that the resulting description is too verbose for the readers to grasp the main characteristics of the camera quickly.

Due to such limitations of gold standard and subjectivity of the quality, the evaluation is hard to standardize. An alternative is to compare the results by each proposed method with the ones made by a professional group. For example, online shopping websites or review sites provide professional reviews written by editors, such as CNET's "editor's review" (Liu et al. 2007). While it may not be the absolute standard for evaluation, the author in the professional group is considered credible enough with the required background knowledge, so the quality of the review may be guaranteed. Researches on developing a systematic comparison method are definitely called for.

9.5.2 Privacy Issues on Metadata

We have seen that metadata are useful in inferring the quality of the content because of their easy and consistent accessibility. However, it is important to recognize the privacy matters around sharing user's personal information with others. Metadata such as time, place, and location can potentially violate people's privacy. Anonymizing user's ID and location with encrypted codes could prevent potential damage, but a more systematic policy for privacy should be developed.

9.5.3 Falsified/Spam Content

Free and easy publication of the content may also produce falsified or spam content. For example, one may deliberately write positive reviews for a particular product in order to increase the sales. Or one could duplicate the review for a particular product and post them several times in order to gain reputation on the website. Unlike e-mail spam, which is mostly automatically filtered, user-generated content spams such as spam reviews are much harder to detect even manually. Jindal and Liu (2008) classified spam reviews into three types as follows:

1. *Untruthful opinions*: those that are deliberately written for the purpose of promoting the products or damaging the reputation of the products
2. *Reviews on brands only*: those that do not comment on the products themselves but only on the brands
3. *Nonreviews*: advertisement or other irrelevant reviews (e.g., question-answering)

They regarded detection of the second and third types of reviews as a classification problem with two classes "spam" and "nonspam" for each type. They utilized the supervised learning methods such as logistic regression, SVM, and naïve Bayes classification. They reported that the logistic regression model is the most effective. As for the first type of review, they focused on predicting harmful spam reviews (e.g., negative spam reviews for good quality product and positive spam reviews for bad quality product) by model building with duplicated review data. However, the levels of duplication are varied, and their study represented only an initial investigation. Further studies on the analysis of complicated types of spam review as well as their detection methods are necessary. In addition, other types of content such as web forums or blogs should also be studied.

9.6 CONCLUDING REMARKS

In this chapter, we described recent work on quality detection that is motivated to promote user-generated content in an effective way. We examined the notion of quality from the perspective of the types of content and the methods to measure such quality based on the characteristics of the content. We also introduced two major methods of determining the quality, or the crowd-based metric and the gold standard model. We examined the two major proposals to evaluate the effectiveness of each method: ranking-based correlation analysis and binary classification by well-known classifiers. Utilizing metadata features gives a robust result in accuracy because of the ease of extraction, but such information is hard to acquire for newly posted contents, which may be overcome by content-based features. This leads to the performance results such that the hybrid approach of utilizing both metadata features and content-based features outperforms any one of the feature-based approaches. We also introduced several applications that use quality assessment, such as filtering the contents with low quality, summarization of the content, recommendation of the content, and predicting the rating of content. With the rapid growth of user-generated content and extended types of content, researches on quality detection must deal with domain adaptation and scalability issues, along with a more stable evaluation mechanism and gold standard model. Finally, privacy issues and falsified or spam contents must be seriously dealt with.

REFERENCES

Agichtein E., C. Castillo, D. Donato, A. Gionis, and G. Mishne. 2008. Finding high-quality content in social media, *Proceedings of WSDM'08*, February 11–12, Palo Alto, CA, pp. 183–193.

Chang, C.-C. and C.-J. Lin. 2001. LIBSVM: A library for support vector machines. *ACM Transactions on Intelligent Systems and Technology*, 2(3), 1–27, 2011. Software available at http://www.csie.ntu.edu.tw/~cjlin/libsvm (accessed on January 7, 2012).

Cohen, J. 1960. A coefficient of agreement for nominal scales. *Educational and Psychological Measurement* 20: 37–46.

Dellarocas, C., N. F. Awady, and X. M. Zhangz. 2007. Exploring the value of online product ratings in revenue fore-casting: The case of motion pictures. Working Paper, Robert H. Smith School Research Paper.

Esuli, A. and F. Sebastiani. 2006. SentiWordNet: A publicly available lexical resource for opinion mining. *Proceedings of LREC*, May 24–26, Genoa, Italy, pp. 417–422.

Ghose, A. and P. Ipeirotis. 2007. Designing novel review ranking systems: Predicting the usefulness and impact of reviews. *Proceedings of ICEC*, August 19–22, Minneapolis, MN, ACM, New York, pp. 303–309.

Hall, M., E. Frank, G. Holmes, B. Pfahringer, P. Reutemann, and I. H. Witten. 2009. *The WEKA Data Mining Software: An Update*, SIGKDD Explorations, Vol. 11, Issue 1.

Hoang, L., J.-T. Lee, Y.-I. Song, and H.-C. Rim. 2008. A model for evaluating the quality of user-created documents. *Proceedings of the 4th AIRS*, LNCS 4993, Springer-Verlag, Berlin, Germany, pp. 496–501.

Huang, S., D. Shen, W. Feng, and Y. Zhang. 2009. Discovering clues for review quality from author's behaviors on e-commerce sites. *Proceedings of ICEC*, August 12–15, Taipei, Taiwan, ACM, New York, pp. 133–141.

Huang, K. Q., Q. Wang, and Z. Y. Wu. 2006. Natural color image enhancement and evaluation algorithm based on human visual system. *Computer Vision Image Understanding* 103(1): 52–63.

Jindal, N. and B. Liu. 2008. Opinion spam and analysis. *Proceedings of the 1st WSDM*, February 11–12, Stanford, CA, ACM, New York, pp. 219–229.

Joachims, T. 1999. Making large-scale SVM learning practical. In: *Advances in Kernel Methods: Support Vector Learning*, B. Schölkopf and C. Burges and A. Smola (eds.). MIT-Press. Software available at http://svmlight.joachims.org/ (accessed on January 7, 2012).

Kim, S.-M., P. Pantel, T. Chklovski, and M. Pennacchiotti. 2006. Automatically assessing review helpfulness. *Proceedings of EMNLP*, July 22–23, Sydney, Australia, Association for Computational Linguistics, Stroudsburg, Pennsylvania, pp. 423–430.

Liu, J., Y. Cao, C.-Y. Lin, Y. Huang, and M. Zhou. 2007. Low-quality product review detection in opinion summarization. *Proceedings of EMNLP-CoNLL*, June 28–30, Prague, Czech Republic, Association for Computational Linguistics, Stroudsburg, Pennsylvania, pp. 334–342.

Lu, Y., P. Tsaparas, A. Ntoulas, and L. Polanyi. 2010. Exploiting social context for review quality prediction. *Proceedings of WWW*, April 26–30, Raleigh, NC, ACM, New York, pp. 691–700.

O'Mahony, M. P. and B. Smyth. 2009. Learning to recommend helpful hotel reviews. *Proceedings of RecSys*, October 23–25, ACM, New York, pp. 305–308.

Pedro, J. S. and S. Siersdorfer. 2009. Ranking and classifying attractiveness of photos in Folksonomies. *Proceedings of WWW*, April 20–24, Madrid, Spain, ACM, New York, pp. 771–780.

Reza, F. 1961. *An Introduction to Information Theory*. McGraw-Hill, New York, 1994. ISBN 0-486-68210-2.

Siersdorfer, S., S. Chelaru, W. Nejdl, and J. S. Pedro. 2010. How useful are your comments?: Analyzing and predicting YouTube comments and comment ratings. *Proceedings of WWW*, April 26–30, Raleigh, NC, ACM, New York, pp. 891–900.

Stone, P. J., D. C. Dunphy, M. S. Smith, and D. M. Ogilvie. 1966. *The General Inquirer: A Computer Approach to Content Analysis*. MIT Press, Cambridge, MA.

Tsagkias, M., M. Larson, and M. Rijke. 2009. Exploiting surface feature for the prediction of podcast preference. *Proceedings of ECIR*, April 6–9, Tenlouse, France, LNCS 5478, Springer-Verlag, Berlin, Germany, pp. 473–484.

Tsagkias, M., M. Larson, W. Weerkamp, and M. Rijke. 2008. Podcred: A framework for analyzing podcast preference. *Proceedings of WICOW*, October 30, Napa Valley, CA, ACM, New York, pp. 67–74.

Tsur, O. and A. Rappoport. 2009. RevRank: A fully unsupervised algorithm for selecting the most helpful book reviews. *Proceedings of the 3rd ICWSM*, San Jose, California, Association for the Advancement of Artificial Intelligence, Palo Acto, California, pp. 154–161.

Wanas, N., M. El-Saban, H. Ashour, and W. Ammar. 2008. Automatic scoring of online discussion posts. *Proceedings of WICOW*, October 30, Napa Valley, CA, ACM, New York, pp. 19–25.

Weimer, M., I. Gurevych, and M. Mühlhäuser. 2007. Automatically assessing the post quality in online discussions on software. *Proceedings of the ACL Demo and Poster Session*, June 25–27, Prague, Czech Republic, Association for Computational Linguistics, Stroudsburg, Pennsylvania, pp. 125–128.

Yang, Y. and J. Pedersen. 1997. A comparative study on feature selection in text categorization. *Proceedings of 14th ICML*, July 8–12, Nashville, TN, pp. 412–420.

Zhang, Z. and B. Varadarajan. 2006. Utility scoring of product reviews. *Proceedings of CIKM*, November 5–11, Arlington, VA, ACM, New York, pp. 51–57.

10 Semantic Web
Ontological Engineering for Knowledge Services

Aviv Segev

CONTENTS

10.1 CONTEXT

10.1.1 RELATED WORK ON CONTEXT

Context has been researched from many aspects, including the aspects of artificial intelligence, natural languages, conversations, formalism of knowledge, goal planning, human expertise in context, knowledge representation, and expert systems.

McCarthy (1987) in his work *Generality in Artificial Intelligence* mentioned some of the main problems existing in the field. The formalization of the notion of context was defined as one of the main problems. McCarthy argued that a most general context does not exist.

Consequently, the formalization of context and a formal theory of introducing context as formal objects were developed (McCarthy and Buvac, 1997). Context was introduced as an abstract mathematical entity with properties useful in artificial intelligence. The abstract definition of context was developed in the Cyc project in the form of microtheories (Guha, 1991). The formal theory of context was used to resolve lexical ambiguity and reason about disambiguation (Buvac, 1996).

The blackboard model of problem solving arose from the Hearsay speech understanding systems (Erman et al., 1980). These ideas were then extended into the standard blackboard architecture in Hearsay-II (HS-II). The blackboard model has proven to be popular for AI problems, and in the years since HS-II, a variety of blackboard-based systems have been developed. HS-III was developed to integrate alternative representations. HS-III had a context mechanism that allowed the integration of knowledge to resolve uncertainty.

Blackboard architectures have been used for interpretation problems such as speech understanding (Lesser et al., 1975), signal understanding (Carver and Lesser, 1992), and image understanding (Williams et al., 1977) and for planning and control (Hayes-Roth, 1985).

Blackboard architecture will be implemented in the context recognition model. The different attributes of the current "world state" are translated into text and added in turn to the blackboard. The data represented in the blackboard model serve as the input to the context recognition algorithm.

10.1.2 INFORMATION SEEKING AND INFORMATION RETRIEVAL

Information seeking is the process in which people turn to information resources in order to increase their level of knowledge in regard to their goals (Modica et al., 2001). Information seeking has influenced the way modern libraries operate (using instruments such as catalogs, classifications, and indexing) and has affected the World Wide Web in the form of search engines.

Although the basic concept of information seeking remains unchanged, the growing need for the automation of the process has called for innovative tools to assign some of the tasks involved in information seeking to the machine level. Thus, databases are extensively used for the efficient storage and retrieval of information. In addition, over the years techniques from the realm of information retrieval (IR) (Salton and McGill, 1983) were refined to predict the relevance of information to a person's needs and to identify appropriate information for a person to interact with. Finally, the use of computer-based ontologies (Smith and Poulter, 1999) was proposed to classify the available information based on some natural classification scheme that would permit more focused information seeking.

Valdes-Perez and Pereira (2000) developed an algorithm based on the concise all pairs profiling (CAPP) clustering method. This method approximates profiling of large classifications. Use of hierarchical structure was explored for classifying a large, heterogeneous collection of web content (Dumais and Chen, 2000). Another method involves checking the frequency of the possible key phrases of articles using the Internet (Turney, 2002). However, this method is based on an existing set of keywords and uses the Internet for ranking purposes only.

There is an extensive body of literature and practice in the area of information science on ontology construction using tools such as a thesaurus (Aitchison et al., 1997) and on terminology rationalization (Soergel, 1985) and matching of different ontologies (Schuyler et al., 1993). In the area of databases and information systems, many models were proposed to support the process of semantic reconciliation, including the SIMS project (Arens et al., 1996), SCOPES (Ouksel and Naiman, 1994), dynamic classification ontologies (Kahng and McLeod, 1996), COIN (Moulton et al., 1998), and CoopWARE (Gal, 1999), to name a few. Ontology construction can be seen as a manual effort to define relations between concepts, while context recognition attempts to identify, in this case automatically, instances of a given situation that could be related to a concept or concepts in the ontology framework.

10.1.3   CONTEXT RECOGNITION

One context recognition approach addressed the creation of taxonomies from metadata (in XML/RDF) containing descriptions of learning resources (Papatheodorou et al., 2002). Following the application of basic text normalization techniques, an index was built, observed as a graph with learning resources as nodes connected by arcs labeled by the index words common to their metadata files. A cluster mining algorithm is applied to this graph and then the controlled vocabulary is selected statistically. However, a manual effort is necessary to organize the resulting clusters into hierarchies. When dealing with medium-sized corpora (a few hundred thousand words), the terminological network is too vast for manual analysis, and it is necessary to use data analysis tools for processing.

Therefore, Assadi (1998) employed a clustering tool that utilizes specialized data analysis functions and clustered the terms in a terminological network to reduce its complexity. These clusters are then manually processed by a domain expert to either edit them or reject them.

Several distance metrics were proposed in the literature and can be applied to measure the quality of context extraction. Prior work had presented methods based on IR techniques (Rijsbergen, 1979) for extracting contextual descriptions from data and evaluating the quality of the process. Motro and Rakov (1998) proposed a standard for specifying the quality of databases based on the concepts of soundness and completeness.

The method allowed the quality of answers to arbitrary queries to be calculated from overall quality specifications of the database. Another approach (Mena et al., 2000) is based on estimating loss of information based on navigation of ontological terms. The measures for loss of information were based on metrics such as precision and recall on extensional information. These measures are used to select results having the desired quality of information.

10.1.4   WEB CONTEXT EXTRACTION MODEL

Several methods were proposed in the literature for extracting context from text. A class of algorithms was proposed in the IR community, based on the principle of counting the number of appearances of each word in a text, assuming that the words with the highest number of appearances serve as the context. Variations on this simple mechanism involve methods for identifying the relevance of words to a domain, using methods such as stop-lists and inverse document frequency. For illustration purposes, a description is provided of a context recognition algorithm that uses the Internet as a knowledge base to extract multiple contexts of a given situation, based on the streaming in text format of information that represents situations.

A *context descriptor* c_i from domain DOM is defined as an index term used to identify a record of information (Mooers, 1972). It can consist of a word, phrase, or alphanumerical term. A weight $w_i \in R$ identifies the importance of descriptor c_i in relation to the information. An example is a descriptor $c_1 =$ Address and $w_1 = 42$. A *descriptor set* $\{\langle c_1, w_1 \rangle\}_i$ is defined by a set of pairs, descriptors, and weights.

Each descriptor can define a different point of view of the concept. The descriptor set eventually defines all the different perspectives and their relevant weights, which identify the importance of each perspective.

The context is obtained by collecting all the different viewpoints delineated by the different descriptors. A *context* $C = \left\{ \left\{ \langle c_{ij}, w_{ij} \rangle \right\}_i \right\}_j$ is a set of finite sets of descriptors, where i represents each context descriptor and j represents the index of each set. For example, a context C may be a set of words (hence DOM is a set of all possible character combinations) defining textual information and the weights can represent the relevance of a descriptor to the information. In classic IR, $\langle c_{ij}, w_{ij} \rangle$ may represent the fact that the word c_{ij} is repeated w_{ij} times in the textual information.

The context extraction algorithm is adapted from Segev et al. (2007a). The input of the algorithm is defined as tokens extracted from textual information. The sets of tokens are extracted as sentences or parsed sets of words, for example, *Get Domains By Zip*, as described in Figure 10.1. Each set of tokens is then

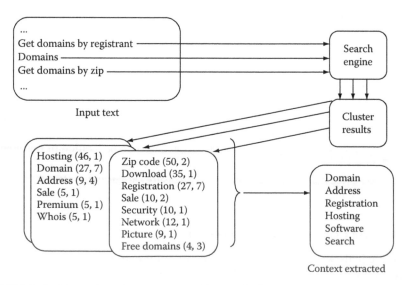

FIGURE 10.1 Example of the context extraction method.

sent to a web search engine and a set of descriptors is extracted by clustering the web pages search results for each token set.

The web pages clustering algorithm is based on the CAPP clustering method (Valdes-Perez and Pereira, 2000). This method approximates profiling of large classifications. It compares all classes pairwise and then minimizes the total number of features required to guarantee that each pair of classes is contrasted by at least one feature. Then each class profile is assigned its own minimized list of features, characterized by how these features differentiate the class from the other features.

Figure 10.1 shows an example that presents the results for the extraction and clustering performed on tokens Get Domains By Zip. The context descriptors extracted include {⟨Zip Code, (50, 2)⟩, ⟨Download, (35, 1)⟩, ⟨Registration, (27, 7)⟩, ⟨Sale, (15, 1)⟩}, ⟨Security, (10, 1)⟩, ⟨Network, (12, 1)⟩, ⟨Picture, (9, 1)⟩, ⟨Free Domains, (4, 3)⟩. A different point of view of the concept can be seen in the previous set of tokens Domains where the context descriptors extracted include {⟨Hosting, (46, 1)⟩, ⟨Domain, (27, 7)⟩, ⟨Address, (9, 4)⟩, ⟨Sale, (5, 1)⟩, ⟨Premium, (5, 1)⟩ ⟨Whole, (5, 1)⟩}. It should be noted that each descriptor is accompanied by two initial weights. The first weight represents the number of references on the web (i.e., the number of returned web pages) for that descriptor in the specific query. The second weight represents the number of references to the descriptor in the textual information (i.e., for how many name token sets was the descriptor retrieved). For instance, in the previous example, Registration appeared in 27 web pages and 7 different name token sets in the text referred to it.

The algorithm then calculates the sum of the number of web pages that identify the same descriptor and the sum of the number of references to the descriptor in the text. A high ranking in only one of the weights does not necessarily indicate the importance of the context descriptor. For example, high ranking in only web references may mean that the descriptor is important since the descriptor widely appears on the web, but it might not be relevant to the topic of the text (e.g., *Download* descriptor in Figure 10.1). To combine values of both the web page references and the appearances in the text, the two values are weighted to contribute equally to the final weight value.

For each descriptor, c_i, the number of web pages refer to it, defined by weight w_{i1}, and the number of times it is referred to in the text, defined by weight w_{i2}, are measured. For example, *Hosting* might not appear at all in the original textual information, but the descriptor based on clustered web pages could refer to it twice in the text and a total of 235 web pages might be referring to it. The descriptors that receive the highest ranking form the context. The descriptor's weight, w_i, is calculated according to the following steps:

- Set all n descriptors in descending weight order according to the number of web page references:

$$\left\{ \left\langle c_i, w_{i1} \right\rangle_{1 \leq i1 \leq n-1} \middle| w_{i1} \leq w_{i1+1} \right\}$$

Current references difference value, $D(R)_i = \left\{ w_{i1+1} - w_{i1}, 1 \leq i1 \leq n-1 \right\}$

- Set all n descriptors in descending weight order according to the number of appearances in the text:

$$\left\{ \left\langle c_i, w_{i2} \right\rangle_{1 \leq i2 \leq n-1} \middle| w_{i2} \leq w_{i2+1} \right\}$$

Current appearances difference value, $D(A)_i = \left\{ w_{i2+1} - w_{i2}, 1 \leq i2 \leq n-1 \right\}$

- Let M_r be the maximum value of references and M_a be the maximum value of appearances:

$$M_r = max_i \left\{ D(R)_i \right\}$$

$$M_a = max_i \left\{ D(A)_i \right\}$$

- The combined weight, w_i, of the number of appearances in the text and the number of references in the web is calculated according to the following formula:

$$w_i = \sqrt{\left(\frac{2 * D(A)_i * M_r}{3 * M_a} \right)^2 + (D(R)_i)^2}$$

The context recognition algorithm consists of the following major phases: (1) selecting contexts for each set of tokens, (2) ranking the contexts, and (3) declaring the current contexts. The result of the token extraction is a list of tokens obtained from the textual information. The input to the algorithm is based on the name descriptor tokens extracted from the textual information. The selection of the context descriptors is based on searching the web for relevant documents according to these tokens and on clustering the results into possible context descriptors. The output of the ranking stage is a set of highest ranking context descriptors. The set of context descriptors that have the top number of references, both in number of web pages and in number of appearances in the text, is declared to be the context and the weight is defined by integrating the value of references and appearances.

Figure 10.1 provides the outcome of the web context extraction method for a DomainSpy web service textual description (see bottom right part). The figure shows only the highest ranking descriptors to be included in the context. For example, *domain, address, registration, hosting, software*, and *search* are the context descriptors selected to describe the DomainSpy service.

10.2   ONTOLOGIES

Ontologies have been defined and used in various research areas, including philosophy (where it was coined), artificial intelligence, information sciences, knowledge representation, object modeling, and most recently, e-commerce applications. In his seminal work, Bunge defines ontology as a world of systems and provides a basic formalism for ontologies (Bunge, 1979). Typically, ontologies are represented using description logic (Borgida and Brachman, 1993) (Donini et al., 1996), where subsumption typifies the semantic relationship between terms, or frame logic (Kifer et al., 1995), where a deductive inference system provides access to semistructured data.

Recent work has focused on ontology creation and evolution and in particular schema matching. Many heuristics were proposed for the automatic matching of schemata (e.g., Cupid (Madhavan et al., 2001), GLUE (Doan et al., 2002), and OntoBuilder (Gal et al., 2005b), and several theoretical models were proposed to represent various aspects of the matching process (Madhavan et al., 2002; Melnik, 2004; Gal et al., 2005a).

The realm of information science has produced an extensive body of literature and practice in ontology construction, for example, Vickery (1966). Other undertakings, such as the DOGMA project (Spyns et al., 2002), provide an engineering approach to ontology management. Work has been done in ontology learning, such as Text-To-Onto (Maedche and Staab, 2001), thematic mapping (Chung et al., 2002), OntoMiner (Davulcu et al., 2003), and TexaMiner (Kashyap et al., 2001), to name a few. Finally, researchers in the field of knowledge representation have studied ontology interoperability, resulting in systems such as Chimaera (McGuinness et al., 2000) and Protègè (Noy and Musen, 2000).

The present model of ontology is based on Bunge's terminology. The aim is to formalize the mapping between contexts and ontologies and provide an uncertainty management tool in the form of concept ranking. When experimenting with the model, the assumption is that an ontology given is designed using any of the tools mentioned earlier.

An *ontology* $O = (V,E)$ is a directed graph, with nodes representing concepts (things in Bunge's terminology; Bunge, 1977, 1979) and edges representing relationships (see Figure 10.2 [top] for a graphical illustration). A single concept is represented by a name and a context C. The relationship of context and ontology is the focus of the next section.

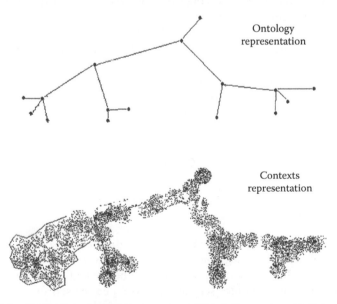

FIGURE 10.2 Contexts and ontology concepts. (From Segev, A. and Gal, A., *J. Data Semantics*, 9, 113, 2007a. With permission.)

10.3 CONTEXTS TO ONTOLOGIES

The relationships between ontologies and contexts can be modeled using topologies as follows. A *topological structure* (topology) in a set X is a collective family $\vartheta = (G_i | i \in I)$ of subsets of X satisfying

1. J finite; $\Rightarrow \bigcup_{i \in J} G_i \in \vartheta$

2. $J \subset I$; $\Rightarrow \bigcap_{i \in J} G_i \in \vartheta$

3. $\varnothing \in \vartheta, X \in \vartheta$

The pair (X, ϑ) is called a *topological space* and the sets in ϑ are called *closed sets*. A context is now defined to be a closed set in a topology, representing a family ϑ of all possible contexts in some set X with the subset relation $\subseteq$. X is a set of sets of pairs $\langle c, w \rangle$, where c is a word (or words) in a dictionary and w is a weight. Note that ϑ is infinite since descriptors are not limited in their length and weights are taken from some infinite number set (such as the real numbers).

The topology is defined by the following subset relation on the context: $\forall C_a \exists C_b$ such that $C_a = \left\{ \left\{ \langle c_{ij}, w_{ij} \rangle \right\}_i \right\}_j \subseteq C_b = \left\{ \left\{ \langle c_{kp}, w_{kp} \rangle \right\}_k \right\}_p$. Stating that for each context there exists another context that includes the existing context. Identity between contexts is defined as follows: $C_a = C_b$ if $c_{kp} = c_{ij}$, $w_{kp} = w_{ij}$, $\forall k, p$. Contexts are identical if all descriptors and their matching weights are identical.

The empty set and X are also contexts. Contexts as sets of descriptor sets are closed under intersection and union.

Contexts were previously defined as closed sets. Next the notion of order of contexts can be defined using a directed set. A *directed set* is a set S together with a relation $\geq$, which is both transitive and reflexive, such that for any two elements $a, b \in S$, there exists another element $c \in S$ with $c \geq a$ and $c \geq b$. In this case, the relation $\geq$ is said to "direct" the set.

A specific directed set is defined using contexts. A context directed set is formally defined by

$$C_0 = \{\varnothing\}$$

$$C_n = \left\{ DS_i, DS_i \cup DS_n \middle| \forall DS_i \in C_{n-1} \right\}$$

The definition is illustrated in Figure 10.3. The different descriptor sets can be viewed as a collection in a bag. One descriptor set DS_1 is randomly selected. Let context C_1 define all the descriptor sets that can be created out of one given context—this is only one descriptor set. Let context C_2 be the sets of descriptors that can be created from two given descriptor sets. Context C_2 contains three descriptor

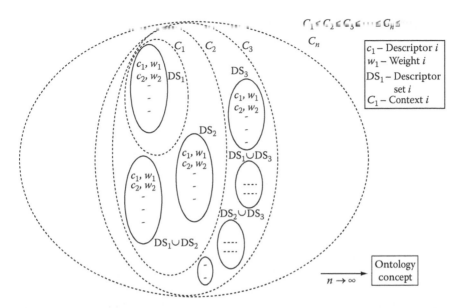

FIGURE 10.3   Context sets converging to an ontology concept. (From Segev, A. and Gal, A., *J. Data Semantics*, 9, 113, 2007a. With permission.)

sets: DS_1 from the previous context, DS_2, which is another descriptor set selected, and the union of both descriptor sets, therefore, $C_1 \leq C_2$. It is possible to continue and build this directed set by adding another descriptor set to C_2 forming a new context C_3, where $C_1 \leq C_3$ and $C_2 \leq C_3$. This process of creating the directed set can continue indefinitely.

This directed set forms a sequence where $C_1 \leq C_2 \leq C_3 \leq \cdots \leq C_n \leq \cdots$.

Whenever a directed set contains contexts that describe a single topic in the real world, such as school or festival, the aim is to ensure that this set of contexts converges to one ontology concept v, representing this topic, that is, $C_n \rightarrow_{n \rightarrow \infty} v$. In topology theory, such a convergence is termed an accumulation point, a point that is the limit of a sequence, also called a *limit point*. Figures 10.2 (bottom) and 10.3 illustrate ontology concepts as points of accumulation. The concept can be viewed as delineating a growing set of descriptors forming the context. The borders outline all of the separate descriptors sets that belong to a specific concept. An overlap between descriptors belonging to different concepts is possible, similar to dynamic taxonomies (Sacco, 2000).

To demonstrate the creation of an ontology concept, let a context be a set containing a singleton descriptor set {Mathematik, 2}. If another singleton descriptor set of {Musik, 2} is added, a new context, which contains three descriptor sets, is formed: {{Mathematik, 2}, {Musik, 2}, {Mathematik, 2, Musik, 2}}. As the possible sets of descriptors describing documents increase, there is increasing coverage of the accumulation point. The directed set composed of these contexts becomes more descriptive. It is possible to converge to an ontology concept, such as *Long Day School*, defined by a set, to which the context set belongs. Basically, the accumulation point forms the context that includes all the descriptor sets required to define a concept.

With infinite possible contexts, is it possible to ensure the existence of ontology concepts to which these contexts converge? The answer is yes. According to the topological definitions, contexts were defined as a subset of a topological space. All of the subsets forming the contexts were defined to be closed sets. According to Kelley (1969), the following theorem holds in regard to closed sets:

Theorem 10.1 *A subset of a topological space is closed if and only if it contains the set of its accumulation point.*

According to this theorem, any subset of contexts, being closed sets, will necessarily include an accumulation point. With a finite set of descriptor sets, when each time another descriptor set is added, an accumulation point, which includes all of the descriptors forming the ontology concepts, will be reached. However, the aforementioned theorem guarantees that even if there are an infinite number of descriptors sets, an accumulation point, which will also be a context, will eventually be reached. This context will include all of the descriptor sets defining the concept.

The model proposed in Segev and Gal (2007a) employs topological definitions to delineate the relationships between contexts and ontologies. A context is a set of descriptors and their corresponding weights. A directed set is a relation of contexts that includes all of their possible unions of sets of descriptors. An ontology concept is the accumulation point of the directed set of contexts.

10.4   WEB SERVICES

10.4.1   RELATED WORK ON WEB SERVICES

In recent years, the use of services to compose new applications from existing modules has gained momentum. Web services are autonomous units of code, independently developed and evolved. The web service description language (WSDL) (Christensen et al., 2001) is used as the de facto standard for service providers to describe the interface of the web services, that is, their operations and input and output parameters. Therefore, web services lack homogeneous structure beyond that of their interface. Heterogeneity stems from different ways to name parameters, define parameters, and describe internal processing. This heterogeneity encumbers straightforward integration between web services.

Web service registries such as universal description, discovery, and integration (UDDI) were created to encourage interoperability and adoption of web services. However, UDDI registries have some major flaws (Platzer and Dustdar, 2005). UDDI registries either are made publicly available and contain many obsolete entries or require registration. In either case, a registry stores only a limited description of the available services.

Semantic web services were proposed to overcome interface heterogeneity. Using languages such as ontology web language for services (OWL-S) (Ankolekar et al., 2001) and WSDL semantics (WSDL-S) (Akkiraju et al., 2005), web services are extended with an unambiguous description by relating properties such as input and output parameters to common concepts and by defining the execution characteristics of the service. The concepts are defined in web ontologies (Bechhofer et al., 2004), which serve as the key mechanism to globally define and reference concepts. Formal languages enable service composition, in which a developer uses automatic or semiautomatic tools to create an integrated business process from a set of independent web services.

Service composition in a heterogeneous environment immediately raises issues of evaluating the accuracy of the mapping. As an example, consider three real-world web services, as illustrated in Figure 10.4. The three services—distance between zip codes (A), store IT contracts (B), and translation into any language (C)—share some common concepts, such as the code concept. However, these three services originate from very different domains. Service A is concerned with distance calculation and uses the zip codes as input, service B defines CurrencyCode as part of the IT contract information to be stored, and service C uses a ClientCode as an access key for users. It is unlikely that any of the services will be combined into a meaningful composition. This example illustrates that methods based solely on the concepts mapped to the service's parameters (as in Paolucci et al., 2002) may yield inaccurate results.

```
<s:element minOccurs="0" maxOccurs="1"name="Zip_Code_1" type="s:string"/>
    <s:element minOccurs="0" maxOccurs="1" name="Zip_Code_2" type="s:string"/>
...
    <s:element minOccurs="1" maxOccurs="1" name="CalcDistTwoZipsMiResult"
type="s:double"/>
```

(a)

```
<s:element minOccurs="0" maxOccurs="1" name="PayeName" type="s:string"/>
<s:element minOccurs="0" maxOccurs="1" name="PayePaymentType" type="s:string"/>
<s:element minOccurs="1" maxOccurs="1" name="PayeAmount" type="s:double"/>
<s:element minOccurs="0" maxOccurs="1" name="CurrencyCode" type="s:string"/>
```

(b)

```
<s:element minOccurs="0" maxOccurs="1" name="ClientCode" type="s:string"/>
<s:element minOccurs="0" maxOccurs="1" name="UserName" type="s:string"/>
<s:element minOccurs="0" maxOccurs="1" name="Password" type="s:string"/>
```

(c)

FIGURE 10.4 Service tagging is misleading: an example. (a) Returns distance in miles or kilometers given two zip codes. (b) Store IT contracts. (c) Translation into and out of any language. (From Segev, A. and Toch, E., *IEEE Trans. Services Comput.*, 2(3), 210, 2009. With permission.)

Segev and Toch (2009) aim at analyzing different methods for automatically identifying possible semantic composition. Two sources for service analysis were explored: WSDL description files and free textual descriptors, which are commonly used in service repositories. Three methods for web service classification are investigated for each type of descriptor: term frequency/inverse document frequency (TF/IDF) (Salton and McGill, 1983), context-based analysis (Segev et al., 2007a), and a baseline method. Contexts are defined as a model of a domain for a given term, which is automatically extracted from a fragment of text. Contexts are created by finding related terms from the web. Unlike ontologies, which are considered shared models of a domain, contexts are defined as local views of a domain (Segev and Gal, 2007a). Therefore, contexts may be different for two fragments of information, even though their domain might be the same. The definition of context used here extends the definition of context in ubiquitous computing, which employs context as any information that can be used to characterize the situation of an entity (Dey, 2000). In many fields, context is used to describe the environment in which a service operates. In this definition, it is used to describe the related set of linguistic terms of a given text.

Segev and Toch (2009) propose a context-based approach to the problem of matching and ranking semantic web services for composition. First, the use of service classification, a process that matches a service to a set of concepts, representing its affinity with a given domain, is proposed. For example, consider the services in Figure 10.4. The context of service A would be a set of geographical terms (such as address, city, and distance). Therefore, it would be classified to a set of concepts taken from a geographical ontology. Service B would be classified to a business transaction ontology and service C to a computer system ontology. Second, the classification and context information is used to improve the process of service composition, ruling out compositions of unrelated services. Given a suggested composition between the number of services, the context overlap between the services is analyzed. The overlap is used to rank the probability of the composition.

Figure 10.5 depicts the stages of the categorization process, including the different methods evaluated. The assumption is that each web service is described using a textual description, which is part of the metadata within UDDI registries, and a WSDL document describing the syntactic properties of the service interface.

Three methods are examined for the service classification analysis: TF/IDF, web context extraction, and a baseline for evaluation purposes. The baseline method is a simple reflection (identity function) of the original bag of tokens, extracted from the service descriptions to a bag of tokens representing sets of words. The basic data structure used by all the methods is a ranked bag of tokens, which is processed and updated in the different stages. The results of the service analysis process are used by the TF/IDF and web context extraction methods. After the different analysis methods were applied, the final categorization is achieved by matching the bag of tokens to the concept names of each of the ontologies.

The field of web service composition is very active. However, most approaches require clear and formal semantic annotations to formal ontologies (Paolucci et al., 2002; Akkiraju et al., 2005; Klusch et al., 2005; Oh, 2006). Since most services that are currently active in the World Wide Web do not contain any semantic annotations,

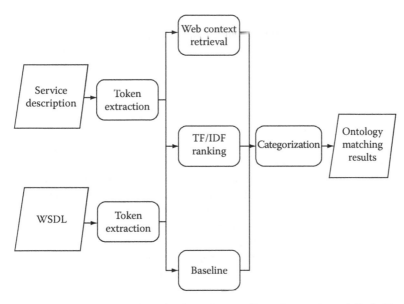

FIGURE 10.5 Web service categorization process. (From Segev, A. and Toch, E., *IEEE Trans. Services Comput.*, 2(3), 210, 2009.)

finding methods that enable composition without semantic annotation is a necessity. Initial work has been done in discovering services directly by querying syntactic web services through their WSDL documentation (Toch et al., 2005; Vouros et al., 2008). Segev and Toch (2009) provide an analysis of different ways for extracting information from syntactic web services and using this information in the context of composition, rather than web service discovery.

The field of automatic annotation of syntactic web services contains several works relevant. Patil et al. (2004) presented a combined approach toward automatic semantic annotation of web services. The approach relies on several matchers (string matcher, structural matcher, and synonym finder), which are combined using a simple aggregation function. Duo et al. (2005) presented a similar method, which also aggregates results from several matchers.

Oldham et al. (2004) showed that using a simple machine learning (ML) technique, namely, naïve Bayesian classifier, improves the precision of service annotation. ML is also used in a tool called Assam (Heß et al., 2004), which uses existing annotation of semantic web services to improve new annotations. While ML effectively improves the efficiency of the semantic annotation, the corpus size used for learning is small, as WSDL documents contain very little text. The approach in Segev and Toch (2009) is complementary to ML methods, as it suggests and provides further information, in the form of textual descriptions and web context. This information can be used by learning methods to improve annotations.

Another relevant field is search engines for syntactic web services. Work by Platzer and Dustdar (2005) and Dong et al. (2004) present search engines for WSDL documents. The search engines use a multitude of IR techniques, including vector space representation, TF/IDF, and text clustering. The main drawback of applying these techniques to WSDL is the relatively short content of a WSDL document, which limits the precision and recall of the search engine.

More recently, several works suggested using information about the web service composition to provide a better annotation process. Bowers and Ludäscher (2005) proposed to explore the relation between input and output parameters of the same operation to infer the semantics from the parameters. If the semantics of the input parameter is known and the logic of the operation is known, then the semantics of the output parameter can be inferred automatically.

Belhajjame et al. (2008) suggest using information about the composition (the term workflow is used in their work) in which the service is used. The composition structure reveals operational constraints between parameters of different operations and can be used to support or disqualify annotations. The aforementioned work by Bowers and Ludäscher (2005) shows the potential of using external information for improving annotations. Segev and Toch (2009) share a similar vision, arguing for the utilization of external information. However, the intention is to produce domain-specific semantic annotation rather than operational semantics. Therefore, the web and public ontologies, rather than the workflow or procedural description of the web services, are used as information resources.

Context-based semantic matching for web services composition has become a focus of interest. An initial prior work describes a context mediator that facilitates semantic interoperability between heterogeneous information systems (Sciore et al., 1994). A recent work presents a context-based mediation approach (Mrissa et al., 2007) that was used to solve semantic heterogeneities between composed web services.

10.5 BOOTSTRAPPING ONTOLOGIES

Ontologies are used in an increasing range of applications, notably the semantic web, and essentially have become the preferred modeling tool. However, the design and maintenance of ontologies is a formidable process (Noy and Klein, 2004; Kim et al., 2005). Ontology bootstrapping, which has recently emerged as an important technology for ontology construction, involves automatic identification of concepts relevant to a domain and relations between the concepts (Ehrig et al., 2005).

Previous work on ontology bootstrapping focused on either a limited domain (Zhang et al., 2006) or expanding an existing ontology (Castano et al., 2007). In the field of web services, registries such as the UDDI have been created to encourage interoperability and adoption of web services. A registry only stores a limited description of the available services. Ontologies created for classifying and utilizing web services can serve as an alternative solution. However, the increasing number of available web services makes it difficult to classify web services using a single domain ontology or a set of existing ontologies created for other purposes. Furthermore, a constant increase in the number of web services requires continuous manual effort to evolve an ontology.

The web service ontology bootstrapping process described here is based on the advantage that a web service can be separated into two types of descriptions: (1) the WSDL describing "how" the service should be used and (2) a textual description of the web service in free text describing "what" the service does. This advantage allows bootstrapping the ontology based on WSDL and verifying the process based on the web service free text descriptor.

The ontology bootstrapping process is based on analyzing a web service using three different methods, where each method represents a different perspective of viewing the web service. As a result, the process provides a more accurate definition of the ontology and yields better results. In particular, the TF/IDF method analyzes the web service from an internal point of view, that is, what concept in the text best describes the WSDL document content. The web context extraction method describes the WSDL document from an external point of view, that is, what most common concept represents the answers to the web search queries based on the WSDL content. Finally, the free text description verification method is used to resolve inconsistencies with the current ontology. An ontology evolution is performed when all three analysis methods agree on the identification of a new concept or a relation change between the ontology concepts. The relation between two concepts is defined using the descriptors related to both concepts.

This approach can assist in ontology construction and reduce the maintenance effort substantially. The approach facilitates automatic building of an ontology that can assist in expanding, classifying, and retrieving relevant services, without the prior training required by previously developed approaches.

10.5.1 BOOTSTRAPPING ONTOLOGY MODEL

The bootstrapping ontology model proposed in Segev and Sheng (2011) is based on the continuous analysis of WSDL documents and employs an ontology model based on concepts and relationships (Gruber, 1993). The innovation of this proposed bootstrapping model centers on (1) the combination of the use of two different extraction methods, TF/IDF and web-based concept generation, and (2) the verification of the results using a free text description verification method by analyzing the external service descriptor. These three methods are utilized to demonstrate the feasibility of the model. It should be noted that other more complex methods, from the field of ML and IR, can also be used to implement the model. However, the use of the methods in a straightforward manner emphasizes that many methods can be "plugged in" and that the results are attributed to the model's process of combination and verification. Segev and Sheng (2011) integrated these three specific methods since each method presents a unique advantage—internal perspective of the web service by the TF/IDF, external perspective of the web service by the web context extraction, and a comparison to a free text description, a manual evaluation of the results, for verification purposes.

The overall bootstrapping ontology process is described in Figure 10.6. There are four main steps in the process. The token extraction step extracts tokens representing relevant information from a WSDL document. This step extracts all the name labels, parses the tokens, and performs initial filtering.

The second step analyzes in parallel the extracted WSDL tokens using two methods. In particular, TF/IDF analyzes the most common terms appearing in each web service document and appearing less frequently in other documents. Web context extraction uses the sets of tokens as a query to a search engine, clusters the results according to textual descriptors, and classifies which set of descriptors identifies the context of the web service.

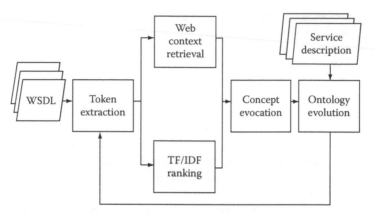

FIGURE 10.6 Web service ontology bootstrapping process. (From Segev, A. and Sheng, Q.Z., *IEEE Trans. Services Comput.*, 2011. With permission.)

The concept evocation step identifies the descriptors that appear in both the TF/IDF method and the web context method. These descriptors identify possible concept names that could be utilized by the ontology evolution. The context descriptors also assist in the convergence process of the relations between concepts.

Finally, the ontology evolution step expands the ontology as required according to the newly identified concepts and modifies the relations between them. The external web service textual descriptor serves as a moderator if there is a conflict between the current ontology and a new concept. Such conflicts may derive from the need to more accurately specify the concept or to define concept relations. New concepts can be checked against the free text descriptors to verify the correct interpretation of the concept. The relations are defined as an ongoing process according to the most common context descriptors between the concepts. After the ontology evolution, the whole process continues to the next WSDL with the evolved ontology concepts and relations. It should be noted that the processing order of WSDL documents is arbitrary.

The main contributions of this work are as follows:

- On a conceptual level, an ontology bootstrapping model, a model for automatically creating the concepts and relations "from scratch," is introduced.
- On an algorithmic level, an implementation of the model in the web service domain is provided, using integration of two methods for implementing the ontology construction and a free text description verification method for validation using a different source of information.

10.6 APPLICATIONS

10.6.1 MEDICAL DIAGNOSTIC ASSISTANCE

This section presents a web-based technique of integrating context recognition and computer vision and demonstrates how this method can be implemented. Usually document analysis focuses on the text part of a document, but Segev et al. (2007b) propose an idea of text understanding by understanding image first, since image can constitute a rich source of information. This idea is based on the assumption that the accuracy of computer vision is high enough to provide a useful hint for context recognition, since an inaccurate computer vision system might also mislead the overall context recognition.

The integration method (Segev et al., 2007b) yields improved results in comparison to the separate use of context recognition or TF/IDF methods. Additionally, use of state of the art as opposed to simple computer vision algorithms can improve the results.

The main advantage of the model for the integration of computer vision into context recognition is its use of the web as a knowledge base for data extraction. The information provided by the computer vision model complements and augments the context recognition process by reducing the number of incorrect diagnoses.

To analyze information consisting of both text and images, a model of the integration of both methods is described in Figure 10.7. The input is separated into text and image. The next step implements a context recognition model for textual analysis and a computer vision model for image analysis. Then the vision is integrated into context, yielding conceptual output. For example, in the field of medicine, the model input can be a medical case study and the model output is a list of words that represent major symptoms or possible diagnoses and these words are checked against the solutions in the medical case studies.

The main advantage of both the web context method and the integrated computer vision and web context over the TF/IDF is the ability to identify a symptom or cause of death that does not appear in the text itself. While the latter has to work within the limits of the original case study text, the context analysis method goes out to the web, using it as an external judge and returning keywords that are deemed relevant, although they were not originally specified in the case description.

The advantage of the integrated computer vision and web context model compared to the web context model can be seen in the reduction of the false

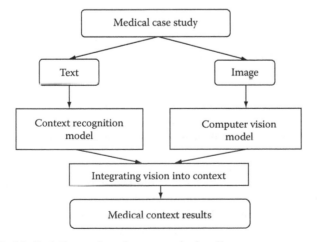

FIGURE 10.7   Medical diagnostic assistance method outline.

positive results. Although the web context by itself in most cases returns the correct results, the ranking of the result is not always high in the result list. The computer vision results allow the identification of which context results should receive higher ranking and consequently the model identifies the correct diagnosis or relevant symptom.

The model achieved high results in both identifying diagnoses that include the identification of the correct diagnosis and identifying symptoms for correct diagnosis. A possible implementation of the model could include a decision support system for a physician analyzing a case. Alternatively, an implementation of the model could be used as a second opinion tool for the patient or his family. Since the model in most cases supplies a list of diagnoses, including the correct diagnosis, a physician would be able to receive an extended list and rule out the incorrect diagnoses.

10.6.2 MULTILINGUAL DECISION SUPPORT SYSTEMS

Experiences in developing information systems have shown it to be a long and expensive process. Therefore, once a generic information system has been developed, it is the aim of the developer to make it as portable as possible and the aim of users to deploy it with minimum effort. In some cases, such deployment requires the change of language, which affects the users' interface as well as the internal decision-making processes. This section focuses on applications in which a language transfer serves as a main obstacle in adapting an information system to users' needs.

As a case in point, consider e-government applications in the European Union. The European Union puts effort into homogenizing its governance procedures to allow easy interoperability. Yet it does so without committing to a single language. On the contrary, the European Union values the preservation of local culture (including language). In such applications, the development of an information system that is monolingual will result in low portability and high deployment costs and therefore multilingual information systems seem to be more appropriate.

Recent advances in information system development suggest the use of ontologies as a main knowledge management tool. Ontologies model the domain of discourse and may be used for routing data, controlling the workflow of activities, assisting in semantic annotation of both data and queries, etc. To take advantage of these recent advances, an ontology-based model for multilingual knowledge management in information systems was proposed in Segev and Gal (2008). The mechanism is based on a single ontology, whose concepts can have multiple representations (i.e., concept names) in various languages. While such solutions already exist (e.g., in Protégé), it is argued that they are insufficient. On the one hand, a single global ontology is preferred over local ontologies when it comes to interoperability. On the other hand, mere translation of ontological concepts from one language to another is insufficient to fully represent differences that may arise from the change of language. Such differences may result in concept ambiguity and generally in underspecification of semantic meaning (Gal and Segev, 2006).

To compensate for ontology underspecification, multilingual ontologies can be supported with a lightweight mechanism, dubbed context. Contexts serve in the literature to represent local views of a domain, as opposed to the global view of an ontology (Gruber, 1993). While the specific representation of contexts vary, one may envision a context, as an example, to be represented by a set of words, possibly associated with weights, reflecting some notion of importance. Contexts, in this solution, are associated with ontological concepts and specified in multiple languages. Therefore, contexts aim at conveying the local interpretation of ontological concepts, thus assisting in the resolution of cross-language and local interpretation ambiguities.

To summarize, the main contributions are as follows:

- The knowledge management model is based on the relationships between ontologies and contexts, thus supporting effective portability and deployment of multilingual information systems.
- The high degree of flexibility this model provides is translated into procedures for the deployment and querying of a multilingual information system.
- The feasibility of the model is demonstrated using an implementation and deployment in the context of a European e-government project.

10.6.2.1   Ontologies, Contexts, and Multilingual Knowledge Management

Now a model for multilingual knowledge management using ontologies and context is described. A common definition of an ontology considers it to be "a specification of a conceptualization" (Gruber, 1993), where conceptualization is an abstract view of the world represented as a set of objects. An *ontology* $O = (V,E)$ is a directed graph, with nodes representing concepts (vocabulary or things; Bunge, 1977, 1979) associated with certain semantics and relationships (Russell and Norving, 2003). For example, in e-government a concept can be *public service* with a relation *includes* to a concept *activity of public administration* and a relation *responsibility* to a concept *local spatial management strategic plan*.

Each descriptor c can be considered to be a different point of view of some concept $v \in V$. A descriptor set then defines different perspectives and their relevant weight, which identifies the importance of each perspective. For example, an ontology concept *local spatial management strategic plan* can be represented by descriptors such as ⟨Immovables, 40⟩, ⟨Building, 25⟩, ⟨Infrastructure, 20⟩, etc. It can now be assumed that each descriptor set represents a different language and then a context is a multilingual representation of a concept.

The model associates an ontology concept with a name and a context. A multiple-name support mechanism is extended and multiple-context support is proposed in a similar fashion. A concept is associated with multiple contexts. Segev and Gal (2007a) defined a context algebra that is closed under the union operator and therefore multiple contexts are in themselves a context, each in a different language. Figure 10.8 provides a schematic illustration of the model for multilingual knowledge management. Four ontology concepts are displayed: *public service*, *citizen*, *activity of public*, and *local spatial*. Each one has concept names also in French,

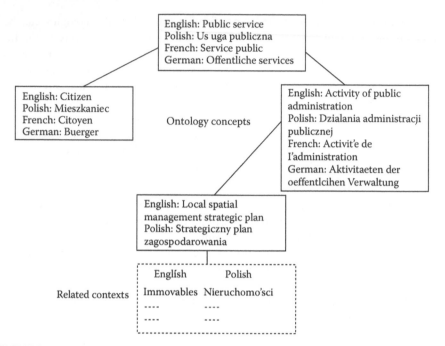

FIGURE 10.8 Multilingual ontology example. (From Segev, A. and Gal, A., *Decision Support Syst.*, 45, 567, 2008. With permission.)

German, and Polish. For the local spatial concept, a set of contexts represents the local perspective of the concepts in both English and Polish.

10.6.3 MULTILINGUAL CRISIS KNOWLEDGE REPRESENTATION

10.6.3.1 Crisis Ontology

In the quest to identify frameworks, concepts, and models for crisis ontologies, the term "open ontology" was addressed by Di Maio (2007). "Open ontology" refers to a given set of agreed terms, in terms of conceptualization and semantic formalization, that has been developed based on public consultation and that embodies, represents, and synthesizes all available valid knowledge thought to pertain to a given domain and necessary to fulfill a given functional requirement.

The *Sphere Handbook* (Sphere Project, 2004) is designed for use in disaster response and may also be useful in disaster preparedness and humanitarian advocacy. It is applicable in a range of situations where relief is required, including natural disasters and armed conflict. It is designed for use in both slow- and rapid-onset situations, rural and urban environments, developing and developed countries, anywhere in the world. The emphasis throughout is on meeting the urgent survival needs of people affected by disaster, while asserting their basic human right to life with dignity.

Analysis of the *Sphere Handbook* index, displayed in Figure 10.9, indicates that it meets many requirements of open ontology. Thus, the current index can be defined

cooking
  fuel supplies 158, 159, 234, 235-6
  environmental impact 123, 159
  stoves 234, 235
  utensils
    access 163, 164
    initial needs 233, 242
  water supplies 64
coordination
  food aid 109, 113-14
  health services 255, 261-3, 263-4
  information exchanges 30, 33-4, 35
  shelter programmes 209-10
crude mortality rates (CMR)
  baseline 260-1
  calculations 301
  documentation 32-3, 259, 271
  maintenance 259, 260
cultural practices
  data gathering 38
  housing 207, 219, 220, 221, 222, 240
  normality 291, 293

  initial assessments 92
  on site soak pits 87-8
  planning 86, 87
  slopes 88, 218
  surface topography 216, 218
  surface water 86
drugs
  donated 266
  essential lists 266, 268
  management 269
  reserve stocks 280

earthquakes, injuries 257, 286
eating utensils 233-4
employment
  food production 128-30
  remuneration 128, 129-30, 131
environment
  erosion 228
  impact
    fuel supplies 123, 234, 235, 242
    settlements 227-9, 241
  protection 13, 227-8

FIGURE 10.9 Index of humanitarian charter and minimum standards. (From Segev, A., *Int. J. Inform. Syst. Crisis Response Manage.*, 1(2), 16, 2009. With permission.)

as an Index Ontology. Generic top-level requirements for an open ontology according to Di Maio (2007) include

- Declaring what high-level knowledge (upper-level ontology) it references. The Index Ontology primary concepts can be identified by the outer level keywords in the index. These keywords serve as a high-level framework defining the primary topics of the Crisis Ontology.
- The ontology allows reasoning/inference based on the index. For example, according to Figure 10.9 the concept *fuel supplies* is related to the class of *cooking* and also related to the concept *impact*, which is related to the concept *environment*. It is also related to the concept *vulnerable groups*. The relational index structure supplies the initial structure of the index ontology.
- Natural language queries can be supported by simple string matching of words from the query against the index ontology concepts. The request to receive relevant information appearing in Figure 10.10, which shows a blog entry posted by a New Orleans resident, displays an example of a textual natural language query that could be analyzed using the index ontology. Simple string matching between the text and the Index Ontology can identify relevant topics such as *food/water/medicine* and *personal hygiene*, which appear in the index. The relevant page numbers of the index topics can supply immediate relevant information delivered in response to the query in any of the aforementioned topics. These could include a short description and possible values required to maintain minimal standards in areas such as *personal hygiene*. A simple web interface

> Right now, it's a matter of survival. There are three important
> aspects to surviving this: you need food/water/medicine, you
> need personal protection, and you need the means to conduct
> personal hygiene in such a way that you're not creating more
> of a problem than you're solving. For any media out there
> reading this, it would be very helpful for you to post guidelines
> for survivalist hygiene.

FIGURE 10.10 Sample blog posting during Katrina Crisis—August 20, 2005. (From Segev, A., *Int. J. Inform. Syst. Crisis Response Manage.*, 1(2), 16, 2009. With permission.)

could support an online connection between the blog and the index ontology, allowing immediate response.

- Use of the index ontology supplies an easy-to-understand mechanism with which most users are familiar. The skills required to utilize the ontology are minimal and can be implemented by any ontology tool, such as Protégé (Noy and Musen, 2000) or Topic Maps Ontopia (Pepper, 1999).
- The "high-level knowledge" represented by the index ontology can easily be linked to classes representing required actions such as status updates, e-mail notification of current crisis situation, resources required for the survivors, and critical locations where immediate intervention is required. The current ontology representation already includes values that can be represented as properties such as measuring acute malnutrition in children under 5 years and other age groups.
- The implementation of the ontology is independent of any ontology language. It can be implemented in any currently used ontology language such as OWL/DARPA agent markup language (DAML) and due to its simplicity can be implemented by alternative ontology languages such as topics, associations, and occurrences (TAO) of topic maps.
- The adoption of an index ontology allows a flexible approach to ontology creation and adoption. As the following section describes, the ontology can be expanded using additional index ontologies or alternatively direct links to information on the web.
- Finally the basic ontology and the knowledge it represents are already defined in multiple languages, allowing multiple viewpoints of similar information in multiple languages. Furthermore, it allows information in multiple languages to be directed to identical ontology concepts.

10.6.3.2 Ontology Design

This section presents the ontology design process. Section 10.6.3.2.1 shows how concepts are extracted from predefined research presented in a book or online documentation to construct the ontology layout. Section 10.6.3.2.2 displays how to extract the concept relations. Section 10.6.3.2.3 depicts how the ontology can be expanded and similar documents based on similar concepts can be added to the ontology. Section 10.6.3.2.4 shows how the ontology can function in a multilingual environment.

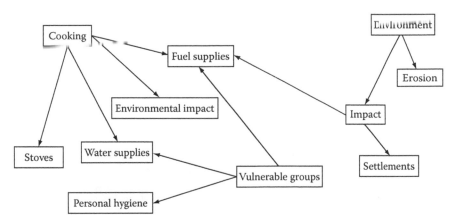

FIGURE 10.11 Sample of the extracted index ontology. (From Segev, A., *Int. J. Inform. Syst. Crisis Response Manage.*, 1(2), 16, 2009. With permission.)

10.6.3.2.1 Extracting the Ontology Layout

Based on the *Sphere Handbook* index (Sphere Project, 2004), an initial ontology can be constructed using existing hierarchical and semantic relations. Furthermore, data linking to additional information can be stored as class properties. Figure 10.11 displays a sample of the index ontology created from the *Sphere Handbook* index (Figure 10.9). The class defined as *cooking* is defined as a superclass of four subclasses: *fuel supplies, environmental impact, water supplies,* and *stoves*. However, *fuel supplies* is a subclass of two additional classes: *vulnerable groups* and *impact*. Similarly, *water supplies* is a subclass of both *cooking* and *vulnerable groups*. The properties of the class *personal hygiene* can match the class with additional information regarding hygiene in the *Sphere Handbook*, such as full description pages or relevant values. Additionally, external information extracted from other resources can be matched with the extracted index ontology.

10.6.3.2.2 Extracting the Concept Relations

The ontology concept relations can be extracted in a similar technique, using the book index. The binary relation is defined as the chapter title shared by each of two concepts. For example, in the *Sphere Handbook*, for each two concepts appearing in the index ontology, the chapter title that connects the two can be defined as the relation.

Figure 10.12 displays an example of the relations of the cooking concept with another four concepts. In the example, it can be seen that the relation of *tools and equipment and lighting* describes both *cooking* and *fuel supply* and *cooking* and *stoves*. The relation that can be automatically extracted in this case supplies an appropriate description.

10.6.3.2.3 Expanding the Ontology

The ontology can be expanded using external information from other resources such as additional data based on books or websites. For example, the Wikipedia website for hygiene includes index information that could be added to the current

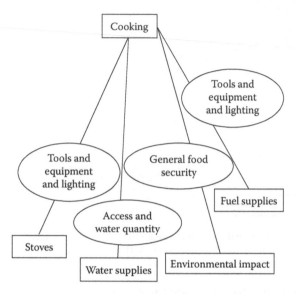

FIGURE 10.12 Ontology concept relations based on document sections. (From Segev, A., *Int. J. Inform. Syst. Crisis Response Manage.*, 1(2), 16, 2009. With permission.)

```
                    Hygiene
        Contents
        1. Personal hygiene
        2. Food and cooking hygiene
        3. Medical hygiene
        4. Personal service/served hygiene
        5. History of hygienic practices
             5.1 Europe
        6. Grooming
        7. Hygiene certification
        8. Academic resources
```

FIGURE 10.13 Possible concepts expansion based on Wikipedia indexing. (From Segev, A., *Int. J. Inform. Syst. Crisis Response Manage.*, 1(2), 16, 2009. With permission.)

index ontology using similar class definitions. Figure 10.13 displays index information from the Wikipedia hygiene index that can be used as concepts for possible ontology expansion. Notice that the concept *personal hygiene* is a subclass of *hygiene* according to this definition. Figure 10.14 displays the ontology expansion based on the Wikipedia hygiene entry. Alternatively, additional index books considered fundamental in the field can be added to the ontology. For example, *the Merck Manual of Medical Information* (Beers, 2003) index can be used for medical class expansion.

There are multiple approaches to merging ontologies such as the formal concept analysis described in Stumme and Maedche (2001). Possible merging operations for the ontology engineer are presented in Noy and Klein (2004). Furthermore,

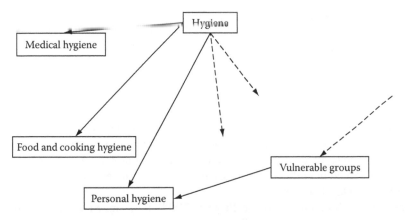

FIGURE 10.14 Ontology expansion based on Wikipedia. (From Segev, A., *Int. J. Inform. Syst. Crisis Response Manage.*, 1(2), 16, 2009. With permission.)

Segev and Gal (2007b) proposed using (machine-generated) contexts as a mechanism for quantifying relationships among concepts. Using this model has an advantage since it provides the ontology administrator with an explicit numeric estimation of the extent to which a modification "makes sense." The present research adopts the method of expanding the ontology based on context mechanism.

10.6.3.2.4 Multilingualism in Crisis Management

As aforementioned, an ontology-based model for multilingual knowledge management in information systems has been proposed in Segev and Gal (2008). The unique feature was a lightweight mechanism, dubbed context, which is associated with

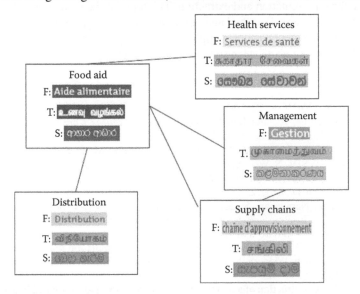

FIGURE 10.15 Sample of the extracted multilingual ontology. (From Segev, A., *Int. J. Inform. Syst. Crisis Response Manag.*, 1(2), 16, 2009. With permission.)

ontological concepts and specified in multiple languages. The contexts were used to assist in resolving cross-language and local variation ambiguities. The technique (described in Section 10.6.3.1) can be adopted to build an ontology where each concept can be represented in multiple languages.

The technique presented here is different from the previous model since it requires the ability to create and modify the ontology in real time as the crisis arises and continues to evolve. This requirement necessitates having a basic predefined multilingual ontology while allowing the expansion of the ontology according to the crisis circumstances and the addition of other languages within the crisis time limitations. The technique can be adopted to build an ontology where each concept can be represented in multiple languages and can be expanded for use in crises, such as the Boxing Day Tsunami.

The *Sphere Handbook* (Sphere Project, 2004) is designed for use in disaster response and was translated into 37 languages. Thus, it supplies a top-level ontology that can be used concurrently in multiple languages. Since each high-level index ontology concept is represented in multiple languages, there is faster ontology adaptation in crisis situations. A sample of a multilingual ontology in English, French (F), Tamil (T), and Sinhala (S) is presented in Figure 10.15.

10.7 CONCLUSION

This chapter has presented a review of knowledge management that uses context and ontology. The knowledge analysis was initially based on extracting the relevant context. Next, the ontology was provided as an outline for representing the framework for organizational knowledge. Mapping from context to ontology is a tool for linking knowledge for representation and extraction. The topic of the matching and composition of web services was described and bootstrapping ontologies for web services was discussed. Knowledge management applications were presented in the fields of medical analysis, multilingual decision support systems, and crisis response systems.

REFERENCES

Aitchison, J., Gilchrist, A., and Bawden, D. (1997). *Thesaurus Construction and Use: A Practical Manual*, 3rd edn. Aslib, London, U.K.

Akkiraju, R., Farrell, J., Miller, J., Nagarajan, M., Schmidt, M. T., Sheth, A., and Verma, K. (2005). WSDL-S Web Service Semantics, W3C Candidate Recommendation, http://www.w3.org/Submission/WSDL-S/ (accessed on January 5, 2012).

Ankolekar, A., Martin, D., Zeng, Z., Hobbs, J., Sycara, K., Burstein, B., Paolucci, M., Lassila, O., Mcilraith, S., Narayanan, S., and Payne, P. (2001). DAML-S: Semantic markup for web services. In *Proceedings of International Semantic Web Workshop (SWWS'01)*, July 29–August 1, Stanford, CA, pp. 411–430.

Arens, Y., Knoblock, C. A., and Shen, W. (1996). Query reformulation for dynamic information integration. In Wiederhold, G. (Ed.), *Intelligent Integration of Information*. Kluwer, Boston, MA, pp. 11–42.

Assadi, H. (1998). Construction of a regional ontology from text and its use within a documentary system. In *Proceedings of the International Conference on Formal Ontology and Information Systems (FOIS-98)*. IOS, Amsterdam, the Netherlands.

Bechhofer, S., Harmelen, F. van, Hendler, J., Horrocks, I., McGuinness, D., Patel-Schneider, P., and Stein, L. (2004). OWL Web Ontology Language Reference, W3C, W3C Candidate Recommendation, http://www.w3.org/TR/owl-ref/ (accessed on January 5, 2012).

Beers, M. (Ed.). (2003). *The Merck Manual of Medical Information*, 2nd edn. Merck Research Laboratories, Whitehouse Station, NJ.

Belhajjame, K., Embury, S. M., Paton, N. W., Stevens, R., and Goble, C. A. (2008). Automatic annotation of web services based on workflow definitions. *ACM Transactions Web*, 2(2), 1–34.

Borgida, A. and Brachman, R. J. (1993). Loading data into description reasoners. In *Proceedings of the 1993 ACM SIGMOD International Conference on Management of Data*, May 26–28, Washington, DC. ACM Press, New York, pp. 217–226.

Bowers, S. and Ludäscher, B. (2005). Towards automatic generation of semantic types in scientific workflows. In *Proceedings International Workshop Scalable Semantic Web Knowledge Base Systems (SSWS'05)*, November 20, 2005, New York, pp. 207–216.

Bunge, M. (1977). *Treatise on Basic Philosophy: Vol. 3: Ontology I: The Furniture of the World*. D. Reidel Publishing Co., Inc., New York.

Bunge, M. (1979). *Treatise on Basic Philosophy: Vol. 4: Ontology II: A World of Systems*. D. Reidel Publishing Co., Inc., New York.

Buvac, S. (1996). Resolving lexical ambiguity using a formal theory of context, semantic ambiguity and underspecification. CLSI Lecture Notes, pp. 1–24.

Carver, N. and Lesser, V. (1992). Blackboard systems for knowledge-based signal understanding. In Oppenheim, A. and Nawab, H. (Eds.), Symbolic and knowledge-based signal processing, Prentice Hall, Englewood Cliffs, NJ, pp. 205–250.

Castano, S., Espinosa, S., Ferrara, A., Karkaletsis, V., Kaya, A., Melzer, S., Moller, R., Montanelli, S., and Petasis, G. (2007). Ontology dynamics with multimedia information: The BOEMIE evolution methodology. In *Proceedings of the International Workshop on Ontology Dynamics (IWOD'07), held with the Fourth European Semantic Web Conference (ESWC'07)*, June 7, 2007, Innsbruck, Austria.

Christensen, E., Curbera, F., Meredith, G., and Weerawarana, S. (2001). WSDL Web Services Description Language, W3C Candidate Recommendation, http://www.w3.org/TR/2001/NOTE-wsdl-20010315 (accessed on January 5, 2012).

Chung, C. Y., Lieu, R., Liu, J., Luk, A., Mao, J., and Raghavan, P. (2002). Thematic mapping from unstructured documents to taxonomies. In *Proceedings of the 11th International Conference on Information and Knowledge Management (CIKM)*, November 4–9, McLean, VA, pp. 608–610.

Davulcu, H., Vadrevu, S., and Nagarajan, S. (2003). Ontominer: Bootstrapping and populating ontologies from domain specific websites. In *Proceedings of the First International Workshop on Semantic Web and Databases*, September 7–8, Berlin, Germany, pp. 259–276.

Dey, A. K. (2000). Providing architectural support for building context aware applications. PhD thesis. Georgia Institute of Technology, Atlanta, GA.

Di Maio, P. (2007). An open ontology for open source emergency response system. Open Source Research Community.

Doan, A., Madhavan, J., Domingos, P., and Halevy, A. (2002). Learning to map between ontologies on the semantic web. In *Proceedings of the 11th International Conference on World Wide Web*, Honolulu, HI. ACM Press, New York, pp. 662–673.

Dong, X., Halevy, A., Madhavan, J., Nemes, E., and Zhang, J. (2004). Similarity search for web services. In *Proceedings International Conference on Very Large Data Bases*, August, Toronto, Canada, pp. 372–383.

Donini, F. M., Lenzerini, M., Nardi, D., and Schaerf, A. (1996). Reasoning in description logic. In Brewka, G. (Ed.), *Principles on Knowledge Representation, Studies in Logic, Languages and Information*. CSLI Publications, Stanford, CA, pp. 193–238.

Dumais, S. and Chen, H. (2000). Hierarchical classification of web content. In *Proceedings of SIGIR, 23rd ACM International Conference on Research and Development in Information Retrieval*, July 24–28, Athens, Greece, pp. 256–263.

Duo, Z., Juan-Zi, L., and Bin, X. (2005). Web service annotation using ontology mapping. In *Proceedings IEEE International Workshop Service-Oriented System Engineering (SOSE'05)*, October 20–21, Beijing, China, pp. 243–250.

Ehrig, M., Staab, S., and Sure, Y. (2005). Bootstrapping ontology alignment methods with APFEL. In *Proceedings of Fourth International Semantic Web Conference (ISWC'05)*, November 6–10, Galway, Ireland.

Erman, L., Hayes-Roth, F., Lesser, V., and Reddy, D. R. (1980). The hearsay II speech understanding system: Integrating knowledge to resolve uncertainty. *Computing Surveys*, 12(2), 213–253.

Gal, A. (1999). Semantic interoperability in information services: Experiencing with CoopWARE. *SIGMOD Record*, 28(1), 68–75.

Gal, A., Anaby-Tavor, A., Trombetta, A., and Montesi, D. (2005a). A framework for modeling and evaluating automatic semantic reconciliation. *VLDB Journal*, 14(1), 50–67.

Gal, A., Modica, G., Jamil, H. M., and Eyal, A. (2005b). Automatic ontology matching using application semantics. *AI Magazine*, 26(1), 21–31.

Gal, A. and Segev, A. (2006). Putting things in context: Dynamic eGovernment re-engineering using ontologies and context. In *Proceedings of the 2006 WWW Workshop on E-Government: Barriers and Opportunities*, Edinburg, U.K.

Gruber, T. R. (1993). A Translation approach to portable ontologies. *Knowledge Acquisition*, 5(2), 199–220.

Guha, R. V. (1991). Contexts: A formalization and some applications. Doctoral dissertation. Stanford University, Stanford, CT (STAN-CS-91-1399-Thesis).

Hayes-Roth, B. (1985). A blackboard architecture for control. *Artificial Intelligence*, 26, 251–321.

Heß, A., Johnston, E., and Kushmerick, N. (2004). ASSAM: A tool for semi-automatically annotating semantic web services. In *Proceedings International Semantic Web Conference*, November 7–11, Hiroshima, Japan, pp. 320–334.

Kahng, J. and McLeod, D. (1996). Dynamic classification ontologies for discovery in cooperative federated databases. In *Proceedings of the First IFCIS International Conference on Cooperative Information Systems (CoopIS'96)*, June 19–21, Brussels, Belgium, pp. 26–35.

Kashyap, V., Dalal, S., and Behrens, C. (2001). Professional services automation: A knowledge management approach using LSI and domain specific ontologies. In *Proceedings of the 14th International FLAIRS Conference (Florida AI Research Symposium), Special track on AI and Knowledge Management*, May 21–23, 2001, Keywest, FL.

Kelley, J. (1969). *General Topology*. American Book Company, New York.

Kifer, M., Lausen, G., and Wu, J. (1995). Logical foundation of object-oriented and frame-based languages. *Journal of the ACM*, 42, 741–843.

Kim, D., Lee, S., Shim, J., Chun, J., Lee, Z., and Park, H. (2005). Practical ontology systems for enterprise application. In *Proceedings of 10th Asian Computing Science Conference (ASIAN'05)*, December 7–9, Kunming, China.

Klusch, M., Fries, B., Khalid, M., and Sycara, K. (2005). OWLS-MX: Hybrid semantic web service retrieval. In *Proceedings First International AAAI Fall Symposium Agents and the Semantic Web*, November 4–6, Arlington, VA.

Lesser, V., Fennell, R., Erman, L., and Reddy, D. R. (1975). Organization of the Hearsay II speech understanding system. *IEEE Transactions on Human Factors in Electronics*, ASSP-23, 11–24.

Madhavan, J., Bernstein, P. A., Domingos, P., and Halevy, A. Y. (2002). Representing and reasoning about mappings between domain models. In *Proceedings of the Eighteenth National Conference on Artificial Intelligence and Fourteenth Conference on Innovative Applications of Artificial Intelligence (AAAI/IAAI)*, July 28–August 1, 2002, Edmonton, Alberta, Canada, pp. 80–86.

Madhavan, J., Bernstein, P. A., and Rahm, E. (2001). Generic schema matching with Cupid. In *Proceedings of the International Conference on Very Large Data Bases (VLDB)*, July 28–August 1, 2002, Rome, Italy, pp. 49–58.

Maedche, A. and Staab, S. (2001). Ontology learning for the semantic web. *IEEE Intelligent Systems*, 16, 72–79.

McCarthy, J. (1987). Generality in artificial intelligence. *Communication of ACM*, 30, 1030–1035.

McCarthy, J. and Buvac, S. (1997). *Formalizing Context, Computing Natural Language*. Stanford University, Stanford, CT, pp. 13–50.

McGuinness, D. L., Fikes, R., Rice, J., and Wilder, S. (2000). An environment for merging and testing large ontologies. In *Proceedings of the Seventh International Conference on Principles of Knowledge Representation and Reasoning (KR2000)*, April 12–15, Breckenridge, CO.

Melnik, S. (Ed.) (2004). *Generic Model Management: Concepts and Algorithms*. Springer, Heidelberg.

Mena, E., Kashyap, V., Illarramendi, A., and Sheth, A. P. (2000). Imprecise answers in distributed environments: Estimation of information loss for multi-ontology based query processing. *International Journal of Cooperative Information Systems*, 9(4), 403–425.

Modica, G., Gal, A., and Jamil, H. M. (2001). The use of machine-generated ontologies in dynamic information seeking. In *Proceedings of the Sixth International Conference on Cooperative Information Systems (CoopIS 2001)*, September, Trento, Italy.

Mooers, C. (1972). *Encyclopedia of Library and Information Science*, Vol. 7, Ch. Descriptors. Marcel Dekker, New York, pp. 31–45.

Motro, A. and Rakov, I. (1998). Estimating the quality of databases. *Lecture Notes in Computer Science*, 1495, 298.

Moulton, A., Madnick, S. E., and Siegel, M. (1998). Context mediation on Wall Street. In *Proceedings of the Third IFCIS International Conference on Cooperative Information Systems (CoopIS'98)*, August 20–28, New York. IEEE-CS, New York, pp. 271–279.

Mrissa, M., Ghedira, C., Benslimane, D., Maamar, Z., Rosenberg, F., and Dustdar, S. (2007). A context-based mediation approach to compose semantic web services. *ACM Transactions Internet Technology*, 8(1), 4.

Noy, N. F. and Klein, M. (2004). Ontology evolution: Not the same as schema evolution. *Knowledge and Information Systems*, 6(4), 428–440.

Noy, F. N. and Musen, M. A. (2000). PROMPT: Algorithm and tool for automated ontology merging and alignment. In *Proceedings of the 17th National Conference on Artificial Intelligence (AAAI-2000)*, Austin, TX, pp. 450–455.

Oh, S. C. (2006). Effective web-service composition in diverse and large-scale service networks. PhD dissertation. University Park, State College, PA.

Oldham, N., Thomas, C., Sheth, A. P., and Verma, K. (2004). Meteor-s web service annotation framework with machine learning classification. In *Proceedings International Workshop Semantic Web Services and Web Process Composition (SWSWPC'04)*, July, San Diego, CA, pp. 137–146.

Ouksel, A. M. and Naiman, C. F. (1994). Coordinating context building in heterogeneous information systems. *Journal of Intelligent Information Systems*, 3(2), 151–183.

Paolucci, M., Kawamura, T., Payne, T., and Sycara, K. (2002). Semantic matching of web services capabilities. In *Proceedings of International Semantic Web Conference*, June 9–12, Sardinia, Italy, pp. 333–347.

Papatheodorou, C., Vassiliou, A., and Simon, B. (2002). Discovery of ontologies for learning resources using word-based clustering. In *Proceedings of the World Conference on Educational Multimedia, Hypermedia and Telecommunications (ED-MEDIA 2002)*, June 24–29, Denver, CO, pp. 1523–1528.

Patil, A., Oundhakar, S., Sheth, A., and Verma, K. (2004). Meteor-s web service annotation framework. In *Proceedings 13th International Conference World Wide Web (WWW'04)*, May 17–22, New York, pp. 553–562.

Pepper, S. (1999). Navigating haystacks, discovering needles. *Markup Languages: Theory and Practice*, Vol. 1, No. 4. MIT Press, Cambridge, MA.

Platzer, C. and Dustdar, S. (2005). A vector space search engine for web services. In *Proceeding of Third European Conference of Web Services (ECOWS'05)*, November 14–16, Vaxjo, Sweden, pp. 62–71.

Rijsbergen, C. J. (1979). *Information Retrieval*, 2nd edn. Butterworths, London, U.K.

Russell, S. and Norving, P. (2003). *Artificial Intelligence: A Modern Approach*, 2nd edn. Prentice Hall, Upper Saddle River, NJ.

Sacco, G. (2000). Dynamic taxonomies: A model for large information bases. *IEEE Transactions Knowledge Data Engineering*, 12(2), 468–479.

Salton, G. and McGill, M. J. (1983). *Introduction to Modern Information Retrieval*. McGraw-Hill, New York.

Schuyler, P. L., Hole, W. T., and Tuttle, M. S. (1993). The UMLS (Unified Medical Language System) metathesaurus: Representing different views of biomedical concepts. *Bulletin of the Medical Library Association*, 81, 217–222.

Sciore, E., Siegel, M., and Rosenthal, A. (1994). Using semantic values to facilitate interoperability among heterogeneous information systems. *ACM Transactions Database Systems*, 19(2), 254–290.

Segev, A. (2009). Adaptive ontology use for crisis knowledge representation. *International Journal of Information Systems for Crisis Response and Management*, 1(2), 16–30.

Segev, A. and Gal, A. (2007a). Putting things in context: A topological approach to mapping contexts to ontologies. *Journal of Data Semantics*, 9, 113–140.

Segev, A. and Gal, A. (2007b). Puzzling it out: Supporting ontology evolution with applications to eGovernment. In *Proceedings of IJCAI Workshop on Modeling and Representation in Computational Semantics*, January 6–12, Hyderabad, India, pp. 45–51.

Segev, A. and Gal, A. (2008). Enhancing portability with multilingual ontology-based knowledge management. *Decision Support Systems*, 45(3), 567–584.

Segev, A., Leshno, M., and Zviran, M. (2007a). Context recognition using internet as a knowledge base. *Journal of Intelligent Information Systems*, 29(3), 305–327.

Segev, A., Leshno, M., and Zviran, M. (2007b). Internet as a knowledge base for medical diagnostic assistance. *Expert Systems with Applications*, 33(1), 251–255.

Segev, A. and Sheng, Q. Z. (2011). Bootstrapping ontologies for web services. *IEEE Transactions on Services Computing* (in print).

Segev, A. and Toch, E. (2009). Context-based matching and ranking of web services for composition. *IEEE Transactions on Services Computing*, 2(3), 210–222.

Smith, H. and Poulter, K. (1999). Share the ontology in XML-based trading architectures. *Communications of the ACM*, 42(3), 110–111.

Soergel, D. (1985). *Organizing Information: Principles of Data Base and Retrieval Systems*. Academic, Orlando, FL.

Sphere Project (2004). Humanitarian charter and minimum standards in disaster response. *The Sphere Project*. Stylus Publishing, Geneva, IL.

Spyns, P., Meersman, R., and Jarrar, M. (2002). Data modelling versus ontology engineering. *ACM SIGMOD Record*, 31(4), 12–17.

Stumme, G. and Maedche, A. (2001). Ontology merging for federated ontologies on the semantic web. In *Proceedings of the International Workshop for Foundations of Models for Information Integration*, September 16–18, Viterbo, Italy.

Toch, E., Gal, A., and Dori, D. (2005). Automatically grounding semantically-enriched conceptual models to concrete web services. In *Proceedings of ER—International Conference on Conceptual Modeling*, Delcambre, L., Kop, L., Mayr, H., Mylopoulos, J., and Pastor, O. (Eds.), *ER*, Springer, New York, pp. 304–319.

Turney, P. (2002). Mining the web for lexical knowledge to improve keyphrase extraction: Learning from labeled and unlabeled data (Tech. Rep. No. ERB-1096; NRC #44947). National Research Council, Institute for Information Technology, Washington, DC.

Valdes-Perez, R. E. and Pereira, F. (2000). Concise, intelligible, and approximate profiling of multiple classes. *International Journal of Human Computer Studies*, 53, 411–436.

Vickery, B. C. (1966). *Faceted Classification Schemes*. Graduate School of Library Service, Rutgers, the State University, New Brunswick, NJ.

Vouros, G. A., Dimitrokallis, F., and Kotis, K. (2008). Look ma, no hands: Supporting the semantic discovery of services without ontologies. In *Proceedings International Workshop Service Matchmaking and Resource Retrieval in the Semantic Web (SMRR)*, October 27, Karlsruhe, Germany.

Williams, T., Lowrance, J., Hanson, A., and Riseman, E. (1977). Model-building in the VISIONS system. In *Proceedings of IJCAI-77*, August 22–25, Cambridge, MA, pp. 644–645.

Zhang, G., Troy, A., and Bourgoin, K. (2006). Bootstrapping ontology learning for information retrieval using formal concept analysis and information anchors. In *Proceedings of 14th International Conference on Conceptual Structures (ICCS'06)*, July 16, Aalborg, Denmark.

11 Acquiring Knowledge from Subject Matter Experts

Nick Milton

CONTENTS

11.1   INTRODUCTION

Every knowledge service needs knowledge. In most cases, the knowledge is obtained by interviewing subject matter experts, that is, people who have many years of experience in the relevant subject areas. There is often a strong correlation between the quality of the knowledge service and the quality of the knowledge. So the knowledge acquisition activities that take place during the development of a new knowledge service are crucial to its quality and usefulness. A successful project, therefore, needs to involve people who can acquire knowledge in an efficient and effective manner.

These people must have a good understanding of the issues involved, the options available, and the best methods to use for a particular situation.

This chapter provides guidelines for people who need to acquire knowledge from subject matter experts. The guidelines are described in terms of 10 key questions that should be considered during the knowledge acquisition phase of a project: What is the project aiming to achieve? What knowledge should I collect? What is my strategy? Should I use a particular methodology? How should I interview the experts? How should I analyze and represent the knowledge? How do I ensure that the knowledge is complete, consistent, and correct? What documentation and other resources should I produce? Does the knowledge need to be updated in the future? What did I learn from the project?

There is a lot of ground to cover in a single chapter so some areas are glossed over with only a short description. This is especially true for the more technical aspects such as knowledge analysis and knowledge modeling. A detailed explanation of these areas can be found in Milton (2007).

11.2 WHAT IS THE PROJECT AIMING TO ACHIEVE?

Imagine two scenarios for the development of a new knowledge service. In the first scenario, you become involved in the project after the knowledge service has been designed. This means you are able to read a specification that describes the architecture, technology, objectives, and uses of the new knowledge service. This provides you with some of the essential information required to define your project activities.

Let us now consider a second scenario in which your activities start before the knowledge service has been designed, and you only have some initial and rather vague ideas about its architecture, technology, objectives, and uses. In this scenario, you must become involved in the design activities or work in parallel with the design activities to find the information you need for knowledge acquisition. If this second scenario is the case, then the following sections provide some guidelines for these early stages of knowledge acquisition.

11.2.1 END USERS AND HOW THEIR ACTIVITIES WILL CHANGE

One of the most important things to know at the start of the knowledge acquisition phase is the uses to which the knowledge will be put. You need to find out who will use the new knowledge service and how it will change their activities. If you do not know this or you only have a rough idea, then you need to obtain this information by talking to three types of people:

1. The other people who are working on the development of the knowledge service to find out what it will do, who will use it, and how the users will be helped
2. The subject matter experts whose job involves communicating and using their expertise within the situation being addressed
3. Some of the people who will be using the knowledge service to find out what they already do and how things can be improved

Rather than just talking to these people, it is a good idea to interview them using one of the special interview techniques described later. For example, you may want to use a semistructured interview where you have a set of preprepared questions. In addition to asking questions, you could use a process mapping technique in which you create diagrams that represent the way people perform their activities. Sending out questionnaires is not a good idea as this rarely provides the depth of information you need, and few people are likely to respond.

11.2.2 BENEFITS TO THE ORGANIZATION

If the new knowledge service is to be used within an organization, it is important to understand how the organization as a whole will benefit from its introduction and use. Is the knowledge service aiming to reduce the time taken to do something? Is the knowledge service aiming to reduce costs, such as by using fewer resources? Is the knowledge service aiming to increase the quality of a product or service? Talking to experts and practitioners is a good way to obtain answers to these questions. You should also try to talk to higher-level managers to understand what they want from the new knowledge service and how it aligns with the organization's strategies and other improvement initiatives.

11.2.3 ENVISIONING THE FINAL PRODUCT

When you know who will use the knowledge service, how they will use it, and the benefits it will provide, you can start to build a vision of the knowledge service and of the knowledge that needs to be collected and implemented. This will provide you with ideas about the types of knowledge you need to acquire and how to structure the knowledge. Here are some possibilities:

- The final product is an expert system that performs complex activities normally performed by expert practitioners. In this case, you will be collecting a wealth of detailed knowledge on how the experts perform their activities and you will structure it in the correct format for use in the expert system.
- The final product is an intelligent database that allows users to find information quickly and easily. In this case, you will need to collect the vocabulary that people use, how the terms in the vocabulary are related to one another, and descriptions for each of the terms. You will need to structure it in a way that will make searching and browsing easy for human users.
- The final product is a help desk that provides users with detailed information on the activities they need to perform. In this case, you will need to capture details of the information people use to perform their activities and how that information is manipulated to provide answers to questions or provide solutions to problems.

11.2.4 IS THE PROJECT PROCESS-FOCUSED OR CONCEPT-FOCUSED?

During the early stages of a project, it is always useful to decide if the project is process-focused or concept-focused as this will affect the strategies and techniques you use to collect and package the knowledge for implementation. Process-focused means the knowledge to acquire is about the things that experts do (i.e., tasks, activities, and processes). Concept-focused means the knowledge is about the concepts (i.e., things) that the expert knows about.

In a process-focused project you will need to identify the tasks involved, the order in which they are performed, the circumstances under which a task is and is not performed, the information that is needed for each task, and the information that is created by each task.

In a concept-focused project, you will be capturing knowledge of concepts such as physical objects, materials, people, organizations, regulations, issues, or ideas. You will find out how they are related to one another and perhaps also find out about the properties of the concepts. If you are developing an ontology for use in a semantic technology system, then you will almost certainly be doing a concept-focused project.

Some projects are partly process-focused and partly concept-focused. If this is the case, then you should decide which type of knowledge fit into which category and choose the best method to capture and model the knowledge for each type of knowledge (see Sections 11.6 and 11.7).

11.3   WHAT KNOWLEDGE SHOULD I COLLECT?

For some projects, the aims are very clear and you will have little choice in deciding what knowledge to collect and the level of detail required to represent it. On other projects, you will have much more flexibility to choose, especially when the goals of the project are at a high level or are described in very general terms. Whichever is the case, there will always be some decisions to make on what knowledge to collect because you can never capture all of the possible knowledge that sits in the heads of the subject matter experts (as it is too vast and detailed).

There are two terms you should know when talking about the knowledge to collect. The first term is "scope." Scope means the knowledge to be collected (as distinct from all the other knowledge that could be collected). For example, the scope of the project might be a detailed description of a particular business process, such as the design and manufacture of a car component. The scope of another project might be the knowledge used by experts for their day-to-day activities, such as lawyers or medics.

Sometimes, the project is a pilot project with the aim of demonstrating the feasibility of creating a full knowledge service. If you are working on a pilot project then less time and money are available, so the scope must be narrower, such as focusing on a particular part of a business process or a small area of law or medicine.

The other term to be aware of is "scoping." Scoping is used to describe that part of a knowledge acquisition project that defines the scope. Scoping involves the initial stages of knowledge capture and is often the first time you meet and interview the subject matter experts. Scoping usually involves three activities: (1) conducting scoping interviews with subject matter experts, (2) analyzing the content of the interviews, and (3) conducting a scoping meeting.

11.3.1 Scoping Interviews

The main aims of scoping interviews are to

- Identify the domain at a high level so you know what possible areas to select for the scope
- Identify which areas of knowledge will be easy to acquire and which will be harder
- Start to understand the jargon and terminology used by the subject matter experts
- Identify the resources that you can use to collect the knowledge, such documentation (reports, books, and websites), and the people to interview
- Establish a rapport with the subject matter experts

All of these are important for scoping but the first is the most important. To map out the domain, try to identify the different types of things involved. It is often useful at this stage to create a concept tree, that is, a taxonomy of the domain (see Section 11.7.2.1).

As covered later, all interviews are best done in a one-to-one, face-to-face interview. Sometimes, meetings with a group of people are useful but not for exploring detailed knowledge. It is often better to see people individually, so their opinions are not affected by hearing what other people say. As with all interviews, you should take an audio recording so you do not miss anything and you can focus on listening rather than spending time writing copious notes.

When interviewing a subject matter expert during scoping, you should ask a mix of general questions to get the breadth of the knowledge and use other questions or techniques to assess the relative importance of the knowledge for the project.

11.3.2 Scoping Analysis

After the scoping interviews, listen to the audio recordings and type notes as you listen. If you have asked the right questions, you should be able to form a picture of what knowledge will be essential to collect, what knowledge will be useful, and what knowledge will be irrelevant. You should also have an idea of the knowledge that will be easy to collect and the knowledge that will be harder to collect. Knowledge is easy to collect if it is already documented; there are plenty of exerts and the experts are used to talking about their knowledge with nonexperts. Knowledge is hard to collect if it is deeply buried in a person's head and the person is very adept at using the knowledge without even thinking about what knowledge is being used.

You should try to decompose the domain into different pieces of knowledge, so you can assess which should be in scope and which should be out of scope. If the project is process-focused, then break the domain into the different tasks that are performed. If the project is concept-focused, then break it down into a set of concepts such as issues, documents, products, or subject areas. It is a good idea at this stage to represent the knowledge you have collected using a diagram such as a tree or process map.

		Ease of developing the knowledge service		
		Easy	Medium	Hard
Benefit of the knowledge service	Very beneficial	Corporate trustee restructuring	Leveraged finance	General lending securitization
	Medium beneficial	Real estate finance insolvency	Equity capital markets	Debt capital markets asset leasing
	Less beneficial	Investment funds	Derivatives trade finance	Taxation

FIGURE 11.1 Example of a scoping grid.

A scoping grid is a useful way of representing what you think should be in and out of scope and why. A scoping grid shows a breakdown of the domain into individual pieces of knowledge and how each is rated against the difficulty of acquiring the knowledge (along one axis) and the benefits of acquiring the knowledge (along the other axis). Figure 11.1 shows an example of a scoping grid. In this example, "corporate trustee" and "restructuring" are the most favored areas for the scope because they have been rated as easy to develop and highly beneficial.

To position a piece of knowledge on a scoping grid requires rating it against the two dimensions along the axes. This can be done by asking the experts to simply rank them. A more detailed technique is to use a set of criteria for each dimension and ask the experts to rate each piece of knowledge against each criterion, then take a weighted average to arrive at the rating (i.e., some of the criteria are more important, hence have more weight, than other ones).

11.3.3 SCOPING MEETING

A scoping meeting is good way to tell people what you have found and discuss your proposed scope. The aim is to decide what the scope should be and to ensure that all the key people are happy with your proposals. You should invite the project manager and a representative sample of your main subject matter experts. Other people to invite are the managers of the subject matter experts (who are often the sponsors/funders of the project) and maybe some members of the development team. Prepare a set of slides that describe (1) what you did, (2) your aims, (3) a sample of the knowledge models you have created such as trees and diagrams, (4) your scoping ideas and your scoping grid, and (5) questions for the meeting to discuss.

11.4 WHAT IS MY STRATEGY?

When the scope is decided, you need to develop a strategy for capturing, analyzing, modeling, and communicating the knowledge. The strategy should identify who to see, when to see them, how long to see each person for, how you will analyze the knowledge, and how you will structure the knowledge using knowledge models.

Simple logistics will affect your strategy such as limitations on when you can see the experts and the distance you have to travel to conduct the interviews. Another important factor when defining the strategy is the way you will structure the knowledge. This will help to define what you need to cover during the interviews with the subject matter experts and will provide a framework for creating the different knowledge models. Such a framework is called a meta-model (or sometimes an ontology).

11.4.1 META-MODELS

You should try to define one or more meta-models as part of your strategy. Each meta-model should consist of several high-level concepts and the relationships between them. For example, you might define the following as your high-level concepts: *system*, *procedure*, *trigger*, *data*, *report*, *file*, and *summary data*. A good way to show the meta-model is as a concept map (see Section 11.7.2.2) on which you show the high-level concepts as nodes and show the relationships between them as arrows. Figure 11.2 shows an example of this. In this example, the high-level classes are shown in the rectangular shapes and the arrows between them show the relationships.

On some projects, the meta-model will also include the properties of each of the high-level concepts as a list of attributes (see Section 11.7.1.2).

During the knowledge capture interviews, you will collect the different subtypes of each high-level concept (class) and identify which ones relate to which other ones using the relationships defined in the meta-model. For example, if you were using the meta-model shown in Figure 11.2, you would identify all of the *systems* and all of the *procedures* and say which system is associated with which procedure. If attributes are involved, you will need to find the values that are related to each attribute, and which values are associated with each concept (see Section 11.7.1.3).

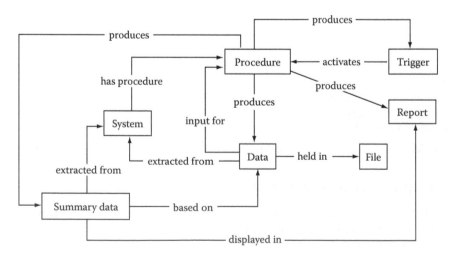

FIGURE 11.2 Example of a meta-model.

11.4.2 SPECIFIC AIMS

You should already have defined the aims of the project as a whole when envisaging the final product. You should also try to define a set of aims for the knowledge acquisition. For example, one aim might be that you are capturing knowledge in a particular area for a particular purpose. Or that you are capturing knowledge to demonstrate that a certain technique can be used to capture a certain type of knowledge. This latter aim is particularly relevant in a pilot project where it is often important to demonstrate that the methods being used will be effective when scaled-up to a full project.

11.5 SHOULD I USE A PARTICULAR METHODOLOGY?

A number of methodologies exist that provide a way of approaching and performing the knowledge acquisition phase of a project. The three main methodologies are CommonKADS, MOKA, and the 47 Steps method. Each of these methodologies covers the initial stages of knowledge acquisition for defining the aims and scope of a project. They also cover the types of meta-model you might want to use and give ideas on capturing the knowledge during interviews and in defining the knowledge base for use during implementation.

Each methodology has a particular emphasis and particular strengths and weaknesses, and you should consider these when you are deciding which one to use for your project. You do not need to use a predefined methodology, but it is useful to know about them so that it can influence the ways in which you work and provide ideas and pointers to improve your own methods and working practices.

11.5.1 COMMONKADS

CommonKADS is focused on capturing and modeling knowledge for use in expert systems, that is, intelligent systems, that aim to emulate the way in which subject matter experts reason about and solve a problem. CommonKADS defines several high-level models of the way people solve problems. These include planning, monitoring, and design using constraint satisfaction.

One of the strongest ideas in CommonKADS is that the generic model is used as a meta-model to drive all acquisition and modeling activities CommonKADS also includes methods for deciding what project to perform and what knowledge to capture. It also presents a formal way of representing the captured knowledge. A full account of CommonKADS can be found in Schreiber et al. (2000).

11.5.2 MOKA

MOKA stands for "methodology oriented to KBE applications," where KBE stands for knowledge-based engineering. MOKA, as the name suggests, is focused on KBE applications, which are intelligent systems that operate as add-ons or within CAD systems (i.e., computer-aided design systems) to help design engineers create designs for new components. Such systems can reduce design times drastically if carefully selected and efficiently developed.

MOKA defines six phases for a KBE development project covering the early stages of project definition, through capture and modeling, to packaging for implementation. MOKA is influenced by CommonKADS but provides a simpler modeling method in two stages: an informal modeling stage that primarily uses forms as templates to be filled in for each of the main concepts and a formal modeling stage that uses diagrammatic representations. MOKA defines five high-level concepts to use, which are entities, constraints, activities, rules, and illustrations. A full account of the MOKA methodology can be found in Stokes (2001), and an overview of knowledge-based engineering can be found in Milton (2008).

11.5.3   47 STEPS

Where CommonKADS and MOKA focus on certain types of knowledge acquisition project, the 47 Steps method is more generally applicable to all types of knowledge acquisition project. As the name suggests, it provides 47 steps to perform from the beginning of a project to its conclusion, covering aims, scoping, interviews, modeling, and preparation for implementation. Its emphasis is less on defining particular meta-models and modeling formats, and more on capturing knowledge from subject matter experts and using general modeling methods. A full account of it can be found in Milton (2007). (*Note*: the 47 Steps methodology was developed by the author of this chapter, so comes highly recommended!)

11.6   HOW SHOULD I INTERVIEW THE EXPERTS?

A common mistake made by people who are new to knowledge acquisition is to think that it is all about meeting subject matter experts, talking to them, writing notes, and then writing up what has been found in a document such as a report for use during implementation. As this chapter suggests, the stages leading up to the interviews are just as important as the interviews themselves. When planning the interview and deciding on the way to capture the knowledge, you should be very clear on the aims of the final product, who will use it, how it will be used, the scope of the knowledge you need to capture, and your strategy. This will provide you with a full context and a clear vision for the end product that will help you to focus the subject matter experts on the project at hand. It is essential that you do your very best to help the expert to stay within that focus and not to drift away from it and talk about interesting but irrelevant topics.

It is tempting to think of an interview as just a question and answer session. This is not the case. On many projects, interviews also involve special techniques that make the expert do things to expose their knowledge or to think about it in unusual ways. Because experts find it difficult, and sometimes impossible, to fully describe their knowledge to the level of detail you require, you must help them in various ways. One way is to approach the same piece of knowledge from different angles using different techniques or different types of question. This will help the expert to recall and describe the knowledge in an efficient way.

11.6.1 Types of Interviews

There are three basic types of interviews: unstructured, semistructured, and structured.

Unstructured interviews are interviews in which you have no predefined questions, and so it is an informal discussion. It is not recommended except for finding out broad aspects of the expert's knowledge and as an aid to designing an interview plan. For example, you might perform an unstructured interview prior to your main interviews as a 10 min chat over the phone or as a 15 min chat at the end of a more structured session.

Semistructured interviews are questioning sessions in which you use a set of preprepared questions to focus and scope what is covered. This type of interview also involves asking unprepared supplementary questions for clarification and probing based on the responses the expert has given to the planned questions. The planned questions should be designed carefully and sent to the expert before the interview. This gives the expert time to think about the questions you will be asking and about the kind of responses to give. Always take two printed sheets of the questions to the interview—one for you and one for the expert (as experts can often lose track of what they are talking about). The printed questions are useful for guiding the expert back to the topic if they stray too far away. If this happens, politely interrupt by saying something like "Can I stop you there, as I think we are drifting a little from the question."

A structured interview is an interview in which you use techniques that make the expert do things rather than just talk about things. These techniques fall into two categories: (1) techniques that involve the expert performing special tasks that reveal their knowledge and (2) techniques that involve a visual representation of the knowledge such as in a diagram or matrix.

In summary, there are three types of interview techniques: unstructured, semistructured, and structured. Unstructured and semistructured interview techniques are used to capture broad knowledge without too much detail. Semistructured and structured interview techniques are used to capture detailed knowledge in specific topic areas. Structured interview techniques are particularly useful for capturing tacit knowledge, as discussed next.

11.6.2 Ways of Acquiring Tacit Knowledge

Tacit knowledge is the subconscious knowledge used by experts and which they often do not know that they use. Since tacit knowledge is so deeply ingrained in a person's thought processes, experts are rarely sure when and how they use tacit knowledge and so find it hard, if not impossible, to describe. It is, therefore, hard to capture tacit knowledge by just asking questions. For this reason, other techniques are required.

As with any type of knowledge, it is important to have a clear idea of what knowledge you want to capture and to keep the expert focused on the knowledge you need. Preparation and planning are important. You should select the right interview technique, make the best use of the resources available, and forewarn the expert by sending an interview plan before the session. To capture the deep, detailed knowledge, you should

- Approach the knowledge from different angles
- Ask the expert to do things (not just say things)
- Represent the knowledge in different ways
- Prompt and probe the expert (but take care not to put forward your own ideas)
- Follow a cycle of interviewing, modeling, and validation (so that you capture the tacit knowledge over several interview sessions)
- Combine knowledge from multiple experts

Techniques that are useful for exposing tacit knowledge are limited data task, card sorting, triadic elicitation, protocol analysis, and twenty questions. Brief descriptions of these are given in the next section alongside some of the other techniques that can be used with experts.

11.6.3 Techniques for Acquiring Different Types of Knowledge

Detailed accounts of various interview techniques can be found in Milton (2007), Cooke (1994), Hart (1984), Hoffman et al. (1995), and Shadbolt (2005). A brief description of each of the main techniques follows.

11.6.3.1 Laddering

Laddering is a questioning technique for finding categories and subcategories of concepts. You repeatedly ask questions such as "What sort of thing is …?" "What types of … are there?" "What other types of … are there?" "What distinguishes a … from a …?" It can be useful to sketch a tree diagram as the expert answers each question.

11.6.3.2 Limited Data Task

Limited data task is a questioning technique to explore knowledge of complex tasks that have lots of factors and inputs. Before the interview, identify a complex task to be explored. At the interview, ask the expert "If you were to perform this task, but only had three pieces of information, what would these be?" After the expert has responded, ask "If you had three more pieces of information, what would these be?" etc. This technique is useful for focusing the expert on essential knowledge and priorities, and the strategies they use.

11.6.3.3 Card Sorting

Card sorting is used to elicit properties (attributes and values) for a set of concepts of the same type. Before the interview, you should identify the set of concepts and write their names on separate cards. At the interview, ask the expert to sort the cards into piles, so that the cards in each pile are similar in some way. After the expert has sorted the cards, ask him/her to give a name to each pile (each name is a value). To elicit more attributes and values, ask the expert to repeatedly sort the cards until he/she cannot sort anymore. An interesting variant is to use objects or pictures for the sorting instead of cards.

11.6.3.4   Triadic Elicitation

Triadic elicitation is often used alongside card sorting to probe for further properties. You should collect the cards together and randomly select three of them. Place these three cards on the table so two are next to each other, and the other is further way. Ask the expert "In what way are these two similar to each other but different from this one." These steps can be repeated for other groups of three randomly selected cards. The answers given by the expert provide further properties, which can then be used for card sorting on all the cards.

11.6.3.5   Protocol Analysis

Protocol analysis (also known as Commentary) is used to capture detailed knowledge of how the expert performs a complex task. Before the interview, select a task that the expert can perform at the interview or in a convenient location. Ask the experts to provide a running commentary as they perform the task. It is essential to take an audio recording or even a video so you can analyze what was said/done after the interview. There are several variations: (1) the expert imagines that they are doing the task rather than doing it for real, (2) the expert commentates as they watch a video of the task, and (3) a second expert does the commentary as the first expert performs the task.

11.6.3.6   Twenty Questions

Twenty questions is used to explore the expert's tacit knowledge by asking him/her to perform an unusual task. Ask the expert to play a game of twenty questions so that he/she will ask questions to you to deduce an answer. Each question the expert asks must have an answer that is "yes" or "no." You do not need to know the correct answers, but that does not matter as you can answer randomly. It is the questions asked by the expert that are the important point. These questions will often expose strategies and conceptual knowledge that may be difficult to capture using other techniques.

11.6.3.7   Process Mapping

Process mapping is used to capture knowledge of a key activity performed by the expert. Draw diagrams with the expert to show the main tasks involved in the activity, their order (using "followed by" arrows), any decision points, and the inputs (resources) and outputs (results) of each task. Each main task on a diagram can be further explored by drawing a separate diagram with the same format, that is, it shows its subtasks and how they relate to one another using "followed by" arrows, decision points, inputs, and outputs.

11.6.3.8   Concept Mapping

Concept mapping is used to capture knowledge of how concepts relate to one another. Draw diagrams with the expert to show a set of concepts as nodes and add arrows between pairs of them to represent the relationships between them. Label the arrows so that the relationships are explicit and clear. Some of the relationships will be specific to the domain of interest, and others will be general such as "part of," "made of," "uses," "creates," "requires," "influences," and "supports."

11.6.3.9 Teach Back

With Teach back, you teach the experts about their expertise by explaining to them what you have learnt in the previous interviews. Comments and corrections made by the experts will help reveal misunderstandings and clarify terminology.

11.6.3.10 Mixing Techniques at an Interview

An interview will often contain a mixture of different techniques. For example, a 90 min interview might contain 30 min of semistructured interview questions, 15 min of card sorting, 10 min of triadic elicitation, 5 min of limited data task, and 30 min of process mapping.

11.6.4 RIDING THE HORSE

It may seem a strange analogy, but in many ways interviewing an expert is like riding a horse. Some experts are like a stubborn horse that you have to motivate and get to move. Other experts are like bucking broncos uncontrollably moving from one topic to another. With such an expert, you have to exercise control with all of the skills and techniques at your disposal. Some experts will be used to moving in a certain direction and you will have to use your different techniques to move their attention and thinking processes to other directions.

Whatever the type of horse, it has all the power (i.e., the knowledge) and you must guide it to move where you want it to move (based on the scope and project aims). Trying to move more than one horse at the same time is a difficult job, so try to have one-to-one interviews whenever possible.

When things work well, rider and horse will form into a single unit and you will be working as a team. That is what good interviewing is all about—teamwork between you and the subject matter expert.

11.7 HOW SHOULD I ANALYZE AND REPRESENT THE KNOWLEDGE?

After an interview, you need to analyze and model the knowledge. Analysis means breaking something down so it can be examined. In this case, you are breaking down the things the expert said to identify the key pieces of knowledge. These pieces are called knowledge objects. When you have identified the knowledge objects, you will create knowledge models to represent the ways in which the knowledge objects are related to one another.

11.7.1 KNOWLEDGE ANALYSIS AND KNOWLEDGE OBJECTS

If you have taken an audio recording of the interview, you will be able to have it transcribed so you have a full textual record of the interview. Analysis can then be performed by highlighting a word or section of text with a marker pen. If you do this, it is a good idea to use different colored marker pens for different classes, for example, yellow for tasks, blue for physical objects, and green for issues. If you do not have a transcript, then you can do the same thing using the notes you have taken during the interview or notes you have made afterward when listening to the recording.

There are four main types of knowledge objects to look for during analysis: concepts, attributes, values, and relationships.

11.7.1.1 Concepts

Most of your knowledge objects will be concepts. These are the things in the domain, such as physical objects, abstract ideas, documents, roles, people, organizations, objectives, and tasks. They are equivalent to the nouns in a sentence. If you are doing a process-focused project, then tasks (activities, and processes) will be one of the main types of concept to look for.

11.7.1.2 Attributes

Attributes are a general property of a set of concepts. For example, "weight," "nationality," "efficiency," and "creation date" are all attributes. They do not give the particular weight or particular nationality of a concept, but provide a general description of what properties the concept possesses. In general, a class of concepts, such as vehicles, animals, or documents, will have the same set of attributes.

11.7.1.3 Values

Values give the specific property of a concept, such as its actual weight or specific nationality. Values come in two varieties: quantitative and qualitative. Quantitative values use a number and a set of units to describe the property, for example, 5 kg or 32 GHz. Qualitative values are adjectives that describe a specific property, for example, heavy, green, important, Russian, or difficult.

11.7.1.4 Relationships

Relationships (also known as relations) represent the ways in which pairs of concepts are related to one another. For example, the "is a" relationship says what class something belongs to, for example, *France—is a—Country*, or *Pig—is a—Mammal*. Another important relationship is "part of" to show composition, for example, *France—part of—Europe*, or *Tail—part of—Pig*. Some other useful relationships are "requires," "followed by," "produces," "uses," "made of," "has role," "has constraint," and "connected to." The choice of which relationships to use should be determined by the domain you are dealing with. That said, almost all projects will use the "is a" relationship and many will use "part of" (or its inverse "has part"). If you have a process-focused project, then look to use relationships such as "followed by," "produces," "input for," and "performs." If you are creating an ontology, then you may need to restrict the number of relationships to a small number of basic ones, such as "is a," "has synonym," and "related to."

11.7.2 Knowledge Models

Once you have a collection of concepts, you can start to say how they are related to one another and to say what properties each one has. To do this, you will be creating a set of knowledge models, each one showing a particular aspect of the knowledge. Building knowledge models is an essential aspect of knowledge acquisition.

Knowledge models are structured representations of knowledge, such as diagrams, grids, and forms. They are helpful in several ways:

- They help you to think clearly, be organized, and be analytical.
- They help you to validate the knowledge with experts.
- They help you to communicate and use the knowledge.
- They will be of help in the future as an aid to maintaining the knowledge service.
- They help in the reuse of the knowledge on other projects.

11.7.2.1 Trees

A tree is a hierarchical diagram that contains nodes connected by links. Each node is a box with text that represents a concept. Each link is a line that represents a relationship between a pair of concepts.

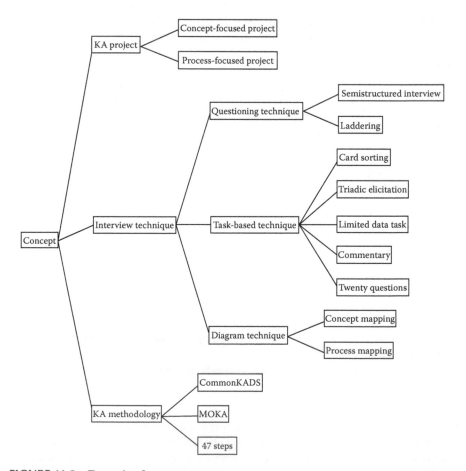

FIGURE 11.3   Example of a concept tree.

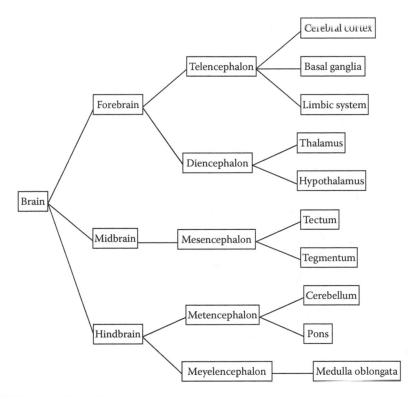

FIGURE 11.4 Example of a composition tree.

The most important type of tree is a concept tree. The thing that makes a concept tree special is that every link is the "is a" relation. By restricting a concept tree to just this one relation means it is used to represent a taxonomy (i.e., a classification structure) of the concepts in the domain of interest, that is, how each concept fits into a hierarchy of classes and subclasses. Figure 11.3 shows an example of a concept tree.

Another important tree is a composition tree on which all the links are the "has part" relation. Such a tree can be used to show how a concept (such as physical component, a document, or a task) breaks down into its component parts. Figure 11.4 is an example of a composition tree showing the parts and subparts of the human brain.

Trees can include more than one type of relation, but be careful not to overload a tree with too many different relations as it will soon become messy and difficult to read. A tree should be a clear hierarchy. Use a concept map if you want to show a more tangled view of concepts and how they are related using many different relations.

11.7.2.2 Concept Maps

A concept map is a diagram that shows nodes connected by arrows. Each node represents a concept such as a product, a person, or a document, and each arrow represents a relation between two concepts. It is important to label each arrow to show the relationship it represents. An example of a concept map is shown in Figure 11.5.

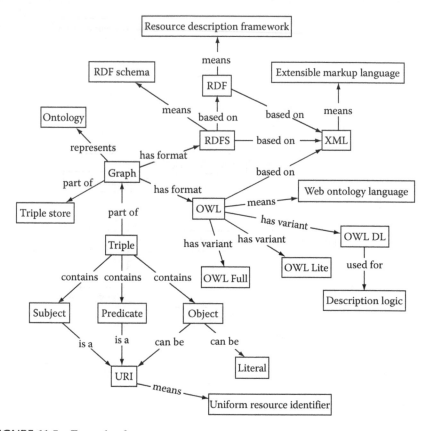

FIGURE 11.5   Example of a concept map.

Concept maps are useful when you want to show information that cannot be presented in a hierarchy using a tree. It can sometimes be useful to draw a concept map on a sheet of paper with a subject matter expert as part of an interview (see Section 11.6.3.8).

11.7.2.3 Process Maps

A process map is similar to a concept map, but has tasks (also known as processes and activities) as the main nodes on the diagram. The other types of nodes are usually resources, people, and decision points, although the exact format will depend on the modeling requirements. A decision point is a special node on a diagram that shows when a task is performed. For example, you might have a decision point called "is the material plastic?" The arrows from a decision point will be labeled with the answer "yes" and "no" and will lead to different tasks.

Figure 11.6 shows an example of a process map. In this example, the tasks are shown in the oval shapes. Items of data that are required to perform a task and/or are produced by a task are shown in rectangular shapes. There are two decision points, which are shown as diamond shapes with a question mark. The box symbol denoting "diagram tool" is a resource required by one of the tasks. Note that the arrows are not

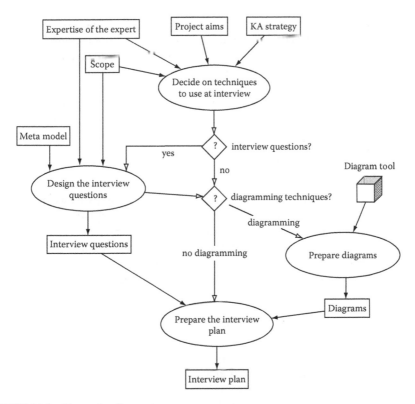

FIGURE 11.6   Example of a process map.

labeled with the name of the relationship (as on a concept map) since their meaning is obvious from the context. Although Figure 11.6 is shown in monochrome, process maps often use colors to denote different classes (node color) and different relationships (arrow color).

A process map is a very useful way of representing how people perform tasks and how information produced in one task is used as an input to another task. Hence, a process map is a vital knowledge model to use when you are working on a process-focused project. Drawing a process map from scratch or reviewing a preprepared version is a good way to help a subject matter expert explain processes in a clear and unambiguous way (see Section 11.6.3.7).

11.7.2.4   Frames

A frame is a grid that shows the properties of an individual concept. The width of the grid is always two columns. The left hand column shows the attributes that are relevant to the concept and the right hand column shows the values that correspond to the attributes.

11.7.2.5   Rules

Rules are a more complex way of representing knowledge than other knowledge models. They are statements that show how knowledge objects link together. A rule is always of the form "IF something THEN something," where the first "something" is called the antecedent and is a condition or a set of conditions, and the second "something" (after the THEN) is called the consequent and is an action or inference, or set of actions and inferences. For example, consider the rule: "IF car has 3 wheels AND width of car is less than half the length THEN car is unstable." In such a rule, the properties of a car are linked together to form a richer form of knowledge than in a frame, which just lists attributes and values. Rules are very useful when you want to represent the detailed reasoning processes that an expert uses to solve a problem or make a decision.

11.7.2.6 Matrices

A grid-format called a matrix can be used to show relationships between concepts and to show the attributes and values of concepts. There are two main forms of matrix. A relationship matrix shows one set of concepts along the x-axis and another set along the y-axis. The cells in the matrix show which concepts are linked by a particular relationship. The other main form of matrix is an attribute matrix that shows the attributes and values (along one axis) for a set of concepts (along the other axis).

11.7.2.7 Forms

A form is a table that describes all the information known about a particular concept. It is similar to a frame but shows more than just attributes and values. It also includes the relationships that exist between the concept and other concepts, and sometimes includes paragraphs of text and images that describe information about the concept in a less formal way. A form is sometimes referred to as an annotation page because it includes these annotated notes, and it is often used as the basis for the main descriptions in a knowledge books and a knowledge web (see Section 11.9).

11.8   HOW DO I ENSURE THE KNOWLEDGE IS COMPLETE, CONSISTENT, AND CORRECT?

The success of a project usually depends on three factors: (1) how well you understood the requirements, (2) how effectively you captured the knowledge, and (3) how well the knowledge was embedded in the deliverable. To succeed on the second of these factors means not only capturing the knowledge in as efficient a manner as possible, but also taking steps to ensure the knowledge you have captured is complete (for the project requirements), consistent (there are no contradictions), and correct (is an error-free representation of what the subject matter experts believe to be true). Ensuring these key success factors requires an ongoing process of validation and cross-validation, and the testing of a prototype deliverable.

11.8.1   VALIDATION

Validation means showing the knowledge you have collected back to the subject matter expert from whom you collected the knowledge. By doing this, the expert can check if it is correct and complete. Knowledge models are an important way to convey the knowledge back to the expert because they are clear, unambiguous, and easy to read. If you have produced good knowledge models, such as a concept tree and a process map, then the expert should be able to spot any gaps and mistakes quickly and easily. You can also use the teach back technique, which was described earlier.

11.8.2 CROSS-VALIDATION

Cross-validation is similar to validation but it means showing the knowledge you have collected to another subject matter expert rather than the one from whom you collected the knowledge. The same techniques can be used for validation. Ideally, you should use a mixture of validation and cross-validation as you capture the knowledge, that is, do not leave it to the end of the interviews to do the validation and cross-validation but do it in an ongoing manner during the interview phase.

11.8.3 RESOLVING INCONSISTENCIES

If you have interviewed more than one subject matter expert, then you might find that the experts disagree on certain things and there are inconsistencies in the knowledge. If this is the case, the best way to resolve this is to (1) identify inconsistencies during interviews and cross-validation sessions with individual experts, (2) compile a list showing the areas of disagreement, and (3) conduct a special session with several experts to go through each area of disagreement and reach a consensus on what knowledge should be used in the project.

11.8.4 CREATING AND TESTING A PROTOTYPE

Before you complete your interviews with the subject matter experts, you should always try to create some kind of prototype of the end deliverable so it can be assessed with a sample of the end users and shown to the experts. This will identify further areas of knowledge that need to be captured. In particular, assessing a prototype can show places where more detailed knowledge is required and places where the level of detail is about right.

You should try to create and assess a prototype as soon as you can so it will give you enough time to complete your knowledge capture activities. If you cannot create a full prototype, then a partial prototype is better than nothing. On some projects, it will be impossible or impractical to create a prototype. In these cases, a useful substitute is a description of the final system as a set of slides with some idea of the knowledge content and the ways it will be used in the end product.

11.9 WHAT DOCUMENTATION AND OTHER RESOURCES SHOULD I PRODUCE?

How you describe the knowledge you have captured will depend on the project deliverable and how the knowledge is to be used in it. Generally speaking, there are two types of project. In one type, your knowledge will be available to the end users directly. In this case, the knowledge is being shared from the experts to the users via the project deliverable. In the other type of project, the knowledge is embedded in the final system in some kind of database or programming code so that the system can do clever things with it. The amount of documentation you produce will differ depending on which type of project you have and who will do the implementation of the knowledge (you or someone else). It will also depend on how much effort has been allocated in the project plan for the documentation activities.

11.9.1 KNOWLEDGE BOOKS

A knowledge book is a document that describes the knowledge captured during the project. Its structure will normally reflect the way you analyzed and modeled the knowledge. If you have a process-focused project, then the knowledge book will usually be structured in terms of processes (tasks, activities), starting with a description of the high-level processes and then going into more detail for each of the subprocesses. For a concept-focused project, each subsection of the document will usually focus on a single concept and describe its properties and its relationships to other concepts.

All your knowledge models should be included in the knowledge book, so that a typical knowledge book will have quite a lot of diagrams such as trees, process maps, and concept maps.

There are two main purposes for a knowledge book: (1) to provide the people who will be implementing the knowledge with a description of the knowledge and (2) to document the knowledge for future uses, for example, when someone wants to reuse the knowledge for another purpose or wants to update or improve the knowledge service.

11.9.2 KNOWLEDGE WEBS

A knowledge web is a collection of web pages that describes the knowledge you captured using the same kind of structure as a knowledge book. It is useful to think of a knowledge web as an electronic version of a knowledge book. Rather than have a subsection in a knowledge book that describes a particular concept, you have a web page that describes the concept.

The advantage of having the information in a web format stems from the fact that knowledge is highly interrelated and so navigation from one topic to another (or from one concept to another) can be achieved with hyperlinks. A disadvantage of a web format is that it can take a lot longer to create, although tools are available (such as PCPACK) that automate the creation of a knowledge web assuming you have created knowledge models and associated notes in a database (Milton 2008). Search is obviously easier with a knowledge web than with a knowledge book, and diagrams such as trees and process maps can have hotspots on their nodes that provide hyperlinks to associated web pages.

11.9.3 KNOWLEDGE BASES AND ONTOLOGIES

As mentioned in the previous section on knowledge webs, an efficient way to store the knowledge you have captured is in an electronic database. A database that contains knowledge from an expert in the form of knowledge models and associated notes is called a knowledge base. Ideally, you will use a special tool to create and edit a knowledge base. An important feature of such a tool is that it can store the knowledge in a format that will be easy to translate into other formats, for example, to make it easy to create a knowledge web and to allow some auto-creation of programming code. Another important feature is that it can be used to display and edit the knowledge using various knowledge models, such as trees, process maps, concept maps, and frames. A tool called PCPACK has been developed for exactly such a purpose. Another tool, Protégé, can also be used, although it is limited in the way you can create and edit diagrams such as trees, concept maps, and process maps.

There are two types of knowledge base. The first type is generally used by people to find out what knowledge has been captured. The second type is used by a software system to do something clever that requires expert knowledge. An ontology is the name used to describe this second type of knowledge base. The essential difference between a knowledge base and an ontology is that the former is more detailed in its descriptions and visual representations (so it can be understood by a person), and the latter is more formal in the way it represents the knowledge (so it can be used by computer).

11.9.4   PSEUDO CODE

Some projects will require rules to be written that represent the reasoning processes taken by experts when they solve problems or take decisions. When the rules you have captured are implanted into the knowledge service, they will be coded using a programming language. To let the programmers know what to code, and to document the knowledge that underlies the code, it is important to describe the rules in a semiformal way that is independent of the programming format. This can be referred to as a type of pseudo code, that is, it is format similar to a programming language but is more readable by people.

11.10   DOES THE KNOWLEDGE NEED TO BE UPDATED IN THE FUTURE?

The world changes—technologies change, working practices change, products change, and organizations change in all manner of ways. This means that the knowledge you capture on a project is likely to become less and less useful over time. If your project requires that the end deliverable lasts for many years, then the knowledge will need to be maintained and updated. If so, what are you going to do now to allow those changes to be made in an efficient and effective manner? Another question for you: Is the knowledge you have captured as complete, correct, and relevant as required by the end users, or does it need updating on a regular basis so that the system as a whole can be improved? If so, you will need a mechanism for the knowledge to be updated.

11.10.1   FEEDBACK FROM USERS

If the users of the knowledge service are people, they will have opinions on how good the system is and how good the knowledge is that has been incorporated into it. If possible, you should provide a way for their opinions and ideas to be fed back to you or someone else on the project team so that improvements can be identified and assessed for inclusion in a future version of the knowledge service. If the end users are reading about knowledge on a web page, then one possibility is to include a way for the user to provide comments, which can be saved in a database or e-mailed to the appropriate person. Other methods for getting feedback are to hold workshops, perform interviews, or send questionnaires to people.

11.10.2   MAKING MAINTENANCE EASIER

There are certain things you can do now that will make maintenance in the future an easier task. When capturing the knowledge, find out from the subject matter experts how long they expect the knowledge to last and which parts are likely to change over the coming years. They may already know about new technologies or new working methods that are going to be introduced or could be introduced. Second, create clear descriptions of the knowledge so that is easy for the people who will be making the changes to see what knowledge was captured and allow them to locate the places that need updating. Using a structured knowledge base that can be viewed using different knowledge models is a good way to do this. If your knowledge is to be embedded in an intelligent software tool or a semantic system, then provide a clear record of the knowledge that was captured in a format that can be understood by people so they can see what knowledge went into the system. Third, think about building special procedures or functionality into the knowledge service so that it can warn people when a review of the knowledge is required.

11.11   WHAT DID I LEARN FROM THE PROJECT?

Acquiring knowledge from subject matter experts is a skilled activity. The more you do it, the better you become. It is important to capture your increasing experience and the lessons you learn on each project in an explicit way. Although this takes a little more time and effort, it is worth it as it will prevent you from forgetting what you have learnt and it will allow other people to benefit from your growing expertise.

11.11.1   LESSONS LEARNT REVIEWS

A lessons learnt review is a meeting in which a group of people who have worked together on a project get together to share what they thought about the project and how things can be improved next time a similar project is performed. These reviews work best when a facilitator from outside the project team leads the review and guides people to be open and think constructively about the improvements that can be made. Such reviews can take place at the end of any project, and so can be used at the end of a project that has involved knowledge acquisition. The facilitator will usually document the findings of the review and make recommendations for changes that can and should be made.

11.11.2 Changing the Methodologies and Tools

Whether or not you take part in a formal lessons learnt review, it is important that you review the methods and tools you used on a project and try to think how they can be improved the next time you undertake a similar project. If you have used a standard methodology such as CommonKADS, MOKA, or the 47 Steps method, or you have used your own way of working, then try to think how the methodology can be improved. Maybe you can skip certain activities, or maybe you need to include additional activities. If you are using software tools to help capture and model the knowledge, then think how they can be improved. Perhaps, you have a library of generic models that you use on projects. If not, think about starting to create such a library. Using a generic model that defines basic classes of concepts and a set of basic relationships means you do not have to start with a blank sheet of paper every time.

Think also about your interviewing strategy and technique and how it can be made more efficient and effective. Maybe you need to do more interviews or spend longer on each interview.

A word of warning: Do not forget that every project is different. If things worked well on one project, they might not work so well on another. So try to learn from other people who do knowledge acquisition, as well as learn from your own experience.

11.12 SUMMARY

The main parts of a knowledge acquisition project are the interviews with the subject matter experts. We saw in Section 11.6 that there are a range of techniques that can be used, each one having its particular strengths in capturing certain types of knowledge. But interviews are not the whole story. Just as important is what happens before, in between, and after the interviews.

In the time leading up to your first interview, it is essential that you find out what knowledge to capture—its type, its scope, and its level of detail. Defining this begins by understanding what the project is aiming to achieve (see Section 11.2) and continues into the scoping phase of the project (see Section 11.3). It is also important before the interviews to start thinking how you will structure (model) the knowledge and if you are going to use a particular methodology (see Sections 11.4 and 11.5).

During the time between interviews you will be analyzing and modeling the knowledge elicited from the experts. This means identifying the key knowledge objects such as concepts, relationships, attributes, and values and building these into clear representations called knowledge models (see Section 11.7). Some interviews will focus on validating and cross-validating the knowledge you have captured so you can check that it is complete, correct, and consistent (see Section 11.8).

During and after the interviews, you will produce documentation and other resources that describe what knowledge you have collected and represent it in a format that will be useful for the remainder of the project (see Section 11.9). On some projects, it is useful to think how the knowledge will be updated in the future (see Section 11.10). It is useful on all projects to ask "what did I learn from the project?" and to use this to improve your methods and working practices (see Section 11.11).

REFERENCES

Cooke, N. J. 1994. Varieties of knowledge elicitation techniques. *International Journal of Human–Computer Studies*, 41(6), 801–849.

Hart, A. 1984. *Knowledge Acquisition for Expert Systems*. New York: McGraw-Hill.

Hoffman, R., Shadbolt, N. R., Burton, A. M., and Klein, G. 1995. Eliciting knowledge from experts: A methodological analysis. *Organizational Behavior and Decision Processes*, 62, 129–158.

Milton, N. R. 2007. *Knowledge Acquisition in Practice: A Step by Step Guide*. London, U.K.: Springer.

Milton, N. R. 2008. *Knowledge Technologies*. Milan, Italy: Polimetrica (this book is available as a free e-book at http://eprints.rclis.org/handle/10760/11158).

Schreiber, A. Th., Akkermans, J. M., Anjewierden, A., De Hoog, R., Shadbolt, N., Van De Velde, W., and Wielinga, B. 2000. *Knowledge Engineering and Management: The CommonKADS Methodology*. Cambridge, MA: MIT Press.

Shadbolt, N. R. 2005. Eliciting expertise. In Wilson, J. R., *Evaluation of Human Work*, 3rd edn. London, U.K.: Taylor & Francis, pp. 185–218.

Stokes, M., Ed. 2001. *Managing Engineering Knowledge: MOKA Methodology for Knowledge Based Engineering Applications*. London, U.K.: Professional Engineering Publishing Ltd.

12 Prospects for Applying Fuzzy Extended Logic to Scientific Reasoning

Vesa A. Niskanen

CONTENTS

12.1 INTRODUCTION

The Western natural sciences have a long methodological tradition based on bivalent reasoning and mathematical formulation. The former paradigm adopts an either–or approach to the examination of the phenomena and the latter often assumes a full correspondence between the mathematical and the real world. Despite their fairly successful results in the natural sciences, these traditions have aroused problems when applied to human reasoning or intelligent computer models. The bivalent models are too coarse for representing complex phenomena, and the conventional mathematical models have often been too complicated in any discipline. In particular, these methods have been more or less controversial when applied in the human sciences (social sciences, behavioral sciences, economics, etc.).

From the philosophical standpoint, the foregoing paradigms are mainly promoted in Marxism (as it was applied in the Soviet Union and East European countries) and the positivistic tradition, the latter including logical empiricism, logical positivism, and analytic philosophy (prevailing in the United States, the United Kingdom, and Scandinavian countries). Today, Marxism seems to be more or less outdated, whereas positivism still quite strongly affects on our contemporary Western methodology in the natural sciences as well as in the quantitative research in the human sciences. In a sense, the conventional natural sciences have been the flagships of positivism, whereas in the human sciences it is more or less widely assumed that positivistic methods alone are insufficient. Hence, alternative philosophical approaches to the human sciences have been suggested in phenomenology, hermeneutics, existentialism, critical theory, and postmodernism, inter alia. This more or less loose group of philosophical streams is also referred to as the Geisteswissenschaften (which is originally a German term). In the methodology of the human sciences, the positivistic approach applies quantitative methods and a lot of computing, whereas the Geisteswissenschaften use primarily qualitative methods as well as some quantitative methods (Denzin and Lincoln, 1994; Fetzer, 1993).

In the real world, we often deal with entities and phenomena that are imprecise, complex, complicated, incomplete, inconsistent, or uncertain by nature. These features may arouse serious problems in the conduct of inquiry, particularly in a computer environment. Later, we will mainly focus on the problems of imprecision. We assume that a linguistic expression is imprecise if it, possibly in addition to the clear cases, refers to borderline cases. For example, the expression "young person" is imprecise because it refers to the set of young persons and this set has borderline cases, that is, such persons who are not clearly young but rather more or less young. Other types of imprecision are also available, but we concentrate on this type of linguistic (and semantical) imprecision (Niskanen, 2004).

The quantitative research usually aims to use precise terms, reasonings, hypotheses, theories, models, and scientific explanations in their conduct of inquiry because it assumes that this approach also guarantees the precise nature of the quantitative studies in general. However, this assumption is based on a fallacy because in fact our studies seem to comprise various imprecise constituents. In particular, the human sciences are very challenging in this respect. Hence, the problem arises how well we can cope with imprecision in the scientific research.

We consider the role of imprecise entities in the scientific research from the standpoint of the philosophy of science in the following. We will examine imprecision in scientific reasoning, theory formation, and hypothesis verification. We will also provide meta-level resolutions for dealing with this subject matter, and in this context we apply the ideas of the fuzzy extended logic. Our approach will hopefully open new prospects for considering the central aspects of science and research.

12.2   APPROXIMATE REASONING WITHIN THE FUZZY EXTENDED LOGIC

Fuzzy logic and its approximate reasoning are one of those central topics that have aroused lively debates with the traditional bivalent approaches. We can study reasoning from such standpoints as psychology, physiology, biology, logic, and methodology. We focus on logico-methodological aspects and thus we mainly consider problems of logic and argumentation in the following. Thus, in general, in reasoning we draw conclusions according to the given premises.

If we perform reasoning, we should first specify our arguments or find the available arguments in our object of study. Second, in scientific research it is also important to draw a distinction between arguments, explanations, and descriptions. For example, consider the following statements:

1. Lotfi Zadeh is the inventor of fuzzy systems because he presented first these ideas.
2. Lotfi Zadeh introduced fuzzy systems in order to construct better computer models.
3. Lotfi Zadeh introduced fuzzy systems.

These sentences represent argument, explanation, and description, respectively, and on some occasions, for example, in scientific explanation and theory formation, we use their combinations simultaneously (Niiniluoto, 1984).

Various types of reasoning are available. We focus on theoretical reasoning and hypothesis verification. The former usually applies affirmations and such standard forms of reasoning methods as syllogisms. The latter deals with methods that are used to verify scientific hypotheses, that is, we consider how our hypotheses can be tested, proved, accepted, rejected, confirmed, or disconfirmed.

Reasoning can base on intuitive and informal rules and assessments, but today symbolic representation and formal arguments are used in particular in the bivalent logics. This principle of the formal correctness of reasoning in the manner of bivalent logic has played a central role in the Western scientific outlooks. However, the bivalent logics have encountered semantic problems because our actual reasoning does not correspond with them, and thus the attempts for modeling human reasoning have failed in this tradition. It has even been stated that bivalent logic was sufficiently simple calculus to use in the precomputer age, whereas today we can abandon it because we may apply more applicable systems with the computers (Haack, 1978; Niiniluoto, 1983, 1984; Rescher, 1980).

We can also study reasoning by considering the nature of our premises, in which case the fundamental question is whether they are necessarily true or not. In the former case, we can apply demonstration and in the latter case dialectics. Examples are Euclid's geometry and Socrates's reasoning method, respectively.

If we, in turn, consider the relationship between the premises and the conclusion, a distinction between deductive and inductive reasoning is usually drawn (Rescher, 1980). Deductive reasoning contains nothing in the conclusion that is not already contained in the premises. This idea provides a basis for syntactic validity (or theorem-hood), which is a research object of proof theory. Semantic validity, in turn, means that the conclusion is true whenever all the premises are true. In traditional bivalent logic tautologies are semantically valid, whereas in the fuzzy logic we can also consider whether truth-preserving reasoning with an alternative degree of truth fulfills semantic validity (Haack, 1978; Niiniluoto, 1983).

In inductive reasoning, it is assumed that the conclusions go beyond what is contained in their premises and thus it is regarded as ampliative with respect to our knowledge if the conclusions are true. Unlike deduction, induction is, however, not necessarily truth preserving, and thus it is possible for the premises to be true, but the conclusion nontrue. In this context, the degree of support for the conclusion provides a basis for the concept of inductive strength. Another clear distinction between deduction and induction is that in the former we can add new premises to our premise set and the conclusion still logically follows from this set. In practice, various types of inductive reasoning are available (Haack, 1978; Niiniluoto, 1983).

The fuzzy systems, which apply a specific multivalued logic, seem to mimic the human reasoning fairly well, and they have also resolved many other problems in particular in computer modeling. Conventional bivalent logic and set theory only deal with precise entities, that is, sets and relations, whereas fuzzy systems can also operate with imprecise, fuzzy, entities. If we can cope with imprecision elegantly, we can greatly improve our research methods, and fuzzy systems seem very promising in this respect. In both cases we usually apply standard mathematical tools, in particular in computer environment. Next we consider briefly the basics of fuzzy set theory and logic.

In the case of conventional sets, the crisp sets, each object either belongs or does not belong to a given set. Fuzzy sets, in turn, also allow partial memberships. In mathematical terms, this idea is usually presented with the membership functions, $\mu: E \to [0, 1]$, in which the domain set E is referred to as the reference set and the co-domain, the closed interval, denotes the degrees of membership. In this context, zero and unity denote nonmembership and full membership, respectively. Thus, crisp sets only operate with the values zero and unity, whereas within fuzzy sets objects may have degrees of membership that range from 0 to 1. This revolutionary idea was first suggested by Lotfi Zadeh in 1965, and it opens several interesting prospects for theory formation and model construction (Bandemer and Näther, 1992; Nikravesh et al., 2008a,b; Zadeh, 1975).

Consider the set of young persons. If crisp sets are used, we have to assign such threshold value that draws a distinction between the young and non-young persons. For example,

If the age of person $X \leq 25$ years, he/she is young otherwise he/she is nonyoung

It follows that if, for example, X is 26 years, he/she is already nonyoung because we only have two alternatives. This reasoning is thus counterintuitive because it seems justifiable to assume that a person aged 26 years is also more or less young, that is, we have borderline cases in this context. Hence, in the case of imprecise sets, the fuzzy sets, in which case we have borderline cases, crisp set theory seems inappropriate.

The fuzzy sets take into account borderline cases and such gradual changes that are typical in the real world. We may now assume that till 25 years the persons have full membership to the set of young persons, and then this membership gradually decreases to zero as the age increases. For example, if our reference set is the ages of persons in years, we may establish such degrees of membership to the set of young persons as shown in Figure 12.1.

$\mu_{young}(10) = 1$
$\mu_{young}(25) = 1$
$\mu_{young}(26) = 0.95$
$\mu_{young}(30) = 0.50$
$\mu_{young}(45) = 0$

This idea can also be applied to fuzzy logic, and thus, in addition to the truth values false and true of bivalent logic, we can also use such values for the borderline

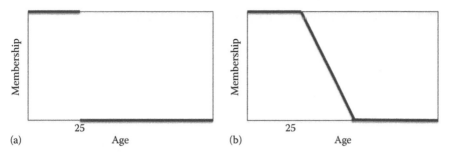

FIGURE 12.1 Degrees of membership to the set of young persons when traditional crisp sets (a) and fuzzy sets (b) are used.

cases as fairly true, neither true nor false, or fairly false. We can also operate with numerical truth values (e.g., from zero to unity) if necessary.

We can now resolve the ancient Sorites paradox, which has been problematic to bivalent logic. Consider this version of the paradox:

A person aged N years is young.
If a person aged N years is young, then a person aged $(N + 1)$ years is also young.
Hence, a person aged $(N + 1)$ years is young.

Our problem now focuses on the second premise, the implication. Is it true or not? If it is true and we apply bivalent logic, we may first reason that

If a person aged 1 year is young, then a person aged 2 years is also young

Next, we may reason that

If a person aged 2 years is young, then a person aged 3 years is young,

and so forth. This chain of reasoning, however, will lead us sooner or later to such counterintuitive conclusions as the persons aged 70 or 100 years are young. Hence, the second premise cannot be true. On the other hand, this premise appears not to be clearly false. If we, in turn, assume that only the persons whose ages are below a given limit point (e.g., 25 years) are young, we encounter the problem of artificial cutoff discussed earlier. The Sorites paradox thus arises in bivalent logic.

In fuzzy logic, which is multivalued by nature, we assume that the truth values of expressions can gradually change between true and false. Within Sorites paradox this means that the truth value of the implication is not true but almost true and thus the truth values of our conclusions approach gradually falsity when age increases. For example,

Truth (The age of a person is N years) = true, if $N \le 25$ years
Truth (The age of a person is 26 years) = almost true
Truth (The age of a person is 30 years) = fairly true
Truth (The age of a person is 40 years) = fairly false
Truth (The age of a person is 50 years) = false

In this manner, we have resolved the Sorites paradox because we can now take into account the gradual decrease of the truth values.

From the mathematical standpoint, one method for assessing the truth values of expressions in fuzzy logic is based on fuzzy similarity relation. Hence, we assess the degree of similarity between a given expression and its true counterpart. The higher the degree of similarity, the closer to truth is our expression. For example,

Truth (John is young) = true, provided that John is in fact young

In this case, we have maximum similarity between the expression young (left) and its true counterpart (right). On the other hand, if minimum similarity (or maximum dissimilarity) prevails, we reason

Truth (John is young) = false, provided that John is in fact old

An example of a truth value between these extremities is

Truth (John is young) = fairly true, provided that John is in fact more or less young

or, alternatively,

Truth (John is 20 years) = fairly true, provided that John is 25

It follows that, unlike in bivalent logic, true is distinct from nonfalse and false is distinct from nontrue. Nonfalse now means anything else but falsity (also including truth), and nontrue includes anything else but truth. We can also establish rules for assigning truth values to negation, compound sentences, and even for quantifiers.

TABLE 12.1
Typical Values Used in Fuzzy Systems

Types of Values	Examples
Precise numerical values and intervals	5, 0.5 [4.5,6]
Approximate numerical values and intervals	About 5, about 0.5, about [4.5,6], from about 4.5 to 6
Precise numerical functions and relations	$x^2 + 2y^3 + 1$, $x = y$
Approximate numerical functions and relations	Approximately $x^2 + 2y^3 + 1$, approximately $x = y$
Precise and approximate linguistic values and relations	Male, negative, small negative, very high, fairly old, not good, young or fairly young, slightly greater than, approximately equal to

Hence, our fuzzy truth evaluation also draws a crucial distinction between fuzz-iness and traditional probability because the probability of John being 20 in the condition that he is 25 would yield zero probability. However, we still have the ever-lasting debate on the distinction between probability, or uncertainty, and fuzziness in the scientific community, although these discussions are quite often based on the misunderstandings of these two concepts. In a nutshell, fuzziness and probability belong to the studies in semantics and epistemology, respectively. On the other hand, we may apply fuzzy probability to uncertain phenomena in which case we operate with such concepts as "probability is approximately 0.5" or "very likely." This issue is discussed in detail later.

Within fuzzy systems, we can operate with both linguistic and numerical values (Table 12.1). One usable method for constructing linguistic values for linguistic vari-ables is to apply the widely used Likert's or Osgood's scales in the manner they are used in the human sciences. For example, given the variable age among persons, we may proceed as follows (Niskanen, 2004):

1. Specify two primitive terms that are antonyms (if possible). *Young* and *old* seem appropriate to age.
2. Specify such linguistic modifiers (adverbs) as *very*, *fairly*, or *more or less*, and construct usable symmetrical scales with the modifiers and primitive terms. Five or seven values are widely used in this context. For example, *young–fairly young–middle-aged (neither young nor old)–fairly old–old*.
3. We can also use negation *not*. For example, *not fairly young*.
4. Use such connectives for compound expressions as *and*, *or*, and *if–then*.
5. If necessary, specify such quantifiers as *all*, *most*, *some*, or *none*.
6. We may also use conventional numerical values, as well as such fuzzy num-bers as *approximately 5* or *approximately between 4 and 6.*

In our studies, given a linguistic variable and the reference set, each linguistic value refers to a certain fuzzy set, and in practice we operate with the corre-sponding membership functions of these sets. Figure 12.2 depicts an example of

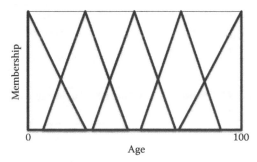

FIGURE 12.2   Tentative membership functions for denoting the linguistic values young, fairly young, middle-aged, fairly old, and old in the context of age.

the membership functions (fuzzy sets), which may be used in calculations with the variable age in a computer environment. Various shapes, triangular, bell-shaped, trapezoidal, etc., for these functions are available. We may specify these functions according to our expertise or empirical data (Bandemer and Näther, 1992; Zadeh, 1975).

Both fuzzy set theory and fuzzy logic were originally designed by Lotfi Zadeh, but today we have thousands of papers and dozens of books in the Globe that consider this topic. We also have thousands of applications in such areas as control, decision making, pattern recognition, data mining, and robotics. It is quite usual that fuzzy systems are integrated with such systems of computational intelligence as neural nets, evolutionary computing, and probabilistic modeling. Computational intelligence is also known as soft computing, in particular when it includes fuzzy systems (Nikravesh et al., 2008a,b).

Recently, Zadeh has established the principles of the extended fuzzy logic, *FLe*, which is a combination of "traditional" provable and "precisiated" fuzzy logic, *FLp*, as well as a novel meta-level "unprecisiated" fuzzy logic, *FLu* (Zadeh, 2009). He states that in the *FLp* the objects of discourse and analysis can be imprecise, uncertain, unreliable, incomplete, or partially true, whereas the results of reasoning, deduction, and computation are expected to be provably valid. In the *FLu*, in turn, membership functions and generalized constraints are not specified, and they are a matter of perception rather than measurement. In addition, in the *FLp* we use precise theorems, classical deducibility, and formal logic, whereas the *FLu* operates with informal and approximate reasoning (Figure 12.3). The *FLe* stems from Zadeh's previous theories on information granulation, precisiated language, and computing with words, as well as on the theory of perceptions (Zadeh, 1975, 1996, 1997, 1999, 2002, 2006, 2009).

Zadeh's ideas mean that we can apply both traditional bivalent-based and novel approximate validity, definitions, axioms, theories, and explanations, inter alia. The central

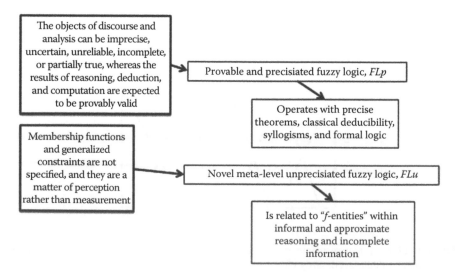

FIGURE 12.3   Zadeh's extended logic, the *FLe*.

concept in Zadeh's *f*-validity is the notion of truth. We can now operate with the degrees of truth, and we may apply the aforementioned meta-rule that the closer a given statement is to its true counterpart, the higher is its degree of truth.

In addition to model construction, we can also apply the foregoing truth evaluation principle to theory formation, hypothesis verification and scientific explanation, inter alia. These examinations, in turn, are based on semantical validity of reasoning, and within the *FLe* this means that, instead of traditional *p*-validity, we use approximate reasoning and other approximate entities, the *f*-entities. These *f*-entities are approximate counterparts of the corresponding traditional constructions. For example, according to Zadeh (2009, p. 3177): "A simple example of a *f*-theorem in *f*-geometry is: *f*-medians of *f*-triangle are *f*-concurrent. This *f*-theorem can be *f*-proved by fuzzification of the familiar proof of the crisp version of the theorem."

Hence, *f*-theorems are approximate counterparts of the corresponding traditional theorems and *f*-validity and *f*-theorems are examples of Zadeh's impossibility principle. This principle informally states that in an environment of imprecision, uncertainty, incompleteness of information, conflicting goals, and partiality of truth, *p*-validity is not, in general, an achievable objective (Figure 12.4). In practice, approximate reasoning thus operates with approximate premises and conclusions. Figure 12.5 provides us with a general picture of the potentiality of the *FLe*.

The well-known syllogism, the fuzzified Modus Ponens, for example, corresponds with Zadeh's *FLe* approach. Its traditional bivalent form is

A
If A, then B
Thus, B

For example,

John is 20 years
If John is 20 years, then he is young
Thus, John is young

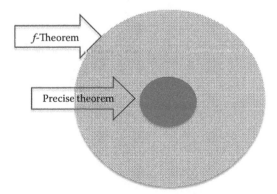

FIGURE 12.4   Zadeh's syntactic *f*-validity yields approximate theorems, which locate in the neighborhood of their corresponding precise or true theorems.

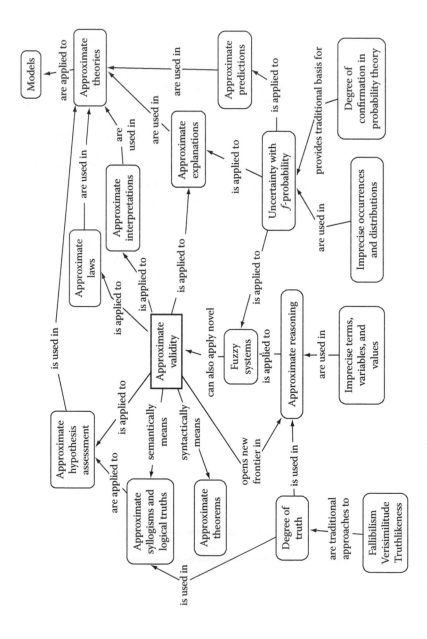

FIGURE 12.5   Role of the *FLe* in the conduct of inquiry.

In the fuzzified version the first premise need not to be identical with the antecedent of the second premise, and thus we use the Modus Ponens in a form

A1
If A, then B
Thus, B1

If statement A1 is now approximately identical with statement A, we can draw the
conclusion B1, and this is approximately identical with B. If A1 = A, our syllogism is
identical with the traditional Modus Ponens and B1 = B. An example of approximate
reasoning is

John is 22 years
If John is 20 years, then he is young
Thus, John is approximately young

In other words, when truth values are applied, we assess the truth value of the conclusion on the basis of the truth values of the premises.

In this reasoning, the second premise, the fuzzy implication, plays an essential
role. Several versions of implications are available, and we usually presuppose that
Truth (antecedent) ≤ Truth (consequent) if the implication is true, otherwise the
implication is nontrue. Sometimes implication's correspondence with our intuition
can be problematic, and no fully satisfactory resolution is available thus far.

If we apply the degrees of truth, we may establish the meta-rule that

In a true implication it holds that Truth (antecedent) ≤ Truth (consequent).
Otherwise, the higher the truth value of the antecedent compared to the
truth value of the consequent, the lower the degree of truth of the implication is obtained.

Hence, if the antecedent is true and the consequent is false, we obtain a false implication,
whereas a true antecedent and fairly true consequent would yield a fairly true outcome.

If numerical degrees of truth from zero to unity are used, Lukasiewicz's implication provides us with one resolution (Haack, 1978):

Truth (if A, then B) = min(1,1-Truth(A) + Truth(B))

Now, within the Modus Ponens, we assess the truth value of the consequent
according to the truth values of the antecedent and implication. Consider again the
example on the fuzzy Modus Ponens mentioned earlier. If we operate with the fuzzy
truth values, we may reason

- John is 22 years (true counterpart)
- If John is 20 years, then he is young (true implication)
- Truth (John is 20 years) = fairly true, provided that John is 22 (antecedent's
  truth value)
- Thus, consequent's truth value, Truth (John is young), is between fairly true
  and true, which means that John is approximately young

We can also apply nontrue or approximate implications to Modus Ponens in which case we draw even more approximate conclusions. Thus, for example, given a non-true implication, we may reason that

A (true)
If A, then B (nontrue)
Thus, B1

and B1 ≈ B even though the first premise is identical with the antecedent. The role of non-true and approximate implications should be studied more because then we can better consider and model the approximate interrelationships between the phenomena.

The fuzzy Modus Ponens is a powerful tool in the fuzzy rule–based systems, which comprise if–then rules, because it enables us to reduce the number of rules considerably (Bandemer and Näther, 1992; Takagi and Sugeno, 1986). Hence, a rule base of a few dozen fuzzy rules may replace thousands of ordinary rules (Figure 12.6). For example, consider a heater that controls the room temperature with the fuzzy rules:

1. If the temperature is approximately ≤0°C, then use the heater with 100% power.
2. If the temperature is approximately 10°C, then use the heater with 75% power.
3. If the temperature is approximately 15°C, then use the heater with 50% power.
4. If the temperature is approximately ≥20°C, then use the heater with 0% power.

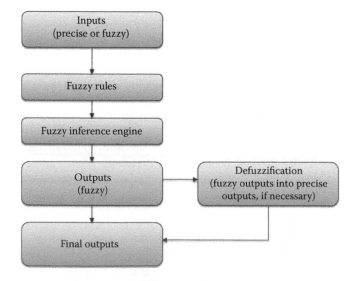

FIGURE 12.6 Flowchart for constructing a fuzzy system. Fuzzy sets and rules for the system are generated according to the knowledge of experts or empirical data.

We notice that we do not have rules for each possible temperature, but only a few typical cases, "landmarks," are given. Thanks for the fuzzy Modus Ponens, we can nevertheless obtain an appropriate heating power for each temperature by applying fuzzy reasoning that is analogous to interpolation. We thus assume that each rule is an implication in the fuzzy Modus Ponens syllogism and our input temperature is the other premise:

> The temperature is $x1°C$.
> If the temperature is approximately $x°C$, then use the heater with $y\%$ power.
> Thus, use the heater with $y1\%$ power.

The closer the given input temperature is to the antecedent of implication, the higher importance, the "firing strength," this rule will have in our reasoning. We evaluate each rule in this manner, and our conclusion, the output, is in a sense the weighted average of the consequents of the rules (sometimes this type of reasoning module is referred to as the inference engine). For example, given the input "the temperature is 14°C," the third and second rules have fairly high and fairly low firing strengths, respectively, whereas the rest of the rules seem irrelevant in this case. Hence, the power of the heater is assigned to be between 50% and 75%, and the output is closer to the lower limit because the third rule is more important. In practice, a more specific or precise output value will be calculated and usually in a computer environment (Bezdek and Pal, 1992; Takagi and Sugeno, 1986).

Unlike in the traditional rule-based systems, the fuzzy systems only need a few central rules, and then we can yield outputs for any given inputs. All our rules are fired simultaneously with various firing strengths. Another advantage is that, instead of mathematical formulas, we may apply user-friendly linguistic rules in our system construction.

Another widely used syllogism is the Modus Tollens, which is traditionally having the form

> If A, then B
> not B
> Thus, not A

We may now reason that

> If John is less than 25 years, then he is young
> John is not young
> Thus, John is not less than 25 years, that is, John is at least 25 years

Its approximate form in the *FLe* would be

> If A, then B
> B1
> Thus, A1

This syllogism is identical to the traditional case when B and B1 have maximal dissimilarity (e.g., B and B1 are antonyms). In other words, when it holds that Truth(B) is false, provided that B1. Then the conclusion, A1, is having the maximal dissimilarity with A. Otherwise, the lower the degree of dissimilarity between B and B1 (the higher the degree of similarity), the higher the degree of truth of the conclusion. For example, if the truth values of the implication are assigned as earlier,

If John is less than 25 years, then he is young (true)
John is fairly young (true)
Thus, Truth (John is less than 25 years) is between false and fairly true

This is due to the fact that in the fuzzy Modus Tollens the Truth (John is young), provided that he is fairly young, is fairly true, and hence, because the implication is true, its antecedent, also our conclusion, should be between false and fairly true. In this context, we can also apply approximate implications in the manner of the approximate Modus Ponens mentioned earlier, and then we will draw even more approximate conclusions.

Fuzzy systems applied to approximate reasoning can resolve problems that are superb to bivalent logics. They can overcome the Sorites paradox and model conveniently such challenging phenomena of the real world that are problematic to traditional approaches. The developments within the *FLe*, in turn, seem to open new prospects at a more general methodological level. The *FLe* also seems to have connections to such traditional approaches as fallibilism, scientific realism, the theory of verisimilitude, and the theory of truthlikeness (Gadamer, 1975; Hempel, 1965; Hintikka and Suppes, 1970; Niiniluoto, 1987; Popper, 1959, 1963). These aspects are considered in the following sections.

12.3   THEORIES AND APPROXIMATION

In the conduct of inquiry, we aim to formulate our considerations and results in an abstract form, and these formulations are referred to as scientific theories. In general, our theories have four tasks, viz., they enable us to understand, explain, and manipulate the phenomena under consideration as well as we can make predictions on these phenomena.

The term *theory* has various interpretations. For example, in its original meaning, which stems from the Greek word *theoria* (view or consideration), theories are general pictures or ideas formulated according to our mental and rational acts. Second, in particular, in the natural sciences, theories comprise concepts that are interconnected by laws (e.g., Einstein's relativity theory or Darwin's evolution theory). Hence, in this sense, theories are sets of propositions, and we can formulate them in a linguistic form by using appropriate sets of sentences. If theories only comprise qualitative, comparative, or quantitative concepts, we can refer to them as qualitative, comparative, and quantitative theories, respectively. Later, we mainly adopt the latter interpretation when we consider theories (Hempel, 1966; Nagel, 1961; Niiniluoto, 1984).

If we consider the nature of the linguistic concepts in theories, these constitute concepts and such interrelationships between these expressions as *x is associated with y*, *x is part of y*, *x is the cause of y*, *x follows y*, *x contradicts y*, *x is an intervening condition for y*, and *x is a property of y*. The traditional aim in the quantitative theory formation has been to replace comparative or qualitative terms with their quantitative counterparts (metrization) in order to make theoretical concepts precise and maximally objective. For example, in the case of the "laws,"

1. Water boils when it is hot (qualitative)
2. Water boils when it is hotter than boiling alcohol (comparative)
3. Water boils at 100°C (quantitative)

The statements 1 and 3 have in this sense the lowest and highest rankings in the light of usability, respectively. The reason for this is that the third law is expressed in a precise (and numerical) form. We have the same goal in statistics in which we aim to use quantitative variables because they are at least at the level of the interval scale of measurement (Hempel, 1966; Nagel, 1961; Niiniluoto, 1984).

However, the paradigm on precise concepts seems to be outdated because the *FLe* also operates well with the imprecise entities. Consider the fuzzy rule, or law, for the business world:

It is profitable to buy cheap and sell expensive.

It is very usable rule, even though it comprises imprecise concepts. Thanks for the *FLe*, we can also computerize the traditional qualitative research to a great extent, which has not been possible thus far.

Theories consist of linguistic entities, whereas models comprise physical or abstract objects that constitute set-theoretical structures. With models we can apply our theories to practice, and they can often be simplified or idealized versions of the corresponding theories. Typical models can be representative (e.g., scale models), theoretical (e.g., mathematical models), or imaginary (e.g., tooth fairy model in dental care) by nature. However, the distinction between theories and theoretical models, in particular, can be problematic in practice (Hempel, 1966; Nagel, 1961; Niiniluoto, 1984).

If we consider the structures of theories from the standpoint of theory formation and their role in the conduct of inquiry, we usually have to study the relationship between the theories and the real world. Hence, our theories may have truth values and their contents are more or less expected to correspond with the facts of the real world (scientific realism, see the following). One alternative, but less adopted, approach primarily presupposes that the theories should be good in practice, whereas their truth values have a minor interest (instrumentalism).

If our theories comprise laws or law-like statements, these can refer directly to the observations and phenomena of the real world or to non-observable entities.

In both cases, the laws can express either coexistence or succession of the entities or phenomena. For example,

The boiling point of water is 100°C

For all real numbers, it holds that $a + b = b + a$

These represent empirical and non-empirical laws of coexistence, respectively.

The laws of succession, in turn, are applied to dynamical and causal systems. Examples of these are

If water is heated up to 100°C, it will boil (empirical)

For all real numbers, it holds that if $a < b$ and $b < c$, then $a < c$ (theoretical)

We can also draw a distinction between the universal, or deterministic, and probabilistic laws. The former expresses general and the latter statistical or probable regularities in the real world (Hempel, 1966; Nagel, 1961; Niiniluoto, 1984). Examples of these are

- In Europe, women live longer than men
- The average age of all the cells in an adult person's body is about 10 years
- Frustration increases more or less aggression
- Rich persons are often more or less conservative

By virtue of the *FLe*, we have today a well-established computerized method to consider the degrees of truth of our laws and theories (Niskanen, 2008, 2009b). This procedure can be carried out by using the meta-rule:

The more similar our theory or law is to its true counterpart, the higher degree of truth can be assigned to it.

In this sense, we may assume that such prevailing theories as the quantum theory in physics or evolution theory in biology are not necessarily completely true yet, but they are only approximately true, that is, they may still include constituents that are only more or less close to truth. However, if we are optimistic, we may attain their corresponding true counterparts in the future. In practice, we can use the traditional testing and confirmation methods when we attempt to approach a true law or theory.

The *FLe* also opens new prospects for the idealized laws in science. We know that many laws or models are formulated in an idealized or simplified form, which means that they only approximately correspond with the real world. Hence, they are in fact nontrue. For example, in statistics we often assume that a given sample or population is normally distributed, but in practice these objects are only approximately normally distributed. Our calculations are nevertheless based on the corresponding "ideal" mathematical distribution. Hence, the mathematical distribution we use is nontrue in practice and our statistical reasoning is also more or less fallacious.

The traditional methods often resolve this discrepancy by adding some appropriate constituents to the original law or model. For example, in statistics we can add various correction coefficients or even new terms to our models. Then we expect that we will obtain a law or model that is sufficiently true.

Zadeh's ideas described earlier provide an alternative approach because we may oper ate with approximate laws, models, and theories in an intelligible manner. His ideas can be generalized by assuming that we construct a comprehensive system that provides us with methods for approximate reasoning, explanations, and theories (Niskanen, 2008, 2009b; Zadeh, 2009). Second, the *FLe* can also be applied to uncertain entities, in particular, to the probabilistic laws. In this context, the fuzzy probability theory plays an essential role, and, thus, we can use both imprecise probability distributions and random variables (see the following). Third, we can apply the *FLe* to the estimation of the degree of lawlikeness. Then we can consider the distinction between the nomothetic and accidental generalizations in which case such criteria as their ability to explain and predict phenomena or cope with counterfactual conditions are relevant.

It seems that the *FLe* maintains scientific realism. This philosophical outlook presupposes that the theories should correspond with the real world, and thus we can also consider their truth values. Hence, the true theories correspond fully with the real world and their concepts refer to existing entities. The *FLe* also promotes the interaction between the human perceptions, precisiated language, and theories as well as it assumes that theoretical concepts refer to existing entities and the truth assessments of theoretical sentences play an essential role. As was mentioned earlier, in practice, it is often possible that we first formulate a non-true theory, and then we may gradually, usually with trial and error, attain its true version (Niiniluoto, 1983; Popper, 1959). Within the *FLe*, we still have to continue our work for developing an intelligible scientific language for this task.

The scientific realism also seems to be the prevailing approach in the scientific community, and it is thus widely presupposed that theories should be true and informative (Hintikka and Suppes, 1970). The latter criteria, in turn, comprises such features as clarity and generality. In addition, it usually seems reasonable to aim at simplicity.

Hence, in practice we have six central stages in our theory formation (Niiniluoto, 1984):

1. We consider the role of our scientific tradition, metaphysical commitments, foreknowledge, or research paradigms in our theory formation. In particular, these aspects are often emphasized in the qualitative research. For example, we can consider the existence of fuzzy entities or the relationship between our perceptions and the real world within the *FLe*.
2. We make certain conventions or restrictions for our object of research. For example, we can establish that the framework of the *FLe* is used and fuzzy sets represent imprecise terms and their degrees of membership range from zero to unity.
3. We collect relevant facts and empirical observations.
4. We make hypothetical assumptions, interpretations, and generalizations according to the outcomes at the third stage. These expressions may also be in an approximate form.
5. We formulate first possible idealized versions, and these provide a basis for our final theory. We also attempt to find simple final resolutions. Here, again, the initial versions may be approximate by nature. Our final versions may also be approximate.
6. We consider the consistency of our theory with the other theories.

Zadeh's *FLe* may thus enhance our theory formation in such areas as in expressing theoretical concepts, applying approximate theories and the idea of lawlikeness, formulating idealizations, and examining deducibility and the truth of our postulations. These aspects, in turn, are related to other important topics based on approximation, and more specific aspects are considered later.

12.4 APPROXIMATE REASONING, PROBABILISTIC CONSTITUENTS, AND HYPOTHESES

The conceptions on probability stem from the problems of uncertainty and ignorance, and thus principally from epistemology. The formal theories on probability have their origin in the mathematical models for gambling in the Renaissance period. Later Kolmogorov, in particular, provided a basis for modern mathematical calculus of probability theory, and this approach is still adopted in the mainstream studies (Niiniluoto, 1983, 1984).

The mathematical, or physicalistic or objective, tradition presupposes that probability is independent of the human knowledge and thus only dependent upon physical properties assigned to the occurrences. This approach is closely related to modality because the role of possible events have been considered in this context (Niiniluoto, 1983, 1984).

This tradition comprises frequency and propensity interpretations. The former assumes that probability is the limit value of event's relative frequency in a large number of trials, and this interpretation is widely used in the modern probability theory and statistics. The propensity approach, in turn, assumes that probability is such numerical "disposition" that yields certain relative frequencies or measures for unique events.

On the other hand, the epistemic approaches, such as the logical and subjective interpretations, presuppose that probability is dependent upon our knowledge and ignorance. In the former case, we assume that probability is the relationship between a hypothesis and its evidence and thus it is related to inductive reasoning, that is, logical probabilities are thus rational degrees of belief of hypotheses according to the evidence (Carnap, 1962; Von Wright, 1957).

The epistemic subjective theories also apply the idea on the degree of belief but now these degrees are mainly based on human rational assessments. Thus, these degrees may vary among the evaluators. On some occasions, subjective theories are combined with the utility theory and Bayesian decision theories.

The epistemic approaches allow us to consider the probabilities of the hypotheses and theories. Then we may apply truth as an epistemic utility, that is, the hypothesis with the highest truth value should always be accepted (however, tautologies would then be the best candidates).

Truth and (epistemic) probability were already combined in the Ancient Greek and Latin, and this was due to the ambiguity of the term probability. In that period of time probability referred to both uncertainty and truth, and thus still in certain languages the word for probability actually refers to truth (e.g., in German the word for probability is Wahrscheinlichkeit, i.e., "truth-looking"). Today the distinction should be clear because epistemic probability and truth are the issues of epistemol ogy and linguistic semantics, respectively (Carnap, 1962). Imprecision, or fuzziness, and uncertainty are also sometimes confused for similar reasons although the former expression usually belongs to semantics and the latter to epistemology (c.f. also an example mentioned earlier). Possibility, in turn, as used within fuzzy systems, is related both to probability and modality (Niiniluoto, 1987).

According to the *FLe*, the idea on the fuzzy probability generally means approximate probability variables and approximate values of these variables (Grzegorzewski et al., 2002; Niskanen, 2010a; Zadeh, 2002, 2009). For example,

- The probability that John's age is 20 is approximately 0.95
- The probability that John's age is 20 is very high
- The probability that John's age is approximately 20 is very high
- The probability that John is young is very high
- The probability that "the probability of John being very young is fairly high" is high (second-order probability)

This approach seems to be usable to both physicalistic and epistemic probability. In the physicalistic case, we apply such approximate probability distributions as approximate normal distributions. Thus, even precise values of random variables yield approximate probabilities (Figure 12.7).

Statistical tests apply much physicalistic probability and random distributions when we consider the acceptance of the null and alternative hypotheses according to the tests of significance (Guilford and Fruchter, 1978; Niskanen, 2010; Zar, 1984). We accept the null hypothesis if the value of our test variable does not deviate too much from the "usual" case, otherwise, we reject it and accept the alternative hypothesis. In practice, we operate with the p value (level of significance, $0 \leq p \leq 1$) in which case we consider the rejection of the null hypotheses if the p value is sufficiently small. In other words, the p value is our risk to draw an erroneous conclusion if we reject the null hypothesis.

Traditionally, the statistical hypothesis verification is based on bivalent reasoning, and thus the rejection of one hypothesis automatically means the acceptance of the other. Hence, formally we reason that if the p value is greater than a given limit point, we accept the null hypothesis, otherwise we reject it. The usual threshold values for p are .05 or .01 (5% and 1% levels of significance, respectively; Figure 12.8).

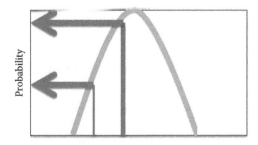

FIGURE 12.7   Approximate probability function maps precise values to imprecise values and imprecise values to even more imprecise values.

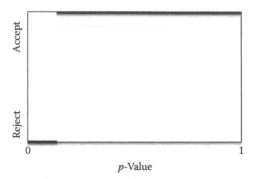

FIGURE 12.8   Acceptance of the null hypothesis in traditional statistics.

For example, when the *t*-test is applied to two independent samples, we have the null hypothesis that there is no difference in the means between the groups, whereas the alternative hypothesis asserts that this difference prevails (the two-tailed case).

In practice, we nevertheless take into account the borderline cases when the acceptance of the null hypothesis is considered. For example, we may pay special attention to the p values that are in the close neighborhood of $p = .05$ in order to avoid erroneous conclusions. Hence, we actually apply approximate reasoning and probability. The *FLe* can make this reasoning more formal and informative if we operate with the degrees of acceptance and rejection in this context. Then, we can establish the meta-rule that the smaller the p value, the lower the risk of error for rejecting the null hypothesis (i.e., the higher the degree of rejection for the null hypothesis). Simultaneously, then it also holds that the higher the degree of accepting the alternative hypothesis (Figure 12.9). We can even construct a fuzzy inference engine including this type of fuzzy rule set for this task.

Possible approximate statistical hypotheses are

- In Europe, the heights of the males are approximately normally distributed
- The Swedish males are slightly taller than the Italian males
- Most Swedish males are very likely fairly tall

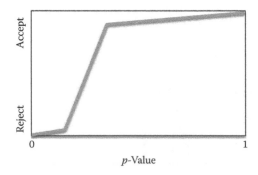

FIGURE 12.9   Tentative curve for the degree of acceptance of the null hypothesis when approximate reasoning is applied to statistics.

- The people in the United States usually drink coffee quite much
- There is a fairly high positive correlation between the grades in mathematics and physics at school
- One-dollar bills fairly likely contain particles of cocaine

In addition to the statistical research, in the scientific explanations we may assume that approximate explanations approximately explain the phenomenon under consideration (Hempel, 1965; Niskanen, 2009a; Rescher, 1970; Von Wright, 1971). In this context, we can also operate with probabilistic or statistical statements. Examples of probabilistic and statistical explanations are

Fact: When tossing the coin in the real world, only approximately 50% will be heads

If we now ask why about 50% of these outcomes are heads, we can provide an approximate explanation that it is due to this approximate frequency probability

or

Most Swedish men are tall because, according to the statistics, their average height is approximately 180 cm

Another approach to the verification of hypotheses with probability statements is the epistemic standpoint. Then, we consider this problem in the light of inductive reasoning and we assume that the probability for accepting a hypothesis is based on our evidence. Thus, this examination is closely related to decision making under uncertainty.

If we aim at true hypotheses and theories, we can consider the inductive support and degree of confirmation for our hypotheses. In the former case, we apply such epistemic utilities as the truth, simplicity, or information contents, whereas in the latter case their true logical consequences play an essential role. In both cases, our examination stems from the problems of truth and reasoning (Carnap, 1962; Von Wright, 1957).

In general, hypotheses play essential role in the conduct of inquiry and our task is to both establish and verify them. In practice, our hypotheses are assumptions or assertions on the phenomena under consideration and our task is to assign truth values to these statements according to such evidence as facts, experiments, observations, and scientific reasoning.

We are still unable to fully explain how our original research hypotheses are usually invented or discovered because they usually stem from our insights, that is, they stem from the researcher's context of discovery and inventions. However, on some occasions we can apply such "automated" procedures as induction, abduction, or reasoning by analogy in this context.

The verification of hypotheses, in turn, is based on the available evidence and scientific reasoning. In the quantitative research, we usually provide hypotheses for describing, predicting, or explaining the phenomena. In the qualitative research, the interpretation hypotheses are also essential.

From the logical standpoint, the hypotheses can be singular, general, or probabilistic statements, that is, they can refer to singular objects or occurrences, regularities, or more or less uncertain events. Hence, in addition to observational statements the hypotheses can deal with the phenomena at a general level or include purely theoretical concepts. To date, the hypothesis verification has usually based on conventional bivalent reasoning, but we can also apply the *FLe* to this subject matter.

The mainstream conventional hypothesis verification usually comprises three reasoning methods. First, John Stuart Mill's method of difference, which, in a nutshell, can be expressed as (Niiniluoto, 1983)

If A_1 and ... and A_n and B, then D
If A_1 and ... and A_n and C, then not D
Thus, B is the cause of D

This reasoning means that we consider two sets of occurrences that contain similar occurrences except one, that is, B in the former and C in the latter set are dissimilar. If now the occurrence D will take place in the former, but not in the latter conditions, we may conclude that B causes D. According to Mill, this method, which is an example of eliminative induction, can be used when we establish competitive hypotheses in a given context.

Second, we can apply the disjunctive syllogism

A or B
Not A
Thus, B

which is used to eliminate the irrelevant hypotheses. Third, we use the foregoing the Modus Tollens syllogism.

Mill's principle and the disjunctive syllogism only provide us with the general guidelines for excluding the false hypotheses and thus we also need specific methods to carry out the verification in practice. Since our hypotheses may also be nonempiric by nature, we are not always able to rely on observations in this task.

An example of a more specific method is the hypothetico-deductive method that is widely used in the quantitative research, and it applies the Modus Tollens syllogism (Hintikka and Suppes, 1970; Niiniluoto, 1983; Popper, 1959, 1963). This method assesses hypotheses in an indirect manner, and thus we can also use nonobservable hypotheses. Given now the hypothesis, H, and its observable or testable and deducible consequence, C, the truth of C is first assessed according to our knowledge, experiments, and observations. If this evidence is inconsistent with C, we conclude that our hypothesis is false. Hence, we apply the traditional Modus Tollens in the form

If hypothesis H holds, then consequence C
The consequence C is inconsistent with the evidence
Thus, hypothesis H is false

For example, consider the hypothesis "the liquid in this bottle is water." We can thus deduce, inter alia, that

If this liquid is water, then it boils at 100°C at the sea level
This liquid does not boil at 100°C
Thus, our hypothesis is false and we reject it

According to the disjunctive syllogism, in turn, this hypothesis is excluded from the set of possible true hypotheses.

However, the problem arises if our previously mentioned evidence corresponds with C because then the traditional Modus Tollens will not provide us with any resolution. Hence, instead of deduction, we must now apply induction. In this case one traditional method is to apply the idea on confirmation and thus, when our evidence corresponds with the consequence C, our hypothesis is more or less confirmed, but not verified. Sufficient confirmation, in turn, will lead to the truth or acceptance of the hypothesis. For example, reconsidering the aforementioned hypothesis,

If this liquid is water, then it boils at 100°C at the sea level
This liquid boils at 100°C
Thus, our hypothesis is more or less confirmed (but not verified yet)

In fact we have now adopted the hypothetico-inductive approach, and we operate with the epistemic probability statements in this context.

Due to its bivalent nature, this traditional approach is unable to specify the idea on confirmation in a satisfactory manner because it can only provide us with either–or type resolutions. The *FLe*, in turn, could provide a better resolution if we apply the foregoing fuzzy Modus Tollens in this context (Niskanen, 2010). We could now apply the approximate hypothetico-deductive method when the consequences of the hypotheses are consistent with the evidence. The general form of our reasoning would thus be (Figure 12.10)

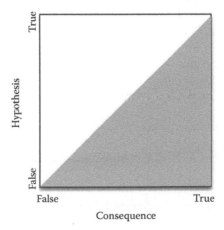

FIGURE 12.10 Fuzzy Modus Tollens approach to hypothetico-deductive method in hypothesis verification.

If hypothesis H, then consequence C (true implication)
Truth (consequence C) is T according to the evidence (T is from false to true)
Thus, Truth (hypothesis H) $\leq$ T, that is, it is from false to T

For example, consider again our hypothesis on water mentioned previously. We may now reason that

If this liquid is water, then it boils at 100°C at the sea level
This liquid boils at 100°C
Thus, the truth value of our hypothesis is from false to true

Consider now the case

If this liquid is water, then it boils at 100°C at the sea level
This liquid boils at 99°C
Thus, the truth of our hypothesis is from false to fairly true

Hence, hypothesis verification based on the *FLe* enables us to acquire at least some information on the truth of our hypothesis, and we usually aim at maximally informative and true hypotheses (even though these aspects are in inverse relation to one another). In general, we may establish here the meta-rule that the more our consequences approach falsity, the closer our hypothesis is also to falsity.

The *FLe* also allows us to assume that the implication itself in the approximate Modus Tollens syllogism is nontrue. For example, if this implication is only fairly true, we can establish that even the false consequence does not necessarily lead to mere false hypotheses. Equally the truth value of the consequence close to true may already lead to conclusion that the truth value of our hypothesis is between false and true. In general, we may thus assume that the non-true implications cause more "dispersion," granulation, or imprecision to our conclusions, and loose reasoning links of this type are typical, in particular, in the human sciences in which we usually operate with noisy data and the complex interrelationships between the variables.

We can also apply the idea on the degree of confirmation in a fuzzified manner in this context in which case our extreme values for the hypotheses would be "full disconfirmation" and "full confirmation." For example,

If hypothesis H, then consequence C
Consequence C if fairly true
Thus, hypothesis H has more or less high degree of confirmation

The degree of confirmation is related to aforementioned degree of belief in epistemic probability because this approach examined the relationship between the hypotheses and their evidence.

Since we adopted the philosophical meta-level approach, we do not provide any concrete methods or tools for applying the foregoing ideas. In practice, the reader may utilize the vast literature on fuzzy system construction for this purpose (Nikravesh et al., 2008a,b). Typical methods for this task comprise the basics of fuzzy logic, application of the extension principle, construction of fuzzy rule–based systems, and computer modeling with such software as MATLAB®'s Fuzzy Logic Toolbox™. Our goal, in turn, has been to open up new vistas of scientific thinking with the novel and revolutionary reasoning.

12.5 CONCLUSIONS

The mainstream Western scientific outlook has maintained bivalent logic >2000 years. Despite its elegant appearance and syntactic structure, it has quite limited applicability to the problems of the real world. The crucial problem is that bivalent logic often provides us with too coarse theories and models although this approach aims at precision. We also encounter this problem in traditional mathematical modeling because mathematics base on bivalent reasoning. In addition, mathematical modeling presupposes deep knowledge on mathematics in order to construct and understand these models. The qualitative research in the human sciences, in particular, has been very challenging to the foregoing methods, and in fact more or less inadequate results have only been achieved in certain areas.

Fuzzy systems may provide some resolutions to these problems because, thanks for their revolutionary and multivalued nature, we can take into account better the gradual changes in the real word, as well as the imprecision that is essential in the human behavior and reasoning. In addition, these systems allow us to use simpler linguistic systems, and they are more user-friendly than the complex mathematical systems in our research work.

The fuzzy extended logic opens new meta-level prospects for the conduct of inquiry. It enables us to adopt a novel and innovative approach to such subject matters as scientific reasoning, theory formation, model construction, hypothesis verification, and scientific explanation. Hence, in addition to the prevailing approaches, which often aim at bivalency and precision, we may also apply their corresponding approximate versions. Prior to the fuzzy systems, approximate entities were only considered in an informal manner although their existence was already recognized in the scientific community, but the fuzzy extended logic provides us with formal and usable tools and methods for this task.

We considered approximate reasoning, theory formation, model construction, and hypothesis verification mentioned earlier by applying Zadeh's fuzzy extended logic. We showed that the approximate reasoning corresponds well with the phenomena of the real world and it can also resolve problems that are superb to the traditional methods. Application of approximate reasoning, in turn, enables us to consider in a plausible manner problems of imprecision and gradual changes in scientific research in general. Since we examined these subject matters from the standpoint of the philosophy of science, we only provided certain meta-level resolutions. In any case, further studies are also expected in this field.

If the novel patterns of thought characteristic of fuzzy systems and the fuzzy extended logic are adopted, our approach will open up new vistas of scientific thinking, and we can also apply more versatile methods and tools in the conduct of inquiry.

ACKNOWLEDGMENT

I express my great gratitude to Prof. Lotfi Zadeh, University of California, Berkeley, for his inspiration and encouragement when examining his work on fuzzy systems and the fuzzy extended logic.

REFERENCES

Bandemer, H. and Näther, W. *Fuzzy Data Analysis*. Kluwer, Dordrecht, the Netherlands, 1992.

Bezdek, J. and Pal, S. *Fuzzy Models for Pattern Recognition*. IEEE Press, New York, 1992.

Carnap, R. *The Logical Foundations of Probability*. The University of Chicago Press, Chicago, IL, 1962.

Denzin, N. and Lincoln, Y. *Handbook of Qualitative Research*. Sage, Thousand Oaks, CA, 1994.

Fetzer, J. *Philosophy of Science*. Paragon House, New York, 1993.

Gadamer, H.-G. *Truth and Method*. Sheed & Ward, London, U.K., 1975.

Grzegorzewski, P. et al. (Eds.). *Soft Methods in Probability, Statistics and Data Analysis*. Physica Verlag, Heidelberg, Germany, 2002.

Guilford, J. and Fruchter, B. *Fundamental Statistics in Psychology and Education*. McGraw-Hill, London, 1978.

Haack, S. *Philosophy of Logics*. Cambridge University Press, Cambridge, U.K., 1978.

Hempel, C. *Aspects of Scientific Explanation and Other Essays in the Philosophy of Science*. The Free Press, New York, 1965.

Hempel, C. *Philosophy of Natural Science*. Prentice Hall, Englewood Cliffs, NJ, 1966.

Hintikka, J. and Suppes, P. (Eds.). *Information and Inference*. Reidel, Dordrecht, the Netherlands, 1970.

Nagel, E. *The Structure of Science*. Routledge & Kegan Paul, London, U.K., 1961.

Niiniluoto, I. *Tieteellinen päättely ja selittäminen (Scientific Reasoning and Explanation)*. Otava, Keuruu, Finland, 1983.

Niiniluoto, I. *Johdatus tieteenfilosofiaan (Introduction to the Philosophy of Science)*. Otava, Keuruu, Finland, 1984.

Niiniluoto, I. *Truthlikeness*. Reidel, Dordrecht, the Netherlands, 1987.

Nikravesh, M., Kacprzyk, J., and Zadeh, L. (Eds.). *Forging New Frontiers: Fuzzy Pioneers I. Studies in Fuzziness and Soft Computing*, Vol. 217. Springer, Heidelberg, Germany, 2008a.

Nikravesh, M., Kacprzyk, J., and Zadeh, L. (Eds.). *Forging New Frontiers: Fuzzy Pioneers II. Studies in Fuzziness and Soft Computing*, Vol. 218. Springer, Heidelberg, Germany, 2008b.

Niskanen, V. A. *Soft Computing Methods in Human Sciences, Studies in Fuzziness and Soft Computing*, Vol. 134. Springer Verlag, Berlin, Germany, 2004.

Niskanen, V. A. Meta-level prospects for scientific theory formation by adopting the information granulation approach. *Proceedings of the IMS'08 Conference*, Sakarya University, Sakarya, Turkey, 2008, pp. 33–41.

Niskanen, V. A. Application of Zadeh's impossibility principle to approximate explanation. *Proceedings of the IFSA'09 Conference*, Lisbon, Portugal, 2009a, pp. 352–360.

Niskanen, V. A. Fuzzy systems and scientific method—Meta-level reflections and prospects. In R. Seising (Ed.), *Fuzzy Set Theory—Philosophy, Logics, and Criticism*. Springer, Heidelberg, Germany, 2009b, pp. 51–82.

Niskanen, V. A. A meta-level approach to approximate probability. In R. Setchi et al. (Eds.), *Lecture Notes in Artificial Intelligence*, Vol. 6279, pp. 116–123. Springer, Heidelberg, 2010.

Niskanen, V. A. Application of approximate reasoning to hypothesis verification. *Journal of Intelligent & Fuzzy Systems*, 21(5), 331–339, 2010.

Popper, K. *The Logic of Scientific Discovery*. Hutchinson, London, U.K., 1959.

Popper, K. *Conjectures and Refutations*. Routledge & Kegan Paul, London, U.K., 1963.

Rescher, N. *Scientific Explanation*. The Free Press, New York, 1970.

Rescher, N. *Empirical Inquiry*. The Athlone Press, London, U.K., 1980.

Takagi, T. and Sugeno, M. Fuzzy identification of systems and its applications to modeling and control, *IEEE Transactions on Systems, Man and Cybernetics*, 15(1), 116–132, 1986.

Von Wright, G. H. *Explanation and Understanding*. Cornell University Press, Cornell, NY, 1971.

Von Wright, G. H. *The Logical Foundations of Probability*. Blackwell, Oxford, U.K., 1957.

Zadeh, L. Fuzzy logic and approximate reasoning. *Synthese*, 30, 407–428, 1975.

Zadeh, L. Fuzzy logic = Computing with words. *IEEE Transactions on Fuzzy Systems*, 2, 103–111, 1996.

Zadeh, L. Toward a theory of fuzzy information granulation and its centrality in human reasoning and fuzzy logic. *Fuzzy Sets and Systems*, 90(2), 111–127, 1997.

Zadeh, L. From computing with numbers to computing with words—From manipulation of measurements to manipulation of perceptions. *IEEE Transactions on Circuits and Systems*, 45, 105–119, 1999.

Zadeh, L. Toward a perception-based theory of probabilistic reasoning with imprecise probabilities. *Journal of Statistical Planning and Inference*, 105(2), 233–264, 2002.

Zadeh, L. From search engines to question answering systems? The problems of world knowledge, relevance, deduction and precisiation. In E. Sanchez (Ed.), *Fuzzy Logic and the Semantic Web*. Elsevier, Amsterdam, the Netherlands, 2006.

Zadeh, L. Toward extended fuzzy logic—A first step. *Fuzzy Sets and Systems*, 160, 3175–3181, 2009.

Zar, J. *Biostatistical Analysis*. Prentice Hall, Englewood Cliffs, NJ, 1984.

13 Personalized Knowledge Service for Smart Cell Phones Based on Usage

Conceptual Model and Intention Prediction Algorithm

Meira Levy, Lior Rokach, Bracha Shapira, and Peretz Shoval

CONTENTS

13.1 INTRODUCTION

Smart cell-phone (SCP) providers are interested to facilitate their customers' cell-phone usage in order to combat potential resistance to adopt new technology, specifically when the customers upgrade their cell phones to new advanced ones or when new SCP applications are introduced in the market. Research (e.g., Chen et al. 2009) showed that users' perceived ease of use (PEOU) and perceived usefulness (PU) have a direct and positive effect on users' attitude, which leads to positive effect on behavioral intention (BI) to use the cell phone. In addition, users' self-efficacy affected BI and is a strong antecedent of PEOU. These findings suggest that SCP adoption relies heavily on how users perceive the technology and whether they consider themselves capable of using it (self-efficacy). In addition, Chen et al. (2009) posit that it is important to design training programs so that trainees will develop positive attitudes toward technological innovations and about their own capabilities to use these technologies. Such training programs, regarding the SCP, may include successive stages where in each stage people will be trained with only selected and essential features of the SCP, for improving their self-efficacy beliefs, and only then more advanced features will be introduced.

One of the common practices for interacting with customers is websites with personalized services that are mainly focused on marketing (Koufaris 2002; Jiang 2009). Several studies in this domain are based on the flow theory (Csikszentmihalyi 1988) that deals with the psychological state of mind within an activity, and how actors feel with the environment they act upon. The flow theory refers to how people concentrate and take control over the environment while filtering out irrelevant thoughts and has been studied in the context of information technologies, computer-mediated environments, and marketing (Hoffman and Novak 1996; Koufaris 2002; Jiang 2009). Jiang (2009) revealed the impact of content relevance on customers' positive attitude toward websites, specifically when goal specificity is high.

While these studies focused on emotional and cognitive components of customers such as intrinsic enjoyment, perceived control, and attention focus as metrics for the online consumer experience, our study aims at understanding how personalized web service, which recommend on product's tutorial that is relevant to users, can help overcome product usage barriers. Former studies that deal with personalized services were concerned with either commercial or learning systems. Personalized commercial services are based on technology that provide content and services tailored to customers based on customer information and customer preferences and behavior within the provided service (Adomavicius and Tuzhilin 2005). Personalized learning systems may provide knowledge either according to the learner's knowledge, the learner's location and needs in that location, and the learner mobility (Cui and Bull 2005; El-Bishouty et al. 2007).

Our research focuses on SCP knowledge personalized service that is based on users' profiles constructed both from their demographic information and from their usage of the cell phones. Since the SCP usage is continuously tracked, the customer's profile is refined allowing an enhanced and dynamic personalized service to the users.

Following Jiang's (2009) findings, we suggest a SCP personalized knowledge service (SCP-PKS) conceptual framework, which provides relevant content based on the user's profile. This service can enhance the user's experience within the SCP knowledge website by providing relevant knowledge that matches his or her goals, toward overcoming SCP usage barriers and extending customer satisfaction. Since the users are first introduced with knowledge regarding familiar goals, this can better influence their PEOU, PU, and self-efficacy that according to Chen et al. (2009) may have positive effect on their attitude and BI to use the SCP.

The SCP-PKS conceptual framework is based on a task model (TM) framework that models and tracks SCP application usage. The TM framework is based on the concept of viewing an application as a set of goals that a user can achieve, and accordingly realizing what the user is required to do in order to achieve any of these goals. The TM has two perspectives: (a) user tracking—tracking the user's activities for the purpose of usage mining; for this purpose, all the usage scenarios of an application are modeled, hence representing technical low-level modeling and (b) goals prediction—tracking and realizing the user's goal while using an application to enable predicting the user's intention; for this purpose, a conceptual structure of the service is modeled, hence, representing high-level modeling.

During the SCP usage tracking, the various goals of the applications are mapped according to user's usage patterns (i.e., the most utilized goals) and user's stereotype, hence creating the linkage between user stereotypes and application's goals. The proposed SCP-PKS utilizes this linkage for offering personalized navigation within a general help service. Once the user enters the application's help service, a navigation map is presented, leading the user to the most relevant help instructions or tutorials for him or her according to his or her stereotype. The general help service provides help suggestions to all the application's goals; the user can override the SCP-PKS map and navigate independently, for example, when a user would like to learn about a new service that he/she has not used yet.

The rest of this chapter is structured as follows: First, the personalization theoretical background is discussed, with emphasis on personalized recommendations and navigation systems. Next, a theoretical background regarding the TM framework and the developed generic TM and TM development tool that served as the infrastructure for exploiting the users' profiles are explained, as well as an intention prediction algorithm that enables to realize users' intended goals while utilizing their phones. Following, the SCP-PKS conceptual framework is presented. Finally, we summarize and discuss future work plan.

13.2 PERSONALIZATION

Personalization in consumer market is aimed at matching and recommending potential products to customers according to their preferences, goals, and past usage behavior (Adomavicius and Tuzhilin, 2005; Frias-Martinez et al. 2006), allowing reduction of search efforts of users and mitigating the encumbrance of information overload (Liang et al. 2006). Previous studies that focused on the interaction between recommendation systems and customers dealt with the following perspectives: understanding customers through profile building, hence profile generation and maintenance, and profile exploitation (Montaner et al. 2003); delivering personalized offering based on the knowledge about the product and the customer; and measuring personalization impact (Adomavicius and Tuzhilin 2005). There are several recommendation methods for building customers' profiles. Frias-Martinez et al. (2006) discuss these methods as user-modeling processes of constructing data that represent user interests, goals, and behaviors. In these processes, user models are derived from both observable and unobservable information. The information for the user model is either provided explicitly by the user or implicitly produced by the system, for enabling adaptive personalized service. Wei et al. (2000) classified recommendation approaches according to the type of data and technique used to arrive at recommendation decisions: popularity, content, association, demographics, reputation, and collaboration. For extracting user preferences, there are methods that differ in the way they calculate the users' preferences: by direct interaction with the users, that is, questioning the users' preferences; semi-direct, that is, users' ratings; or indirect, based on tracking users' actions in the recommended environment, that is, hyperlink clicks (Sakagami and Kamba 1997). Several studies examined user characteristics and product features on recommendations (Im and Hars 2001; Hung 2005). Liang et al. (2008) added a semantic expansion approach to recommendation system. However, all the studied recommendation systems are based on users' behaviors within the recommendation system without taking into consideration users' behaviors regarding the recommended product. This has great impact on personalized systems since at the beginning the recommendation systems do not have any information about the user, hence requiring a learning phase and usage of a standard personalization. In addition, machine learning techniques that are used by the personalized systems need to cope with millions of users that a system can have, and be flexible to changes of users' preferences (Frias-Martinez et al. 2006). The SCP-PKS framework presented in this research addresses these problems since the user models are created and dynamically changed before using the service, which allow the personalized system to skip a learning phase.

13.2.1 PERSONALIZED NAVIGATION

Personalized navigation is a mechanism for presenting users with an environment that better suits the way in which they interact with the environment. It is part of personalized interface of web pages within e-commerce sites, typically dealing with issues such as customization of colors, folders, content organization (i.e., news and weather), and content updating frequencies (Frias-Martinez et al. 2006). Current personalized navigation systems aim at improving a user's awareness of resources available to them on the Internet and in intranets, while minimizing the effort required to discover those (Budzik et al. 2002). Budzik et al. describe a system named Watson that proactively retrieves documents from online repositories that may be relevant to users in the context of a specific task. They further extend this system with communities of practice components that enable users to discover information from users with similar goals and interests, and communicate with them both synchronously and asynchronously. The users' current tasks are derived automatically from their behavior in software tools, hence enabling to provide them useful resources in the context of the work that they are performing. Godoy and Amandi (2008) propose a method to derive semantically enhanced contexts that point out the information that is more likely to be relevant according to the user's profile and the current browsing activity. They developed agents that act as browsing assistants who treat web browsing as a cooperative search activity, tracking users' behavior on the web for learning users' preferences and providing real-time display of recommendations. While most personalized navigation systems track the users in the same environment where the recommendations are displayed, our proposed PKS utilizes the already tracked cell-phone usage patterns for various users and goals as well as analyzed user profiles' goals and preferences that are stored in the smart knowledge base. The logical infrastructure for tracking the cell-phone usage is a TM that defines the structures and the possible activities that can be performed by users of interactive application installed in the device, as will be discussed in the next section. The proposed knowledge service does track the usage patterns of the users in the environment for refining the smart knowledge base of the system.

13.3   TASK MODEL AS THE PERSONALIZATION INFRASTRUCTURE

13.3.1   What Is a Task Model?

A TM is a logical description of activities that can be performed by users of interactive application (i.e., software systems that run on computing devices and involve user and system interaction) who are willing to achieve certain goals (Chen et al. 2009). A TM can enhance the design of interactive applications as well as track users' behaviors while using such applications. Mori et al. (2002) describe TMs as follows: "The basic idea is to identify useful abstractions highlighting the main aspects that should be considered when designing effective interactive applications"; "Task models play a particularly important role because they indicate the logical activities that an application should support"; "Tasks can range from a very high abstraction level (such as deciding a strategy for solving a problem) to a concrete, action oriented level (such as selecting a printer). User task model (how users think that the activities should be performed) and the system task model (how the application assumes that activities are performed) correspond closely."

PaternÒ (1999) and Mori et al. (2002) elaborate on the use of TMs: improve understanding of the application domain; record the result of interdisciplinary discussion; support effective design; enable requirement analysis—where, through a task analysis, designers identify requirements that should be satisfied in order to perform tasks effectively; design interactive applications; support usability evaluation; identify task efficiency or analyze logs of user interactions with the support of TMs; and support the user during a session and documentation. Mori et al. (2002) claim that TMs represent the intersection between user interface design and more formal software engineering approaches like UML that lacks the consideration of the user interface.

13.3.2   Generic Task Modeling Framework

The generic task modeling method, developed in this research, enables representing cell-phone applications and tracks actual use of the applications by users, thus enabling the SCP providers realizing the users' activities, learns about the efficiency of use of the cell-phone applications, predicts users' intentions of use (thus avoiding erroneous use), and also predicts interests in future SCP applications. The generic TM has two representations: passive and active. While the passive TM represents the available/possible usage scenarios of applications, the active TM enables tracking the users' actual activities while operating the cell-phone applications. In addition, the TM has two perspectives as described earlier in Section 13.1: (a) user tracking—tracking the user activities for the purpose of usage mining; for this, all the usage scenarios of an application are modeled, hence representing technical low-level modeling and (b) goals prediction—tracking and realizing the user's goal while using an application to enable predicting the user's intention.

13.3.2.1 Generic Task Model

The task modeling method presented in this chapter is part of the SCP research project aimed at tracking the activities of users of SCP while using the various applications included in their cell-phone devices. The TM of each application included in such cell phone is represented in a relational database. We distinguish between passive and active TM: (a) passive TM—includes definitions of each of the applications, including their goals, tasks, screens, and possible transitions between them and (b) active TM—includes the users' activities when operating/using the cell-phone applications.

The SCP TM has several perspectives:

- Passive perspective—static models of available applications on the devices.
- Goals perspective—static models of the conceptual structure of the service, hence goal hierarchies and linkages between screen inputs with goals.
- Active perspective—recording of the user activities. The captured screen representations are linked to the static models.
- Reuse perspective—the TM modeling enables reuse of existing models. For this purpose, a special mechanism of inheritance is implemented in the TM modeling tool.
- Simulation perspective—the TM model includes graphic representations of the TM screens. This feature enables running a simulation of a TM scenario with the TM Designer tool.

13.3.2.2 TM Designer Tool

TM Designer is a graphical software tool, presented in Figure 13.1, aimed to model TMs of SCP applications. The tool enables the modeler to define a meta-model for an application, and to create a static instance of the meta-TM. The architecture of the tool consists of three layers:

- Model—defines the metadata of the model and the graphical definitions, hence their notations. We use a graphical modeling framework that automatically generates code for the graphics and the elements in the model. The elements in the model match the elements in the TM syntax and TM database entities.
- Logic—code that supports extended features such as import and export.
- Runtime environment—used by the modeler of an application. In the runtime environment, there are three views: (a) the canvas, which is mapped to the main element that are currently modeled; (b) the tool palette, which enables dragging elements to the canvas; and (c) the properties view, which presents the properties of the selected element in the canvas.

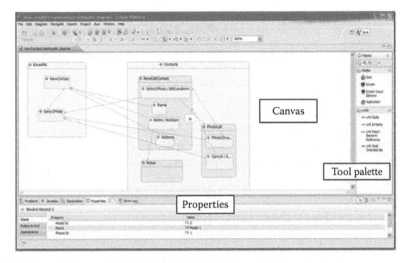

FIGURE 13.1   Example of a diagram in TM Designer.

13.3.2.3   TM Infrastructure

The aforementioned TM and TM Designer tool are part of a TM infrastructure presented in Figure 13.2. The TM infrastructure is activated in the three main phases of the system's life cycle, namely, modeling, tracking, and analysis. In the modeling phase, a human modeler models SCP applications using the TM Designer tool. The modeler defines all the screens, the screen input elements, and the goals of the applications. In addition, a unique signature is automatically captured from the client

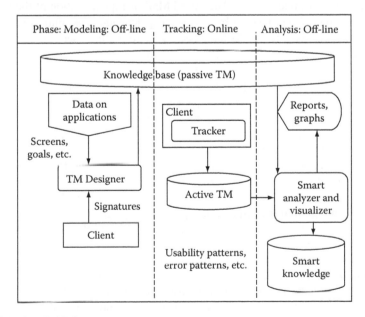

FIGURE 13.2   TM infrastructure.

for each of the screens, representing the physical items of the screen. This process is done off-line by a modeler. The output of the modeling phase is the passive TM, which is recorded and sent to the TM database.

The tracking phase is performed online when the user operates his or her SCP. The relevant passive TM is first downloaded to the user's cell phone. A special tracker module records the users' activities and captures the users' screens, input elements in each screen (SIEs), and goals, generating the active TM database.

The analysis phase is activated by corporate managers. This phase uses the passive and the active TMs and activates intelligent analysis modules identifying usage patterns, error patterns, intention prediction, usability hotspots and more, generating the smart knowledge base.

13.4 USER INTENTION PREDICTION

Prediction of the intentions of users of a system can improve the services provided to the users by, for example, adjusting the interface to their needs. Intention prediction is the basis for predicting user's intentions while performing a sequence of actions. Intention prediction is also known as intention recognition and plan recognition. Predicting the next step of the user within a sequence of interactions can be categorized as a problem with a sequential problem space, which is commonly solved using sequence learning algorithms (Sun and Giles 2001).

Intention prediction has become a very active research area. One example is the Lumiere Project (Horvitz et al. 1998) that is aimed at leveraging methods for reasoning under uncertainty about the goals of software users using Bayesian user models. The models can be employed to infer user's needs by considering a user's background, actions, and queries. They used Markov representation of the temporal Bayesian user-modeling problem by considering dependencies among variables at adjacent time periods. Few works have been done to predict user's intentions while surfing the web. Chen et al. (2002) used a modified Naïve Bayes classifier algorithm to support incremental learning for modeling the user's intended action on a computer. Sun et al. (2002) presented a method for predicting the user's browsing intention based on the web page sequences she had previously visited. The proposed method employs a multistep dynamic n-gram model. TaskPredictor (Shen et al. 2006) is a machine learning system that attempts to predict the user's current activity. First feature selection, a threshold for making classification decisions, and Naïve Bayes are applied to decide whether to make a prediction. Then, a discriminative model (linear support vector machine) is applied to make the prediction itself. In this work, they treated the task prediction problem as a traditional supervised learning problem and ignored the sequential aspect of the problem. Several projects have performed research intelligent wheelchairs. One of which (Taha et al. 2008) presents a technique to predict the wheelchairs users' intended destination at a larger scale (and not only locations immediately surrounding the wheelchair). The system relies on minimal user input obtained from a standard wheelchair joystick—together with a learned partially observable Markov decision process (POMDP).

Although there exist methods for predicting user's intentions, they do not refer to personal aspects such as user characteristics. It might be helpful to refer to the difference between the users characteristics in order to predict the next step of the user in a certain application. The goal of this module is to create an intention prediction model that is automatically trained according to the user's fixed characteristics (e.g., demographic data) and user's sequences of actions in the system. The model is based on hidden Markov model (HMM) (Rabiner 1989) and built hierarchically to increase the accuracy of prediction. We assume that a system consists of a set of goals the user would like to accomplish defined in the TM, for example, in an e-mail application; the goals might include "send e-mail" or "add contact." Each goal can be accomplished by performing a sequence of tasks.

13.4.1 HMM FOR USER INTENTION PREDICTION

For this research, we surveyed methods of sequence learning including neural networks, dynamic Bayesian networks, different Markov models, and others. The sequence learning algorithm we selected for our problem is HMM (Rabiner 1989). The reasons for selecting this well-known algorithm are the ability to calculate the probability of sequences given in the model very efficiently and the existence of an efficient training algorithm (the Baum–Welch algorithm). As far as we know, no research until now tried to increase the accuracy of prediction using user's profile.

A HMM is one type of dynamic Bayesian network (Friedman et al. 1998). An HMM is a stochastic process with an underlying stochastic process that is not observable (it is hidden), but can only be observed through another set of stochastic processes that produces the sequence of observed symbols. HMM can be viewed as a specific instance of a state space model in which the latent variables are discrete.

Our approach for predicting users' intentions is based on the TM, and takes into consideration the differences in usage behavior between users. We create intention prediction tree with HMM in each node to increase the accuracy of prediction. Figure 13.3 presents the proposed process schematically. The left side in Figure 13.3 specifies the generation of the models used for intention prediction based on a profile-driven HMM tree for each possible goal. The input for this process is a set of user's action sequences divided into goals and user's attributes. The output of the process is a tree for each goal based on different usage behaviors of users in attaining this goal and user attributes. The right side of the figure presents the prediction process. The inputs for the process are the goal trees generated on the models generation phase, a specific user session (i.e., a sequence of actions), and the attributes of a user. The output is the predicted goal that the specific user was trying to accomplish in this sequence (i.e., the goal with the highest probability estimation).

An illustration of the obtained tree is provided in Figure 13.4. As user characteristics for this illustration, demographic data is used, but it can be any other static data about the users. Each graph in a tree represents transition probabilities between screens (actions) in a goal. For simplicity only four screens are needed to accomplish a goal, but there is no limitation for number of screens in the model. In the root graph, transition probabilities between screens for all users are presented. The graphs in the next level represent division of the users by gender (one of the users' characteristics). The left child stores transition probabilities for males and the right child stores transition probabilities for females. The graphs in the lower level in

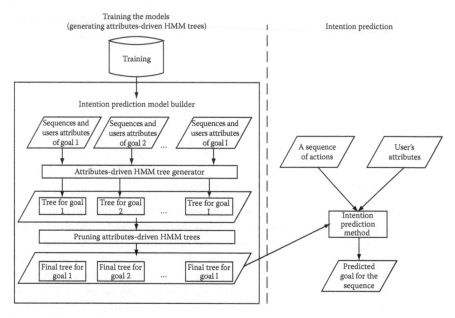

FIGURE 13.3 Diagram of the proposed attributes-driven HMM tree–based method for intention prediction.

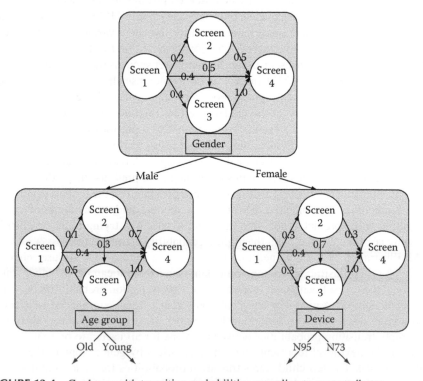

FIGURE 13.4 Goal tree with transition probabilities according to user attributes.

the tree (not shown in the figure) divide the users from the parent level by another attribute from the users' characteristics. After building the model using training sequences, the profile-driven intention prediction tree can provide more knowledge to the company that wishes to improve the users' interface. If a young male uses the system, the company can use the model to predict the next step this user intends to do and provide him a user's interface that fits better to his usage and intentions.

13.4.2   HELP SUGGESTIONS BASED ON THE USER'S PROFILE

The induced users' profiles can facilitate the help offering during the mobile usage. The scenario is as follows: while the user activates the mobile, the system tracks and analyzes the users' operations and tries to figure out what are the user's desired goals according to the usage pattern and the specific user's profile. In case the user fails to accomplish a goal (i.e., by choosing the return or cancel operations), the system will provide help suggestions according to the goals that were analyzed by the intention prediction algorithm regarding the specific user's profile that match activity patterns that were tracked by the system.

13.4.3   EXPERIMENTAL STUDY

In order to evaluate its performance and examine the potential of the proposed framework and the profile-driven HMM tree algorithm, we conducted a field study. The profile-driven HMM algorithm and the HMM algorithm were evaluated on a real data obtained from a mobile application. The examined mobile e-mail application has four different goals (tasks). Forty-eight users have participated in this field study. The users were given tasks (goals) to perform using a mobile phone. They were aware that data was being collected about them. Goals included "configure account settings," "read/send an e-mail," "send a picture to an e-mail address," and "add a contact to address book." They have totally conducted 1652 sessions (sequences). We also received demographic data about the users and from this data four attributes were used to differentiate between users in order to build each goal tree (age, mobile provider, customer type, and payment type). The actions users performed (usage data) with their device were recorded and transformed to input for the intention prediction algorithm.

13.4.3.1   Experimental Process

In this experiment, the true positive rate and false positive rate for each task were calculated. In order to estimate these rates, the collected dataset was randomly divided 10 times into a training set and test set. The training set comprises 70% of the dataset and the test set comprises 30%.

Using the training set, the models for the profile-driven HMM trees were built. Following this stage, the trees were tested using the test set. Each sequence on the test set was divided into subsequences by action steps (only first action, then two first actions in the sequence, etc.).

The input for the test was these subsequences of the sequence together with the user's attributes. The output of the test is a goal that the algorithm predicts as the goal the user will do with the highest probability. Table 13.1 presents the true positives

TABLE 13.1
Goal Identification Results

Task	True Positive (%)	True Negative (%)
Configure account settings	98	0
Read/send an e-mail	94	3
Send a picture to an e-mail address	100	0
Add a contact to address book	96	3

and true negatives for each task separately after completing 50% of the sequence. The results are very encouraging. In fact in most of the goal we achieved a true positive rate of greater than 95%.

13.5   SMART CELL-PHONE PERSONALIZED KNOWLEDGE SERVICE

The SCP-PKS utilizes the analyzed user stereotypes stored in the smart knowledge base for recommending navigation patterns within a knowledge service site that targets new users of SCP. The knowledge service serves as the cell phone's tutorial where each user is motivated to learn first his or her most relevant goals, according to his or her already identified usage behavior, and then get acquainted with more new features. In case the user is familiar with the presented recommendations, he or she can navigate according to his or her interests. The tutorials can be based on the help data that is included in the TM, since the TM holds, for each goal, short and detailed explanations about what and how to do to achieve it.

13.5.1   SMART CELL-PHONE PERSONALIZED KNOWLEDGE SERVICE ARCHITECTURE

The SCP TM and the organizational DB are the infrastructure for the smart knowledge, but are not part of the SCP-PKS, which includes several layers and presented in Figure 13.5. The organizational DB is utilized for retrieving the user's personal information according to which the users' stereotypes are constructed. Since the data is aggregated in relation to users' stereotype and not user's specific data the privacy is kept.

The Smart Knowledge base is constructed during the tracking of the SCP usage. The most relevant parts in the Smart Knowledge base are the users' stereotypes and the goals that these users' stereotypes activate. The Smart Knowledge base enables the Users' Stereotype Analyzer to relate a specific user to a specific users' stereotype for realizing the user's most relevant goals. Next, the Knowledge Navigation Recommender can extract and present to the user the required tutorials from the Tutorials Database. The Web Handler takes care of presenting the tutorials to the user in a menu-driven format that first presents the most relevant goals for the specific user, for enhancing the user navigation and information search in the site. The Interaction Layer Database aggregates all the interactions that the user has performed within the knowledge service site. The Interaction Database (DB) keeps all these users' interactions for understanding the users' preferences within the knowledge

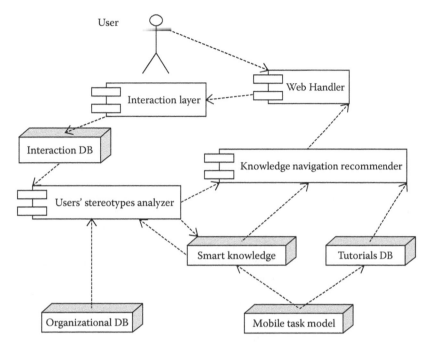

FIGURE 13.5 Smart cell-phone personalized knowledge service architecture.

service environment. If the user did not enter the proposed navigation patterns, these patterns will not be suggested to him or her in the next site browsing session, but rather new patterns that may interest the user. The information in the Interaction DB is also utilized by the Users' Stereotype Analyzer for refining the Smart Knowledge base, since the information exhibit the goals that a user was interested in and that may interest other users who belong to the same user's stereotype. The Interaction Layer is responsible for handling the Interaction DB and mining the database for revealing new information such as new relevant goals for various users' stereotypes.

13.5.2 ILLUSTRATION OF THE SMART CELL-PHONE PERSONALIZED KNOWLEDGE SERVICE

The SCP personalized service begins by generating TMs for various SCP applications. For example, the application e-mail is modeled according to the generic TM structure and the TM database includes all the e-mail application's goals, screens, and how to activate the SCP in order to achieve a specific goal (i.e., to send a message). Then, when a user starts using the SCP, the system tracks the users' activities and with the intention prediction algorithm recognizes which goals are the most prominent for specific users' stereotypes.

In our example, the system identified that educated users, hence users with occupation such as lawyer or engineer, in the age range of 25–55, are utilizing the e-mail option of the cell phone, whereas uneducated users in this age range are using mainly sending text messages via the SMS option. Therefore, when the lawyer enters the

SCP-PKS site, the Users' Stereotype Analyzer relates the lawyer to a specific stereotype, based on retrieving the user's personal details from the organization DB and matching them to existing stereotypes in the Smart Knowledge base. According to the analyzed lawyer's stereotype, which is "educated at the age range of 25–55," the Knowledge Navigation Recommender sends the Web Handler the tutorials that are associated with the most utilized goals for that stereotype, such as the e-mail application. The Web Handler presents the lawyer tutorials that relate to the e-mail application in the SCP-PKS site and enables the lawyer to navigate to other tutorials according to his or her interests (e.g., the camera application, music downloads, and more). The Interaction Layer tracks the site for identifying the lawyer's usage patterns for realizing the goals that may interest the same identified stereotype and saving these patterns in the Interaction DB. In case the lawyer is interested in new goals, such as taking a picture with the camera, that were not identified in the Smart Knowledge base, the Users' Stereotype Analyzer will update the Smart Knowledge base, so when a user with a similar user's stereotype will connect to the SCP-PKS site, the tutorial of this goal—in our case, taking a picture with the camera—will be presented too. The same site will present to another user, for example, an uneducated grocery owner in the age range of 25–55, first the tutorial that relates to text messages (SMS) rather than the e-mail tutorial. In case the uneducated user will be interested in more sophisticated applications, the user's stereotype-related goals will be updated and these tutorials will be presented to other users with the same stereotype.

We believe that such personalized service can have a positive effect on new cell-phone users, both regarding their BI to use the cell phone and also regarding their self-efficacy. In particular, having a service that can train them efficiently with relevant information enables them to start using the cell phone immediately after purchasing it. This follows the flow theory that fosters filtering out irrelevant information that distracts users' attention. In addition, their first good experience with the cell phone may develop positive attitudes and encourage them to use more advanced capabilities of the cell phone.

13.6 SUMMARY AND FUTURE WORK

We presented a novel architecture of a SCP-PKS that will facilitate new SCP users overcoming usage barriers, in particular when replacing or upgrading their cell phones. The architecture is based on a generic TM that enables modeling various SCP applications and a smart knowledge base that is created during tracking and analyzing usage patterns of users of the SCP. In addition, we introduced an improved algorithm for intention prediction that is based on HMM. We used users' characteristics to build hierarchical intention prediction model for analyzing the users' intentions while using the phone, allowing connecting users' stereotypes and intended goals. Since the smart knowledge base holds the most relevant goals of the SCP services for different stereotypes of users, each user can be offered first with the most relevant tutorials, in addition to other tutorials that relate to other existing cell-phone services. Tracking the usage patterns of the suggested service will help identifying additional interests of SCP users for enhancing further tutorials offering for the various SCP users' stereotypes.

Our research aims to investigate the suggested conceptual framework on real SCP applications for realizing its benefits to users, and to learn how the recommended navigation patterns that are based on users' stereotypes enable users to overcome usage barriers and augment the new SCP usage.

REFERENCES

Adomavicius, G. and Tuzhilin, A. 2005. Personalization techniques: A process-oriented perspective, *Communications of the ACM*, 48(10), 83–90.

Budzik, J., Bradshaw, S., Fu, X., and Hammond, K. 2002. Supporting on-line resource discovery in the context of ongoing tasks with proactive software assistants. *International Journal of Human Computer Studies*, 56(1), 47–74.

Chen, Z., Lin, F., Liu, H., and Liu, Y. 2002. User intention modeling in web applications using data mining. *World Wide Web*, 5(3), 181–191.

Chen, J.V., Yen, D.C., and Chen, K. 2009. The acceptance and diffusion of the innovative smart cell phone use: A case study of a delivery service company in logistics. *Information & Management*, 46(4), 241–248.

Csikszentmihalyi, I.S. 1988. *Optimal Experience: Psychological Studies of Flow in Consciousness*. Cambridge University Press, Cambridge, U.K.

Cui, Y. and Bull, S. 2005. Context and learner modeling for the mobile foreign language learner. *SYSTEM*, 33(2), 353–367.

El-Bishouty, M.M., Ogata, H., and Yano, Y. 2007. PERKAM: Personalized knowledge awareness map for computer supported ubiquitous learning. *Educational Technology & Society*, 10(3), 122–134.

Frias-Martinez, E., Magoulas, G., Chen, S., and Macredie, R. 2006. Automated user modeling for personalized digital libraries. *International Journal of Information Management*, 26, 234–248.

Friedman, N., Murphy, K., and Russell, S. 1998. Learning the structure of dynamic probabilistic networks. In *Proceedings of the Fourteenth Conference on Uncertainty in Artificial Intelligence UAI'98*, July 24–26, University of Wisconsin Business School, Madison, WI.

Godoy, D. and Amandi, A. 2008. Exploiting user interests to characterize navigational patterns in web browsing assistance. *New Generation Computing*, 26, 259–275.

Hoffman, D.L. and Novak, T.P. 1996. Marketing in hypermedia computer-mediated environments: Conceptual foundations. *Journal of Marketing*, 60(3), 50–117.

Horvitz, E., Breese, J., Heckerman, D., Hovel, D. et al. 1998. The Lumiere Project: Bayesian user modeling for inferring the goals and needs of software users. In *Proceedings of the Fourteenth Conference on Uncertainty in Artificial Intelligence*, Madison, WI, pp. 256–265.

Hung, L.P. 2005. A personalized recommendation system based on product taxonomy for one-to-one marketing online. *Expert Systems with Applications*, 29(2), 383–392.

Im, I. and Hars, A. 2001. Finding information just for you: Knowledge reuse using collaborative filtering systems. In *Proceedings of the Twenty-Second International Conference on Information Systems*, New Orleans, LO, pp. 349–360.

Jiang, C. 2009. Does content relevance lead to positive attitude toward websites? Exploring the role of flow and goal specificity. In *Proceeding of AMCIS 2009*. San Francisco, CA.

Koufaris, M. 2002. Applying the technology acceptance model and flow theory to online consumer, behavior. *Information Systems Research*, 13(2), 205–223.

Liang, T.P., Lai, H.J., and Ku, Y.C. 2006. Personalized content recommendation and user satisfaction: Theoretical synthesis and empirical findings. *Journal of Management Information Systems*, 23(3), 45–70.

Liang, T.P., Yang, Y.F., Chen, D.N., and Ku, Y.C. 2008. A semantic-expansion approach to personalized knowledge recommendation. *Decision Support Systems*, 45(3), 401–412.

Montaner, M., Lopez, B., and Mosa, L.D.A. 2003. A taxonomy of recommender agents and the Internet. *Artificial Intelligence Review*, 19, 285–330.

Mori, G., Paterno, F., and Santoro, C. 2002. CTTE: Support for developing and analyzing task models for interactive system design. *IEEE Transactions on Software Engineering*, 28(8), 797–813.

PaternÒ, F. 1999. *Model-Based Design and Evaluation of Interactive Applications*. Springer Verlag, Berlin, Germany.

Rabiner, L.R. 1989. A tutorial on hidden Markov models and selected applications in speech recognition. *Proceedings of the IEEE*, 77(2), 257–286.

Sakagami, H. and Kamba, T. 1997. Learning personal preferences on online newspaper articles from user behaviors. *Computer Networks and ISDN Systems*, 29, 1447–1455.

Shen, J., Li, L., Dietterich, T.G., and Herlocker, J.L. 2006. A hybrid learning system for recognizing user tasks from desktop activities and email messages. In *Proceedings of the 11th International Conference on Intelligent User Interfaces (IUI'06)*, Sydney, New South Wales, Australia. ACM, New York, pp. 86–92.

Sun, X., Chen, Z., Liu, W., and Ma, W.Y. 2002. Intention modeling for web navigation. In *Proceedings of the 11th World Wide Web Conference (WWW)*, May 7–11, Honolulu, Hawaii.

Sun, R. and Giles, C.L. 2001. Sequence learning: From recognition and prediction to sequential decision making. *IEEE Intelligent Systems*, 16(4), 67–70.

Taha, T., Valls, M.J., and Dissanayake, G. 2008. POMDP-based long-term user intention prediction. In *Proceedings of the IEEE 2008 International Conference on Robotics and Automation (ICRA'08)*, May 19–23, Pasadena, CA.

Wei, J., Bressan, S., and Ooi, B.C. 2000. Mining term association rules for automatic global query expansion: Methodology and preliminary results. In *Proceedings of the First International Conference on Web Information Systems Engineering*, Hong Kong, China, pp. 366–373.

Part III

Human Networks in Knowledge Services

14 Design of Knowledge Service Model and Approaches for Professional Virtual Community in Knowledge-Intensive Industries

Yuh-Jen Chen and Meng-Sheng Wu

CONTENTS

14.1   INTRODUCTION

With the burgeoning growth of service-oriented knowledge economy, knowledge has become an important resource for enterprises seeking to enhance competitive advantage while service is a critical value for enterprises to push the economic growth. Thus, knowledge-intensive industries (KII) have become a trend nowadays for industrial development (Bryson and Rusten 2005; Chen 2009; Rodwell and Teo 2004).

Enterprise activities in KII are typically highly creative. By performing and accomplishing each enterprise activity, the domain professional knowledge and experiences involving various ideas such as service innovation or service value added are used. Therefore, effective knowledge management is needed to rapidly accumulate the knowledge assets of enterprises and enhance the efficiency of KII.

Enterprise knowledge management can be implemented as either a systematization strategy or personalization strategy (Hansen et al. 1999; Nonaka and Takeuchi 1995). The systematization strategy manages explicit knowledge and enhances the spread and distribution of explicit knowledge through information systems. The personalization strategy allows an expert to share the other experts' own tacit knowledge (empirical knowledge) through cooperation and communication. Tacit knowledge, which is associated with an enterprise's value, is generally hidden inside personal mental models. The inability to transform tacit knowledge into organizational knowledge (explicit knowledge) results in the loss of important intellectual assets when knowledge workers leave a firm.

A professional virtual community is an interactive platform for enterprise experts to create and share empirical knowledge in KII (Pan and Leidner 2003; Wenger 1998; Wenger et al. 2002). Expert discussions on the platform typically create high volumes of useless information and empirical knowledge. Therefore, managing and sharing useful contents of knowledge discussion have become important issues for knowledge service management in a professional virtual community.

The number of studies of virtual communities has increased rapidly in recent years (Chang 2002; Chang et al. 2008; Li and Wu 2010; Lin and Hsueh 2006; Rouibah and Ould-Ali 2007). For instance, Chang (2002) proposed a concept-mapping method to retrieve the most suitable response to a learner's question from a discussion in a web-based forum. Lin and Hsueh (2006) developed a knowledge-map creation and maintenance mechanism that dynamically manages community documents for communities of practice on the Internet. Rouibah and Ould-Ali (2007) designed a virtual engineering community that supports concurrent product development for geographically distributed partners and then developed the web-based workspace that allows different partners to share engineering data. Chang et al. (2008) constructed a journal publishing community to promote effective document and information sharing in a web-based coursework environment. Li and Wu (2010) applied sentiment analysis and text mining techniques in developing an integrated approach to analyze text data in online forums and detect and forecast hotspots. However, these recent studies focused mainly on managing and searching for explicit knowledge from documents and information in virtual community. That is, they lacked a complete solution for managing and sharing empirical knowledge from professional knowledge and experiences. Therefore, the empirical knowledge requirements of experts in a professional virtual community cannot be satisfied, such that experts cannot create services that meet customer demands.

Hence, this study designs a knowledge service model and approaches to support professional virtual community in KII and effectively assist KII enhancing service innovative abilities. To accomplish this objective, the following tasks are performed: (i) propose a knowledge service model for professional virtual community and (ii) design the approaches for enabling knowledge service technology, including knowledge extraction, knowledge verification, knowledge reasoning, knowledge adaptation, and expert recommendation and consultation.

14.2 DESIGN OF KNOWLEDGE SERVICE MODEL FOR A PROFESSIONAL VIRTUAL COMMUNITY

This section first defines the KII and analyzes its characteristics. Empirical knowledge for a professional virtual community in KII is then modeled. Based on this empirical knowledge model, a knowledge service model for a professional virtual community is finally designed.

14.2.1 DEFINITION AND CHARACTERISTIC ANALYSIS FOR KII

KII are a service value chain of high-value knowledge, which have been established by utilizing cooperation modes as well as by combining resources from science, engineering, and academia. KII use innovative operational modes and technology application techniques to pursue the innovations of product, brand management, operation mode, and service through conducting technologies, Internet, professional knowledge, and services. KII comprise business services, communication services, financial services, educational services, legal consultation services, distribution services, and health services.

Based on the definition of KII, it has the following characteristics:

1. *Knowledge oriented*: In KII, the performance and accomplishment of each enterprise activity rely highly on utilization of domain knowledge and experience to ensure that business models operate normally. Consequently, customer satisfaction and enterprise competitiveness increase.

2. *Knowledge expertise*: KII are a service value chain of high-value knowledge. The establishment and supply of professional services are generally based on professional knowledge. Therefore, knowledge expertise must be checked when collecting knowledge to create quality services that meet customer requirements.

3. *Knowledge innovation*: Service innovation is the primary goal for KII. To provide excellent services to customers, enterprises generally enhance their service innovative abilities via knowledge innovation, which relies on empirical knowledge exchange and communication.

4. *Knowledge value added*: Collaboration is an important strategy for increasing competitive advantage of KII. Thus, the scope of required knowledge has been extended from "point" mode into "plane" mode, such that completed knowledge value-added services can be provided to customers.

14.2.2 Empirical Knowledge Modeling for a Professional Virtual Community

Empirical knowledge in a professional virtual community can be exchanged and shared effectively via interaction among experts, and the range of this empirical knowledge is extensive. Thus, empirical knowledge exchanged and shared among experts in a professional virtual community is modeled (Figure 14.1). In understanding empirical knowledge, experts first focus on a discussion topic for exchanging and sharing empirical knowledge. Then, the definition, purpose, and implementation steps for the topic are shared and understood, respectively. Meanwhile, experts can use the document search function in a virtual community to search for and refer to

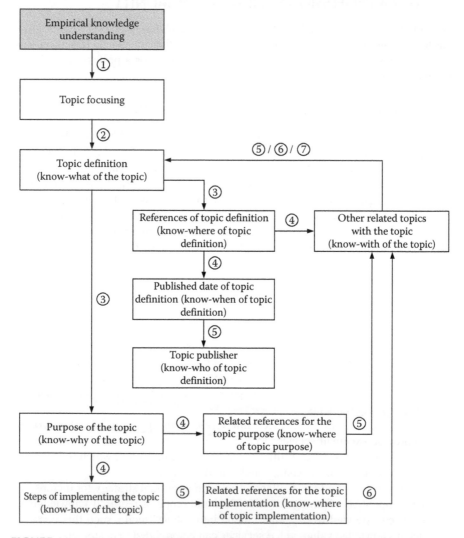

FIGURE 14.1 Empirical knowledge model for a professional virtual community.

other related definitions, purposes, and implementation steps for the topic. Related discussion topics can also be identified in these reference documents. Moreover, publishing dates and publishers would be referred while experts understand the topic definitions.

14.2.3 KNOWLEDGE SERVICE MODEL DESIGN FOR A PROFESSIONAL VIRTUAL COMMUNITY

Based on the aforementioned characteristics of KII and the empirical knowledge model for a professional virtual community, the knowledge service management model for a professional virtual community is designed. As shown in Figure 14.2, the business model of a KII must be performed and achieved by knowledge workers using empirical knowledge to satisfy customer groups from different service activities. However, to enhance customer satisfaction effectively, knowledge workers generally use a professional virtual community to create and share each other's empirical knowledge, thereby enhancing themselves knowledge innovative abilities to create quality services that meet

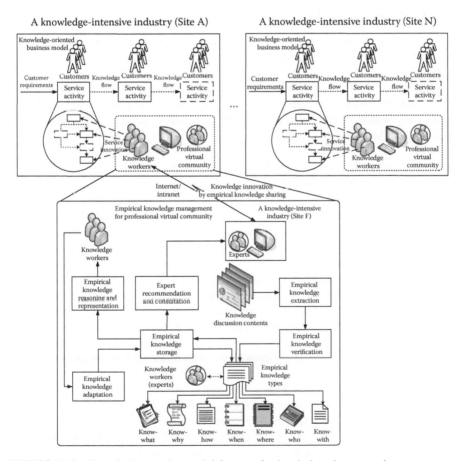

FIGURE 14.2   Knowledge service model for a professional virtual community.

customer needs. In a professional virtual community, knowledge discussion contents from expert discussions are important sources for accumulating an enterprise's empirical knowledge. This valuable empirical knowledge can be managed and reused effectively through knowledge extraction, verification, storage, reasoning, adaptation, and expert recommendation and consultation to achieve the goal of service innovation for KII.

14.3 DESIGN OF APPROACHES FOR ENABLING KNOWLEDGE SERVICE TECHNOLOGY IN A PROFESSIONAL VIRTUAL COMMUNITY

According to the knowledge service model designed in Section 14.2.3, this section designs the approaches for enabling knowledge service technology in a professional virtual communities. These approaches include knowledge extraction, knowledge verification, knowledge reasoning, knowledge adaptation, and expert recommendation and consultation. Each approach is introduced as follows.

14.3.1 APPROACH FOR KNOWLEDGE EXTRACTION

In the expert discussion in a professional virtual community, high-volume and precious empirical knowledge hidden behind the knowledge discussion contents should be effectively extracted to be managed and shared.

The procedure of knowledge extraction (Figure 14.3) involves "topic classification," "domain dictionary construction," "ontology-based topic empirical knowledge model construction," "topic concept extraction and representation," as well as "path establishment between topic concepts," as described in the following:

1. *Topic classification*: Topics for discussion documents in a professional virtual community are inducted. All responding documents to the same creator are treated as the documents under the same topic so as to combine them into one document unit. The document unit set obtained through the topic classification then becomes the main source for discovering the empirical knowledge evolution course.
2. *Domain dictionary construction*
   a. *POS tagging*: Based on the document unit set, keywords for related discussion domains are determined and parts of speech (POS) for each keyword is analyzed by conducting sentence breaking and word tagging.
   b. *Term-pair combination*: In executing the sentence breaking and the word tagging for document unit set, domain keywords in documents may be broken owing to the ability to use a corpus with different domains or an insufficient corpus. In this case, accurate domain keywords cannot be extracted. Hence, these segmented keywords must be recombined through the term-pair combination to ensure the accuracy of domain keywords.
   c. *Domain keyword filtering*: The term frequency-inverse document frequency (TF-IDF) can assist experts in filtering the obtained domain keywords to ensure the accuracy of a domain dictionary.

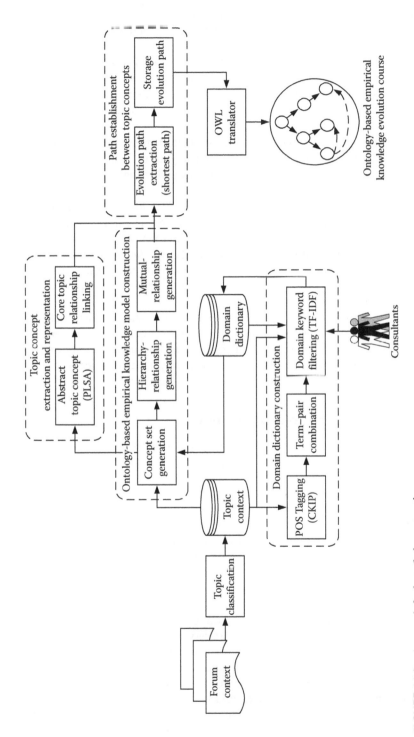

FIGURE 14.3 Approach for knowledge extraction.

3. *Ontology-based topic empirical knowledge model construction*
   a. *Concept-set generation*: According to the inducted document units and the extracted domain dictionary, the concept set of discussion domain is analyzed by performing formal concept analysis (FCA).
   b. *Hierarchy-relation generation*: In this subpart, the mutual inclusion relation between the concept set and the document set is analyzed to establish a hierarchical classification structure, ultimately forming an empirical ontology.
   c. *Mutual-relation calculation*: The strength degrees for similarity and intersection between concept sets are analyzed to facilitate the analysis of evolution course for concept expanding and changing during experts' discussion process.
4. *Topic concept extraction and representation*: The document units and the domain dictionary are utilized as the data sources. The most common discussion topic concepts and their related domain keywords in a discussion domain are then identified based on probabilistic latent semantic analysis (PLSA). Additionally, the document units from part (1) are classified according to their topic concepts to analyze the discussion topics hidden inside of the document units effectively.
5. *Path establishment between topic concepts*: The shortest path between topic concepts is constructed through the shortest path searching method to present the evolution course of topic expanding and changing during discussion among community experts, ultimately predicting accurately the required empirical knowledge for community members.

14.3.2 Approach for Knowledge Verification

The knowledge verification is important not only to facilitate the logic accuracy of empirical knowledge in the knowledge repository, but also to provide knowledge workers with decision support based on accurate knowledge.

This section first designs the ontology-based empirical knowledge verification model. Then, the verification procedure for ontology-based empirical knowledge is designed based on the knowledge verification model.

The ontology-based empirical knowledge verification model consists mainly of empirical knowledge verification and empirical knowledge validation (Figure 14.4), as described in the following:

1. *Verification of empirical knowledge*: This task involves conducting concept matching and relation comparison for the ontology-based empirical knowledge model, that is, source ontology, as obtained from the empirical knowledge extraction and the verified empirical knowledge, that is, target ontology, in the empirical knowledge repository. The concept matching method matches the concept similarity and determines the synonym and homograph. Next, relations including "The Same" (high similarity) and "Similar To" (medium similarity) are built based on the

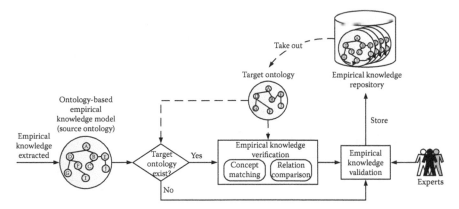

FIGURE 14.4   Ontology-based empirical knowledge verification model.

concept similarity. If no similar concept is available after calculating the concept similarity, then the concept itself must be marked as an independent concept such as concept E in Figure 14.5. This is in order to achieve the construction of relations between the ontology-based empirical knowledge model and the verified empirical knowledge in the empirical knowledge repository (Figure 14.5). The relation comparison method searches for all possible logic paths (Figure 14.6) for the verified ontology empirical knowledge inside of the empirical knowledge repository and then verifies its logical structure to pave the way for empirical knowledge validation.

2. *Validation of empirical knowledge*: Based on logical errors from the verification of empirical knowledge, domain experts must determine and modify these logical errors and, then, store modified empirical knowledge into the empirical knowledge repository as a reference for logical judgments for future empirical knowledge verification in a professional virtual community.

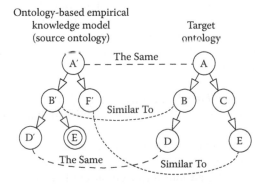

FIGURE 14.5   Relation construction between ontology-based empirical knowledge.

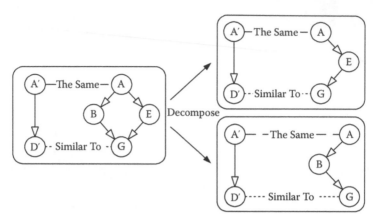

FIGURE 14.6   Logic path decomposition.

According to the designed ontology-based empirical knowledge verification model, the procedure of ontology-based empirical knowledge verification is designed to detect logical conflict problems generated from the priority difference in time for new and old knowledge models. The procedure thus provides experts with verification and modification for empirical knowledge (Figure 14.7). The ontology-based empirical knowledge verification procedure consists mainly of concept matching, relation comparison, and empirical knowledge validation. Meanwhile, concept matching involves concept mapping, concept locating and relation renewing, and relation labeling, while relation comparison involves path finding, relation and path combination, and relation consistency checking. Empirical knowledge validation comprises concept modification and relation modification.

14.3.3 APPROACH FOR KNOWLEDGE REASONING

Knowledge workers can describe and inquire the encountered problems in the topic discussion to match and obtain related empirical knowledge and solutions by using the knowledge reasoning in order to satisfy their empirical knowledge requirements.

The empirical knowledge reasoning includes six portions, namely, multilayer empirical knowledge representation model design, empirical knowledge concept schema design, OWL-based empirical knowledge ontology establishment, reasoning rules design for single-layer empirical knowledge, reasoning rules design for cross-layer empirical knowledge, and empirical knowledge reasoning procedure design. They are discussed as follows:

1. *Design of ontology-based multilayer empirical knowledge representation model*: This study first divides empirical knowledge into four different layers of "know-what," "know-why," "know-how," and "know-with" based on the empirical knowledge characterization such as hierarchical, descriptive, causal, procedural, and relational. The conceptual model of ontology-based multilayer empirical knowledge representation is then designed by using ontology techniques, as shown in Figure 14.8.

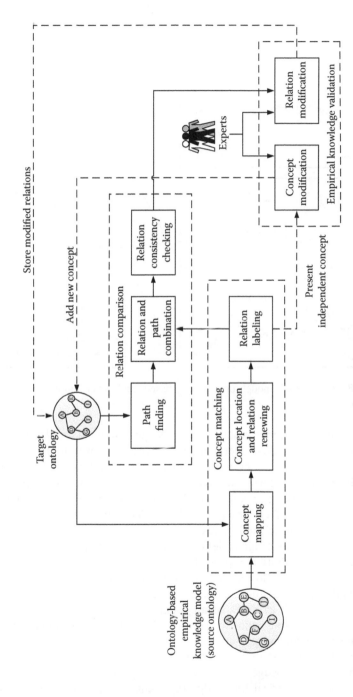

FIGURE 14.7   Approach for ontology-based empirical knowledge verification.

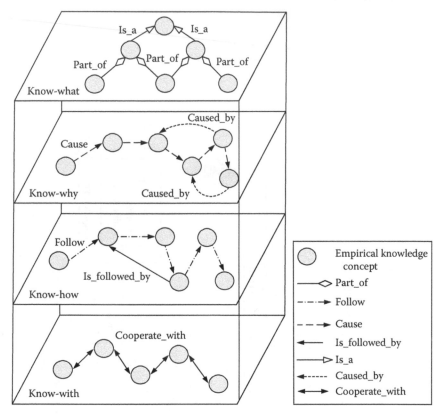

FIGURE 14.8 Conceptual model of ontology-based multilayer empirical knowledge representation.

In the figure, descriptive empirical knowledge is defined as a know-what layer that identifies class, hierarchy, layer, and composition of empirical knowledge; in addition, the relationships include the taxonomy relationship "Is_a" and the partonomy relationship "Part_of." Casual empirical knowledge is treated as a know-why layer that explains the causality and is a consequence of empirical knowledge. Therefore, the relationships are defined as "Cause" and "Caused_by," respectively. Moreover, procedural empirical knowledge is defined to a know-how layer that describes the operational activity and procedure of an event; the relationships are "Follow" and "Is_followed_by." Finally, the know-with layer mainly presents the relational empirical knowledge that records the relationships among operational activities of an event. The relationship is therefore defined as "Cooperate_with" to represent the collaboration among empirical knowledge concepts.

2. *Design of ontology-based empirical knowledge concept schema:* For the empirical knowledge concept in the designed conceptual model of ontology-based empirical knowledge representation, this study adopts the

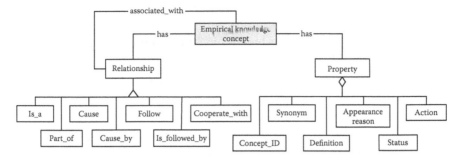

FIGURE 14.9   Empirical knowledge concept schema.

object-oriented method (Bhalla 1991; Martinez et al. 2008; Xu et al. 2007) to design an concept schema of empirical knowledge, as shown in Figure 14.9. The empirical knowledge concept schema consists of three elements of concept, property, and relationship. Each element is described as follows:

a. *Concept*: Concept refers to the name of the basic unit that can constitute an empirical knowledge ontology to express a visible or invisible object.

b. *Property*: Empirical knowledge properties include "Concept_ID," "Definition," "Synonym," "Appearance Reason," "Status," and "Action."

c. *Relationship*: Relationships between empirical knowledge concepts include "Is_a," "Part_of," "Cause," "Cause_by," "Follow," "Is_followed_by," and "Cooperate_with."

3. *Establishment of OWL-based empirical knowledge ontology*: According to the aforementioned conceptual model of ontology-based empirical knowledge representation and the ontology-based empirical knowledge concept schema, an OWL-based empirical knowledge ontology is developed in the following based on the basic axiom and constraint of OWL DL (Horrocks et al. 2003; Hsu and Kao 2005; Motik et al. 2005) and OWL language, thus paving the way for reasoning rules design of ontology-based empirical knowledge. Table 14.1 lists the relevant symbols for establishing OWL-based empirical knowledge ontology. In the table, OWL-based empirical knowledge ontology is composed of different OWL constructs. Each OWL construct can be represented by a different DL Syntax. Each OWL construct is detailed as follows:

a. *rdf:subClassof*: the hierarchical relationship between two concept classes.

b. *owl:ObjectProperty*: the object relationship between two concept objects.

c. *owl:TransitiveProperty*: the inheritance relationship between two concepts. In the relationship, the inheritance inherits the characteristics from the original concept that had been inherited.

d. *owl:SymmetricProperty*: the relationship symmetry between two concepts.

TABLE 14.1

Relevant Symbols for Establishment of OWL-based Empirical Knowledge Ontology

OWL Construct	DL Syntax	Relationship of Empirical Knowledge Ontology
rdf:subClassof	$\subseteq$	Is_a
owl:ObjectProperty	$P \subseteq P_i$	Part_of, Cause, Follow, Cooperate_with
owl:TransitiveProperty	$P^+ \subseteq P$	
owl:SymmetricProperty	$P_i \equiv P_i^-$	Cooperate_with
owl:inverseOf	$P \equiv P^-$	Has_part, Caused_by, Is_followed_by, etc.
OWL Construct	**DL Syntax**	**Property of Empirical Knowledge Ontology**
owl:DatatypeProperty	$U \subseteq U_i$	Appearance reason, status, action
owl:minCardinality	$\geq n\ P$	Appearance reason
owl:maxCardinality	$\leq n\ P$	Status, action

    e. *owl:inverseOf*: the reversibility between concept relationships.

    f. *owl:DatatypeProperty*: the data property of a concept.

    g. *owl:minCardinality*: constraint on the minimum of a property value.

    h. *owl:maxCardinality*: constraint on the maximum of a property value.

4. *Design of ontology-based reasoning rules for single-layer empirical knowledge*: This study describes the ontology-based reasoning rules for single-layer empirical knowledge based on the aforementioned OWL-based empirical knowledge ontology and related methods of knowledge reasoning, description logics, and OWL DL basic axiom and constraint (Horrocks et al. 2003; Hsu and Kao 2005; Motik et al. 2005), as shown in Table 14.2. The empirical knowledge reasoning rules in the table comprise four different types of know-what, know-why, know-how, and know-with.

5. *Design of ontology-based reasoning rules for cross-layer empirical knowledge*: According to the aforementioned OWL-based empirical knowledge ontology and the related methods of knowledge reasoning, description logics, and OWL DL basic axiom and constraint (Horrocks et al. 2003; Hsu and Kao 2005; Schulz and Hahn 2005), the relationships among empirical knowledge in different layers are analyzed based on the ontology-based multilayer empirical knowledge representation model. Additionally, the ontology-based reasoning rules for cross-layer empirical knowledge are designed by using ontology language OWL DL. According to Table 14.3, the ontology-based reasoning rules for cross-layer empirical knowledge involve three categories. They are, that is, "know-what layer vs. know-why layer," "know-what layer vs. know-how layer," and "know-what layer vs. know-with layer," respectively.

6. *Design of empirical knowledge reasoning scenario*: According to the previously designed results, the empirical knowledge reasoning scenario is constructed, as shown in Figure 14.10.

TABLE 14.2
Ontology-Based Reasoning Rules for Single-Layer Empirical Knowledge

Single-Layer Empirical Knowledge Type	Relationship	Reasoning Rule
Know-what layer	Is_a	• (?A rdfs:subClassOf ?B)∧(?B rdfs:subClassOf ?C) ⇒ (?A rdfs:subClassOf ?C)
	Part_of Has_part	• ∀D,E,F: Part-of(D,E)∧Part-of(E,F) ⇒ Part-of(D,F) • Has-part ≡ Part-of⁻¹ (inverse) • ∀D,E: Part-of(D,E) ⇔ Has-part(E,D)
Know-why layer	Cause Caused_by	• (?Cause rdf:type owl:TransitiveProperty)∧(?G ? Cause ?H)∧(?H ? Cause ?I) ⇒ (?G ? Cause ?I) • Cause ≡ Caused_by⁻¹ (inverse) • (?Cause owl:inverseOf ? Caused_by)∧(?H ? Cause ?I) ⇒ (?I ? Caused_by ?H)
Know-how layer	Follow Is_followed_by	• (?Follow rdf:type owl:TransitiveProperty)∧(?J ? Follow ?K)∧(?K ? Follow ?L) ⇒ (?J ? Follow ?L) • Follow ≡ Is_followed_by⁻¹ (inverse) • (?Follow owl:inverseOf ? Is_followed_by)∧(?K ? Follow ?L) ⇒ (?L ? Is_followed_by ?K)
Know-with layer	Cooperate_with	• (?Cooperate_with rdf:type owl:TransitiveProperty)∧(?M ? Cooperate_with ?N)∧(?N ? Cooperate_with ?O) • (?M ? Cooperate_with ?O) • (?Cooperate_with rdf:type owl:SymmetricProperty)∧(?N ? Cooperate_with ?O) ⇒ (?O ? Cooperate_with ?N)

Notes: Parameters A, B, C, etc. denote the concept names of empirical knowledge. Symbol "∧" represents the logical operator AND. Symbol "⇒" means reasoning.

TABLE 14.3
Ontology-Based Reasoning Rules for Cross-Layer Empirical Knowledge

Cross-Empirical Knowledge Type	Relationship	Reasoning Rules
Know-what layer vs. know why layer	Part_of Cause	• X ⊆ ∃ Cause.Y∧Y ⊆ ∃ Part-of. Z ⇒ X ⊆ ∃ Cause. Z
Know-what layer vs. know-how layer	Part_of Follow	• U ⊆ ∃ Follow.V∧V ⊆ ∃ Part-of. W ⇒ U ⊆ ∃ Follow. W
Know-what layer vs. know-with layer	Part_of Cooperate_with	• R ⊆ ∃ Cooperate_with.S∧S ⊆ ∃ Part-of. T ⇒ R ⊆ ∃ Cooperate_ with. T

Notes: Parameters X, Y, Z, etc. denote the concept names of empirical knowledge. Symbol "∧" represents the logical operator AND. Symbol "⇒" implies reasoning. Symbol "⊆" implies a subset. Symbol "∃" implies an existential quantifier.

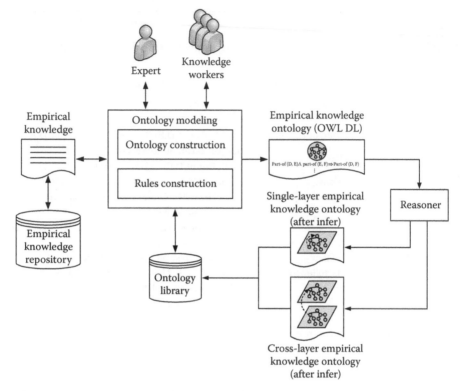

FIGURE 14.10   Empirical knowledge reasoning scenario.

14.3.4   APPROACH FOR KNOWLEDGE ADAPTATION

According to knowledge workers' past usage behaviors, the suitable domain ontology can be adapted to satisfy every knowledge worker's knowledge requirements, and ultimately to enhance the reuse value of knowledge.

This section first designs three models, including the model for adaptation factor analysis, the context analysis model for users, and the personalized adaptation model. Based on these models, a procedure of domain ontology adaptation is then designed. Each model is described as follows.

14.3.4.1   Adaptation Factor Analysis Model for Domain Ontology

Various context requirements and motivations generate diverse application interfaces and services in a user information system (Kwon 2009; Lee 2007). Hence, this study treats the usage context as the basis of use requirement of domain ontology.

Traditional use requirement of domain ontology must rely on the active input of users for judging the system (Hong et al. 2009). Although domain ontology can accurately judge user requirements, there is a lack of intelligent and real-time qualities for system operations. This produces error intention of use requirement to users who are not familiar with a certain application domain. Conversely, the usage history log records behavioral characteristics of users in an active, real time, and objective way (Hong et al. 2009; Komlodi et al. 2007). Therefore, this study induces the individual requirement and behavior of users through exploring literatures on the usage history log and using these results to conduct adaptation factors for domain ontology.

Information retrieval (IR) (Komlodi et al. 2007; Sufyan Beg 2005; Tian et al. 2009) often uses the usage history log as a search quality measure (SQM) to determine the search result satisfaction of users. This method analyzes the visited content and browsing time of users to judge whether the user is suitable for, or familiar with, that content. Then, how the user has used that content and the usage frequency are recorded to judge the importance of that content. Web mining (Khasawneh and Chan 2006; Lia and Zhong 2004) analyzes and computes the non-idle state of the user, and adjusts the weights of travel time and use action to validate that the content can meet use requirements. In marketing management (Chen et al. 2009), the model of RFM customer value analysis uses three elements as the value measurement of customers to determine valuable customers and their preferences. These three elements include recent consuming time (recently, R), consuming frequency during a certain period (frequency, F), and consuming amount during a certain period (monetary amount, M), respectively.

Based on the exploration of literatures given earlier, this study induced five ontology adaptation factors to facilitate the design of an adaptation factor analysis model for domain ontology, as listed in Figure 14.11.

Based on the adaptation factors for domain ontology from Table 14.4, the adaptation factor analysis model for domain ontology is designed, as shown in Figure 14.11. This proposed model has the following elements:

1. *Feature extraction*: Based on user features in user profiles, the concept of entropy (Chen et al. 2009) is employed to compute the information gain of each user feature. Through the method, the user features with higher identification can be found to describe the various user features. These user features are then stored in the user features repository to facilitate the establishment of a usage context analysis model for users.

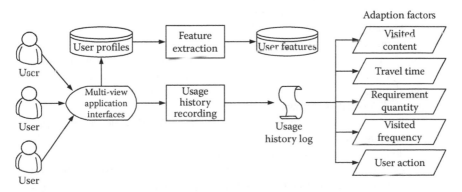

FIGURE 14.11   Adaptation factor analysis model for domain ontology.

TABLE 14.4

Important Adaptation Factors for Domain Ontology from Literature

Adaptation Factor	Explanation
Visited content (VC)	All visited concept nodes from user
Travel time (TT)	User's attentive time to each concept node
Requirement quantity (RQ)	The number of instances that are mapped to each visited concept node
Visited frequency (VF)	The visited times to each concept node
User action (UA)	Usage of the knowledge content that is mapped to the visited concept node from user

2. *Usage history recording*: The ontology usage histories of users are first derived from their application interfaces. Subsequently, their usage history logs are created to calculate the values of adaptation factors for domain ontology. These values are the significant basis for adapting domain ontology.

14.3.4.2   Usage Context Analysis Model of Users

Due to a diverse usage context in the requirements of users (Figge 2004; Lee et al. 2005), this study established the usage context analysis model for domain ontology users to effectively represent the use requirements of users, as shown in Figure 14.12. The usage context analysis model includes three portions, namely, user preference analysis, use requirement-based concept clustering, and user pattern formation. Each portion is explained as follows:

1. *User preference analysis*: The user preference analysis involves two steps of usage, history tracing and log path mapping, introduced as follows:
   a. *Usage history tracing*: This step obtains the preference of users to certain concept nodes in the domain ontology by tracing the use action of users and figuring out the values of adaptation factors from the usage history log.
   b. *Log path mapping*: This step maps the results from tracing usage history with the concept nodes from the domain ontology to transfer use action into a sequence formed with concept nodes of domain ontology. Then, this sequence is stored in the log paths of users to facilitate clustering for use requirement-based concepts.
2. *Use requirement-based concept clustering*: The use requirement-based concept clustering includes three steps, namely, path association computing, domain ontology modeling, and use requirement range setting. Each step is described as follows:
   a. *Path association computing*: Based on the acquired sequence in log path mapping, the association support and the confidence among concepts in the domain ontology are obtained using the association rules of data mining.
   b. *Domain ontology modeling*: Based on the results from path association computing as well as path length and path depth of domain ontology, the domain ontology is modeled through social network representation

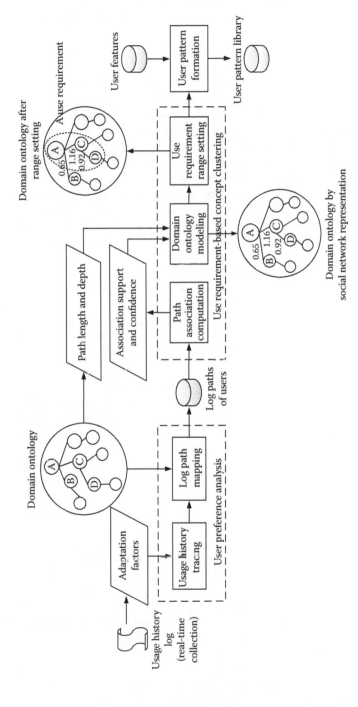

FIGURE 14.12 Usage context analysis model for domain ontology users.

(Forte et al. 2008). This social network–based domain ontology representation can obviously represent one specific concept node cluster and
the relation strengths between these concepts, while the user is using
the domain ontology.
  c. *Use requirement range setting*: According to concepts in the social net-
     work–based domain ontology, hierarchical clustering (Hsu et al. 2007; Tan
     et al. 2006) is conducted to cluster these concepts. The clustering results are
     treated as the use requirement range for domain ontology users. However,
     this use requirement range changes based on the usage history log to
     promptly present the variances in use requirement of domain ontology.
3. *User pattern formation*: The user pattern describes the requirement detail
   of using the domain ontology. The results from the use requirement-based
   concept clustering are used to match with user features to form a user pattern. Subsequently, the association rules determine the support degree of
   user pattern formation. Finally, the support degree of user pattern formation
   is used to evaluate if the user pattern meets the threshold value of user pattern formation. The qualified user pattern is stored in the user pattern library.

14.3.4.3 Personalized Adaptation Model

Based on the aforementioned adaptation factor analysis model for domain ontology
and the usage context analysis model, this study establishes the personalized adaptation model to help users obtain an appropriate user pattern, as shown in Figure 14.13.
This personalized adaptation model includes two portions of user confirmation and
pattern recommendation. Each portion is discussed as follows:

1. *User confirmation*: The user confirmation judges whether the user is a new
   user according to the user history log and the user profile. If the user is a
   new user, the user preference analysis is directly performed. Conversely, the
   user profile is downloaded from the user profiles repository to conduct user
   preference analysis to generate a user log path.
2. *User pattern recommendation*: The user pattern recommendation comprises
   two steps of pattern mapping and pattern projection, introduced as follows:
   a. *Pattern mapping*: Pattern mapping matches user log paths with the user
      pattern library to find out similar user patterns. Subsequently, the con-
      fidence values from similar user patterns are ranked.
   b. *Pattern projection*: By utilizing the results of user preference analy-
      sis and user patterns, the preference degree and requirement range of
      domain ontology usage for the individual user are represented, respec-
      tively. Through combining with the domain ontology, the personalized
      domain ontology is projected and recommended to users. This person-
      alized domain ontology is adopted to build or update the user profile.
      Based on the user profile, the support degree of user pattern formation
      can be calculated to consider if this user pattern meets the threshold
      value of user pattern formation. If the user pattern satisfies the thresh-
      old value, the qualified user pattern is stored in the user pattern library
      as the recommended basis of user pattern for all users.

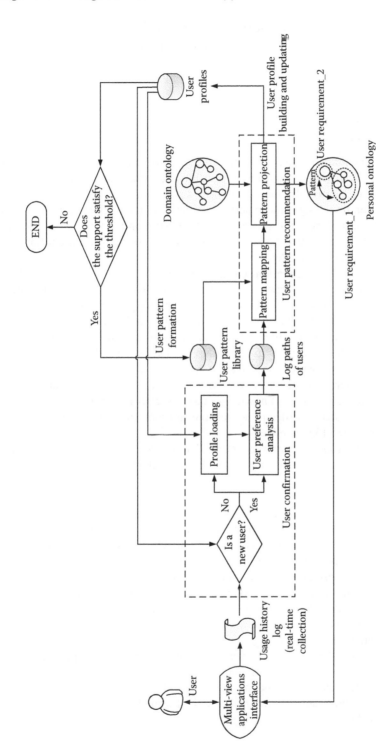

FIGURE 14.13   Personalized adaptation model.

14.3.4.4 Domain Ontology Adaptation Approach

By integrating previously mentioned adaptation factor analysis, model for usage context analysis, and personalized adaptation model, the approach of domain ontology adaptation is established, as shown in Figure 14.14.

14.3.5 Approach for Expert Recommendation and Knowledge Consultation

The expert recommendation can facilitate requesters of empirical knowledge quickly and correctly searching for appropriate consultative experts in a professional virtual community in order to solve their encountered problems.

This section first designs the expert recommendation model for empirical knowledge consultation. Then, the procedures of registration/withdrawal for empirical knowledge and consultative expert recommendation are designed according to the expert recommendation model.

The expert recommendation model for empirical knowledge consultation includes registration/withdrawal of empirical knowledge and consultative expert recommendation, as shown in Figure 14.15. Ontology is employed to establish experts' profile and sharable empirical knowledge and, then, store them in the metaempirical knowledge library via registration of empirical knowledge to encourage consultative experts to share empirical knowledge. Moreover, empirical knowledge is withdrawn by empirical knowledge withdrawal that provided from consultative experts. Ontology is utilized by consultative expert recommendation to achieve matchmaking of knowledge according to requirements of empirical knowledge requester in order to search out appropriate consultative experts for consulting empirical knowledge.

Based on the proposed expert recommendation model, the procedures of registration/withdrawal for empirical knowledge and consultative expert recommendation are described as follows:

1. *Procedure of empirical knowledge registration and withdrawal*: Figure 14.16 shows the procedure for registration and withdrawal of empirical knowledge, which includes four main steps of input of consultative expert's profile and sharable empirical knowledge, registration of empirical knowledge, withdrawal of consultative expert's profile and sharable empirical knowledge, and withdrawal of empirical knowledge. Meanwhile, registration of empirical knowledge involves personal ontology construction, concept name and synonym matching, concept nodes clustering, and a new global ontology generation.

2. *Procedure of consultative expert recommendation*: The procedure of consultative expert recommendation comprises three steps: requested knowledge parsing, concept matching of empirical knowledge, and attribute name set matching of practical knowledge items, as shown in Figure 14.17.

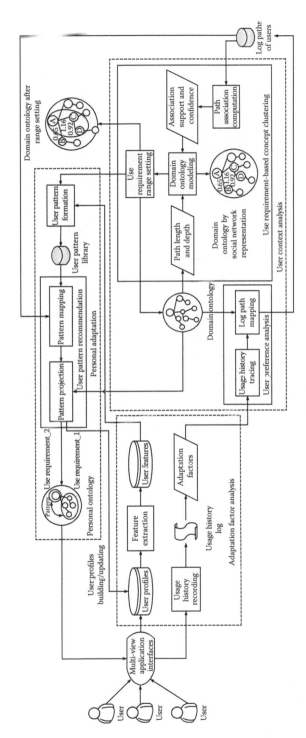

FIGURE 14.14   Approach for domain ontology adaptation.

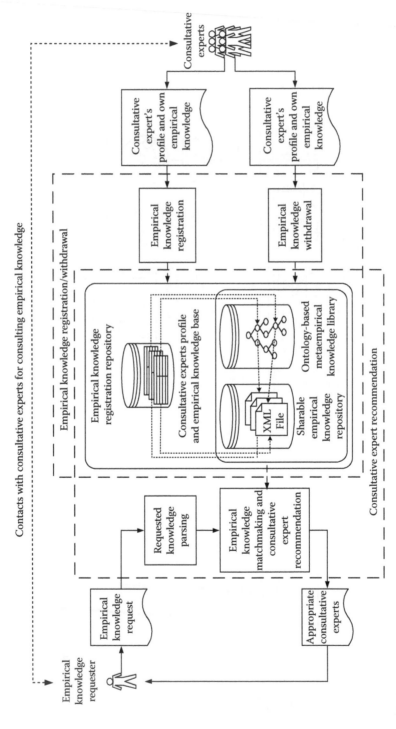

FIGURE 14.15 Ontology-based expert recommendation model for empirical knowledge consultation.

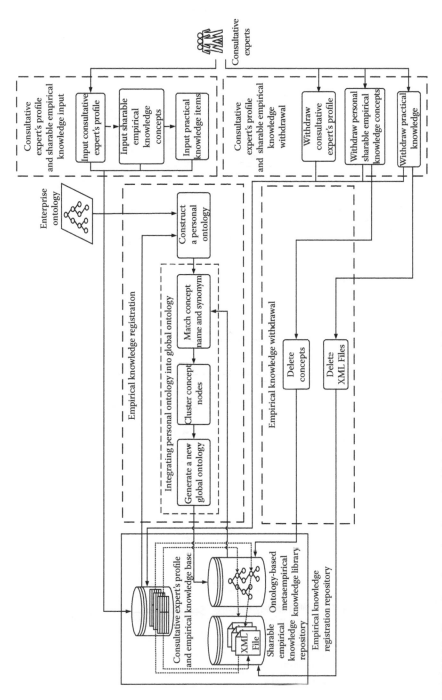

FIGURE 14.16 Procedure for registering and withdrawing empirical knowledge.

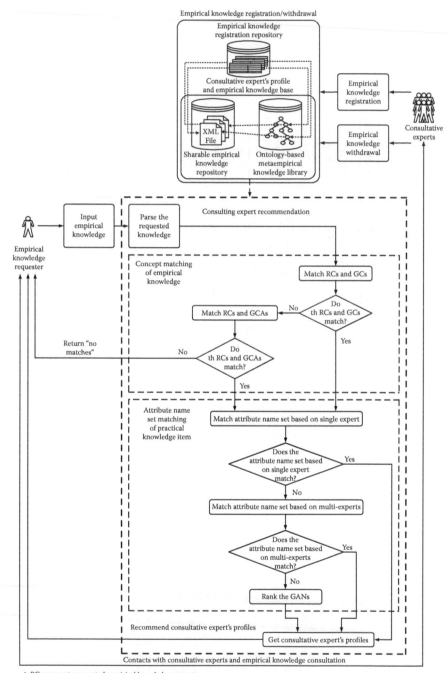

- RC represents requested empirical knowledge concept
- GC represents global empirical knowledge concept
- GCA represents practical knowledge item attribute of global empirical knowledge concept
- GAN represents attribute name set of global practical knowledge item

FIGURE 14.17 Procedure for consultative expert recommendation.

14.4 CONCLUSIONS

This study first defined KII and analyzed its characteristics. Then, empirical knowledge from a professional virtual community was identified and modeled. According to the definition, characteristics, and empirical knowledge of KII, the knowledge service model for a professional virtual community was designed. Moreover, the approaches for enabling knowledge service technology for a professional virtual community were designed.

The results and contributions of this study are summarized as follows:

1. *Knowledge service model for professional virtual community*: This study proposed a knowledge service model for professional virtual community, which can be an important reference model for implementing the knowledge service management in other learning organizations.
2. *Approaches for enabling knowledge service technology*: The designed approaches can effectively extract, verify, store, reason, and adapt the empirical knowledge from knowledge discussion contents in a professional virtual community, to assist enterprises in fulfilling the knowledge service.

ACKNOWLEDGMENT

The authors would like to thank the National Science Council of the Republic of China, Taiwan, for financially supporting this research under Contract Nos. NSC98-2221-E-327-039 and NSC99-2221-E-327-036.

REFERENCES

Bhalla, N. 1991. Object-oriented data models: A perspective and comparative review. *Journal of Information Science* 17: 145–160.

Bryson, J. R. and Rusten, G. 2005. Spatial divisions of expertise: Knowledge intensive business service firms and regional development in Norway. *The Service Industries Journal* 25: 959–977.

Chang, C. K. 2002. An architectural system for retrieving messages to respond dangling questions in learning forum. *Computers and Education* 39: 51–64.

Chang, C. K., Chen, G. D., and Li, L. Y. 2008. Constructing a community of practice to improve coursework activity. *Computers and Education* 50: 235–247.

Chen, Y. M. 2009. Determinants of industry performance: Region vs. country effects in knowledge-intensive service industries. *The Service Industries Journal* 29: 297–316.

Chen, Y. L., Kuo, M. H., Wu, S. Y., and Tang, K. 2009. Discovering recency, frequency, and monetary (RFM) sequential patterns from customers' purchasing data. *Electronic Commerce Research and Applications* 8: 241–251.

Figge, S. 2004. Situation-dependent services? A challenge for mobile network operators. *Journal of Business Research* 57: 1416–1422.

Forte, M., Souza, W. L., and Prado, A. F. 2008. Using ontologies and web services for content adaptation in ubiquitous computing. *The Journal of Systems and Software* 81: 368–381.

Hansen, M. T., Nohria, N., and Tierney, T. 1999. What's your strategy for managing knowledge? *Harvard Business Review* 77: 106–116.

Hong, J., Suh, E. H., Kim, J., and Kim, S. Y. 2009. Context-aware system for proactive personalized service based on context history. *Expert Systems with Applications* 36: 7448–7457.

Horrocks, I., Patel-Schneider, P. F., and van Harmelen, F. 2003. From SHIQ and RDF to OWL: The making of a web ontology language. *Web Semantics: Science, Services and Agents on the World Wide Web* 1: 7–26.

Hsu, C. C., Chen, C. L., and Su, Y. W. 2007. Hierarchical clustering of mixed data based on distance hierarchy. *Information Sciences* 177: 4474–4492.

Hsu, I. C. and Kao, S. J. 2005. An OWL-based extensible transcoding system for mobile multi-devices. *Journal of Information Science* 31: 178–195.

Khasawneh, N. and Chan, C. C. 2006. Active user-based and ontology-based web log data preprocessing for web usage mining. In *Proceedings of the 2006 IEEE/WIC/ACM International Conference on Web Intelligence*, Hong Kong, China, pp. 325–328.

Komlodi, A., Marchionini, G., and Soergel, D. 2007. Search history support for finding and using information: User interface design recommendations from a user study. *Information Processing and Management* 43: 10–29.

Kwon, O. 2009. A social network approach to resolving group-level conflict in context-aware services. *Expert Systems with Applications* 36: 8967–8974.

Lee, W. P. 2007. Deploying personalized mobile services in an agent-based environment. *Expert Systems with Applications* 32: 1194–1207.

Lee, I., Kim, J., and Kim, J. 2005. Use contexts for the mobile Internet: A longitudinal study monitoring actual use of mobile internet services. *International Journal of Human–Computer Interaction* 18: 269–292.

Li, N. and Wu, D. D. 2010. Using text mining and sentiment analysis for online forums hotspot detection and forecast. *Decision Support Systems* 48: 354–368.

Lia, Y. F. and Zhong, N. 2004. Web mining model and its applications for information gathering. *Knowledge-Based Systems* 17: 207–217.

Lin, F. R. and Hsueh, C. M. 2006. Knowledge map creation and maintenance for virtual communities of practice. *Information Processing and Management* 42: 551–568.

Martinez, C. J., Campbell, K. L., Annable, M. D., and Kiker, G. A. 2008. An object-oriented hydrologic model for humid, shallow water-table environments. *Journal of Hydrology* 351: 368–381.

Motik, B., Sattler, U., and Studer, R. 2005. Query answering for OWL-DL with rules. *Web Semantics: Science, Services and Agents on the World Wide Web* 3: 41–60.

Nonaka, I. and Takeuchi, H. 1995. *The Knowledge-Creating Company*. Oxford, U.K.: Oxford University Press.

Pan, S. L. and Leidner, D. E. 2003. Bridging communities of practice with information technology in pursuit of global knowledge sharing. *The Journal of Strategic Information Systems* 12: 71–88.

Rodwell, J. J. and Teo, S. T. T. 2004. Strategic HRM in for-profit and non-profit organizations in a knowledge-intensive industry. *Public Management Review* 6: 311–331.

Rouibah, K. and Ould-Ali, S. 2007. Dynamic data sharing and security in a collaborative product definition management system. *Robotics and Computer-Integrated Manufacturing* 23: 217–233.

Schulz, S. and Hahn, U. 2005. Part-whole representation and reasoning in formal biomedical ontologies. *Artificial Intelligence in Medicine* 34: 179–200.

Sufyan Beg, M. M. 2005. A subjective measure of web search quality. *Information Sciences* 169: 365–381.

Tan, P. N., Steinbach, M., and Kumar, V. 2006. *Introduction to Data Mining*. New York: Addison Wesley, pp. 515–523.

Tian, X., Du, X. Y., Hu, H., and Li, H. H. 2009. Modeling individual cognitive structure in contextual information retrieval. *Computers and Mathematics with Applications* 57: 1048–1056.

Wenger, E. 1998. *Communities of Practice, Learning, Meaning, and Identity.* Cambridge, U.K.: Cambridge University Press.

Wenger, E., McDermott, R. A., and Snyder, W. 2002. *Cultivating of Communities of Practice.* Boston, MA: Harvard Business School Press.

Xu, W. L., Kuhnert, L., Foster, K., Bronlund, J., Potgieter, J., and Diegel, O. 2007. Object-oriented knowledge representation and discovery of human chewing behaviors. *Engineering Applications of Artificial Intelligence* 20: 1000–1012.

15 Social Network Analysis of Knowledge-Based Services

Meeyoung Cha and Minsu Park

CONTENTS

15.1 INTRODUCTION

Just a few years ago, social media was still considered a new trend. A number of websites such as YouTube and Flickr were used as a private tool for social networking, mainly used for uploading and sharing user-generated videos and photos. However, the mass adoption of Facebook, Twitter, and a number of other real-time online social networks (OSNs) over the past few years secured the role of social media as a valuable source of information. Through the development of smartphone technology, more and more people are adopting smartphones and are connected to the Internet all the time. Thereby, OSNs have gained massive number of registered users, allowing social media to serve a real-time platform for information sharing anytime and anywhere.

The OSNs allow individuals to become news generators or citizen journalists by capturing events in images, texts, and videos without delay. A huge growth of OSNs affects a range of social activities such as social campaigns, news broadcastings, and business. More recently, OSNs have played an important role in social demonstrations and public protests such as in the Iran election in 2009 and Egyptian revolution in 2011. With the advent of OSNs, individuals can easily attract their followers through social interaction. For these reasons, the data embedded in OSNs are drawing the attention of various applied service providers, including marketers, stock market analysts, health care providers, etc. (Bollen et al., 2011a,b).

OSN analysis also has its basis in the broad fields of network analysis and graph theory. Network analysis has been applied widely in mathematics, physics, anthropology, biology, communication studies, economics, geography, information science, organizational studies, social psychology, sociolinguistics, and so on. The power of network analysis is derived from its unique standpoint, which assumes that the network structure provides opportunities or constraints to individual actors. Network analysis produces an important view, where the relationships of individual actors and connections between other actors within network are more important than the attributes of individuals themselves. This approach has turned out to be useful for explaining many real-world phenomena.

In this chapter, we introduce some of the basic concepts about graph theory and focus on two specific roles of OSNs, namely, as information broadcast system and as a sentiment capturing system. By introducing observations from analysis of recent social network research, we hope the readers grasp some of the fundamental ideas in social network analysis of knowledge-based services.

15.2 GRAPH-BASED ANALYSIS

Graph theory, which is the study of network structure, produces a possibility to build a networked structure that describes characteristics of real-world phenomena in a legible and neat form called graphs. Many problems of practical interest can be represented by graphs. For example, graphs are used to represent networks of communication, transportation system, data organization, and society. Graph theory provides a set of abstract concepts and methods for the analysis of graphs. In this section, we introduce some of the basic concepts about graph-based analysis, which are important in social network analysis.

15.2.1 BASIC DEFINITIONS OF GRAPHS

A social network consists of a structure made up of individuals (or organizations) called *nodes* that are linked or connected by one or more specific types of interdependency, such as friendship, kinship, common interest, financial exchange, dislike, sexual relationship, relationship of beliefs, knowledge, or prestige. Social network analysis views social relationships in terms of network theory consisting of nodes and *edges* (also called ties, links, or connections). Nodes represent the individual actors within the networks and edges represent the relationships between the actors. The structures of typical social network systems are often very complex. There can be many kinds of edges between the nodes. Research in a number of academic fields

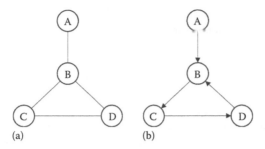

FIGURE 15.1 Two types of graphs: (a) an undirected graph on 4-node and (b) a directed graph on 4-node. (From Easley, D. and Kleinberg, J., *Networks, Crowds, and Markets: Reasoning about a Highly Connected World*, Cambridge University Press, Cambridge, MA, 2010, p. 22. With permission.)

have shown that social networks operate on many levels, from families up to the level of nations, and play a critical role in determining the way problems are solved, organizations are run, and the degree to which individuals succeed in achieving their goals.

The edges can be expressed as a directed (asymmetric), which has a directivity, or as an undirected (symmetric) line. For example, the Twitter contact network is asymmetric, meaning that if a user A follows another user B, B does not need to follow A back. For instance, the Twitter contact network can be expressed by the directed graph like Figure 15.1b. On the other hand, the Facebook contact network built by mutual relationship can be expressed by the undirected graph like Figure 15.1a. These two toy examples in Figure 15.1 show the difference between directed and undirected graphs. In Figure 15.1, four circles labeled A, B, C, and D represent nodes. Lines between nodes represent edges. Undirected graphs are generally drawn with simple edges, implying that nodes are connected with each other as shown in Figure 15.1a. Directed graphs are generally drawn with edges represented by arrows to express asymmetric relationships as shown in Figure 15.1b (Easley and Kleinberg, 2010).

Because graphs can serve mathematical models of network structure, they are useful. For example, we can see an imaginary social network as a simple depiction using nodes and edges as shown in Figure 15.2. A key is how to draw which nodes are connected to which other nodes. Nodes can be people or groups of people, and

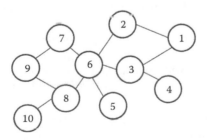

FIGURE 15.2 Drawing of an imaginary 10-node social network graph.

edges can be any kinds of social interaction. In Figure 15.2, for example, we can consider that nodes represent people, and an edge connects two nodes if there is a direct friendship between them.

We can imagine many other classes of graph structures. For instance in communication networks, nodes are computers or other devices that can relay messages and the edges represent direct links along which messages can be transmitted. In information networks, nodes are information resources such as web pages or documents and edges represent logical connections such as hyperlinks, citations, or cross references. The list of areas in which the concept of graphs can be applied is hence broad. Graphs are useful in representing how things are either physically or logically linked to one another in a network structure (Easley and Kleinberg, 2010). We introduce several key concepts that are often used in graph-based analysis in the following. We use the toy example in Figure 15.2 to explain each of these concepts.

1. *Adjacency.* Two nodes are called *adjacent* if they share a common edge (i.e., connected), in which case the common edge is said to join two nodes. In Figure 15.2, for example, nodes 1 and 2 are adjacent (as well as many other pairs of nodes in the figure, such as nodes 3 and 4). Nodes 1 and 6 are not adjacent because they are nodes directly connected. Note that adjacency simply means connection; it does not imply any direction or lack of direction.

2. *Neighborhood and degree.* The *neighborhood* of a node *v* in a graph is the set of nodes adjacent to *v*. The neighborhood does not include *v* itself. For example, the neighborhood of node 6 in Figure 15.2 includes nodes 2, 3, 5, 7, and 8. The degree of a node is the total number of nodes adjacent to that node. We can equivalently define the degree of a node as the cardinality of its neighborhood. For example, the degree of node 6 in Figure 15.2 is 5. In a directed graph, we define degree in the same manner, but it is also important to define *indegree* and *outdegree.* Recall that any directed edge has two distinct ends: an arrowhead and a tail. Each end is counted separately. The sum of head endpoints count toward the indegree of a node and the sum of tail endpoints count toward the outdegree of a node.

3. *Density.* The *density* of a graph $G = (V, E)$ that comprises set of nodes (V) and a set of edges (E) measures how many edges are in set E compared to the maximum possible number of edges between nodes in set V. Density is calculated as follows:

    a. An undirected graph with no loops can have at most $|V| \times (|V| - 1)/2$ edges, so the density of an undirected graph is $2 \times |E|/(|V| \times (|V| - 1))$.

    b. A directed graph with no loops can have at most $|V| \times (|V| - 1)$ edges, so the density of a directed graph is $|E|/(|V| \times (|V| - 1))$.

In Figure 15.2, the graph has a total 11 edges and 10 nodes, and the graph can have 45 maximum possible numbers of edges. Hence, the density of the graph in Figure 15.2 is 0.24 by the simple formula.

4. *Paths and walks.* A *path* $P = (v_1, v_2, v_3, ..., v_k)$ is an ordered list of nodes, where $v_1, v_2, v_3, ..., v_k$ represent the nodes in the path. The first node of a path, v_1, is called the origin and the last node, v_k, is called the destination. Both origin and destination are called endpoints of the path. In Figure 15.2, for example, there are many paths from nodes 1 to 8. One such path is $P_1 = (1, 3, 6, 8)$; another path from 1 to 8 is $P_2 = (1, 2, 6, 7, 9, 8)$. We define a *walk* similarly to a path. However, a walk may visit the same node more than once, but a path must never do so. This implies that every path is a walk, but some walks are not paths (i.e., any walk that visits a node more than once). A cycle is a walk such that the origin and destination are the same; a simple cycle is a cycle that does not repeat any nodes other than the origin and destination. In Figure 15.2, for example, there are many walks from nodes 1 to 8. Our previous examples $P_1 = (1, 3, 6, 8)$ and $P_2 = (1, 2, 6, 7, 9, 8)$ are also walks from nodes 1 to 8. In addition, $W_1 = (1, 2, 6, 3, 4, 3, 6, 7, 6, 8)$ and $W_2 = (1, 3, 1, 3, 6, 8)$ are examples of walks from nodes 1 to 8 that are not paths.

5. *Length, distance, and diameter.* The *length* of a path or a walk is the number of edges that it uses, counting multiple edges multiple times. In Figure 15.2, $(5, 6, 3)$ is a path of length 2 and $(1, 2, 6, 3, 1, 2)$ is a walk of length 5. The *distance* between two nodes x and y is written $d(x, y)$ and defined as the length of the shortest path from x to y. The distance between any node and itself is 0. If there is no path from x to y (i.e., if x and y are in different connected components), then $d(x, y)$ is infinity. The *diameter* of a graph G is the maximum distance between any pair of nodes in G. If G is not connected, then the diameter of G is infinity.

15.3 SOCIAL NETWORKS AS AN INFORMATION BROADCAST SYSTEM

A social network plays a fundamental role as a medium for spreading information, idea, and influence among its members. An idea or innovation will appear—for example, the use of cell phones among college students, the adoption of a new drug within the medical profession, or the rise of a political movement in an unstable society—and it can either die out quickly or make significant inroads into the population (Kempe et al., 2003). If we want to understand the extent to which such ideas are adopted, it becomes important to understand how the dynamics of adoption are likely to unfold within the underlying social network. It is also important to understand the extent to which people are likely to be affected by decisions of their friends and colleagues and the extent to which *word-of-mouth* effects will take hold (Kempe et al., 2003). Achieving such goals has many implications presented by studies of epidemic propagation, information cascades, viral marketing, and others.

In Sections 15.3.1 and 15.3.2, we introduce studies that analyzed popular blog domains including Wordpress and Blogger.com and multimedia contents sharing websites such as YouTube and Flickr to concentrate on understanding propagation characteristics of various forms of contents such as texts, videos, and photos.

In Sections 15.3.3 and 15.3.4, we introduce how information propagate and some of mechanisms for managing online behavior in communities based on research on Twitter, one of the most popular OSNs. Later, we highlight some of the recent studies of OSNs on information propagation to understand the trends of social network analysis. The key trends are moving from measuring properties of information in various networks to monitoring or predicting capturable phenomena on OSNs in real time.

15.3.1 How Blogs Carry Different Types of Information

Social media updates report not only emergencies, but also everyday chats. In particular, blogs play a significant role in today's Internet culture in relation to daily information propagation activities. People discuss political issues, review new products and cultural contents, form communities and special interest groups, share multimedia contents, and so on. To meet needs of diverse groups, blogs have evolved into different forms. Blog can be either a text-based blog or photo blog, video blog, and other kinds. To gain more insights on understanding which factors affect its efficiency for information dissemination and how different types of content affect the shape of the blog graph, we need to analyze structural properties of the blog graph and types of content on the blog.

Cha et al. (2009a,b) presented a number of interesting findings related to characterizing how the structure of the blogosphere influences the patterns of content spreading. First, the network structure of blogs shows a heavy-tailed degree distribution following power law, low reciprocity, and low density. In particular, the low reciprocity indicated a marked discrepancy between indegree and outdegree in blogs. Figure 15.3 shows the indegree and outdegree distribution. The horizontal axis represents the node degree and the vertical axis represents the cumulative number of blogs with degrees (i.e., the number of neighbors) greater than or equal to the degree specified in the horizontal axis. Although the majority of blogs connect only to a few others, certain blogs connect to thousands of other blogs. These high-degree

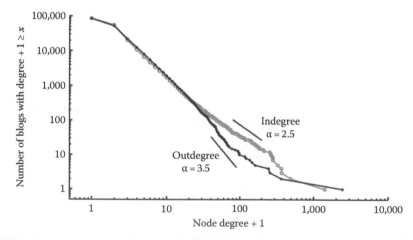

FIGURE 15.3 Degree distribution of the blog graph.

blogs are often content aggregators (high indegree), recommenders (high outdegree), or reputed content producers. In contrast to other OSNs, most links are unidirectional and the network is sparse in the blogosphere. This is because links in typical OSNs represent friendship where reciprocity and mutual friends are expected, while blog links are used to reference information from other data sources. Unlike other OSNs, bloggers likely have a clear preference when they cite a particular topic or an information source (e.g., each user has his own preferred news outlet).

Second, there exists a particular interaction between different blog domains and language groups. The analysis reveals that a significant portion of content referral links spanned different blog domains (e.g., Blogger.com, Wordpress.com, and typepad.com). Blog interactions were not limited by the domain of the blog hosting sites. This result indicates that analyzing the blog graph based on the data from a single blog domain will miss a lot of the rich linkage structure in the blogosphere (Cha et al., 2009a,b). However, authors only found few links between blogs that were written in different languages. When they do occur, links between different languages tended to be unidirectional: unsurprisingly, the most common form is a non-English blog pointing to an English-written blog.

Third, media content spreads according to two broad patterns: flash floods and ripples. Authors conducted analysis to understand how a specific content such as a YouTube video propagates in the blog graph. The two suggested patterns, flash floods and ripples, represent interesting characteristics. Figure 15.4 shows that bloggers tend to rapidly follow up and spread videos of certain topics, but not all other topics. This means that certain topics are more viral than others. On the contrary to this result, bloggers are more relaxed in following up for the other specific categories. The first group includes topical content such as news, political commentary, and opinion. Like flash floods, these types of content spread quickly by the hour and then quickly disappear. This demonstrates the role of blogs as a social medium that helps and influences how opinions form and spread on current issues. The second group includes non-topical content such as music and entertainment. Like ripples, old content that is produced more than a year ago can get rediscovered and again start gaining the attention of bloggers, albeit at a slow rate.

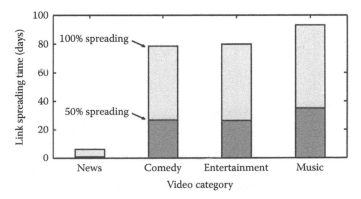

FIGURE 15.4   Time lag in the spread of videos in the blogosphere.

Authors found a list of the most linked websites from the blog dataset. Most of top linked sites on the list were video or photo uploading and sharing services including YouTube and Flickr. This result may reflect the trends that new industries are born or existing industries try to find new ways of doing business utilizing multimedia contents. Probably, this trend will be more pronounced due to the growth of the use of smartphone and microblogs such as Twitter and Facebook. Findings in Cha et al. (2009a,b) open up new research directions and give us chance to develop new business ideas. We introduce how photo-related information propagate on the networks in the following section.

15.3.2 How Photo Favorite Markings Spread on Flickr

OSN sites like Facebook, YouTube, and Flickr have become a popular way to share and disseminate content. Their massive popularity has led to viral marketing techniques that attempt to spread contents, products, and ideas on these sites. Nevertheless, there has been little data publicly available on viral propagation in the real world and few studies have characterized how information spreads over current OSNs.

Cha et al. (2009a,b) collected and analyzed large-scale traces of information dissemination in the Flickr social network. Their analysis, based on crawls of the *favorite markings* of 2.5 million users on 11 million photos, aims at answering three key questions: (a) how widely does information propagate in the social network? (b) how quickly does information propagate? and (c) what is the role of word-of-mouth exchanges between friends in the overall propagation of information in the network? Contrary to viral marketing intuition, they found that (a) even popular photos that own more than a couple of hundred fans do not spread widely throughout the network, (b) even popular photos spread slowly (i.e., over several years) through the network, and (c) information exchanged between friends is likely to account for over 50% of all favorite markings, but with a significant delay at each hop.

Here, we cover two key observations a little bit more deeply. First, Figure 15.5 shows how widely photos were consumed in the network. Authors assumed that if photos spread widely throughout the network then the local and global hotlists

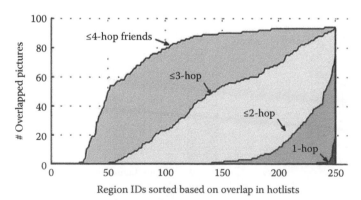

Region IDs sorted based on overlap in hotlists

FIGURE 15.5   Resemblance in local and global hotlist.

(i.e., a set of the most popular photos) will have large overlap as well. The 2.5 million users who have favorite-marked at least one photo were considered to consist the entire network. Then authors selected 250 users (or seed nodes) out of the set of the 2.5 million users and considered each of the 250 selected users' neighborhoods as 250 local regions. The top 100 pictures from the each local and global region were identified and compared. The horizontal axis in Figure 15.5 represents the number of local regions sorted based on overlap in hotlists and the vertical axis represents the number of overlapped pictures between the local and global regions. Based on 1- or 2-hop neighborhoods, hotlists of pictures or the set of most popular photos within the neighborhood were localized. That is, these pictures were not widely available throughout the network. On the other hand, focusing on the 4-hop neighborhood, we see a high overlap in global and local hotlists. This is because this neighborhood covers a large number of users—36% of the entire graph.

This also means that information is reachable within few hops (i.e., small world network). These results mean that most fans or the ones who favorite-mark a photo of a given picture are within a few hops of the picture uploader. For less popular photos, 91% of all fans are within 2 hops of the uploaders. But even for top popular photos, 81% of all fans are within 2 hops of uploaders. It is intuitive that globally less popular photos exhibit strong locality, since these are typically personal photos of family and friends, which are by definition interesting primarily to people pictured in the photo and those who know them. However, it is surprising that popular pictures with >500 fans also show a high level of content locality.

Figure 15.6 shows the long-term trends in popularity growth of pictures and it demonstrates the finding that pictures spread slowly throughout the social network. The horizontal axis represents the age of the photo, or the time since the pictures' uploads. The vertical axis represents the fraction of fans a photo obtained by the given age, out of the total number of fans it obtained at the end of the first year. Many photos do show an active rise in popularity during the first few days after they are uploaded. After the first few (10–20) days, most pictures in aggregate enter a period of steady linear growth. Surprisingly, the steady-growth trend is sustained over extended periods of time—the median growth rate does not show any sign of

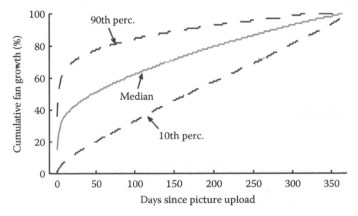

FIGURE 15.6 Popularity growth patterns over a long-term period.

slowing down even after 1 or 2 years. As a result of this steady-growth pattern, the fans of these pictures obtained during the first few days accounts for an ever decreasing fraction of total fans. For a majority of pictures, over 40% of fans were acquired after the first 100 days. Conversely, authors' analysis suggests that Flickr users take a long period of time to find out about interesting pictures. This steady linear growth in popularity cannot be easily explained by traditional information diffusion theories such as *diffusion of innovations* (*Bass diffusion model*), which predict an exponential growth in popularity followed by a saturation or maturity.

In addition, authors found that photo favorite marking is also followed by the power-law degree distribution like blogs degree distribution. However, it shows high reciprocity (68% of social links are bidirectional), which contrasted with the low reciprocity (6% of blog links are bidirectional) seen in blogs (Cha et al., 2009a,b). Gouldner described that relations with little or no reciprocity may occur when power disparities allow one party to coerce the other (Gouldner, 1960). However, it is still unclear if reciprocity is a rule that obtains in specific contexts or if it is a widely shared behavioral norm. While most observers would agree that reciprocity operates at different levels (René, 1989), can reciprocity treated as a culturally induced moral obligation or as the expression of a rational calculus of costs and benefits? The implications of studies about the network structures of blogosphere and Flickr enable us to explore the underlying meanings of reciprocity seen in OSNs data.

15.3.3 GENERAL INTRODUCTION ON TWITTER AS A BROADCAST SYSTEM

Understanding the characteristics of different types of information and how they relate to other aspects of a network is necessary in describing the ability of network as an efficient delivery system. In particular, many researches study online social media and its impact on word of mouth, because consumer-to-consumer interactions help information diffusion more actively and broadly in the area of computer-mediated communication compared to face-to-face communication. By using computer-mediated communications tools such as social networking services (SNS), personal opinions have become an important type of information that is shared among the Internet users. Word of mouth not only plays significant role in marketing but in many other ways is related to information distribution and consumer decision making. Without word-of-mouth effects on OSNs, the diffusion of real-time information such as the recent social movement regarding the Egypt revolution and its protests would not have disseminated rapidly in the network. Twitter, a microblogging service, has emerged as a new medium in spotlight through recent happenings, such as the antigovernment protests in Egypt in 2011 and the U.S. Airways plane crash on the Hudson river in 2009 (Kwak et al., 2010).

Kwak et al. analyzed how directed relationship of users in Twitter set it apart from the existing OSNs. First, authors asked a question, why do people follow each other? Authors guessed the reason in two ways: reflection of offline social relationships and subscription to others' messages. On Twitter, only 22.1% of user pairs follow each other reciprocally. This rate was much lower than Flickr (69%), Yahoo! 360 (84%), and Cyworld (77%). It means that following someone is not similarly considered as being friends as in other OSNs. Therefore, the following relationship on Twitter is

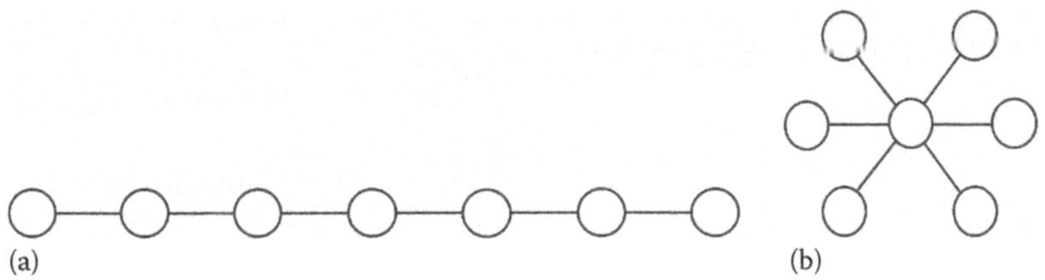

FIGURE 15.7 Examples of a 7-node network: (a) line topology network and (b) star topology network.

not a reflection of an offline social relationship, but represents an active subscription of tweets. Additionally, authors collected CNN Headline News from their Twitter data collection and conducted an analysis to compare the trending topics seen from Twitter against CNN News website. As a result, more than half the topics were already reported from CNN. However, first blow of some news came from Twitter and it shows a live broadcasting nature (e.g., sports matches and accidents). These findings support not only the assumption that Twitter is a medium that allows an active exchange of information but also the assumption that users talk about timely topics on Twitter.

Then, authors analyzed their dataset to determine whether Twitter effectively delivers information on its network or not. Although individual users on Twitter do not have many followers, they reach large public by word of mouth quickly. Consider each circle is an information source and information propagates following the edge in Figure 15.7. Which network is more efficient for propagating information by word of mouth? Intuitively we know that the average path length is an important factor for the efficiency of a network for word of mouth. A network with shorter average path length is more effective for word of mouth. For example, the graph in Figure 15.7b shows that shorter chains connect any two actors in a star topology than in a line topology as in Figure 15.7a.

Information disseminates only from a user to his followers in Twitter. Therefore, the average path length might be longer because of low reciprocity (22.1%) than in other OSNs. However, surprisingly authors observed that the average path length is 4.1 and effective diameter is 4.8 even in the directed Twitter network. For 97% of node pairs, the path length is 6 or shorter. These results are shown in Figure 15.8 and the results indicate that Twitter can be an efficient medium for word of mouth.

Active word-of-mouth phenomenon on Twitter is facilitated by *retweet*, which is to repeat or relay other's tweet thereby allowing it to propagate in the network. Without any retweet, an original tweet can only reach the 1-hop neighbors of the tweet uploader. With the help of retweets, an original tweet can reach farther out in the network. Authors found that the average number of additional readers by a retweet was a few hundred no matter how many followers a user has. It means that a tweet is likely to reach a certain size of audience, once the user's tweet starts spreading via retweets. In addition, retweet process was occurring quickly. Thirty-five percent of all retweets occurred within 10 min of the original tweet and a half of retweets occurred within an hour. This indicates that retweets assist users propagate information widely and quickly.

Such characteristics of Twitter allow us to understand the specific word-of-mouth examples in the real-world situation. OSN can be a great resource of information

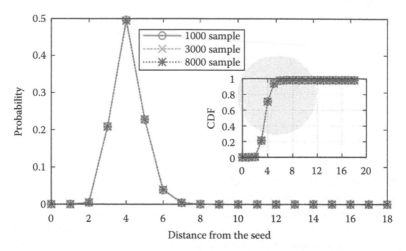

FIGURE 15.8 Hop distance between random pairs of nodes. (From Kwak, H. et al., What is twitter, a social network or a news media? in *Proceedings of International World Wide Web (WWW) Conference*, April 26–30, Raleigh, NC, 2010, p. 3. With permission.)

in particular for spreading breaking news and live events. These real-time communication characteristics are useful in disseminating news on emergency situations. In that case, how does information propagate through networks? Next, we delve deep into the patterns of information sharing and adopting mechanisms related to emergency situations.

15.3.4 POTENTIAL USAGE OF SOCIAL NETWORK SERVICES IN CRISIS COMMUNICATION

The 9.0-scale earthquake that hit Japan on March 11, 2011, at 2:45 PM (local time) was the biggest quake in Japan's history and the fifth biggest recorded across the globe. The subsequent tsunamis caused by the tremors lead additional devastation across three coastal prefectures, wiped two towns off the map, and claimed thousands of lives and displaced more than half a million people. At the time, the mass media reported that Twitter was one of the few functioning communication tools immediately after the earthquake (Acar and Muraki, 2011).

An important characteristic of Twitter is its real-time nature. Twitter is an instantly updated social media site that provides a real-time information network that connects users to the latest information about what their Twitter neighbors generate or find. It can be accessed via the World Wide Web, the mobile phone text messaging system (SMS), or any third-party tools used with smartphones and computers. In the case of the 2011 earthquake, many people tweeted related to the earthquake, which enabled the detection of earthquake occurrence promptly, simply by observing the frequency of tweets.

Most of the tweets in disaster-hit areas were about warnings, help requests, and reports about the incident. Twitter accounts set up by local authorities at the time of the earthquake were particularly useful and were well-followed and retweeted

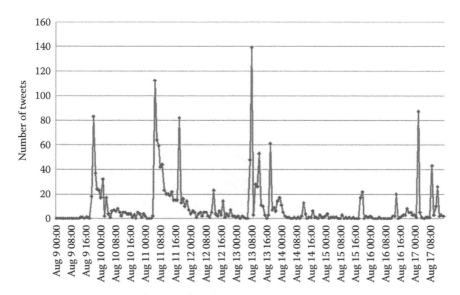

FIGURE 15.9 Number of tweets related to earthquakes in Japan. (From Sakaki, T. et al., Earthquake shakes twitter users: Real-time event detection by social sensors, in *Proceedings of International World Wide Web (WWW) Conference*, April 26–30, Raleigh, NC, 2010, p. 4. With permission.)

extensively, especially when warnings of an imminent tsunami were predicted. Already a wide range of studies suggest that information sharing networks, such as Twitter, can be very useful in times of crisis by quickly and effectively disseminating relevant news. Figure 15.9 shows the number of tweets in relation to the earthquake in Japan. However, we need to find potential solution for a couple of issues before we can effectively utilize social networking tools during disasters (Sakaki et al., 2010).

The biggest problem is the reliability of Twitter updates that were misplaced or lied particularly in calls for help. Another problem is the low signal-to-noise ratio for messages using hashtags. Hashtags are a set of keywords prefixed with the "#" symbol that would normally allow users to filter updates of particular interest. Hashtag misuse (e.g., spammers adopting popular hashtags in their tweets) can lead to difficulties in finding important messages in the areas earthquake hit directly. One more minor problem is that there were too few official updates from the government and the mass media compared to updates from social media.

Related to crisis communication, researchers have paid attention in utilizing social media as a monitoring tool of crisis for an early warning system. Sakaki et al. (2010) conducted research to test the usage of Twitter in monitoring disasters such as earthquake and typhoon. They considered each Twitter user as a sensor and applied Kalman filtering and particle filtering, which are widely used for location estimation in ubiquitous and pervasive computing, for semantic analysis on tweets and spatio-temporal modeling. The particle filter works better than other existing methods in estimating the centers of earthquakes and the trajectories of typhoons.

To obtain tweets on the target event precisely, authors conducted semantic analysis on tweets. For example, users might post tweets such as "Earthquake!" or "Now it

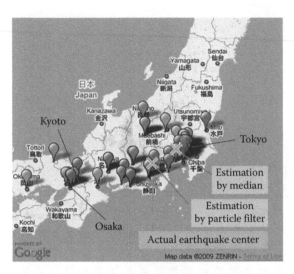

FIGURE 15.10 Earthquake location estimation based on tweets. (From Sakaki, T. et al., *Earthquake shakes twitter users: Real-time event detection by social sensors*, in *Proceedings of International World Wide Web (WWW) Conference*, April 26–30, Raleigh, NC, 2010, p. 7. With permission.)

is shaking," thereby making earthquake and shaking a keyword. However, users might also post tweets such as "I am attending an Earthquake Conference" or "Someone is shaking hands with my boss," in which case the two keywords are no longer relevant to the event. In order to distinguish these cases, the authors prepared training data and devised a classifier using a support vector machine based on features such as keywords in a tweet, the number of words, and the context of words about target events.

The outcome of location estimation from Twitter related to the earthquake in Japan is shown in Figure 15.10. Balloons show the tweets on the earthquake. Lighter colored balloons represents earlier tweets and darker colored balloons represents later tweets. The cross shows the earthquake center. Authors approximately found the actual earthquake center utilizing the particle filter. If the center of the earthquake is in an oceanic area, the estimation is more difficult to locate the center of the earthquake precisely from tweets. And it becomes more difficult to make good estimations in less-populated areas. The great number of Twitter users on the earthquake makes the estimation more precisely. OSNs, gaining a huge amount of active users, give us possibilities to track particular pieces of information (e.g., personal status updates in urgent situation) simultaneously.

In this section, we discussed how social media can be used to design valuable real-time services such as earthquake report and early warning systems. Early warning systems can help us reduce the enormous losses. For example, upon a warning, we can turn off a stove or heater in our house and hide ourselves under a desk or table if we have several seconds before an earthquake actually hits. Current social media will become more useful. Users generate texts that can be analyzed to infer the urgency and tension of the information producers by capturing their tones and emotions. Next section, we will introduce sentiment analysis to understand the possibility for analyzing moods of users that can be captured from user-generated texts in online social media.

15.4 ONLINE SOCIAL NETWORKS AS A SENTIMENT CAPTURING SYSTEM

Web is places where people consume and create content, and interact with other people. So far, we covered a number of studies about common behaviors of millions of users to gain insights into various phenomena of information propagation. The behavior that any user can share and contribute content, express opinions, and link to others reveals a fundamental characteristic in that human behavior. People are curious about others' sentiments and emotions. For instance, many companies struggle to know consumers' opinions about their products, and politicians wonder how people think about their new policy. These are snippet examples. However, studies on those of examples may open up a possibility of developing prediction systems through utilizing social media.

In recent couple of years, a number of researchers and practitioners from social network analysis area show interest on the sentiment detection. Sentiment analysis is generally defined as an application of natural language processing, computational linguistics, and text analytics to identify and extract subjective information in source materials. It aims to determine the attitude of a speaker or a writer with respect to some topic or the overall contextual polarity (e.g., positive or negative) of a document (Pang and Lee, 2008). So, it is called *opinion mining* in the other term. Recently, sentiment analysis has been used not only in discerning users' opinions but also in discovering broad implications. For example, Bollen et al. (2011a,b) measured public moods in OSN, Twitter, and showed that the measured moods from OSN can be utilized to predict the stock price fluctuations by using sentiment analysis. In this section, we introduce the basic concepts of sentiment analysis techniques and discuss some of the interesting studies.

15.4.1 How Does Sentiment Analysis Work?

There are two major techniques used to automate the sentiment analysis process. The most frequently used one in commercial applications is based on linguistic resources and the other is based on machine learning. The first technique is based on predetermined lists of different categories of words such as positive and negative words. The text in question is checked for how many times any of these words appear in it. Here is a simple example: "iPhone is good and useful but expensive." This sentence would be regarded as positive, as it contains two words from the positive list ("good" and "useful") and only one word from the negative list ("expensive"). A slightly more sophisticated approach may use different scores and weights for different words and account for negation (e.g., "not good").

A further step is to take care of larger linguistic units such as phrases and sentences. For example, a pattern "noun phrase + positive verb + brand or product name" may capture phrases like "I love the Samsung Galaxy Tab" and "My friend prefers Apple." But this approach can involve a number of linguistic techniques that are not always robust. One of the well-known problems of sentiment analysis based on linguistic resources is that we cannot always predict the ways sentiments are expressed. Existing sentiment analysis approaches that operate at phrase or sentence level assume that people use normal or standard predictable language. However, this is rarely the case and particularly not in social media, because people use all kinds of new word and slangs to express their feelings. Therefore, extending word dictionaries based on released software programs or developing new dictionaries becomes important.

The second popular technique is based on machine learning. This technique relies on a computer's ability to automatically learn the language that is used for expressing sentiment regardless of the meaning of language is. To do this approach, a machine needs some information to learn from (called a training corpus), which is often composed of a set of examples annotated by humans. Once the machine has learned the examples, it can apply the acquired knowledge to new, unseen documents and classify them into sentiment categories. This approach is not perfect either. It is weak in treating domain dependency. For example, if a machine was trained on a corpus of movie reviews, it will be very inaccurate if applied to reviews of automobiles. It means that one needs to train a machine in all domains it is to be used.

15.4.2 HAPPINESS ASSORTATIVITY

The behaviors and interaction patterns between nodes in a network tend to be correlated with the characteristics of the nodes. The old adage "birds of a feather, flock together" might explain this phenomenon in a clear sense. There are several studies conducted over time indicating that beyond demographic features such as age, sex, and race, even psychological states such as "loneliness," can be assortative in a social network (Bollen et al., 2011a,b). In addition, according to a study of Bollen et al., happier people tend to be linked by their tweets. The study, entitled "Happiness is assortative in online social networks," found that people who are similar, especially those who are generally happy, tend to group themselves together, even on Twitter.

The analysis used standard algorithms borrowed from psychological research to assess the *subjective well-being* of users from their tweets by looking for trends in positive or negative words. By investigating aggregation trends, they found that happier people are more likely to retweet and mention people who are also marked as happy. They also observed the same pattern for people marked as unhappy. To understand this phenomenon better, authors analyzed 129 million tweets of 102,009 users using a sentiment analysis tool, OpinionFinder, to determine sentence-level subjectivity over a 6 month period. They defined the happiness, the subjective well-being (SWB) score $S(u)$ of user u as the fractional difference between the number of tweets that contain positive OpinionFinder terms and those that contain negative terms as follows:

$$S(u) = \frac{N_p(u) - N_n(u)}{N_p(u) + N_n(u)} \tag{15.1}$$

where $N_p(u)$ and $N_n(u)$ represent, respectively, the number of positive and negative tweets posted by user u.

Figure 15.11 is an example of a tweet classified by SWB score. Tweets that have a larger SWB score than the score of 0.5 saliently use the words that indicate positive emotions, which cause description on the objects or context to seem positive as well.

Tweets submitted by high SWB users (>0.5).

So ... nothing quite feels like a good shower, shave and haircut ... love it
My beautiful friend. i love you sweet smile and your amazing soul
i am very happy. People in Chicago loved my conference. Love you, my sweet friends
@ anonymous thanks for your follow I am following you back, great group amazing people

Tweets submitted by low SWB users (<0.0).

She doesn't deserve the tears but i cry them anyway
I' m sick and my body decides to attack my face and make me break out!! WTF :(
I think my headphones are electrocuting me.
My mom almost killed me this morning. I don't know how much longer i can be here.

FIGURE 15.11 Examples of tweets posted by users with very high and very low subjective well-being (SWB) values. (From Bollen, J. et al., *Artif. Life*, 17(3), 237, 2011a. With permission.)

Tweets that have a smaller SWB score than the score of 0.0 relatively show more frequent use of negative words, and the content itself also expresses such negative emotions. Tweet data are evaluated by the SWB score, and authors confirmed assortativity in a certain tweet network through the method.

Authors formally defined pairwise SWB assortativity as follows: For each edge (V_i, V_j) in a particular social network, they extracted the corresponding two SWB values, one for the source node and another for the target node. These values are then aggregated into two vectors, $S(S)$ and $S(T)$, for sources and targets, respectively. The value of the pairwise assortativity, denoted as $Ap(Gcc)$, is then given by the Pearson correlation coefficient of these two vectors.

$$A_P \equiv \rho(S(S), S(T)) = \frac{1}{n-1} \sum_i \left[\left(\frac{S(S_i) - \langle S(S) \rangle}{\sigma(S(S))} \right) \left(\frac{S(T_i) - \langle S(T) \rangle}{\sigma(S(T))} \right) \right] \quad (15.2)$$

The pairwise assortativity is defined in the $[-1, +1]$ interval, with -1 indicating perfect disassortativity, 0 indicating a lack of any assortativity, and $+1$ meaning perfect assortativity. The authors also calculated the neighborhood assortativity of Gcc with regards to SWB by defining two vectors, one for the SWB values of every unique user and another for the average SWB values of their neighborhoods. Figure 15.12 shows the distribution of SWB values of all edges and nodes in Gcc and confirms the observed correlation between the SWB values of connected users in Gcc.

The pairwise assortativity scatterplot in Figure 15.12 represents that there are a significant bimodal distribution of SWB values, clustered within ranges of $[-0.05, 0.05]$ and $[0.1, 0.3]$. Horizontal axis indicates a user's SWB score, and vertical axis indicates another user's SWB score who corresponds to the previous user. The bar located at the right side of Scatterplot showing the scale of the density indicates that the more reddish that the plot becomes, the more frequent that the user appears. So to speak, the users included in the clustered range of $[-0.05, 0.05]$ and $[0.1, 0.3]$ are likely to be connected with the users with similar SWB value. The assortativity values were found to be 0.433^{***} ($N = 2,062,714$ edges) for the pairwise SBW assortativity, statistically significant (p values < 0.001) for the sample sizes.

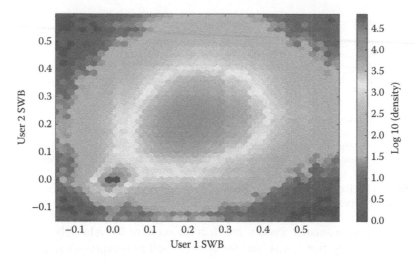

FIGURE 15.12 Scatterplot of the SWB values for users connected in Twitter friends network. (From Bollen, J. et al., *Artif. Life*, 17(3), 237, 2011a. With permission.)

Although the observed correlation is not obviously linear, the clustering pattern indicates that users with SWB values in a particular range are preferentially connected to users within that same range. This result confirms the observation in relation to SWB assortativity. Interestingly, the authors obtained the probability distribution of emotionality values of data cells, and confirmed two peak bimodal distribution where one was in the range [−0.1, 0.1] and another in the range [0.2, 0.4]. Furthermore, the tweeter users in their sample data were included in the positive SWB values in a narrow range [0.1, 0.4]. This result indicates the sociocultural differences in how emotions and mood are expressed on Twitter.

As shown in the result, massive networked datasets from OSNs can reveal much more than just the structural properties. In this study, the authors have uncovered hidden sentimental information from short postings in Twitter. This work demonstrated that even short messages can reveal plenty of sentimental information of their writers but are by no means conclusive and causal reaction. The clustering shown in Figure 15.12 might be explained by the notion of "homophily." In other words, the demographic, behavioral, and structural similarities between nodes might lead to such phenomenon. Another possible explanation is "mood contagion" where there are actually some causal reactions taking place among connected nodes so that they can affect each other over time. The last possibility is the case where people control their expression level of their emotions after watching how their friends react. This might be a particularly the case when the common perception of open expression varies by culture.

This research shows that social networks can convey subtle information not only about the social relationships but also about individuals' emotions. We expect new research using sentiment analysis methods will be able to capture mood contagions on social networks.

15.4.3 Tracking Political Sentiments on Twitter

The social media outlets have been successfully used in the recent presidential campaigns. While television still plays a vital role in political campaigns, social media has rapidly become an equally crucial part of the political campaign. The 2008 U.S. presidential election marked great changes. Nearly 3.1 million people have contributed to democratic nominee Barack Obama's campaign on Facebook, and Obama's supporter network on Facebook was considered one of the most crucial parts of his victory. In the 2008 election, Obama had a unique edge in embracing the power of OSN, but the trend is no longer special. Now many people across the globe who are interested in politics use social media.

Politically engaged online users actively encourage others to vote by posting politics-related status updates, or spread the word about events in a user-friendly format. OSNs can spark specific subgroups of political supporters. Even the traditional news media also pay attention to grasp trends in social media. Users' messages on OSNs can be useful indicators of offline trends. The following study examined the general idea that there might be a connection between political Twitter messages and the offline political processes.

Tumasjan et al. (2010) examined how accurately Twitter informed us about the constituencies' political sentiment. They downloaded >100,000 tweets, containing the name of either at least one of the six major parties or selected prominent politicians in German election period. All collected tweets published between August 13 and September 19, 2009, 1 week prior to the election. Then, they simply extracted tweet sentiment by using Linguistic Inquiry and Word Count (LIWC), which is a text analysis software program that calculates the degree of sentiments in 70 and other dimensions including positive and negative emotions.

In order to analyze the political sentiment of the tweets, authors generated multidimensional profiles of the politicians in their sample using the relative frequencies of LIWC category word counts. Figure 15.13a indicates all leading candidates have a very similar sentiment profile. Only polarizing political characters, such as liberal leader *Westerwelle* and socialist *Lafontaine*, deviate in line with their roles as opposition leaders. Messages mentioning *Steinmeier*, who was sending mixed signals regarding potential coalition partners, are the most tentative. Figure 15.13b shows that positive outweigh negative emotions, except in the case of *Seehofer* who is associated the most with anger. Authors explained that Seehofer irritated many voters with his attacks on desired coalition partner at that time. For *Steinbrück* and *zu Guttenberg*, the money and work issues reflect their roles as finance and economics ministers.

Authors concluded sentiment profiles plausibly reflect many nuances of the election campaign and politicians evoke a more diverse set of profiles than parties. They predicted the result of election based on the political proximity between parties. The prediction was conducted by similarity calculation of sentiment profiles and calculation of relative tweet volume. They found that the mere number of tweets mentioning a political party can be considered a plausible reflection of the vote share and its predictive power even comes close to traditional election polls. This study

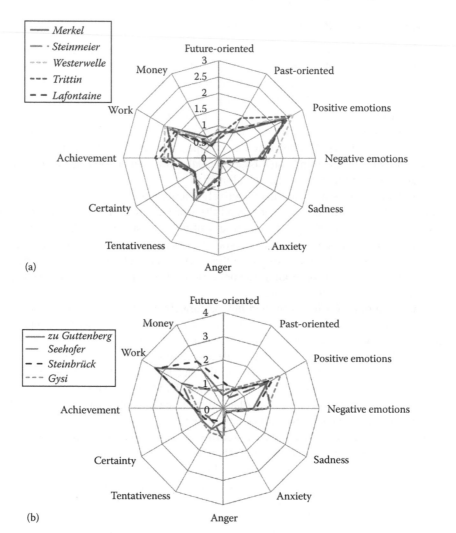

FIGURE 15.13 (a) Profiles of leading candidates and (b) profiles of other candidates. (From Tumasjan, A. et al., Predicting elections with twitter: What 140 characters reveal about political sentiment, in *Proceedings of International AAAI Conference on Weblogs and Social Media (ICWSM) Data Challenge*, May 23–26, Washington, DC, 2010, pp. 4–5. With permission.)

shows that Twitter is used as a forum for political deliberation and online messages on Twitter validly reflect offline political sentiment.

A lot research was being conducted and many approaches were attempted in this field. Nevertheless most of the approaches are still very naive and it seems that the contemporary results can be improved in many ways. An upcoming trend that seems to be followed by solving uncovered issues in introduced studies in this section. Customizing and extending the breadth of the corpus will allow us to build up a more complete resource for social network analysis. Integrating temporal information will allow us to better examine the changing characteristics of sentiments dynamics on the network and to analyze reactions to occurred events. Although sentiment analysis method is still unstable, we can consider the prediction system through the emergence of social network analysis techniques using big data.

15.5 CONCLUSION

In social networks, user interaction is presented and driven by structural rules of the surrounding environment. Individual impacts on their neighbors and the opposite situation might be occurred in a same manner. Each of individuals or group of those of individuals may adjust their behavior to be stable in a society. Through experience, individuals learn to organize particular networks with their identity including specific roles, relations, and activities in the surrounding environment. Studies of social networks try to understand the exact mechanisms of such individual and collective user behaviors.

So far, we had difficulties in analyzing social networks because of lack of massive data. Surveys were often used for data collection, which inherently limits the scale of the study. Hence, it was difficult to examine the interaction patterns and information flows that occur in social networks in great detail. Now, with the advent of the Internet, we are given with a great opportunity to analyze large social networks, because hundreds of millions of people use OSNs in posting updates about situational, personal, and interpersonal cues.

Profile updates in OSNs present unique research opportunities. OSN services such as blogs and microblogs contain individuals who voluntarily post information about themselves—personal thoughts, feelings, beliefs, activities—in a public arena with unlimited access for anyone with an Internet connection.

Previous studies have identified the potential of social media as a communication medium and monitoring sensor system of the offline social networks. Those studies explained how information propagates on particular networks and presented how to predict social phenomena based on using quantitative analysis methods. Throughout this chapter, we have covered some of the basic concepts of social network analysis and introduced key findings in recent research.

While there are many directions social network analysis could expand in the near future, one of the exciting future directions is on adopting qualitative analysis methods. Qualitative analysis methods such as sentiment analysis could be more broadly used to analyze subjects which have not been studied widely in the existing social networks. Health care–related issues are one such topic.

Many people use Internet-based health care communities. About one-third of Americans who go online to research their health currently are known to use OSNs to find fellow patients and discuss their conditions (Elkin, 2008), and 36% of OSN users are known to leverage other consumers' knowledge before making health care–related decisions (Jupiter Research, 2007). OSNs hence hold considerable value to health care organizations. As more people use OSNs for seeking and sharing health-related information, health care organizations could potentially interact with various types of people to inform users of new treatments and care pathways.

While this sounds promising, researchers will need to develop sophisticated methods to treat data before OSN data can be used directly for health care purposes.

First, there is an urgent need to gather richer data from OSNs. In a recent work, Christakis et al. found that an individual is 57% more likely to be obese if he or she has a friend who is obese. In this study, the type of friendship between users was an important variable: persons in closer, mutual friendships had higher effect on each other than those in other types of friendships (Christakis and Fowler, 2007). In order to repeat a similar study using OSNs, one will need detailed demographic and geolocation information about users.

Second, researchers should take care in generalizing an observation that is made from one OSN. This is because environmental factors might affect the data. To overcome such bias, one will need to conduct sensitivity analysis, which addresses the impact of latent factors. Furthermore, findings from qualitative analysis (e.g., sentiment analysis) may not hold over time, because user behaviors and social structures within OSNs change over time.

Despite various challenges, social network analysis is a promising research area and it can be applied to many disciplines. Social network analysis makes massive and fast growing networks data affordable for analysis. We expect to see researchers in various disciplines adopting tools and methods used in social network analysis to make new findings and key observations in various fields like crisis communication, health care systems, businesses, and marketing.

REFERENCES

Acar, A. and Muraki, Y. Twitter for crisis communication: Lessons learnt from Japan's tsunami disaster. *International Journal of Web Based Communities*, 7(3), 392–402, 2011.

Bollen, J., Gonçalves, B., Ruan, G., and Mao, H. Happiness is assortative in online social networks. *Artificial Life*, 17(3), 237–251, 2011a.

Bollen, J., Mao, H., and Zeng, X.J. Twitter mood predicts the stock market. *Journal of Computational Science*, 2(1), 1–8, 2011b.

Cha, M., Antonio, J., Pérez, N., and Haddadi, H. Flash floods and ripples: The spread of media content through the blogosphere. In *Proceedings of the Third International AAAI Conference on Weblogs and Social Media (ICWSM) Data Challenge Workshop*, May 17–20, 2009, San Jose, CA, 2009a.

Cha, M., Mislove, A., and Gummad, K.P. A measurement-driven analysis of information propagation in the flickr social network. In *Proceedings of the Eighteenth International Conference on World Wide Web*, April 20–24, 2009, Madrid, Spain, pp. 721–730, 2009b.

Christakis, N.A. and Fowler, J.H. The spread of obesity in a large social network over 32 years. *New England Journal of Medicine*, 357(4), 370–379, 2007.

Easley, D. and Kleinberg, J. *Networks, Crowds, and Markets: Reasoning about a Highly Connected World*, Cambridge University Press, Cambridge, MA, 2010.

Elkin, N. *How America Searches: Health and Wellness*. Scottsdale, AZ: iCrossing, 2008. http://www.icrossing.com/research/how-america-searches-health-and-wellness.php (accessed on August 1, 2008).

Gouldner, A.W. The norm of reciprocity: A preliminary statement. *American Sociological Review*, 25(2), 161–178, 1960.

Kempe, D., Kleinberg, J., and Tardos, É. Maximizing the spread of influence through a social network. In *Proceedings of the Nineth ACM SIGKDD International Conference on Knowledge Discovery and Data Mining*, August 24–27, ACM Press, Washington, DC, pp. 137–146, 2003.

Kwak, H., Lee, C., Park, H., and Moon, S. What is twitter, a social network or a news media? In *Proceedings of the Nineteenth International Conference on World Wide Web*, April 26–30, Raleigh, NC, pp. 591–600, 2010.

Levy, M. *Online Health: Assessing the Risk and Opportunity of Social and One-to-One Media*, JupiterResearch, 2007.

Pang, B. and Lee, L. Opinion mining and sentiment analysis. *Foundations and Trends in Information Retrieval*, 2(1–2), 1–135, 2008.

René, L. African peasantries, reciprocity and the market: The economy of affection reconsidered. *Cahiers d'études africaines*, 29, 33–67, 1989.

Sakaki, T., Okazaki, M., and Matsuo, Y. Earthquake shakes twitter users: Real-time event detection by social sensors. In *Proceedings of the Nineteenth International Conference on World Wide Web*, April 26–30, Raleigh, NC, pp. 851–860, 2010.

Tumasjan, A., Sprenger, T.O., Sandner, P.G., and Welpe, I.M. Predicting elections with twitter: What 140 characters reveal about political sentiment. In *Proceedings of the Fourth International AAAI Conference on Weblogs and Social Media (ICWSM) Data Challenge*, May 23–26, Washington, DC, pp. 178–185, 2010.

16 Radio Frequency Identification–Enabled Social Networks

Ashraf Darwish, Sapna Tyagi,
Abdul Quaiyum Ansari, and M. Ayoub Khan

CONTENTS

16.1 INTRODUCTION

Online social networks are establishing novel forms of interaction among users; millions of people around the world, young and old, knowingly and willingly use Facebook, Friendster, MySpace, Match.com, LinkedIn, and hundreds of other sites to communicate with friends, dates, and jobs. In doing so, they wittingly reveal personal information to strangers as well as friends. Before moving down, let us study the real statistics of Facebook that will show the impact of social networking on our daily lives as shown in Table 16.1.

TABLE 16.1
Facebook Statistics

Statistics	Study
People on facebook	• More than 500 million active users • 50% of our active users log on to Facebook in any given day • An average user has 130 friends • People spend over 700 billion minutes per month on Facebook
Activity on facebook	• There are over 900 million objects that people interact with (pages, groups, events, and community pages) • An average user is connected to 80 community pages, groups, and events • An average user creates 90 pieces of content each month • More than 30 billion pieces of content (web links, news stories, blog posts, notes, photo albums, etc.) are shared each month
Global reach	• More than 70 translations are available on the site • About 70% of Facebook users are outside the United States • Over 300,000 users helped translate the site through the translation application
Platform	• Entrepreneurs and developers from more than 190 countries build with Facebook platform • People on Facebook install 20 million applications every day • Every month, more than 250 million people engage with Facebook on external websites • Since social plug-ins launched in April 2010, an average of 10,000 new websites integrate with Facebook every day • More than 2.5 million websites have integrated with Facebook, including over 80 of comScore's U.S. Top 100 websites and over half of comScore's Global Top 100 websites
Mobile	• There are more than 250 million active users currently accessing Facebook through their mobile devices • A person that uses Facebook on their mobile devices is twice as active on Facebook as nonmobile users • There are more than 200 mobile operators in 60 countries working to deploy and promote Facebook mobile products

Source: Facebook, 2011, http://www.facebook.com/press/info.php?statistics

If we need information, the Internet offers a wealth of resources. But if you are hunting down a person or a thing, a computer is not much help. After the World Wide Web (the 1990s) and the mobile Internet (the 2000s), we are now heading to the third and potentially most revolutionary phase of the Internet—the "Internet of Things" (IoT), which is a new dimension added to the world of information and communication technologies (ICT), which means a communication from anytime (day/night), anywhere (nearby PC, somewhere else indoor, outdoor, while moving) connectivity for anyone, to anything, for example, between PCs, human to human (H2H), human to thing (H2T), and thing to thing (T2T). Multiple connections create an entirely new dynamic network of networks—an IoT (Simonov et al. 2008). The IoT will create a dynamic network of billions or trillions of wireless identifiable "things" communicating with one another and integrating the developments of pervasive computing, ubiquitous computing, and ambient intelligence (Vermesan et al. 2009). Now we have reached the era where electronic tags promise to link the objects of the real world with the virtual world, thus enabling anytime, anyplace connectivity for anything and for anyone. It refers to a world where physical objects and human beings, as well as virtual data and environments, all interact with each other in the same space and time, which will give new paradigm shift for social networking.

16.1.1 Background of Internet of Things

The IoT is an integrated part of Future Internet and could be defined as a dynamic global network infrastructure with self-configuring capabilities based on standard and interoperable communication protocols where physical and virtual "things" have identities, physical attributes, and virtual personalities; use intelligent interfaces; and are seamlessly integrated into the information network as shown in Figure 16.1 (Vermesan et al. 2009).

According to the predictions

*   We will have hundreds of billions of RFID-tagged objects at approximately five cents per tag by 2015.
*   Today, there are roughly 1.5 billion Internet-enabled PCs and over 1 billion Internet-enabled cell phones. The present "Internet of PCs" will move toward an "IoT" in which 50–100 billion devices will be connected to the Internet by 2020 (Sundmaeker et al. 2010).

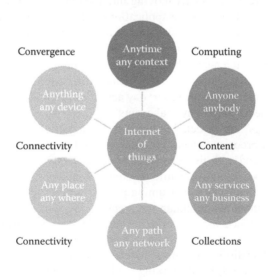

FIGURE 16.1 Internet of things. (From Vermesan et al., Internet of things—Strategic research roadmap, http://ec.europa.eu/information_society/policy/rfid/documents/in_cerp.pdf, 2009.)

According to the IMS research

- There will be over 6 billion cell phones in use around the world, the great majority of which will be Internet connected.
- There are around 2.5 billion TVs in use today; many of these will be replaced with Internet-connected sets, either directly or via a set-top box.
- An increasing proportion of the world's 1.1 billion cars will be replaced over time by models that have Internet connectivity. Chief Futurist for Cisco Systems, Dave Evans, predicts that by 2014, 90% of data will be video.
- By 2050, a computer with the computing power of nine billion brains will be available for $1000. We currently only know 5% of what we will know in 50 years.

IoT, also called sensor network, uses information sensing equipments such as radio frequency identification (RFID), sensor, global positioning system (GPS), and laser scanner to connect things with the Internet based on the given agreement to carry out information interchange and communications, realizing intelligent identification, positioning, tracking, monitoring, and management. IoT involves numerous technologies such as radio frequency (RF), integrated circuit, communications, computer, software, system integration, and Internet (IoT China 2011).

We will be wearing smart clothes, made of smart fabrics, which will interact with the climate control of our cars and homes, selecting the most suitable temperature and humidity levels for the person concerned; smart books of the future will interact with the entertainment system, such as a multidimensional, multimedia hypertext, bringing up on the TV screen additional information on the topic we are reading in real time; and so on.

The technology RFID is rapidly moving into the real world through a wide variety of applications. A glimpse of the applications is discussed in Section 16.2. In this type of RFID system, each physical object is accompanied by a rich tiny tag with a unique number that can be globally accessible and that can contain both current and historical information about object's physical properties, origin, ownership, and sensory context (e.g., the temperature at which a milk carton is being stored). It can also transform the way we perform everyday activities by giving applications current and detailed knowledge about physical events.

In the IoT, "things" are expected to become active participants in business, information, and social processes where they are enabled to interact and communicate among themselves and with the environment by exchanging data and information "sensed" about the environment, while reacting autonomously to the "real/physical world" events and influencing it by running processes that trigger actions and create services with or without direct human intervention. We stand on the verge of an era where devices and objects are networked. There are already some exciting applications that span the technology spectrum. NASCAR.com offers a Java applet-driven virtual dashboard that displays real-time wireless telemetry from cars during a race—registering location, speed, RPM, braking, and more. The race cars have effectively become networked devices. Meanwhile, NASA, with the assistance of GE Medical, is now able to monitor real-time medical data (such as blood pressure, respiration, and heart rate) from its space shuttle astronauts (Meloan 2003) (Table 16.2).

TABLE 16.2
Applications of IoT

Segments	Description
Object to Internet to human	Object-initiated service that results in an e-mail to a human respondent
Human to Internet to object	Human communicates via Internet to activate a control device in the home
Object to Internet to object	Object-activated control service via the Internet that results in an object or system activation, control event, or information update, possibly with a human interface to allow monitoring of events
Objects to dedicated IoT infrastructure to object	Exploiting a dedicated infrastructure and domain features to support a new range of object-oriented applications and services, possibly with human interfaces as appropriate for interactive function

16.1.2 SEGMENTS OF APPLICATIONS FOR THE INTERNET OF THINGS

Following are the various segments where RFID-enabled services are added to make the application alive.

16.2 RFID-ENABLED "IoT": RELATED APPLICATIONS

RFID technology is being used for identification of a wide range of objects such as goods in the stores, books in the libraries, frozen biological samples in research and pharmaceutical laboratories, criminology, museums, as well as for other forms of identification (bracelets in clinics, night clubs, theme parks, ski resorts, biometric passports, train tickets, tracking friends, tags on the cars, etc.). Another application includes RFID tag added to the sole of a shoe interacting with RFID scanner hidden in the floor mat or at the side of a doorway. The RFID reader sends the data obtained from the shoe tag to the local web server that transmits it to a social network in a form of a message about user location. In addition to that, it is being proposed to equip places such as cafes and similar places with buttons similar to Facebook button "Like" that would allow users standing on the same floor mat to press the button and instantly add each other to the lists of friends. The main problem with this idea is that RFID reading devices are not installed in most of the public places.

This "real-life" context can unlock the door to various business, environmental, personal, and social contexts applications. We present some of the RFID applications being used for social networking.

1. RFIDDER was developed at the University of Washington, which is also known as "friend finder" that lets people use data from radio tags to inform their social network where they are and what they are doing. The feature can be used on the web and on a mobile phone, with a connection to the social networking service Twitter. RFIDDER is capable of adding or removing friends in real time, such as a person is hanging in a garden or shopping mall and there he or she finds someone and likes to add to his or her own network.

2. NASDAQ-traded company called VeriChip manufactures RFID chips specifically for use in human beings, the idea being that the chips would provide a quick and reliable way to store and retrieve emergency medical information; VeriChip is also marketed in South America as a way to track kidnapped victims.

3. Another research was done where human-implantable RFID tags merge with the online social networking craze. In this, all the information in your Facebook profile were tucked snugly into a tiny RFID-like chip embedded in the ball of your thumb or tucked in your wrist; RFID-enabled cell phone could beep every time you walked past somebody 2° of separation or less from you or who had the same favorite novel you do or who liked to play scrabble (Grossman 2007).

4. Another experiment was done at the University of Washington with RFID tags equipped with tiny computer chips that store an identification number unique to each tag. Researchers installed about 200 antennas throughout the computer science building that pick up any tag near them every second. The researchers hope to expand the project, funded by the National Science Foundation, to include participation by about 50 volunteers—people who regularly use the building. Volunteers will have the option of removing their data at any point. The system can show when people leave the office, when they return, how often they take breaks, where they go, and who is meeting with whom.

5. The LAB's Personal Digital Diary application detects and logs a person's activities each day and uploads them to a Google calendar. Users can search the calendar to jog their memories about when they last saw someone or how, where, and with whom they spent their time.

6. RFID readers placed around shopping malls and airports could help government agencies collect information about visitors' travel patterns, shopping habits, and relationships.

7. Asics has done campaign in a very innovative way in a New York Marathon, which blended social media with the real world. In this marathon, runners attached RFID tags to their shoes and when they passed readers at specific points of the New York Marathon track, they were played messages from loved ones on giant screens to encourage them along the 26 mile course. This has bought real-world experiences to the online social world.

8. Another application is "poken." The poken makes exchanging social network information in the real world simple and easy. Using RFID and supercool plastic creature tags, the poken allows you to swap ID information with other poken users when you meet them. Poken tags are easy to spot in the open, and that helps encourage the sharing aspect. Poken is an access point for all of your social networks, so when you meet people in the real world, you can use the poken tool to share information and easily make connections on popular social networks like Facebook and MySpace and others.

9. FaceChipz is a new social networking site designed just for kids. Intended primarily for the "tween" set who have outgrown children's websites but have not quite aged into Facebook yet, FaceChipz merges real-world networking with an online component. After purchasing a starter set of five chips, the child has their parent register an account for them on the FaceChipz website. Then the game begins.

The child registers all their chips online using the unique identification code found on the back of each token. When all the chips have been registered, they can be distributed to friends. In return, the child's friends will hand them their FaceChipz. When the exchange is complete, the child returns to the computer to register the new codes from the chips they have collected. The end result is a social network of friends with a physical counterpart in the real world. Parents will appreciate the fact that the FaceChipz network offers a more secure and private environment for their kids than traditional social networking sites like MySpace and Facebook. No strangers can solicit friendship requests here—the child's only online friends are those they have connected with in real life. There is not even a search mechanism for friends to find each other without first trading chips. FaceChipz prepare children for the online world of social networking. All the elements are there that could make FaceChipz a success: collectable tokens, an online element, and parent-friendly company ethics. There is another bonus, too: the chips are cheap. A five pack is just $4.99 at ToysRUs and the one-time site registration fee is only $1.00. By this technology, mom and dad get involved and are made aware of the child's online activities.

16.3 ENABLING TECHNOLOGIES FOR SOCIAL NETWORKS

"Anytime, anywhere, any media" has been for a long time the vision pushing forward the advances in communication technologies (Atzori et al. 2010). Actualization of the IoT concept into the real world is possible through the integration of several enabling technologies. In this section, we discuss the most relevant one, which is RFID system.

16.3.1 RFID TAG TECHNOLOGY

An RFID system consists of RFID reader with antennas, host computer, and transponders or RF tags that the reader recognizes. An RFID tag is uniquely identified by tag ID stored in memory and can be attached to almost anything like watch, thumb, wrist, etc., which is usually of very low dimensions. Hitachi has developed a tag with dimensions 0.4 mm × 0.4 mm × 0.15 mm. RFID tags are categorized into three groups: basic classifications, classification according to the memory, and classification according to the method of communication (Table 16.3).

Such tag IDs are specified through EPC (electronic product code) standard, and as the tagged objects pass within the vicinity of the reader, an antenna broadcasting radio waves on specific frequency, the transponder "wakes up" and sends the chip data to a transceiver (Khoussainova et al. 2008).

16.3.2 ROLE OF EPC IN SOCIAL NETWORKS

EPC is a unique number that identifies a specific item that will be stored on RFID tag's memory. These codes are generic and follow a universal numbering scheme for physical objects. The EPC is capable to identify every single, individual product item, whereas the barcode only identifies the product. The structure of a 96 bit EPC is as follows.

TABLE 16.3
Classification of Tags

Basic Classification	Description
Passive tags	They do not have onboard power supplies and harvest the energy required for transmitting their ID from the query signal transmitted by an RFID reader in the proximity
Semipassive tags	Semipassive RFID uses an internal power source to monitor conditions but requires RF energy transferred from the reader/interrogator similar to passive tags to power a tag response
Active tags	Active RFID tags use an internal power source, such as a battery, within the tag to continuously power the tag and its RF communication circuitry. Active RFID tags are continuously powered, whether in the reader/interrogator field or not, and are normally used when a longer tag read distance is desired
Classification according to the Tag Memory	
Read only	In read only tag, the memory is factory programmed and cannot be modified after its manufacture. Its data remain static, and very small quantity of data, usually 96 bit of information, can be stored
Read–Write	Read–write tags can be read as well as can be written into. Its data can be dynamically altered and it is capable of storing large amount of data ranging from 32 to 128 kB
Classification according to the Communication Technologies	
Induction	This tag uses close proximity electromagnetic field or inductive coupling generally using LF and HF bands, and near field communication
Propagation	This tag uses electromagnetic waves, operates in UHF, and is suitable for far field communication

The first field in the header defines the coding schemes in operation with the remaining bits providing the actual product code. The manager field is responsible for identifying the product manufacturer. The object class defines the product class itself. The serial number is unique for an individual product class. The length of EPC may be of 64, 96, 128, 256, 1K, 4K bits (Khan and Sanjay 2009). The 96 bit EPC belongs to class I generation that identifies 268 million manufactures (228) uniquely. Each manufacturer can have 16 million (224) unique object classes and 68 billion unique serial numbers (236) in each class (Table 16.4).

16.3.3 RFID CONCEPTUAL ARCHITECTURE

We have presented a common architecture illustrating how an RFID application works. Each object is tagged with an RFID chip that contains a unique EPC. In layer 1, data from these tags are collected periodically by RFID readers and sent to RFID middleware in form of tuple <epc, Reader_id, time>.

Instead of storing these massive, dirty, poor-semantic reads directly in repository, the RFID middleware will filter (e.g., eliminate duplicate reads and missing reads) and correlate them with the business context to generate all clean and meaningful events.

TABLE 16.4
96 Bit FPC Format

Header	EPC Manager	Object Class	Serial Number
8 bit	28 bit	24 bit	36 bit

Source: Khan, M.A. and Sanjay, O., SHA-256 based n-bit EPC generator for RFID tracking simulator, in *Proceeding of IEEE International Conference on Advance Computing (IACC)*, March 6–7, Patiala, India, 2009.

These events are then passed to EPCIS through capture interface and stored permanently in its repository or pushed to some applications interested in real-time information. The data can be queried from partners' accessing applications through query interface. Physical layer consist of reader and RFID middleware.

A brief overview of all the layers of EPC architecture (Figure 16.2) is presented in Table 16.5.

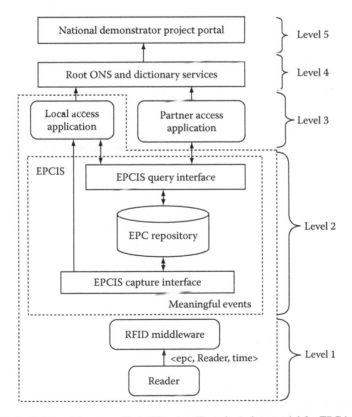

FIGURE 16.2 EPC architecture. (From Nguyen, T. et al., A data model for EPC information services, DEWS, http://www.ieice.org/~de/DEWS/DEWS2007/pdf/m6-3.pdf, 2007.)

TABLE 16.5
EPC Layer Description

Layer	Implementation	Description
1	Tags and readers	EPC data tagged on item are captured by the reader in the form of hexadecimal number
1	RFID middleware	An on-site software component that converts multiple reads to one read adds information such as reader_id, location_id, and timestamp
2	EPCIS	It is responsible for capturing meaningful events; it performs data modeling, maintains database, and provides framework to execute query to obtain the information relevant to business context
3	Local object name services	Each of the industry partners maintains its own repository of product-specific data. The local ONS provides a pointer to the local database
4	Root ONS and dictionary services	The ONS identifies unique numbers for manufacturers and the discovery service points to a particular EPCIS where detailed information can be obtained for a specific item
5	National demonstrator project portal	The portal displays the data collected and stored in the system and allows the queries on the movement of goods through supply chain

16.3.4 RFID-Enabled Architecture for Social Networks

The deployment of a pervasive RFID-based infrastructure in an everyday environment holds the promise of enabling new classes of applications that go beyond tracking and monitoring. As illustrated in Figure 16.3, the system architecture is divided into three layers. The bottom layer that we can call as primitive data processing layer consists of the RFID readers and their associated database RDB (reader database) and the middleware and associated databases (middleware database, MDB). As soon as the tags are detected and captured by the readers, the readings are stored in the RDB from where it is consumed by middleware for further processing like eliminating noise, removing duplicates, filtering, and cleaning. Afterward, the clean data are stored in MDB. Now MDB streams its tag readings, tag reads event (TRE) <Tag_id, Reader_id, Location_id, Timestamp>, to the social networking layer (SNL). The RFID data in the form of <Tag A, Reader X, Location Y, Timestamp T> offer low-level information in terms of tags and antennas. To make the TREs helpful to IoT user, SNL can transform them into more meaningful, high-level information about events that applications and users can directly consume (e.g., Anne is taking coffee in Cafe Coffee Day at New Street road).

This layer can be called as complex events processing layer. In a friend-finder application (Welbourne et al. 2009), the TRE streams can serve to automate sharing of a user's current or historical location, as well as the activities they perform (e.g., having lunch). Finally, in the hospital scenario, TREs can help monitor the location and status of patients, staff members, and equipment (Collins 2004). Furthermore, because such

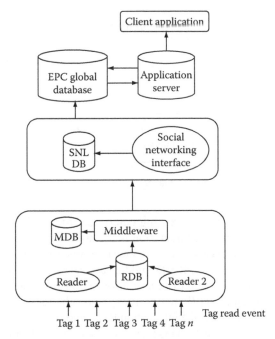

FIGURE 16.3   RFID-enabled architecture for social networks.

high-level events are personal and potentially sensitive, users must be able to precisely control all information disclosure to avoid privacy breaches.

All these streamed, clean, semantically more meaningful events are routed to SNL database. SNL database is further inclined with main RFID-enabled social networking application server, which runs application logic and EPC global database that are responsible in responding to event queries and maintaining all interactions with RFID-enabled client application provisioned by whoever deploys the application.

16.3.5   ROLE OF SOCIAL NETWORKING INTERFACE

SNI (social networking interface) layer should have the ability to analyze information from different RFID readers, collate it, and send it to relevant enterprise information systems or end-user applications. In order to give real-time visibility to tagged objects, RFID data need to be aggregated and transformed into meaningful, actionable information. Current RFID systems, after doing filtering at the middleware based on tag patterns, pass RFID data to existing SNI layer (Figure 16.4), and in turn, SNI layer derives meaningful information that aligns with the business process. This puts too much onus on the enterprise information systems to correlate the event patterns and make sense of the RFID data.

SNI can invoke specific functions like event-pattern matching, event aggregation, and event hierarchy, which are needed when a combination of certain events makes sense to a business process or back-office application. By doing event-pattern matching in the SNI layer, information can be presented in the way an enterprise information system needs it to be, and SNI can invoke specific functions of the system.

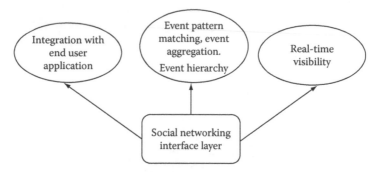

FIGURE 16.4   Social network interface layer.

Integration of RFID data with existing enterprise information systems forms a vital part in RFID deployments. Integration is the ability to communicate the right RFID event or aggregated events or alert to the right process in the right format. It is a complex process today because RFID middleware sends granular RFID events to the existing systems, leaving them to deduce information and route the right event to the right process. When an RFID network expands, changes have to be updated in three places: the middleware, the integration layer, and the processes of concern in the resource management systems. This increases the cost and the time of integration and maintenance.

SNI can simplify the integration process to a great extent by using the combination of RFID data analysis and standard integration adapters (a software component) within AIS to connect to multiple business applications (e.g., ERP, WMS, TMS, and SCM). In other words, integration becomes a simpler process when the logic for extracting meaningful information resides within the AIS layer and the communication to the existing system occurs via standard integration adapters. In this way, expanding the functionality or features for the entire RFID-enabled system will become easier and cost-effective.

The purpose of using RFID technology extends real-time visibility to social networking by giving them intelligent RFID feeds (sending the real-time information to the real-time objects or systems at the right time and in the right format) so they can respond appropriately in real time. SNI should deploy measures so that end-user applications are alerted or notified only when some meaningful event has occurred, and not for every RFID read that happened like Tom is taking coffee break and is presently in cafeteria.

16.3.6   Models Used for RFID-Enabled Social Networking

Data modeling comprises a user-level model, location model, an entity model, and an event model. The data model abstracts away the many technical details and difficulties of an RFID deployment to present applications with data in a form that is easier to work with. The location model hides the details of the RFID infrastructure while capturing an abstract notion of tag location and movement. The entity model allows applications to work with meaningful entities (e.g., people, places, and things).

Finally, the event model defines how entity movements and relationships can map to high-level events and how these events are represented.

16.3.6.1   User-Level Model

High-level events are personal and potentially sensitive; users must be able to precisely control all information disclosure to avoid privacy breaches. As such, we need tools that let users directly control how their RFID data are transformed and disclosed in the RFID-enabled social networking; the tools may be like an application interface available on consumers' mobile sets to manage all tag events like registration of tags and association of tags with personalized messages. Also, the user must be privileged to hide, to delete the activities associated with tags, and to modify the current location of tags. Location-based services (LBS) for mobile phones empower subscribers to use their phones to find information about nearby businesses or services, such as movie theaters, banks, or cafés. They can also use their phones for navigation. For example, one user can receive directions from one location to another, locate another person's mobile phone, and receive updates or alerts about bus delays, traffic jams, or sales at nearby businesses. The user model implies that users exert full control over RFID tags by means of appropriate authentication mechanisms. Objects by default do not respond to network requests. The context decisions concerning when and how the use of tags is appropriate are thus taken by the object owner (Guenther and Spiekermann 2005). All these information can then further be used by SNI to produce meaningful information controlled by the user himself.

16.3.6.2   Location Model

A location can be either a geographic location or a symbolic location such as a warehouse, a shipping route, a surgery room, or a smart box. A change of location of an EPC-tagged object is often signaled by certain RFID readers. The location histories of RFID objects are then transformed automatically from these RFID readings and stored in a location relation in an RFID data store. Location model is defined by the location relation that has the schema: location (timestamp, tag_id, loc_id). Here, the location attribute contains not a reader identifier but a value that is meaningful to applications, which is a real-time place. For example, the tuple (1:10 p.m., 10, room 230) indicates that at 1:10 p.m., the tag with ID 10 was located in room 230. For each unique (tag ID, time) combination, thus, for (10, 1:10 p.m.), the system may store the tuple mentioned earlier but also the tuple (1:10 p.m., 10, room 231), indicating that the tag was in the adjacent room 231.

16.3.6.3   Static Entity Model

The entity model can be divided into person, object, and reader as shown in Figure 16.5.

16.3.6.4   Dynamic Event Model

In this section, we will formalize the semantics and specification for RFID events. In particular, we will discuss temporal RFID events, which are highly temporally constrained and cannot be well supported by traditional ECA (event-condition-action) rule systems. An *event* is defined to be an occurrence of interest in time, which could

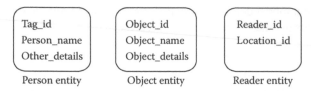

Person entity   Object entity   Reader entity

FIGURE 16.5 Entity model.

TABLE 16.6
RFID Primitive Events

Tagged object "A" enters the location "L" at time "t"
Tagged object "A" reaches the proximity of reader "R" at time "t"
Tagged object "A" leaves the proximity of a Reader "R" at time "t"
Tagged object "A" is near to the tagged object "B" at time "t"
Tagged object "A" leaves the location "L" at time "t"
Tagged object "A" and tagged object "B" are "d" distance apart at time "t"

---

be either a primitive event or complex event. Let us see the stream of RFID primitive events that are encountered in any RFID environment (Table 16.6).

Primitive events occur at a point in time (e.g., "X is at location L") generated during the interaction between readers and tagged objects (Wang et al. 2006). That is, a primitive event is a reader observation, in the format of observation <Tag_id, Reader_id, Timestamp, Location>. Primitive events are generated at random. There is no start_time or end_time associated with the observations (Wang et al. 2006). Complex events are patterns of primitive events and happen over a period of time (e.g., "Anne is taking a coffee break"). RFID-enabled social networking needs implementation of real-time data management that unites the immediacy of in-memory decision making with the historical visibility of a database. Complex event processing deals with the task of processing multiple streams of simple events with the goal of identifying the meaningful events on the fly, within those streams, and comparing with historical data, with the same immediacy of response (Figure 16.6).

Following are the examples of relational model for complex events. The tuple in meeting event (2011-05-03, Ale, Sam, 4.0, room 520) believes with probability 4.0 that a meeting is held between Ale and Sam in room 520. Another example may include

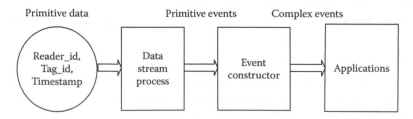

FIGURE 16.6 Dynamic event model.

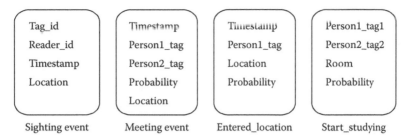

FIGURE 16.7 Complex events. (From Khoussainova, N. et al., Probabilistic event extraction from RFID data (poster), in *Proceedings of the 24th ICDE Conference*, April 7–12, Cancún, Mexico, 2008.)

tuples for Entered_ Location (2011-05-03: 11 a.m., Bob, Café Coffee Day, 4.0), (2011-05-03: 11 a.m., Alice, Café Coffee Day, 4.0), we can assume that Bob and Alice have been to Café Coffee Day with probability 4.0. As another example, if two ENTERED-LOCATION as (2011-05-03: 10 a.m., Alice, room 50, 2.3) and (2011-05-03: 10 a.m., Bob, room 50, 2.5) occurred, they can be composed to form a START-STUDYING event with some probability. For example, START-STUDYING event might be stored in a table with schema START-STUDYING (time, person1, person2, room, probability). An example tuple is (1:10 p.m., Alice, Bob, room 50, 2.0), which represents Alice and Bob are having a study session at 10 a.m. with probability 2.25 (Figure 16.7).

16.4 CHALLENGES AND ISSUES PREVALENT IN RFID-ENABLED SOCIAL NETWORKING

Trust and privacy issue is seen with highest importance in the social acceptance of the IoT. One of the most crucial ways in which RFID could be criminally manipulated is by tracking them illegally. It is theoretically possible for anyone with an RFID reader to read RFID signals and track them. This makes both personal security as well as military information at risk. RFID's potential for tracking consumers and the personal information that these tokens might reveal about their owners have often overshadowed the technical advantages (Garfinkel et al. 2005). Privacy with regard to location and personal data of the user is obviously a concern that needs to be adequately addressed. Everyone and everything will be connected to the network. Eventually, every person and everything with an electronic identification on the planet will be connected to the "IoT." The resulting network traffic will require highly scalable, reliable systems. The reason social network security and privacy lapses exist results simply from the astronomical amounts of information the sites process each and every day that end up making it that much easier to exploit a single flaw in the system. There are a number of users that are extremely vulnerable to privacy threats and eventually identity theft. Those who are necessarily the most tech savvy enjoy the use of social networking to keep in touch with family and friends. Features that invite user participation—messages, invitations, photos, open platform applications, etc.—are often the avenues used to gain access to private information,

TABLE 16.7

Tips to Protect You on SNs

S. No.	Things You Should and Should Not Do while on Social Networking Sites
1	It is a good idea to limit access to your profile. Do not allow strangers to learn everything they can about you
2	Use caution when you click links that you receive in messages from your friends on your social website
3	Do not trust that a message is really from who it says it is from
4	Never post your full name, social security number, address, phone number, full birth date, financial information, or schedule. These will make you vulnerable to identity thieves, scams, burglars, or worse
5	Think twice before posting your photo. Photos can be used to identify you off-line. They can also be altered or shared without your knowledge
6	Do not allow social networking services to scan your e-mail address book
7	Type the address of your social networking site directly into your browser or use your personal bookmarks not from within your e-mail accounts
8	Be selective about who you accept as a friend on a social network
9	Do not post information that makes you vulnerable to a physical attack. Revealing where you plan to meet your friends, your class schedule, or your street address is almost an open invitation for someone to find you
10	Choose your social network carefully
11	Be careful about installing extras on your site

especially in the case of Facebook. A Facebook user with 900 friends and 60 group memberships is a lot more likely to be harmed by a breach than someone who barely uses the site (Roadmap 2008). Table 16.7 gives you a glimpse of some tips to protect you while you are on social networking sites).

Security lapses on social networks do not necessarily involve the exploitation of a user's private information. Take, for example, the infamous "Samy" MySpace XSS worm that effectively shuts the site down for a few days in October 2005. The "Samy" virus (named after the virus' creator) was fairly harmless, and the malware snarkily added the words "Samy Is My Hero" to the top of every affected user's MySpace profile page.

EPC code tracking is also susceptible to denial of service hacks. It is possible to populate the root servers of EPC tracking systems with thousands of illicit requests for data, thereby effectively crippling the system. This is an ongoing concern for businesses planning to implement the EPC system.

Technical improvement can mitigate some privacy concerns, for example, encryption prevents the eavesdropping of transactions (Hancke 2008), but privacy also requires complementary operational regulation or legislation to be enforced (Ayoade 2007). Current security protocols rely on well-known elliptic curve cryptography (ECC), Rivest–Shamir–Adleman (RSA), and advanced encryption standard (AES), but the application of these algorithms to "IoT" is still under research whether they can be successfully implemented with given constrained memory and processor speed. Policy can take the form of governmental recommendation

(Brussels 2009) on privacy and data protection or can be a short set of operating principles protecting end users, such as Garfinkel's "RFID Bill of Rights" (Garfinkel 2002).

16.5 CONCLUSIONS

The RFID is quite an old technology, already used for over a decade in, for example, animal identification chips and electronic door keys. RFID chips contain a unique number or a small amount of data that can be transmitted wirelessly without the need for an onboard battery. In this chapter, we have presented a state-of-art literature on "IoT" for social networks. We have explored many applications currently prevailing all around the world. We have also presented a basic framework for RFID architecture that is needed for social networking. We have also presented various models like user-level model, location model, static event model, and dynamic event model used in social networking scenario. We have also explored complex event processing. We have discussed various securities and privacy issues prevailing in RFID-enabled social networking. We conclude that use of RFID in social networking provides many vital information. The success of RFID-enabled social networking depends on the public acceptance. To make such applications successful, we need to balance between security and privacy aspects. The ideas, concepts, and methods must be communicated and discussed to avoid experiences already made by the retail industry in their pioneer RFID applications, where consumers blocked the adoption of electronic tags. One of the central questions that must be answered is: Who owns the data in networked systems?

REFERENCES

Vermesan O., M. Harrison., and H. Vogt (2009), Internet of things—Strategic research roadmap, http://ec.europa.eu/information_society/policy/rfid/documents/in_cerp.pdf

Atzori L., A. Iera, and G. Morabito (2010), The internet of things: A survey 2010, *Computer Networks*, 54: 2787–2805, Elsevier.

Ayoade J. (2007), Privacy and RFID systems, roadmap for solving security and privacy concerns in RFID systems. *Computer Law and Security Report*, 23(6): 555–561.

Brussels (2009), Commission recommendation on the implementation of privacy and data protection principles in applications supported by radio-frequency identification. Commission of the European Communities Recommendation {SEC(2009) 585, SEC(2009) 586}, http://ec.europa.eu/information_society/policy/rfid/documents/recommendationonrfid2009.pdf (accessed on September 25, 2011).

Collins J. (2004), Hospital gets ultra-wideband RFID, *RFID Journal*, http://www rfidjournal.com/article/view/1088/1/1 (accessed on September 5, 2011).

Facebook (2011), Statistics, http://www.facebook.com/press/info.php?statistics (accessed on September 20, 2011).

Garfinkel S. (2002), An RFID bill of rights, *MIT Technology Review*, October, http://www.technologyreview.in/communications/12953/ (accessed on September 25, 2011).

Garfinkel S., A. Juels, and R. Pappu (2005), RFID privacy: An overview of problems and proposed solutions, *IEEE Security and Privacy*, 3(3): 34–43.

Grossman L. (2007), Tag, you're it. *Times Magazine*, 170(16), http://www.time.com/time/magazine/article/0,9171,1673283,00.html (accessed on September 10, 2011).

Guenther O. and S. Spiekermann (2005), RFID and perceived control—The consumer's view, *Communications of the ACM*, 48(9): 73–76.

Hancke G.P. (2008), Eavesdropping attacks on high-frequency RFID tokens, in *4th Workshop on RFID Security—RFIDSec08*, July 9–11, Budapest, Hungary, pp. 100–113.

IoT China (2011), Research report on IoT, http://fxnewschina.com/currency/2011/08/31/research-report-on-chinese-internet-of-things-industry-2010-2012/ (accessed on August 21, 2011).

Khan M.A. and O. Sanjay (2009), SHA-256 based n-bit EPC generator for RFID tracking simulator, in *Proceedings of IEEE International Conference on Advance Computing (IACC)*, March 6–7, Patiala, India, pp. 988–991.

Khoussainova N., M. Balazinska, and D. Suciu (2008), Probabilistic event extraction from RFID data, in *Proceedings of the 24th ICDE Conference*, April 7–12, Cancún, México, pp. 1480–1482.

Meloan S. (2003), Towards a "Global" internet of things, OSDN, http://java.sun.com/developer/technicalArticles/Ecommerce/rfid/ (accessed on September 12, 2011).

Nguyen T., Y.-K. Lee, R. Huq et al. (2007), A data model for EPC information services, DEWS, http://www.ieice.org/~de/DEWS/DEWS2007/pdf/m6-3.pdf (accessed on September 20, 2011).

Roadmap (2008), Internet of things in 2020: Roadmap for future: INFSO D.4 networked enterprise and RFID INFSO G.2 micro and nanosystem in cooperation with the working group RFID of the ETP EPOSS, version 1.1.

Simonov M., R. Zich, and F. Mazzitelli (2008), RFID, energy, and internet of things http://www.ursi.org/proceedings/procGA08/papers/C11p5.pdf (accessed on September 26, 2011).

Sundmaeker H., P. Guillemin, P. Friess, S. Woelfflé et al. (2010), Vision and challenges for realising the internet of things: Cluster of European Research Project on internet of things. European Research Information society & Media, http://www.internet-of-things-research.eu/pdf/IoT_Clusterbook_March_2010.pdf (accessed on September 26, 2011).

Wang F., S. Liu, P. Liu, and Y. Bai (2006), Bridging physical and virtual worlds: Complex event processing for RFID data streams, in *Advances in Database Technology—EDBT 2006*, March 26–31, Munich, Germany, pp. 588–607. doi:10.1007/11687238_36.

Welbourne E., L. Battle et al. (2009), Building the internet of things using RFID: The RFID ecosystem experience, *IEEE Internet Computing*, 13(3): 48–55.

17 Geosocial Networking Services

Jae-Gil Lee

CONTENTS

17.1 INTRODUCTION

Geosocial networking is a type of social networking in which geographic services and capabilities are used to enable additional social dynamics (Quercia et al., 2010). The location data collected from users allow social networks to connect users to *local* people who share some common interests. With the advent of mobile devices, huge amounts of location data are being collected. Despite of growing concerns on tracking location data, this trend will surely accelerate. At the same time, online social networks (OSNs) are becoming increasingly popular. As of July 2011, Facebook is known to have more than 750 million active users. Twitter rapidly gained worldwide popularity, with 200 million users as of 2011, generating over 200 million tweets and handling over 1.6 billion search queries per day.

Geosocial networking has a lot of useful applications, and three examples are borrowed from Wikipedia (2011). With regard to the third case, there is an interesting piece of news. British Prime Minister David Cameron thinks he has found some culprits to blame in the recent riots that have rocked London and other cities—Facebook and Twitter (from CNN News on August 12, 2011).

- Geosocial networking allows users to interact with people close to their current locations. Popular geosocial applications such as Foursquare and Brightkite enable users to share their locations as well as recommendations for venues. This is the most basic application of geosocial networking.
- In disaster scenarios, geosocial networking allows users to collaboratively disseminate geotag information on hazards and disaster aid activities to develop collective intelligence by assembling individual perspectives. Furthermore, geolocated messages could be used to automatically detect and track potential dangers, for example, an emerging epidemic, for the public.
- Geosocial networking has political applications since it can be used to organize, track, and communicate events and protests. For example, people can tweet using their mobile phones to quickly organize a protest event before authorities can stop it.

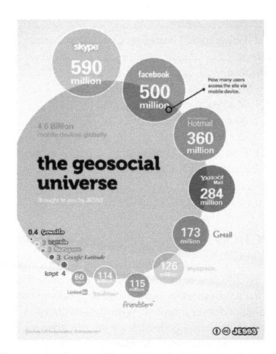

FIGURE 17.1   Geosocial universe. (From Labs, J., The geosocial universe, 2010, http:// jess3.com/geosocial-universe/)

Figure 17.1 is an infographic illustrating and comparing the popularity of different geosocial networking services as of August 2010. Please notice that not all of social networking services in this figure are purely location based. For example, in Facebook or Twitter, users can choose to reveal their location. However, Facebook or Twitter is *not* generally regarded as location-based social networking services. In this chapter, we will discuss only *pure* location-based social network services such as Foursquare.

The rest of this chapter is organized as follows. Section 17.2 explains the most representative location-based social networking services. Section 17.3 explains a few measures for geosocial properties. Section 17.4 presents analytic results for the impact of the geographic distance to information dissemination. Section 17.5 explains two algorithms for detecting communities from geosocial networks. Finally, Section 17.6 summarizes this chapter.

17.2 LOCATION-BASED SOCIAL NETWORKING SITES

17.2.1 FOURSQUARE

Foursquare (http://www.foursquare.com) is a location-based social networking website based on mobile devices. This service is available to users with GPS-enabled mobile phones such as Android phones and iPhones. Native mobile apps are provided for many mobile operating systems including iOS, Android, Windows Phone, and BlackBerry OS. Foursquare is regarded as the leading provider for geosocial networking, and its history goes back to Dodgeball founded in 2000.

Users can check-in at venues using mobile devices as in Figure 17.2a. Checking-in is an act of declaring that he or she is now at a specific venue. Checking-in is possible at preregistered venues, or users are allowed to register new venues. Users can leave a tip and read the tips already written by other users. The history of visited places is maintained in each user's profile as in Figure 17.2b. Users can connect their Foursquare accounts to their Facebook or Twitter accounts: their check-ins are automatically posted to their Facebook or Twitter accounts.

Foursquare incorporates some gaming elements into social networking. Points are rewarded for checking-in at various venues. The concept of the Mayor is devised to encourage users to check-in as many as possible. The Mayor is, basically, the user who has earned the most points at a specific venue (Foursquare.com, 2010): "If a user has checked-in to a venue on more days (meaning only one check-in per day qualifies for calculating mayorship) than anyone else in the past 60 days, the check-ins are valid under Foursquare's time and distance protocols, and they have a profile photo, they will be crowned "Mayor" of that venue, until someone else earns the title by checking in more times than the previous mayor."

Badges are earned when a user fulfills some requirements. For example, Newbie is earned at the first check-in; Adventurer, Explorer, and Superstar are earned when a user has checked-in at 10, 25, and 50 different venues, respectively. The Badge screen in Figure 17.2c shows which badges I have earned. Foursquare has not announced the full list of badges, and there are some very rare and unknown badges.

Like other social networking services, the concept of the friend is incorporated into Foursquare. Users can invite other users to become friends. Users are allowed

(a)   (b)

(c)

FIGURE 17.2   Three screen shots of Foursquare: (a) check-in screen, (b) user profile screen, and (c) badge screen.

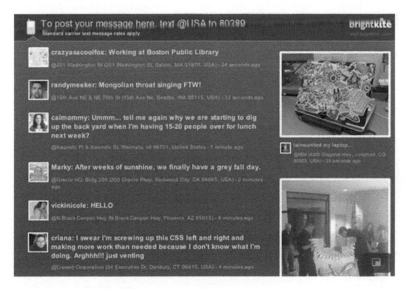

FIGURE 17.3   Screen shot of the Brightkite Wall.

to see only the most recently visited place for each friend. Also, it is allowed to see a friend's badges and a friend's friends for further social networking. When a user sign up for Foursquare, it first recommends possible friends from his or her Facebook, Twitter, or Gmail accounts.

17.2.2   BRIGHTKITE

Brightkite (http://www.brightkite.com) is a location-based social networking website. It provides services very similar to Foursquare, which enables its users to share their locations and meet new people. Brightkite offers a desktop website, SMS tools, mobile phone sites, and native mobile apps. One of the novel features is the Brightkite Wall shown in Figure 17.3. It allows you to turn your monitor into an interactive, live Brightkite display. Actually, the Brightkite Wall is a Flash application showing a feed from a place, user, or keyword.

17.2.3   GOWALLA

Gowalla is a location-based social networking service. It provides functionalities similar to Foursquare or Brightkite. Users check-in at spots in their local vicinity by using native mobile apps or by connecting to the mobile website. Users can create a spot and a trip consisting of a sequence of spots. Some spots and trips are features by Gowalla and are introduced on the home page. Especially, Gowalla released an iPad app on April 2010 (Figure 17.4).

Gowalla also combines gaming elements with social networking. When users check-in at spots, new items will be randomly given to them as rewards. If a user is the first person to check-in at a new spot, the user can opt to drop one of his or her items to become a founder. When a user arrives at a spot where items have been

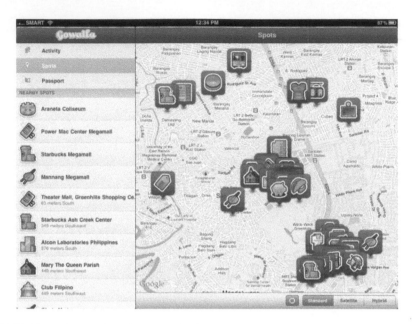

FIGURE 17.4   Screen shot of Gowalla for iPad.

dropped before, then he or she can pick up the item after leaving one of his or her items. That is, users can drop items at spots or swap items if they find something they like. If users keep some items forever, they can do that by placing the items into vaults. Special items have been created for important events such as the 2010 Winter Olympic Games, and promotional items are sometimes linked to real-world prizes.

17.2.4 GOOGLE LATITUDE

Google Latitude (www.google.com/latitude) is a location-based mobile application developed by Google. Actually, Google Latitude is the successor of Dodgeball created by the founder of Foursquare. Latitude allows a mobile phone user to see where his or her friends are right now. A user first sends an invitation to a friend on his or her contacts. After the friend accepts the invitation, the user can see the friend on the map as in Figure 17.5. Each user's current location is mapped on Google Maps. Latitude helps friends and family stay in touch with you by sharing your location with whomever you choose.

For privacy, each user is in full control of how his or her location is provided: an exact location can be allowed or only city-level information can be allowed. In addition, Latitude can be turned off by the user, or a fake location can be manually entered. Google assures that Latitude stores only the most recent location updated by each user on its server.

Now, Latitude provides check-in facilities similar to Foursquare as in Figure 17.6. If a user checks-in at some place, the user can see its profile from Google Place. The detailed information and ratings from other users are provided to the user. Latitude provides automatic check-in/checkout facilities using your current location: when you arrive at a specific place, you will be automatically checked in while you are there.

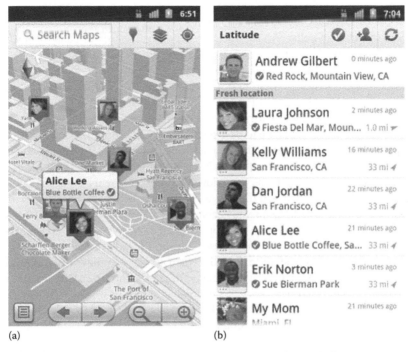

(a)   (b)

FIGURE 17.5   Screen shot of Google Latitude: (a) map screen and (b) list screen.

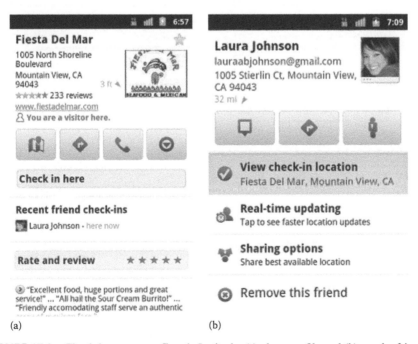

(a)   (b)

FIGURE 17.6   Check-in screen on Google Latitude: (a) place profile and (b) nearby friends.

17.2.5 Yelp

Yelp.com (http://www.yelp.com) is a social networking, user review, and local search website. First of all, Yelp.com provides local search functionalities. For example, you can search "Korean restaurant" in San Francisco. In Figure 17.7a, the list of Korean restaurants is displayed with their basic information, address, and five-star rating. If you choose a specific place, as in Figure 17.7b, you will see the detailed information and user reviews on the place. The reviews are really helpful for deciding whether you will visit the place. Yelp is known to be especially useful for finding good restaurants.

Yelp combines local reviews and social networking service to create a local online community (Dvorak, 2005). Like other OSNs, you can invite people to become friends on Yelp. Once you find the persons you know, you can send them a compliment, send them a private message, and add them as a friend. In addition, you can follow other users just like Twitter, which means you become a fan of them. If you follow other users, you will see their recent reviews in your profile, and their reviews will appear first in Yelp's default ranking whenever you browse a business they have reviewed.

Yelp has the rule for the Elite Squad, which is very similar to the Mayor of Foursquare. The Elite Squad is a way of recognizing the most passionate users who have uploaded many useful reviews. To become eligible for the Elite Squad, users must provide their photo and real name. This concept is adopted to indicate that the user is a trusted author of business reviews.

(a)

FIGURE 17.7   Two screen shots of Yelp.com for local search: (a) local search and (b) user reviews.

(b)

FIGURE 17.7 (continued)

17.2.6 COMMON FEATURES

In geosocial networking, three types of data are intermixed as shown by Figure 17.8:

- Location (geographic) data: the current or past locations of users
- Link data: the friend relationship between users
- Content data: the contents generated by users (e.g., reviews and tips for venues)

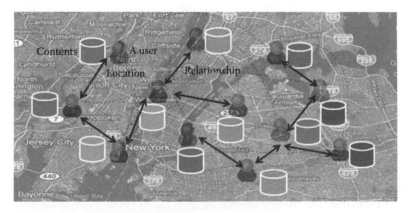

FIGURE 17.8 Three types of data in geosocial networking services.

Since three types of data are intermixed and the amount of such data is huge, we have unprecedented opportunities for analyzing geosocial networking data. In the rest of this chapter, we will survey the recent studies in this direction.

17.3   MEASURES FOR GEOSOCIAL PROPERTIES

17.3.1   DISTANCE METRICS

In order to consider both geographic location and social relationship in OSNs, the need for new metrics has been raised. The main purpose in this direction is to characterize how geographic distance affects social structure. Scellato et al. (2010) proposed two metrics: node locality and geographic clustering coefficient.

The *node locality* is defined by Equation 17.1. Let us consider an undirected geographic social network. Suppose a node i with a particular geographic location has the set Γ_i of its neighbors. The node degree k_i is the number of these neighbors, that is, $k_i = |\Gamma_i|$. β is a scaling factor used for avoiding extremely small node locality. Basically, the node locality represents how much geographically close its neighbors are:

$$\mathrm{NL}_i = \frac{1}{k_i} \sum_{j \in \Gamma_i} e^{-l_{ij}/\beta} \tag{17.1}$$

While the node locality represents how close the neighbors are, the geographic clustering coefficient represents how tightly connected the neighborhood of a node is. The definition of the geographic clustering coefficient is based on that of the clustering coefficient. The clustering coefficient measures the proportion of triangles among the neighbors of a given node. The geographic clustering coefficient attempts to weight triangles differently depending on how close nodes are. For a node i, each existing triangle formed by nodes i, j, and k is assigned a weight w_{ijk} defined by Equation 17.2. Here, Δ_{ijk} is the maximum length among the three links, that is, $\Delta_{ijk} = \max(l_{ij}, l_{ik}, l_{jk})$:

$$w_{ijk} = e^{-\frac{\Delta_{ijk}}{\beta}} \tag{17.2}$$

Then, the *geographic clustering coefficient* is defined by Equation 17.3, which is the average of the weights for differently ordered couples of the neighbors in Γ_i:

$$\mathrm{GC}_i = \frac{1}{k_i(k_i - 1)} \sum_{j, k \in \Gamma_i} w_{ijk} \tag{17.3}$$

17.3.2   ANALYSIS RESULTS

The same authors analyzed four popular OSNs about geosocial properties. The data sets are collected from four popular OSNs: Brightkite, Foursquare, LiveJournal, and Twitter. LiveJournal and Twitter are not location-based OSNs, but users can specify their locations optionally. We will summarize their findings in this section.

17.3.2.1 Small-World Effect

The authors confirm the small-world effect. The sampled average path length is above 4.5 hops in Foursquare, Brightkite, and LiveJournal. It is only 2.77 hops in Twitter. The average (standard) clustering coefficient is 0.207 and 0.256 in Twitter and Foursquare, respectively. LiveJournal scores 0.185, and Brightkite 0.181. These results conform to the small-world effect. The average path length is only a few hops, and the clustering coefficient is higher than that in a randomized network of the same size.

17.3.2.2 Node Locality

In Brightkite, about 40% of users have a node locality higher than 0.90, and in Foursquare, this phenomenon becomes more evident since 25% of users have a node locality close to 1. User of these location-based services shows a high average node locality in overall: Brightkite has an average value of 0.82, while the average goes up to 0.85 in Foursquare.

However, in LiveJournal and Twitter, this effect is weaker. In LiveJournal, 10% of users have a node locality closed to 1, and the average is 0.73 for in-locality and 0.71 for out-locality. In Twitter, the average value is lower than that of other networks: 0.57 for in-locality and 0.49 for out-locality.

These results show location-based services such as Brightkite and Foursquare are characterized by short-range friendship links among users, resulting in a large portion of users with high values of node locality (Scellato et al., 2010). Thus, considering locations in addition to contents shared by users will give us more opportunities for discovering potential friends who live nearby.

17.3.2.3 Geographic Clustering Coefficient

The authors provide the results for the geographic clustering coefficient. The average value is 0.165 in Brightkite, 0.237 in Foursquare, 0.146 in LiveJournal, and 0.108 in Twitter. In the first two data sets, the geographic clustering coefficient is close to the standard clustering coefficient. This indicates that the users within the same cluster are quite geographically correlated. In contrast, in LiveJournal and Twitter, the geographic clustering coefficient is lower than the standard clustering coefficient since there is no strong geographic correlation. Thus, tighter (i.e., smaller) clusters are generated in the first two data sets.

In addition, about 20% in Brightkite and 10% in Foursquare have the coefficient value 1.0. On the other hand, in LiveJournal and Twitter, almost no user has the coefficient value 1.0. Brightkite and Foursquare, which are location-based services, tend to have more geographically confined triangles than other social networks focusing on content production and sharing.

17.4 SOCIAL CASCADES

Scellato et al. (2011) analyzed the impact of geographic locations to the content delivery network (CDN). Content providers such as YouTube allow us to easily access multimedia contents, generating high bandwidth and storage demand on the CDN.

Moreover, dissemination of the contents is accelerated by OSNs such as Twitter and Facebook. Social cascades are observed when users increasingly repost (i.e., retweet) links received from others.

In the work by Scellato et al. (2011), the authors study how geographic information can improve caching of multimedia files in the CDN. To this end, they conduct an extensive study of information dissemination on a geosocial dataset gathered from Twitter. (In Twitter, users can choose to specify their location.) They investigate the extent of social cascades and characterize social cascades over space and time by taking account of users' geographic information.

A *CDN* is a system of networked servers holding copies of data items, placed at different geographic locations (Scellato et al., 2011). The purpose of the CDN is to efficiently deliver contents to clients, for example, providing a video to a client like YouTube. Each request is served by a geographically close server, and the content is moved across servers to optimize the quality of service perceived by users. Of course, it is much better if the files are stored in a server located closely to the requester.

The authors first measure the node locality (Scellato et al., 2010) to see geosocial properties. The average geographic distance of a social link is 5117 km, indicating that Twitter users tend to engage in long-range connections. Nonetheless, there are around 10% of users with the node locality greater than 0.6 and 20% with the node locality greater than 0.5. This indicates that only a minority of Twitter users have geographically limited audiences, whereas most users have a set of followers spreading over large regions.

Instead, the authors show how users are spreading contents mainly over the less frequent shorter connections. A piece of information can disseminate from a user to others like a virus in an epidemics. People share information with their friends who might share the information again and again. This phenomenon is called the *social cascade*. Formally, it is defined that a user B is reached by a *social cascade* about the content c if and only if

- There is another user A who posted the content c
- The user A posted content c before a user B posted it
- There was a social connection from the user A to the user B when A posted c

Then, social cascades are represented as a tree over the geographic social network. The initiator node becomes the root node. Here, a link from a user A to a user B means that B has received some content from A. A link is annotated with two timestamps t_1 and t_2, which are the timestamps when A and B posted a content, respectively. A cascade step is limited by the time threshold 48 h, that is, a cascade step is valid only when two postings have happened within 48 h from each other.

In order to investigate the geographic properties of a social cascade, two measures are defined as follows (Scellato et al., 2011):

- The *geodiversity* of a cascade is the geometric mean of the geographic distances between all pairs of users in the cascade tree.
- The *georange* of a cascade is the geometric mean of the geographic distances between the users and the root user of the cascade tree.

The result of the analysis for the real dataset from Twitter is summarized as follows (Scellato et al., 2011):

- Geographic distance: Here, we are interested in the distribution of the geographic distance between authors of two consecutive tweets in a social cascade. About 10% of cascade steps are less than 1 km, 20% of them shorter than 100 km, and over 30% of them shorter than 1000 km. This result slightly does not conform to the distribution of link lengths of Twitter. Even though less than 5% of the social connections are shorter than 100 km, this faction for the cascade steps increases up to 20%. Content dissemination through social networks is expected to travel over geographically short-range connections rather than over more numerous long-distance links.
- Geodiversity and georange: About 40% of the cascades exhibit geodiversity lower than 1000 km, and about 20% of them lower than 300 km. Thus, even though many cascades reach a broad audience, some of them remain geographically limited. On the other hand, about 90% of the cascades exhibit georange smaller than 1000 km and about 30% of them smaller than 100 km. This indicates that a cascade may take place in a broad region, but each user is still close to the initiator.

It is observed that even the initial locality of the first user is already correlated with the geographic spreading of the cascade. Moreover, including the second user as well makes this correlation even stronger. Thus, the final properties of a cascade can be estimated even from the users involved in the initial stages. Also, the geographic and social properties of the initiator are sufficient to understand whether a cascade will spread locally or globally.

17.5 COMMUNITY DISCOVERY

In recent years, it is hard to deny the booming popularity of social networking sites such as Facebook and MySpace. More people keep joining such social networking sites. Users can add friends and send them messages. Social networks are often modeled as graphs, where nodes are users, and edges are relationships between users. The graphs are very dynamic: users can join or leave, and friends can be added or deleted. Moreover, the size of the graphs is so huge. Facebook is the world's largest social network with over 750 million users (as of July 2011). One of the most active topics is finding communities that share common interests or background, which is really useful for viral marketing.

Social community analysis has been the focus of many studies over the past 80 years (Wasserman and Faust, 1994). One of the earliest studies includes the work by Rice (1927) on the analysis of communities of individuals based on their political biases and voting patterns. Since then, there have been numerous studies on social community discovery. Mostly, these studies were based on user surveys, and the data were manually collected for a small scale of sampled persons.

17.5.1 Quality Measures

Informally, a *community* in a network is a group of nodes with higher cohesion than to the rest of the network. This concept can be formalized in different ways according to the quality measure that an algorithm tries to optimize. The modularity (Newman and Girvan, 2004) and the normalized cut (Shi and Malik, 2000) are the most popular ones, but many other measures also have been proposed in the literature. In this section, we briefly review the two quality measures. Before going into details, we first define the notation for explaining the quality measures in Table 17.1.

17.5.1.1 Modularity

The *modularity* has become quite popular as a way of measuring the goodness of graph clustering. If a division is good, the modularity will be high. In this case, internal connections between vertexes within communities are dense, but there exist only sparse connections between different communities. The idea behind the modularity is that the farther the subgraph corresponding to each community is from a random subgraph (i.e., the null model), the better the discovered community structure is. The modularity Q for a division of the graph into k clusters $\{V_1,\ldots, V_k\}$ is defined by Equation 17.4:

$$Q = \sum_{i=1}^{k} \left[\frac{A(V_i, V_i)}{m} - \left(\frac{degree(V_i)}{2m} \right)^2 \right] \tag{17.4}$$

The modularity of a clustering is defined to be the fraction of the edges that fall within the given communities minus the expected such fraction if edges were distributed at random. The definition of the modularity reveals an inherent trade-off: to maximize the first term, many edges should be contained in clusters, whereas the minimization of the second term is achieved by splitting the graph into many clusters with small total degrees. Notice that the first term is also known as *coverage*.

TABLE 17.1

The Notation for Quality Measures

Notation	Description
V	The set of vertexes in the graph
V_i	A cluster of vertexes
m	The number of edges in the graph
$degree(i)$	The degree of the vertex i
$degree(V_i)$	The total degree of the cluster V_i
$A(i, j)$	The affinity between the vertexes i and j
$A(V_i, V_j)$	The sum of edge affinities between vertexes in V_i and V_j, that is, $A(V_i, V_j) = \sum_{u \in V_i, v \in V_j} A(u, v)$

17.5.1.2 Normalized Cut

The *normalized cut* of a group of vertexes $S \subset V$ is defined by Equation 17.5 (Shi and Malik, 2000). The normalized cut is the sum of affinities of the edges that connect S to the rest of the graph $\bar{S}$, normalized by the total edge affinity of S and that of $\bar{S}$:

$$Ncut(S) = \frac{\sum_{i \in S, j \in \bar{S}} A(i,j)}{\sum_{i \in S} degree(i)} + \frac{\sum_{i \in S, j \in \bar{S}} A(i,j)}{\sum_{j \in \bar{S}} degree(j)} \qquad (17.5)$$

17.5.2 Joint Cluster Analysis

Now, we summarize the algorithms of detecting communities in geosocial networks. These algorithms exploit both attribute data (geographic locations can be one specific type of attribute data) and relationship data.

17.5.2.1 Connected *k*-Center

The closest work to this category is joint cluster analysis done by Ester et al. (2006). They claim that attribute data and relationship data are two principal types of data that represent the intrinsic and extrinsic properties of entities. It is common that both data types carry complementary information. For example, two persons share many common characteristics, but they have not met each other (thus, no social relationship between them). In contrast, two very close friends may have totally different characteristics. Thus, joint cluster analysis of both data types allows us to achieve more accurate results. To this end, they introduce the *Connected k-Center* (*CkC*) problem.

The authors mention an interesting example (Ester et al., 2006): "Market segmentation divides a market into distinct customer groups with homogeneous needs, such that firms can target groups effectively and allocate resources efficiently, as customers in the same segment are likely to respond similarly to a given marketing strategy. Traditional segmentation methods were based on attribute data only such as demographics (age, sex, ethnicity, income, education, religion, etc.) and psychographic profiles (lifestyle, personality, motives, etc.). Recently, social networks have become more and more important in marketing. Ideas and behaviors are contagious. The relations in networks are channels and conduits through which resources flow. Customers can hardly hear companies but they listen to their friends; customers are skeptical but they trust their friends. By word-of-mouth propagation, a group of customers with similar attributes have much more chances to become like-minded. Depending on the nature of the market, social relations can even become vital in forming segments, and purchasing intentions or decisions may rely on customer-to-customer contacts to diffuse throughout a segment, for example, for cautious clients of risky cosmetic surgery or parsimonious purchasers of complicated scientific instruments. The *CkC* problem naturally models such scenarios: a customer is assigned to a market segment only if he has similar purchasing preferences (attributes) to the segment representative (cluster center) and can be reached by propagation from customers of similar interest in the segment."

Definition 17.1 shows the decision version of the *CkC* problem for a constant $k \geq 2$ and a dimensionality $d \geq 1$.

Definition 17.1 (Ester et al., 2006) *Given the number k of centers, a radius constraint $r \in \mathcal{R}^+$, a distance function $\|\cdot\|$, and a graph $G = (V, E)$, where every vertex in V is associated with a coordinate vector $w:V \to \mathcal{R}^d$, decide whether there exist disjoint partitions $\{V_1,\ldots, V_k\}$ of V., that is, $V = V_1 \cup \ldots \cup V_k$ and $\forall 1 \leq i < j \leq k$, $V_i \cap V_j = \varnothing$, which satisfy the following two conditions:*

1. *The induced subgraphs $G[V_1],\ldots, G[V_k]$ are connected (internal connectedness constraint).*
2. *$\forall 1 \leq i \leq k$, there exists a center node $c_i \in V_i$ such that $\forall v \in V_i$, $\|w(v) - w(c_i)\| \leq r$ (radius constraint).*

Intuitively, the problem is to check whether the input graph can be divided into k connected components so that every component is a cluster with radius less than or equal to r, that is, in each cluster, all vertexes are within distance r to the center of the cluster c. Figure 17.9 illustrates this concept, and four clusters can be created according to Definition 17.1.

17.5.2.2 Connected X Cluster

The drawback of the *CkC* problem is that it is usually difficult to determine the number of clusters in advance. To solve this problem, Moser et al. (2007) propose an extension that they call the *Connected X Cluster* (*CXC*) problem. They develop an algorithm of discovering an arbitrary number of *true clusters* in joint cluster analysis without specifying the number of clusters a priori. True clusters are assumed to be compact and distinctive from neighboring clusters in terms of attribute values and to be internally connected in terms of relationship data. Unlike classical clustering methods, the neighborhood of clusters is not defined in terms of attribute data but in terms of relationship data (Moser et al., 2007).

To motivate the assumption mentioned earlier, the authors present motivating examples for community identification and hotspot analysis (Moser et al., 2007): "In the context of criminology, the goal of hotspot analysis is to identify high-crime areas (hotspots) surrounded by areas of relatively low crime rates (coldspots). Attribute data represent the frequencies of certain crime types in a particular area over a given period of time. Relationship data represents connectivity information such as a road network. The road network constrains the space in which offenders move and also plays a major role in the appearance of crime attractors and generators like pubs or needle sharing places. Thus, two high-crime areas, which are close to each other using the Euclidean

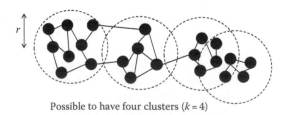

Possible to have four clusters ($k = 4$)

FIGURE 17.9   Example of the *CkC* problem.

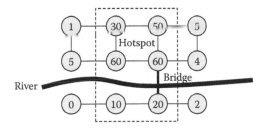

FIGURE 17.10   Hotspot example.

distance, but not directly connected on the road network, have most likely different attractors or generators and should not be merged into one hotspot."

Figure 17.10 illustrates this example. Here, a vertex represents an area, and an edge means the road between two areas. The number (attribute value) in each vertex denotes the number of crimes in the corresponding area. Two parts of the hotspots exist on the different sides of the river. The crime rates of these two hotspots are pretty different. If we considered only attribute values for clustering, two parts would form two separate clusters. However, the two areas are connected by a bridge, and the bridge is the only path between the two areas. When considering two data types, the hotspot will be correctly identified as one cluster.

The authors propose a two-phase algorithm for the *CXC* problem. In the first phase, cluster atoms—basic building blocks—that cannot be split in the second phase are generated. These cluster atoms are further refined iteratively for the second phase. In the second phase, the cluster atoms are merged in the bottom-up fashion as long as the quality measure increases. The Silhouette coefficient is adapted for the joint cluster analysis.

17.6   SUMMARY

In this chapter, we briefly reviewed geosocial networking services. First, we explained popular location-based social networking services. Geosocial networking contains three types of data: location data, link data, and content data. A lot of research efforts have been devoted to fully take advantage of these abundant data. Next, we explained a few measures that consider both geographic distance and social relationship. The impact of the geographic distance to social cascades is also analyzed using these measures. Community discovery is one of the most important analytic operations for social network data. We explained two approaches for the joint cluster analysis that takes account of both attribute data and relationship data. In summary, since geosocial networking services are all interesting and there are a lot of potentials in this direction, we believe many new services and research directions will emerge in the near future.

REFERENCES

Dvorak JC (2005) Two wrongs make a right. *PC Magazine*, pp. 1–2.
Ester M, Ge R, Gao BJ, Hu Z, Ben-Moshe B (2006) Joint cluster analysis of attribute data and relationship data: The connected k-center problem. In: *Proc. 6th SIAM Int'l Conf. on Data Mining (SDM)*, Bethesda, MD.

Foursquare.com (2010) Foursquare help page. URL: http://support.foursquare.com

Labs J (2010) The geosocial universe. URL: http://jess3.com/geosocial-universe/

Moser F, Ge R, Ester M (2007) Joint cluster analysis of attribute and relationship data without a-priori specification of the number of clusters. In: *Proc. 13th ACM SIGKDD Int'l Conf. on Knowledge Discovery and Data Mining (KDD)*, San Jose, CA, pp. 510–519.

Newman MEJ, Girvan M (2004) Finding and evaluating community structure in networks. *Physical Review E* 69(2):026113.

Quercia D, Lathia N, Calabrese F, Lorenzo GD, Crowcroft J (2010) Recommending social events from mobile phone location data. In: *Proc. 10th IEEE Int'l Conf. on Data Mining (ICDM)*, Sydney, New South Wales, Australia, pp. 971–976.

Rice SA (1927) The identification of blocs in small political bodies. *The American Political Science Review* 21(3):619–627.

Scellato S, Mascolo C, Musolesi M, Crowcroft J (2011) Track globally, deliver locally: Improving content delivery networks by tracking geographic social cascades. In: *Proc. 20th Int'l Conf. on World Wide Web (WWW)*, Hyderabad, India, pp. 457–466.

Scellato S, Mascolo C, Musolesi M, Latora V (2010) Distance matters: Geo-social metrics for online social networks. In: *Proc. 3rd Workshop on Online Social Networks (WOSN)*, Boston, MA.

Shi J, Malik J (2000) Normalized cuts and image segmentation. *IEEE Transactions on Pattern Analysis and Machine Intelligence* 22(8):888–905.

Wasserman S, Faust K (1994) *Social Network Analysis: Methods and Applications*. Cambridge University Press, Cambridge, U.K.

Wikipedia (2011) Geosocial networking. URL: http://en.wikipedia.org/wiki/Geosocial_networking

Part IV

High-Performance Knowledge Service Systems

Part IV

High-Performance Knowledge Service Systems

18 Emerging Ubiquitous Knowledge Services

From Mobile Sensing to Ubiquitous Crowdsourcing and Beyond

Uichin Lee, Howon Lee, Bang Chul Jung, and Junehwa Song

CONTENTS

18.1 INTRODUCTION

While traditional knowledge services are mainly delivered by the experts with domain-specific knowledge, recent advances of information and communication technologies (e.g., smartphones and Web 2.0) have dramatically changed the nature of knowledge services. These technologies have facilitated ubiquitous sensing and networked collaboration among people, thus significantly broadening the range of knowledge services and lowering the knowledge transfer barriers. People generate significant amount of content and share it with other users via Web 2.0 services (e.g., YouTube and Twitter). Collaboration over the Internet is common as one can easily create an online project with tens of thousands of participants (called crowdsourcing), for example, Wikipedia, online Q&A services, and Innocentives. Further, sensing happens everywhere, ranging from surveillance cameras on the streets to smartphone sensors (e.g., collecting mobility traces of users) to even human sensors (e.g., sensing social events). This dramatic shift of ubiquitous sensing and collaboration has enabled new types of knowledge services called "ubiquitous knowledge services." The key role of ubiquitous knowledge services is to seamlessly integrate content from various sources at large scales and to derive new values for end users in ways that the contributor of the content did not plan or imagine (Figure 18.1).

To illustrate ubiquitous knowledge services, consider the case of a recent flooding that happened in July 2011 in Seoul, South Korea, which is reported as the heaviest rainfall ever recorded. Due to heavy rainfall in a relatively short period of time, many places were flooded, and the city government was not able to properly track them. Yet people took photos/videos of the local areas and shared them with others via social networking services (e.g., Twitter and Facebook). Given this, various knowledge services were instantly created to help people to make informed decisions; for example, one service collected all the news feeds from social networking and visualized related photos/videos on the map. Another case is delivering traffic information to the drivers. Traditional traffic information systems mostly use expensive

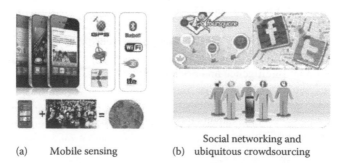

            (a)   Mobile sensing   Social networking and
                                      (b)   ubiquitous crowdsourcing

FIGURE 18.1   (a) Mobile sensing, (b) social networking, and ubiquitous crowdsourcing.

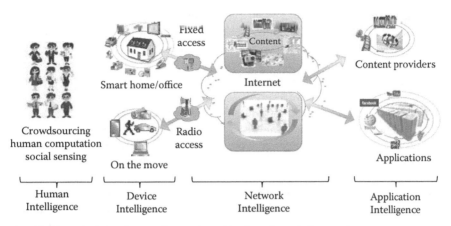

FIGURE 18.2 Driving forces of emerging ubiquitous knowledge services.

sensors for traffic condition estimation (e.g., fixed cameras or induction loops on the roads), and yet, the coverage of those systems is extremely limited due to high installation and maintenance costs. An alternative is to let drivers use their smartphone apps to gather/share the traffic information (Thiagarajan et al., 2009). While individual reports may not be as accurate as traditional approaches, this crowdsourced knowledge service can *collectively* deliver accurate information even with extended coverage and minimal infrastructure costs. Other knowledge services include cooperatively estimating/sharing the schedule of public transits (Thiagarajan et al., 2010) or traffic signals (Koukoumidis et al., 2011). These use cases (and many to come) clearly demonstrate that ubiquitous sensing and collaboration provide unique opportunities for enabling various ubiquitous knowledge services.

As illustrated in Figure 18.2, we claim that the main driving forces of emerging ubiquitous knowledge services are the intelligences on device/network (e.g., smart devices [SDs], mobile sensing, and in-network processing), application (e.g., web, service mashup, and cloud computing), and human (e.g., crowdsourcing, human computation, and social sensing).

- Recent SDs such as smartphones and vehicles are equipped with various sensors (e.g., GPS [Global Positioning System], cameras, and accelerometers) and wireless networking technologies (e.g., Bluetooth, Wi-Fi, and 3G). Sensing capabilities with always-on connectivity enabled a variety of new mobile sensing applications such as people sensing and vehicular sensing. Further, the network infrastructure is becoming more intelligent as it supports various services ranging from simple quality of service control to in-network data processing and application services (e.g., machine to machine and Internet of things [IoT]).
- Application intelligence has facilitated a rapid innovation of application service developments. For instance, service mashup techniques allowed application developers to easily use and combine data, presentation, or functionality from multiple sources to create new services. Cloud computing

enabled ubiquitous, convenient, on-demand network access to a shared pool of configurable computing resources (e.g., networks, servers, storage, applications, and services) that can be rapidly provisioned and released with minimal management effort or service provider interaction.

- The areas of human intelligence include the capability of collaboration among people to achieve a certain goal (e.g., crowdsourcing) and the capability of human computation where humans perform *the computation tasks* that machines cannot do well (e.g., image labeling and natural language processing) or *the perception tasks* by sensing/processing ambient/social environments (e.g., earthquake reports and social events).

In this chapter, we begin by providing a detailed review of these areas to better understand ubiquitous knowledge services with emphasis on device, network, and human intelligences. In particular, we review various mobile sensing and ubiquitous crowdsourcing applications for ubiquitous knowledge services, namely, monitoring personal environmental impact (e.g., CO_2 generation/exposure), analyzing context awareness with sound sensing, sharing personal context over social networking, monitoring road surface (e.g., potholes) and real-time traffic, facilitating fuel-saving driving, and social sensing with Twitter (e.g., weather monitoring and earthquake detection).

Given that the fundamental goal of ubiquitous knowledge services is integrating content (e.g., data, information, and knowledge) from various sources (e.g., sensors, databases, and humans) at large scales and driving new values to end users, we review the existing systems on sensor data collection, retrieval, and service systems, ranging from research prototypes to recent standards such as machine-to-machine (M2M) communications, EPC (electronic product code) global network, and semantic sensor web (SSW). Since our ultimate goal is to design an open platform for ubiquitous knowledge services, we also suggest the desirable features based on our insights from the review, namely, semantic query processing, service-device/network interaction (for cross layer optimization), human computation/sensing, social networking, and agent support. Beyond simple data collection and processing, ubiquitous knowledge services naturally require complex semantic query processing at large scale. Service-device/ network interactions are necessary as we need to optimize query processing over the network of resource-constrained SDs. Social networking and collaboration will enable various crowdsourcing and human computation/sensing projects. We also anticipate that future knowledge services will be more proactive to achieve better user experiences (UXs). For example, user agents perceive the intention of a user and provide the customized knowledge services to the user at any place and any time, with any device even with physical restrictions. Having said that, we finally present our vision platform called the knowledge communication service (KCS) toward the goal of realizing these features.

18.2 FUNDAMENTALS OF SMART DEVICES

An SD is generally referred to as an active computer network device with intelligence (see Figure 18.3). Mark Weiser proposed three basic forms for ubiquitous system devices, namely, tabs, pads, and boards (Weiser, 1999). Tabs are

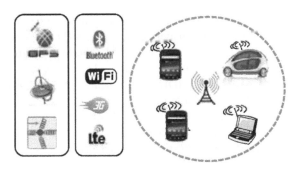

FIGURE 18.3 Examples of SDs.

accompanied or wearable centimeter-sized devices, for example, smartphones; pads are handheld decimeter-sized devices, for example, laptops and iPads; and boards are meter-sized interactive display devices, for example, horizontal surface computers and vertical smart boards. These SDs could be embedded into home appliances (e.g., smart TVs and refrigerators) and vehicles (e.g., SatNav and vehicular sensors). Also, dedicated sensors for monitoring physical environments as in sensor networks can be considered as SDs. Note that recent SDs are equipped with various sensors. Typical sensors include motion sensors such as accelerometer, magnetometer (digital compass), and gyroscope (rotational motion and changes in orientation); an ambient light sensor (brightness of the light), a proximity sensor (presence of nearby objects); a camera/video; a voice sensor; and location sensors. In this section, we review the capabilities of SDs, for example, processing, wireless networking, localization, and multimodal interfaces.

18.2.1 PROCESSING CAPABILITIES OF SMART DEVICES

We review the system configuration of SDs, namely, memory, CPU, and storage capacity. We show that unlike regular computing systems such as PC and laptops, SDs are resource constrained, and thus, the resource constraints must be carefully considered when designing knowledge services.

Most SDs use flash memory to store operating systems and data (Khatib et al., 2007). They typically use the shadowing technique such that during system booting time, the entire code image of an OS and applications is copied from flash memory to DRAM for execution. Low-end SDs such as SatNavs are typically equipped with 64–128 MB DRAM and 1–2 GB of flash memory; high-end SDs such as smartphones and smartpads are equipped with 512–1 GB DRAM and 16–32 GB of flash memory. For instance, Clarion EZD580 SatNav has 64 MB DRAM and 2 GB NAND flash memory; TomTom GO910 has 64 MB DRAM and 1 GB flash memory. iPhone 4 and Samsung Galaxy S smartphones are equipped with 512 MB DRAM and 16–32 GB flash memory; smartpads such as iPad 2 and Galaxy Tab have higher memory of 1 GB. SDs tend to keep the DRAM size minimal because DRAM is power hungry; each DRAM refresh cycle dissipates a few milliwatts per MB (Khatib et al., 2007). Moreover, it is known that for a given workload, there are some threshold values for

DRAM and flash memory sizes such that increasing the size beyond those threshold values will not bring any further performance gain (Douglis et al., 1994). For instance, smartphones have about 512 MB DRAM. The trend is also true in recent multimedia systems. For instance, Apple TV has a 1 GHz Pentium M Crofton processor, 256 MB DRAM, and a 40 G HDD. SDs typically use low-power and low-cost CPUs such as ARM and xScale whose clock speed is around 1 GHz, or low-power/low-cost multimedia processors such as Texas Instruments OMAP 2/3 and RMI Alchemy Au series. Although Intel recently released a high-performance mobile CPU called Atom (Silverthorne 1.6/8 GHz), its performance is about the same as 900 MHz Celeron M.

In summary, we observe that SDs are limited in terms of DRAM and CPU power, by a few orders of magnitude, compared to standard desktop machines or servers. The recent trend suggests that such a computing/memory resource gap will continue to exist between SDs and regular desktop machines in the foreseeable future.

18.2.2 WIRELESS NETWORKING TECHNOLOGIES FOR SMART DEVICES

Popular wireless networking technologies for SDs include cellular networks (3G/4G), wireless local area networks (WLANs), personal/body area networks (PANs and BANs), and vehicular communications (DSRC [dedicated short-range communication]/WAVE [wireless access in a vehicular environment]). The cellular networks are for long-range communications, whereas WLANs and vehicular communications (DSRC/WAVE) are for medium- and short-range communications. PANs/BANs such as Bluetooth and ZigBee are used for only short-range communications. In Figure 18.4, we present the wireless networking standards in terms of range and data rate.

Cellular networks: Cellular systems have been evolving rapidly to support the ever increasing demands of mobile networking. 2G systems such as IS-95 and GSM support data communications at the maximum rate of 9.6 kbps. To provide higher

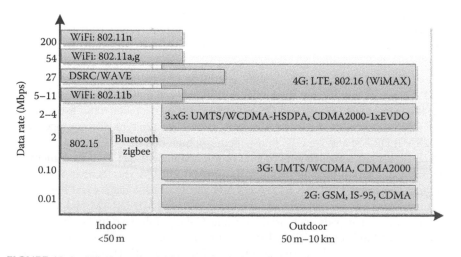

FIGURE 18.4   Wireless networking standards (range vs. rate).

rate data communications, GSM-based systems use GPRS (<171 kbps) and EDGE (<384 kbps), and IS-95-based CDMA systems use 1xRTT (<141 kbps). Now 3G systems support much higher data rate. UMTS/HSDPA provides maximum rates of 144 kbps, 384 kbps, and 2 Mbps under high mobility, low mobility, and stationary environments, respectively. CDMA2000 1xEvDO (Rev. A) provides 3 and 1.8 Mbps for down- and uplinks, respectively. 4G systems such as 3GPP LTE and IEEE WiMAX provide the data rate (tens of Mbps), mobility (<60 km/h), and coverage (<10 km) required to deliver the Internet access to mobile clients, which are in the process of deployment.

WLAN: Wi-Fi supports broadband wireless services. 802.11a/g/n provides up to several hundreds of Mbps and has nominal transmission range of 38 m (indoor) and 140 m (outdoor). Despite its short radio range, its ubiquitous deployment makes Wi-Fi an attractive method to support broadband wireless services. Wi-Fi supports direct peer-to-peer communications called Wi-Fi Direct, a standard that allows Wi-Fi devices to talk to each other without the need for wireless access points (APs) (hot spots). Wi-Fi Direct–enabled devices can signal other devices to make connections and can also form a Wi-Fi Direct Group for device-to-device communications.

Bluetooth and ZigBee: Bluetooth is an always-on, low-power, and short-range hookup for implementing wireless PANs (WPANs). It became the most popular PAN communication device. The concept behind Bluetooth is to support universal short-range wireless capabilities using the 2.4 GHz unlicensed low-power band. Bluetooth allows users to connect up to eight devices by forming a star-shaped cluster, called piconet. The cluster head is called master and the other nodes are called slaves. Two Bluetooth devices within 10 m range can exchange data up to 2 Mbps (Bluetooth v2.0). Readers can find more information about Bluetooth and its applications such as content distribution in the work by Lee et al. (2010). Another WPAN radio is ZigBee, a low-cost, low-power, wireless network standard with data transmission rates of 20–250 kbps. The low cost allows the technology to be widely deployed in wireless control and monitoring applications. Low power usage allows longer life with smaller batteries. ZigBee supports P2P style mesh networking that is simpler and less expensive than Bluetooth.

Vehicular communications: DSRC is a short- to medium-range communication technology operating in the 5.9 GHz range (Jiang et al., 2006). The Standards Committee E17.51 endorses a variation of the IEEE 802.11a MAC for the DSRC link. DSRC supports vehicle speed up to 120 mph, nominal transmission range of 300 m (up to 1000 m), and default data rate of 6 Mbps (up to 27 Mbps). This will enable operations related to the improvement of traffic flow, highway safety, and other Intelligent Transport System (ITS) applications in a variety of application environments called DSRC/WAVE. DSRC has two modes of operations: (1) ad hoc mode characterized by distributed multihop networking (vehicle-vehicle) and (2) infrastructure mode characterized by a centralized mobile single hop network (vehicle gateway). Note that depending on the deployment scenarios, gateways can be connected to one another or to the Internet, and they can be equipped with computing and storage devices, for example, Infostations (Frenkiel et al., 2002).

18.2.3 LOCALIZATION

We review localization techniques that are commonly used by the SDs, namely, GPS, A-GPS (Assisted-GPS), network positioning systems (or cell-tower localization), and Wi-Fi positioning systems (WPSs).

The GPS is a space-based global navigation satellite system that provides location and time information. There are 27 satellites in the orbit, each of which carries highly accurate atomic clocks and is controlled by the ground control station. The orbits are arranged so that at anytime, anywhere on Earth, there are at least four satellites "visible" in the sky. A-GPS receiver's job is to locate three or more of these satellites, figure out the distance to each, and use this information to deduce its own location. This operation is based on a mathematical principle called trilateration, a process of determining absolute locations of points by measurement of distances, using the geometry of circles or spheres. Imagine you are somewhere in Korea and you are totally lost—for some reason, you have absolutely no clue where you are. You find a friendly local and ask, "Where am I?" He says, "You are 70 km from Daejeon." You ask somebody else where you are, and she says, "You are 60 km from Daegu." Now you have two circles that intersect. You now know that you must be at one of these two intersection points. If a third person tells you that you are 100 km from Gwangju, you can eliminate one of the possibilities. You now know exactly where you are.

SDs that use cellular networks mostly have an A-GPS that significantly reduces satellite search space by focusing on where the signal is expected to be. A cellular network service provider has an assist server and position server with GPS that collects satellite information (e.g., time sync, frequency, and visible satellites), which is transmitted over cellular network to speed up location fixing. In practice, however, even with such enhancement, the performance of A-GPS in SDs is less satisfactory. Due to small form factors, SDs mostly use microstrip- or wire-based antenna whose performance is inferior to that of spiral helix antenna that is commonly used in SatNavs.

Another commonly used localization method is cell-tower localization. The overall localization method is very similar to GPS in that a mobile node measures the distances from nearby cell towers (whose positions are known) and trilateration is performed to determine its location. Distance measurement can be done with time of arrival of signals or signal-strength measurement. Unlike GPS, such measurements in urban environments are error prone, and, thus, the error range is much greater than GPS (hundreds of meters on average).

A WPS is based on fingerprinting (e.g., RADAR and Skyhook). WPS uses a fingerprinting technique and is generally composed of two steps: training and positioning phases. In the training phase, the system builds a fingerprint table: for each location, the system collects signal-strength samples from towers and keeps the average for each location. In the positioning phase, the system calculates the distance in signal-strength space between the measured signal strength and the fingerprint DB, for example, by selecting k fingers with the smallest distance and using arithmetic average as an estimated position. One of the major shortcomings is that building the fingerprint table (e.g., wardriving) is costly and the table must be periodically updated as Wi-Fi hot spots tend to change over time.

18.2.4 MULTIMODAL INTERFACES

Basic user interfaces in SDs are the standard graphic-based WIMP (window, icon, menu, and pointing device). SDs also have multiple modalities and are able to interpret information from various sensory and communication channels. For instance, smartphones use touch-based interactions; LG Optimus smartphones allow users to use personalized gestures as shortcuts to commands. Oviatt (2003) defined that multimodal interfaces process two or more combined user input modes (such as speech, pen, touch, manual gesture, gaze, and head and body movements) in a coordinated manner with multimedia system output. They is a new class of interfaces that aim to recognize naturally occurring forms of human language and behavior and which incorporate one or more recognition-based technologies, for example, speech, pen, and vision. Thus, they can provide richer and more natural methods of interactions, for example, gesture, speech, and even all the five senses.

Multimodal interfaces were initially considered as more efficient than existing unimodal interfaces, but experimental results showed that the speedup of task completion is mere 10%, suggesting that speedup should not be considered as a primarily measure (Oviatt, 1997). Instead, multimodal interfaces have been shown to significantly reduce the error rate and improve the reliability; for instance, Oviatt (1997) showed that users made 36% fewer errors with a multimodal interface than with a unimodal interface. People generally prefer to use multimodal interaction due to its expressive power.

18.3 APPLICATION INTELLIGENCE

Application intelligence facilitated a rapid innovation of these application services. Key enabling technologies of the application intelligence include service mashup, cloud computing, and knowledge visualization. There are several other techniques that contribute to the application intelligence such as context awareness, augmented reality, and privacy/security. The application intelligence permits rapid, efficient software development for the developers and rich UX for the users. In this section, we review the key enabling technologies, namely, service mashup, cloud computing, and knowledge visualization.

In general, service mashup means a web page or application that combines data, service, presentation, or functionality from multiple sources to create new services. To make existing data more valuable for personal and professional purposes is very important to provide a better UX, and this is one of the key success factors in the application and content business. In other words, the mashup is a way to create new web applications by combining existing web resources utilizing data and web APIs. Web services are emerging as a major technology for deploying automated interactions between distributed and heterogeneous applications. Even though many service providers recognized its value and have already adopted the service mashup concept, fully realizing the service mashup concept is still challenging, and much work remains to be done. Since services generally face a broader range of heterogeneity compared to data, resolving heterogeneity is the key challenge in the area of service mashup.

Cloud computing delivers hosted services over the Internet. It has enabled ubiquitous, convenient, on-demand network access to a shared pool of configurable computing resources (e.g., networks, servers, storage, applications, and services) that can be rapidly provisioned and released with minimal management effort or service provider interaction (Mell and Grance, 2011). Cloud computing describes a new ecosystem model for IT services based on Internet protocols, and it typically involves provisioning of scalable and virtualized resources, such as storages and applications.

Knowledge visualization aims to improve the creation and transfer of knowledge among people via various computer or non-computer-based visualization methods. While information visualization focuses on the use of computer-supported tools to derive new insights, knowledge visualization concentrates on transferring insights, experiences, attitudes, values, expectations, perspectives, opinions, and predictions with various complementary visualization methods (Burkhard, 2004; Burkhard and Meier, 2004; Eppler and Burkhard, 2005). For an effective transfer and the creation of knowledge via visualizations, following four perspectives should be taken into account, such as function type, knowledge type, recipient type, and visualization type. A function type perspective and knowledge type perspective answer why visualization should be used (aim) and what type of knowledge needs to be visualized (content), respectively. A recipient type perspective points to the different backgrounds of the audience, and the visualization type perspective organizes the main visualization types in accordance with their individual properties.

18.4 UBIQUITOUS SENSING FOR KNOWLEDGE SERVICES

We present ubiquitous sensing applications for SDs; for people sensing, we review knowledge services on personal environmental impact monitoring, sound sensing, and state sharing over social networking, and for vehicular sensing, we review knowledge services on road surface monitoring, real-time traffic monitoring, and fuel-efficient driving.

18.4.1 PEOPLE SENSING–BASED KNOWLEDGE SERVICES

18.4.1.1 Personal Environmental Impact Report

Participatory sensing is to realize distributed data collection and analysis at personal, urban, and global scale. Mobile phones are used to gather data, and web services are used to aggregate and interpret the collected data. Personal Environmental Impact Report (PEIR) is a mobile personal sensing platform that uses mobile handsets to collect/upload data and generates web-based reports about personal environment impacts (Mun et al., 2009). The PEIR system estimates the metrics related to health and wellness: (1) carbon impact: a measure of transportation-related carbon footprint, (2) sensitive site impact: a user's transportation-related airborne particulate matter emissions near sites with populations sensitive to emissions, (3) smog exposure: a user's transportation-related exposure to particulate matter emissions, and

PEIR: Personal environmental impact report

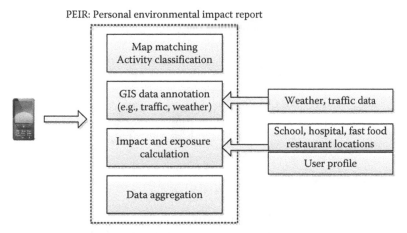

FIGURE 18.5   PEIR system architecture.

(4) fast-food exposure: the time integral of proximity to fast-food restaurants. The ultimate goal of PEIR is for people to take notice of the impact and exposure and to raise individual/community awareness on the potential risks.

Figure 18.5 shows how the PEIR system works. PEIR basically collects the location traces from individuals using the onboard GPS in mobile phones. Once the data are uploaded, raw GPS readings are matched with the map to reduce the GPS errors. Then, a user's activity is classified by processing the GPS readings (e.g., walking and driving). PEIR uses external data sources. For each site visited, it annotates the data with weather and traffic information by utilizing external information sources (e.g., government organizations for weather and transportation). The system utilizes the user profile (e.g., vehicle type) and existing location information of schools, hospitals, and fast-food restaurants. For a given trace, the carbon and sensitive site impacts are estimated using the automobile emission factor (EMFAC) model customized with a user's vehicle type and weather condition. Similarly, the smog exposure can be estimated by incorporating the traffic information. The fast-food exposure can be simply calculated based on the location traces and fast-food locations.

18.4.1.2   Sound Sensing for People-Centric Applications

Sound captured by the microphone of a mobile phone is a rich source of information that helps make inference about the person carrying the phone, the ambient environments, and even the social events. SoundSense is a sound processing application that automatically classifies the sound into several categories of interest (e.g., voice and music) (Lu et al., 2009). SoundSense is also capable of learning ambient sounds considered significant (e.g., vehicle starting sound and vacuum cleaner sound). To this end, it incrementally builds probabilistic models of encountered sounds. The detected sounds are then ranked based on importance and are given to the user to perform human-in-the-loop labeling. This ambient sound logs can be used as a diary or can be used to collect/share the sound information across the city to understand the noise level of an urban environment.

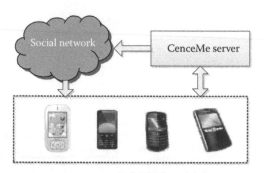

FIGURE 18.6   CenceMe overview.

18.4.1.3   Sensing and Social Networking

Given that mobile phones are equipped with sensors such as GPS and accelerometer, they can be used to create mobile sensor networks to sense information important to people. One example would be using mobile phones to automatically detect/publish the current locations or activities of users. CenceMe is a mobile platform that tracks users (e.g., location and activity) and shares the user information over social networks (Miluzzo et al., 2008). CenceMe's Click Status application displays the status of CenceMe buddies from Facebook (see Figure 18.6). For privacy management, CenceMe provides a privacy setting GUI through which users can enable/disable sensors used for tracking.

CenceMe uses various smartphone sensors to extract user information. Simple detections are performed in the phones, for example, classifying whether the incoming sound is human voice or environmental noise and classifying acceleration data to activities (e.g., sitting, standing, walking, and running). For sophisticated analysis (e.g., conversation, social context, and mobility mode), the relevant sensor data are transmitted to the back-end servers in which the classification operations are performed.

18.4.2   Vehicular Sensing–Based Knowledge Services

18.4.2.1   Road Surface Monitoring

Municipalities around the world spend millions of dollars to keep roads in good ride quality. Ride quality is mainly measured by the pavement roughness of a road surface that is an expression of the surface irregularity. The roughness causes vertical vibration of vehicles and thus reducing ride quality. There are several main sources of roughness: steps, dips, humps, bumps, and defects (crack, potholes). It also causes the stresses and fatigue damages of both vehicle and road structures. Municipalities have been profiling roads using noncontact profiling devices mounted on the vehicles that use GPS and accelerometer/laser sensors. This approach is quite expensive due to the operation costs and limited availability of specialized profiling devices.

Researchers have recently considered using less expensive commodity sensors; for example, smartphones are good sensing platforms as demonstrated by Mohan et al. (2008). Eriksson et al. (2008) developed a pothole detection algorithm that uses

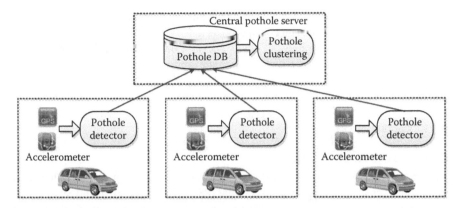

FIGURE 18.7   Pothole patrol system overview.

off-the-shelf GPS and 3-axis accelerometer sensors. For pothole detection, the acceleration data needs to be carefully examined because potholes must be differentiated from other road anomalies (e.g., railroad crossings and expansion joints). The detection algorithm mainly uses the X-Z ratio since potholes influence the acceleration values only on one side of the vehicles (as opposed to both sides in railroad crossings and expansion joints). It also removes anomalies caused by turning, veering, braking, and also samples collected at high speed. As shown in Figure 18.7, a mobile node runs the detection algorithm and returns results to the central server. For robust detection, the server clusters at least k events in the same location. The robustness can be further improved when multiple vehicles participate in the monitoring project.

18.4.2.2   Traffic Monitoring Systems

Most traffic information systems currently analyze data from closed-circuit cameras and sensors installed on the roads to estimate traffic conditions. This information is then shared via radio broadcasts, traffic report on the Internet (e.g., Google traffic), and en route traffic displays and signals. The coverage of those systems, however, is extremely limited (e.g., mainly highways) due to high installation and maintenance costs. To overcome such limitations, researchers have proposed to use vehicle as sensors to collect GPS measurements. This mobile sensor approach greatly extends coverage, thus enabling *street-level traffic flow estimation*. Yoon et al. (2007) systematically showed that various features can be used for accurate traffic flow estimation, for example, average speed, speed distribution, speed changes, and time duration/distribution.

For data collection and sharing, we can consider various wireless connectivity such as roadside Wi-Fi APs, vehicle-to-vehicle communications, and cellular communications (2G/3G). CarTel (Hull et al., 2006) uses roadside Wi-Fi APs, and TrafficView (Dikaiakos et al., 2005) and MobEyes (Lee et al., 2006) use car-to-car communications. Also, cellular communications have been recently considered: TruTraffic by Dash Express's SatNav system and smartphone-based system like VTrack. In particular, TruTraffic collects GPS measurement data from SatNav users using 2G/3G and provide a real-time traffic information service such as dynamic route guidance and congestion notification. If smartphones are used, energy consumption and GPS errors must be considered as in VTrack (Thiagarajan et al., 2009). Adaptive sampling is preferable as GPS is energy hungry. Due to inferior antenna and urban canyon, GPS is not always accurate and even may be unavailable. In this case, alternative location services must be employed (e.g., Wi-Fi and cellular localization). VTrack solves these challenges by employing a novel map matching algorithm that greatly reduces the required number of samples and yet significantly lowers location tracking errors.

18.4.2.3   Fuel-Efficient Driving

Fuel-efficient driving can be realized using participatory sensing. We introduce two participatory sensing systems, namely, GreenGPS (Ganti et al., 2010) and SignalGuru (Koukoumidis et al., 2011). GreenGPS recommends the most fuel-efficient route to the drivers. SignalGuru predicts the traffic signal schedule as in a Green Light Optimal Speed Advisory (GLOSA) system; when heading toward a signalized intersection, drivers can find the optimal speed that minimizes the waiting time at the intersection (thus, saving fuel).

GreenGPS uses grassroots collaboration among users who have onboard OBD-II sensors that are capable of retrieving detailed vehicle statistics in real time (e.g., RPM, fuel consumption, and speed). The fuel consumption depends on multiple parameters such as year, model, and make of a vehicle. However, when the number of participants is small, it is challenging to accurately model fuel consumption particularly (requiring a large number of samples). Instead, GreenGPS builds a generalized prediction model with model clustering, that is, grouping similar items, say the same make and model. The resulting model takes the input parameters such as stop sign count, traffic light count, distance traveled, average speed, and vehicle weight. The experimental results show that an average 10% fuel saving can be achieved with GreenGPS.

SignalGuru also relies on information sharing among drivers. Unlike traditional GLOSA systems where dedicated devices are installed at the intersection (e.g., counting down unit or digital sand watch), SignalGuru is infrastructureless as it uses commodity cameras in smartphones to detect signal changes. The detection module periodically processes the video frames to detect signal changes (e.g., green to yellow to red). The detected information is then disseminated via vehicle-to-vehicle communications. Once enough information is collected, the schedule prediction can be performed. Predicting the schedule of pretimed traffic signals (with fixed schedules) is straightforward. SignalGuru also supports the schedule prediction of traffic-adaptive signals where the schedule dynamically changes on the basis of the current vehicle density (sensed via loop detectors on the roads). The experimental results show that SignalGuru can accurately predict the schedule (less than several seconds of an error), and an average of 20.3% fuel saving can be achieved.

18.4.2.4   Urban Surveillance

Vehicular sensing can be used for proactive urban monitoring services (Lee et al., 2006) where vehicles continuously sense events from urban streets; maintain sensed data in their local storage; autonomously process them, for example, recognizing license plates; and possibly route messages to vehicles in their vicinity to achieve a common goal, for example, to permit police agents to track the movements of specified cars. Vehicular sensing could be an excellent complement to the deployment of fixed cameras/sensors. This completely distributed and opportunistic cooperation among sensor-equipped vehicles has the "deterrent" effect of making it harder for potential attackers to disable surveillance. Another less sensational but relevant example is the need to track the movements of a car, used for a bank robbery, in order to identify thieves, say. It is highly probable that some vehicles have spotted the unusual behavior of thieves' car in the hours before the robbery and might be able to identify the threat by "opportunistic" correlation of their data with other vehicles in the neighborhood. It would be much more difficult for the police to extract that information from the massive number of multimedia streams recorded by fixed cameras.

18.5 UBIQUITOUS CROWDSOURCING FOR KNOWLEDGE SERVICES

The term crowdsourcing was first coined by Jeff Howe in a Wired Magazine article (derived from outsourcing). Crowdsourcing is the act of taking a job traditionally performed by an employee and outsourcing it to the public, hoping to exploit the spare processing power of millions of human brains in the distributed labor network over the Internet (i.e., any people behind their computers who are willing to contribute to the project). Examples of crowdsourcing projects include Wikipedia, social Q&A (e.g., Yahoo! Answer), and Innocentives.* The required human skills of crowdsourcing range from simple (e.g., image labeling and transcribing) to complex (e.g., research and development). Note that simple tasks that human perform better than machines (e.g., image labeling and transcribing) are often called "human computation tasks." The motivation behind crowdsourcing participation largely depends on the types of systems, e.g., monetary rewards in Amazon's M-Turk,† a mechanized labor marketplace, altruism in Wikipedia, and fun in human computation games. Readers can find more detailed information about human computation in the work by Quinn and Bederson (2011).

In recent years, the availability of networked labor forces has significantly improved with SDs due to always-on wireless connectivity. Further, a variety of sensors in SDs (e.g., cameras and accelerometers) enable a new type of crowdsourcing, namely, sensing tasks (e.g., taking photos from streets), which we call ubiquitous crowdsourcing. Sensing activities can be largely classified into two categories: (1) passive sensing task: data are automatically collected on the background without any user intervention and (2) active sensing task: a user needs to perform physical activity for sensing (e.g., taking photos). Participatory sensing that we reviewed in Section 18.3 belongs to this category. An alternative to using sensors is to use a human's five senses and cognitive ability (inference); for example, a flooding/earthquake can be easily detected. Soliciting participation in ubiquitous crowdsourcing can be done in a number of ways, for example, people with the same interest or from the same community as in

---

* http://www.innocentive.com/
† https://www.mturk.com/mturk/welcome

participatory sensing, mechanized labor forces as in Amazon's M-Turk, and people from the social networks as in Twitter. In this section, we illustrate several ubiquitous crowdsourcing applications by reviewing existing systems that utilize mechanized labor forces and social networks.

18.5.1 Location-Based Ubiquitous Crowdsourcing

Location-based tasks (e.g., taking photos, price/inventory checks, and verifying map information) are one of the key services of ubiquitous crowdsourcing, and several online services include mCrowd,* Field Agent,† and Gigwalk.‡ Microsoft Bing has been collecting photos using Gigwalk for panoramic 3-D photosynthesis of businesses and restaurants in Bing Map. mCrowd developed by University of Massachusetts Amherst supports a feature of interfacing with existing labor marketplaces such as Amazon's M-Turk and ChaCha in which location agnostic tasks can be effectively handled.

Given that location-based ubiquitous crowdsourcing is still in its early stage, understanding the UXs is very important. A recent study on the UXs reports the following observations (Alt et al., 2010). People prefer address-based task selection (or map-based visualization). Among all the tasks, photo-taking tasks were most popular mainly due to lower cognitive overhead. People are not willing to make significant efforts on performing tasks as most completed tasks are close to their usual places (e.g., home and work). When searching for tasks, people wish to view the tasks close to their current location. Tasks are usually done after work or during breaks, and the response time varies widely (several minutes to one day or longer). Voluntary tasks (with no monetary rewards) have much lower chances of completion rate as opposed to tasks with monetary rewards.

18.5.2 Human Intelligence and Social Networking

We review ubiquitous knowledge services that exploit crowdsourced sensing and collaboration over social networking tools such as Twitter and Foursquare. Several application services have been proposed by the researchers at University of Buffalo, for example, weather information collection and noise mapping (Dermirbas et al., 2010) and location-based query resolution (Bulut et al., 2011).

Given that the current mobile platforms lack an infrastructure that assists mobile users to perform collaboration and coordination ubiquitously and permits searching/ aggregating the data published by mobile users, Dermirbas et al. (2010) proposed a crowdsourcing architecture based on Twitter, an open social networking system that allows users to share short messages (140 character limit). The key components include Sensweet and Askweet. Sensweet is a smartphone application that uses the smartphone's ability to work in the background for sensing ambient environments without distracting a mobile user. Sensweet also defines a markup language called

---

* http://crowd.cs.umass.edu/
† http://www.fieldagent.net/
‡ http://gigwalk.com/

TweetML to standardize the sensor data format in Twitter. Askweet is a server program that monitors a Twitter account (dedicated for a specific service) for questions, processes the questions, and aggregates the replies to answer the questions. It first attempts to answer the queries using the data available on Twitter (including the data published by Sensweets). If that fails, Askweet finds experts on Twitter to which the queries are forwarded. Dermirbas et al. (2010) demonstrated two application services: crowdsourced weather radar and noise mapping. In the crowdsourced weather radar service, random Twitter users are probed with a question about the current weather; in the noise mapping service, a Sensweet mobile application measures the noise level using the microphone and reports the results via Twitter. In their experiment of collecting responses of NYC weather, they found that weather probing takes reasonable delay (e.g., more than 50% of answers have less than 30min), and Twitter users are more responsive during afternoon and evening.

This work was extended to location-based queries by incorporating the location-based social networking services, e.g., Foursquare (Fatih et al., 2011). Search queries typically fall into two categories: factual, for example, "hotels in Miami," and nonfactual, for example, "cheap, good hotel in Miami." In the case of nonfactual queries, it is well known that traditional search engines perform poorly. Noting that a significant fraction of location-based queries is nonfactual, Fatih, Yilmaz, and Dermirbas extended the aforementioned system by considering the location information of users (from location-based social networking services such as Foursquare) when selecting people to ask questions. Their experimental results showed that the user participation significantly improved when location information was used; e.g., the median response delay was 13min.

Instead of asking Twitter users to perform human computation tasks, Sakaki et al. (2010) showed that Twitter messages posted by users can be automatically analyzed to detect natural events such as earthquakes and typhoons. They proposed a machine learning–based approach using support vector machines for event detection. For earthquake detection, features of interest include statistics (e.g., tweet message length), key words (e.g., shaking, earthquake), and word context (e.g., location, time). Like real sensors, semantic analysis may result in errors. For accurate detection, time-series data are further analyzed to build spatial/temporal models; for example, the frequency of earthquake-related messages exponentially decreases. In the experiments, they analyzed a total of 49,314 tweets over the 1-year period from August 2009 to September 2009 (Sakaki et al., 2010). When the results were compared with the Japan Meteorological Agency (JMA) data, they were able to detect 96% of earthquakes that were stronger than scale 3 or more during the period.

18.6 BUILDING UBIQUITOUS KNOWLEDGE SERVICE PLATFORMS

Accessing remote sensor networks (e.g., sensor data retrieval) is one of the critical requirements of ubiquitous knowledge services. If there is only a single sensor network, we can assign a single node as a sink (or a base station [BS] or gateway) for data collection. When there are multiple wireless sensor networks that are geographically distributed, diverse system architectures for data collection are feasible. A simple approach would be using a centralized server over the Internet. An alternative is to maintain a hierarchical structure (e.g., local servers maintain all the data generated in the local area, and a central server keeps the index for global data retrieval) or to use peer-to-peer storage (e.g., distributed hash tables maintain all the sensor data).

So far, various Internet-based systems were proposed for these purposes, for example, a simple two-tier structure (local vs. back-end servers) such as ArchRock (Woo, 2006), SensorBase (Reddy et al., 2007), and SensorMap (Nath et al., 2007); a semihierarchical structure such as IrisNet (Gibbons et al., 2003) and Global Sensor Networks (GSN) (Aberer et al., 2006); and a fully distributed peer-to-peer structure such as GeoServ (Ahn et al., 2011). In ArchRock and SensorBase, sensor data from a sensor network are aggregated at the local gateway and are published to the front-end server through which users can share the data. In SensorBase, back-end servers (called republishers) further process sensor data to enable sensor data searching. SensorMap is a web portal service that provides mechanisms to archive and index data, process queries, and aggregate and present results on geo-centric web interfaces such as Microsoft Virtual Earth. In IrisNet, each organization maintains database servers for its own sensors, and a global naming service is provided for information access; a similar approach is used in GSN to allow users to query local and remote sensor data sources. GeoServ, a fully distributed peer-to-peer sensor networking platform based on a distributed hash table, provides a location-aware sensor data retrieval scheme for geographic range queries and a location-aware publish-subscribe scheme for multicast routing over a group of subscribed users (Ahn et al., 2011).

Along with these research prototypes, there have been continued efforts on standardization of technologies such as Sensor web enablement (SWE), SSW, M2M communications, and EPCglobal network. Also, several related research projects were recently conducted in Europe such as BRIDGE (building radio frequency identification for the global environment) (BRIDGE, 2009) and SENSEI (2010). These standards and research systems feature various functions ranging from simple sensor data collection to internetworking across multiple domains to semantic sensor data analysis. In this section, we review these activities and provide a detailed comparison of key features. Further, we present our research proposal called the KCS.

18.6.1   M2M

In the past, the communication system mostly considered the scenarios with human interactions, namely, human to human (H2H) such as voice communications and human to machine (H2M) such as remote service access including web browsing. Recently, the industry has been also looking at alternative communication scenario where human interactions are absent, which is called M2M communication or machine type communication (MTC). A service optimized for M2M differs from a service optimized for H2H communications in that it has different market scenarios with a potentially very large number of low-cost sensor nodes with little data traffic per node (3GPP, 2010a,b). Service areas of M2M include security, tracking/tracing, payment, health, remote maintenance control, metering, and consumer devices. For example, in the smart grid area, autonomous and smart monitoring with M2M communications enables utility providers to remotely sense and control their utility equipment, thus obviating the need of laborious manual intervention. Many analysts expect that the M2M market will increase drastically in the near future.

Many groups have actively involved in M2M standardization, e.g., IEEE 802.16p (WiMAX), IEEE802.11 (Wi-Fi), 802.15.4 (ZigBee), GSMA, WiMAX Forum, WFA, OMA, TIA, and CCSA NITS. Here, we overview the proposals of leading standardization groups, namely, 3GPP and ETSI. 3GPP has considered M2M as one of the most promising applications to support in the LTE-Advanced standards (3GPP, 2010b). For successful adoption, 3GPP proposed to improve their cellular network systems as follows (3GPP, 2011): (1) architectural enhancements to support a large number of MTC devices in the networks, (2) architectural enhancements to fulfill MTC service requirements, and (3) support combinations of architectural enhancements for MTC. While 3GPP focused on communication networks, ETSI aims at standardizing more generic service architecture and requirements of M2M (ETSI, 2010). It claimed the following clauses: (1) general requirements describing communication features necessary for the correct establishment of M2M communications; (2) management related to malfunction detection, configuration, and accounting; (3) functional requirements for M2M services such as data collection/reporting and remote control operations; (4) security for M2M device such as authentication, data integrity, and privacy; and (5) naming, numbering, and addressing intended to M2M devices. ETSI also provided the overall end-to-end M2M functional architecture, including the identification of the service entities, the related interfaces, and the relations with the underlying existing networks (ETSI, 2011).

The following are the core system components of ETSI proposals (see Figure 18.8): (1) M2M devices and local area networks, (2) M2M gateway, (3) M2M network core, and (4) M2M applications. M2M applications interact with the service capabilities in the M2M network core to acquire the sensor data of interest. Actual sensor data will be collected via the M2M area network through M2M gateways (or sink nodes). Any local service capabilities in the M2M gateways are registered at the network core for centralized access and control of service capabilities in the M2M network.

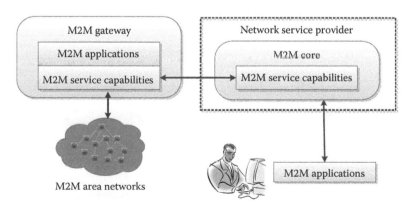

FIGURE 18.8 M2M architecture overview.

18.6.2 SENSOR WEB ENABLEMENT AND SEMANTIC SENSOR WEB

SWE is an initiative of the Open Geospatial Consortium (OGC) to add real-time sensing to the Internet and the web, by building an open standard framework for exploiting web-connected sensors and sensor systems, e.g., flood gauges and web-cams (Botts et al., 2008). SWE has been focused on developing web standards (based on XML) to enable (1) discovery, (2) access, (3) tasking, and (4) alert notification. SWE's standards can be classified into two categories: (1) encoding standards: how to describe sensors and sensor data (i.e., SensorML [sensor model language] and O&M [observations and measurement schema]) and (2) web service standards: how to provide sensor web services (i.e., SOS [sensor observations service], SPS [sensor planning service], and SAS [sensor alert service]). SensorML describes the standard models and XML schema for describing the detailed information about sensors and their processes (e.g., functional models and process chains), including the information needed for sensor discovery. O&M defines the standard models and XML schema for encoding sensor data (e.g., units and features). SOS provides a standard web service interface for querying and retrieving sensor data and sensor system information. SPS is a standard web service interface for configuring the sensors (and testing sensing feasibility), and SAS is a standard web service interface for publishing and subscribing to alerts from sensors. As shown in Figure 18.9, for a given set of sensors, the service catalogs (i.e., the registries of the sensor system) are built based on SensorML and O&M, and the resulting catalogs are used in SPS and SOS (e.g., discovering services, sensors, providers, and sensor data). Once the sensing system has initialized, a user can make a collection request to SPS or directly to the sensors (via SOS).

The SSW is a framework for providing "enhanced meaning to sensor data" to improve situation awareness by augmenting standard sensor languages of the SWE with semantic annotations through which applications can understand and reason over sensor data consistently, coherently, and accurately (Sheth et al., 2008). SSW is based on the semantic web, an evolving extension of the web where semantics

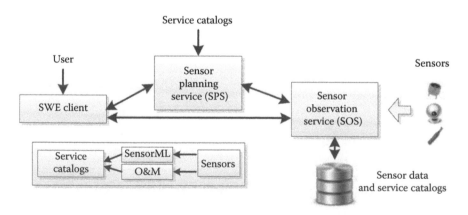

FIGURE 18.9   SWE overview.

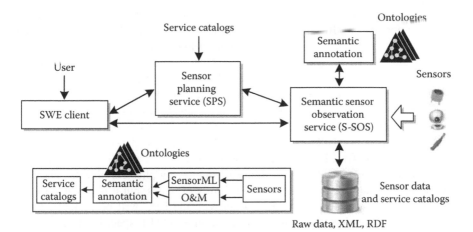

FIGURE 18.10 SSW overview.

(meaning) of information is captured using ontologies, enabling machines to inter-pret and relate data. As shown in Figure 18.10, the key components of the seman-tic web include the Resource Description Framework (RDF) data representation model and the ontology representation languages, namely, RDF schema and web ontology language (OWL). For semantic annotation, SSW uses spatial, temporal, and thematic ontologies along with domain-specific ontologies for representing sensors (e.g., biomedical ontology). Here, annotation spans from raw sensor data to extracted features and detected entities from sensor data. By using semantic web tools, semantic reasoning and computation over semantically annotated data can be performed, including semantic-enabled search, integration, answering complex queries, pattern finding, discovery, visualization, etc. For instance, if a group of sensors provides information about temperature and precipitation, possible road condition can be specified using simple rules (say, an icy road if temperature is below 0°C and it is raining).

18.6.3 INTERNET OF THINGS

The phrase Internet of things (IoT) was coined by the Auto-ID Center with the vision of building a global network called the EPC network that enables tracking items (e.g., using RFID tags and readers) and sharing information over the Internet. Due to recent advances of RFIDs, sensor networks, near field communications (NFC), and SDs, the IoT has been widely accepted as an enabler of ubiquitous sensor networking over the Internet, that is, remotely monitoring the physical world using sensors (or associated services) and acting upon changes in the real world (Santucci, 2010). In fact, such real-time sensing and actuation call for open, scalable, secure infrastruc-ture standards. Some of the recent efforts include Auto-ID's EPC network (and EPC sensor networks), OCG's SWE, ISO/IEC standards, and IEEE 1451 standards (Song and Lee, 2008).

The EPC network basically considers small, inexpensive, high-performance RFID tags and readers. Its goal is to provide the ability to provide item tracking and information sharing to the current supply chain architecture. In the EPC network, all items are uniquely identified using EPC. Information related to each item is stored within a network of databases with each touch point within the supply chain collating information specific to that product. The EPC network also defines a common communication language and standard interfaces, thus guaranteeing interoperability between vendors. As shown in Figure 18.11, the EPC network is composed of the following components: EPC, ID system (RFID tags and readers), EPC application level event (EPCALE), EPC information services (EPCIS), and discovery services. EPC is an identification scheme for uniquely identifying physical objects. The current coding schemes are based on the GS1 family of codes, but it is possible to use any existing coding schemes. An RFID tag with the EPC will be attached to an object, and RFID readers can read out information. EPCALE specifies how the captured RFID data are filtered and grouped to some events. After passing through the EPCALE, the captured RFID data will be stored into the EPCIS's repository. The querying interface of EPCIS allows external applications to access the RFID data. Discovery services enable global access to data associated with a specific EPC (e.g., between different enterprises).

While the EPC network considers passive RFID tags, it can be also extended to include smart tags with communication and processing capabilities (forming an EPC

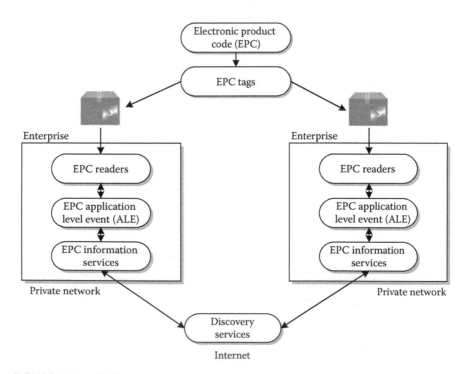

FIGURE 18.11 EPC network overview.

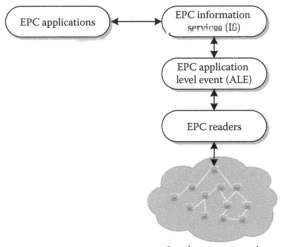

Local sensor network

FIGURE 18.12   EPC sensor network.

sensor network) (Sung et al., 2007). As shown in Figure 18.12, a group of smart tags (or a local sensor network) is connected to the back-end system via a reader (called a BS). It uses IEEE 1451 TEDS (Transducer Electronic Data Sheet) standards for sensor and sensor data description. Various air interfaces are used for communications among smart tags (e.g., Bluetooth, ZigBee, and NFC), but it does not specify sensor networking protocols in a local sensor network. Unlike the original EPC network, the functions of EPCALE are extended to support more complex operations on sensor data (e.g., data processing such as averaging).

Like the EPC sensor network proposal, the BRIDGE project aimed at supporting ambient sensors and sensor-enabled active RFID tags in the EPCglobal network for supply chain monitoring (BRIDGE, 2009). Given that existing RFID-related standards and proposals cannot efficiently support sensor data, BRIDGE incorporates OGC's SWE into the EPC network. These standards share the same functionalities; for example, EPC network's EPCIS/ALE corresponds to SWE's SOS and Discovery Service (ONS) to Catalog. Yet SWE lacks the concept of BS, reader management, and globally unique address, while EPC Network does not provide sensing task planning. In BRIDGE, SWE is extended to include the BS and to interwork with the EPC network, which causes minimal modification to the EPC network standards.

The SENSEI project models wireless sensor and actuator networks into resources and proposes to network "resources" over the Internet (SENSEI, 2010). For instance, a sensor node in the vehicle can provide one or more resources (e.g., engine and speed sensors). SENSEI offers a rendezvous service to allow locating resources of interest (called a resource directory). Moreover, it supports components that enable contextualized information retrieval and interaction, which requires an entity directory. Sensor data are represented based on the O&M standards and can be processed to derive entity-based contextual information. SENSEI is composed of the following components: resource directory, entity directory, execution manager, and semantic query resolver. A user's query is routed to the semantic query resolver, which then access the entity/resource directories to make execution plans. The resulting plans are then submitted to the execution manager, and the final results will be delivered to the user.

18.6.4 FEATURE COMPARISON

The key features of knowledge integration and retrieval systems can be summarized as follows: sensor model and encoding, local area sensor networking, application service support, dynamic multidomain service networking, semantic query processing, service-device/network interaction, human computation/sensing, social networking, and agent support. The feature comparison results can be found in Table 18.1. In SWE, sensor model and encoding are supported via O&M and SensorML, and application services are supported via SOS and SPS. The EPC network was designed to support RFID tags and related services, including multidomain service networking (e.g., different enterprises). The EPC sensor network and BRIDGE extend the EPC

TABLE 18.1
Feature Comparison

	SWE	SSW	EPC-NET	EPC-SN	BRIDGE	SENSEI	M2M ESTI	KCS
RFID support			o	o	o	o	o	o
Sensor model and encoding	o	o		o	o	o	o	o
Local area sensor networks				o	o	o	o	o
Service support	o	o	o	o	o	o	o	o
Dynamic multidomain service networking			o	o	o	o		o
Semantic query processing		o				o		o
Service-device/network interaction								o
Human computation (and sensing)								o
Social networking								o
Agent support								o

network to support active sensors. SENSEI provides "resource" (or service)-based networking and semantic query processing. ESTI's M2M standards basically provide similar features as the EPC sensor network and BRIDGE, but the current M2M standards focused on single domain scenarios.

From our review on SDs and related mobile/people sensing applications, including ubiquitous crowdsourcing, we find the following observations: (1) humans are part of the systems (human computation), (2) existing mobile sensing services are usually optimized due to resource constraints (requiring service-device/network interactions), and (3) when human resources are utilized for knowledge services, exploiting social networks is crucial. Another relevant technique is agent technologies that can automatically collect user information and infer a user's wants and needs to provide personalized services. In the following, we briefly introduce our proposal called the KCS that provides these features by interworking with sensors, SDs, and people.

18.6.5 TOWARD KNOWLEDGE COMMUNICATION SERVICES

The goal of KCS is to perceive the intention of a user and provide the customized knowledge services to the user at any place and any time, with any device even with physical restrictions. To attain high-quality UXs, exploiting device/network, application, and human intelligence is necessary. We envision new system architecture for such ubiquitous knowledge services with the following components (see Figure 18.13): knowledge server (KS), knowledge gateway (KG), and SD. The desired features of the KS include knowledge brokering, virtual user knowledge agents, semantic query processing, and knowledge repository. Knowledge brokering enables content exchanges (e.g., data, information, and knowledge) across KSs in different network/domains (e.g., Verizon, AT&T, and KT) and across various content providers (e.g., Google, Yahoo!, and Naver). The collected data will be stored in a KS's knowledge repository for later access. Virtual user knowledge agents and semantic query processors will permit personalized knowledge services; that is, the user agent will proactively search for content with help of the semantic query processors that gather and analyze all the available data distributed over the network. A KS interacts with KGs in the same domain. A KG has a lightweight KS (e.g., with knowledge brokering, agents, and caching) and supports internetworking with legacy wireless sensor networks including M2M and RFID tags. SDs are equipped with a user knowledge agent, data/information gathering units, and adaptive/flexible user interfaces for efficient human-computer interactions. User knowledge agents basically use the services in a local KG and central KSs for knowledge acquisition and processing. Further, they can use direct device-to-device communications among SDs (including with KGs) to acquire resources from nearby nodes (e.g., data, information, knowledge, and computing power). In this way, the agent can collect and process a user's context information to accurately predict the intension and to provide personalized knowledge services.

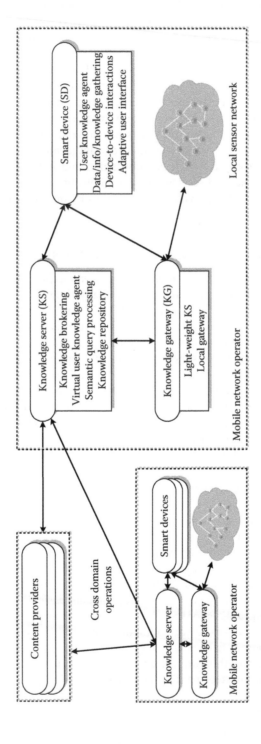

FIGURE 18.13 KCS overview.

18.7 CONCLUSION

We introduced the concept of ubiquitous knowledge services that seamlessly integrate content (e.g., data, information, and knowledge) from various sources (e.g., content providers, sensors, SDs, and humans) at large scales and derive new values for end users. In this chapter, we began with reviewing the key enabling technologies of ubiquitous knowledge services in the areas of device, network, application, and human intelligences. We then reviewed and compared the existing platforms for sensor data collection, retrieval, and application services. Finally, we proposed a set of desirable features of the ubiquitous knowledge service platforms based on our vision of the KCS.

REFERENCES

3GPP. Evolved universal terrestrial radio access (E-UTRA) and evolved universal terrestrial radio access network (E-UTRAN), TS 36.300 V10.0.0, 2010a.

3GPP. Service requirements for machine-type communications; Stage 1, TS22.368 v.10.1.0, 2010b.

3GPP. System improvements for machine-type-communications, TS23.888 v.1.4.0, 2011.

Aberer, K., M. Hauswirth, and A. Salehi. A middleware for fast and flexible sensor network deployment, in *Proceedings of the 32nd International Conference on Very large Data Bases (VLDB)*, Seoul, South Korea, September 12–15, 2006.

Ahn, J. H., U. Lee, and H. J. Moon. GeoServ: A distributed urban sensing platform, in *IEEE/ ACM International Symposium on Cluster, Cloud, and Grid Computing (CCGrid'11)*, Newport Beach, CA, May 23–26, 2011.

Alt, F., A. S. Shirazi, A. Schmidt, U. Kramer, and Z. Nawaz. Location-based crowdsourcing: Extending crowdsourcing to the real world, in *Proceedings of the 6th Nordic Conference on Human-Computer Interaction (NordiCHI): Extending Boundaries*, New York, 2010.

Botts, M., G. Percivall, C. Reed, and J. Davidson. OGC® sensor web enablement: Overview and high level architecture, in F. Fiedrich and B. Van de Walle (eds.) *Proceedings of the Fifth International ISCRAM Conference*, Washington, DC, May 2008.

Bowman, P., J. Ng, M. Harrison, T. M. López, and A. Illic. Sensor based condition monitoring, in *Building Radio Frequency Identification for the Global Environment*, 2009.

BRIDGE. Sensor based condition monitoring, BRIDGE Consortium, 2009.

Bulut, M. F., Y. S. Yilmaz, and M. Demirbas. Crowdsourcing location-based queries, in *IEEE Conference on Pervasive Computing and communication Workshops (PercomW'11)*, pp. 513–518, 2011.

Burkhard, R. and M. Meier. Tube map: Evaluation of a visual metaphor for interfunctional communication of complex projects, in *International Conference on Knowledge Management and Knowledge Technologies (I-KNOW)*, 2004.

Burkhard, R. Learning from architects: The difference between knowledge visualization and information visualization, in *International Conference on Information Visualization*, Washington, DC, 2004.

Demirbas, M., M. A. Bayir, C. G. Akcora, and Y. S. Yilmaz. Crowd-sourced sensing and collaboration using twitter, in *Proceedings of the 2010 IEEE International Symposium on A World of Wireless, Mobile and Multimedia Networks (WOWMOM)*, Buffalo, NY, November 24, 2010.

Dikaiakos, M. D., S. Iqbal, T. Nadeem, and L. Iftode. VITP: An information transfer protocol for vehicular computing, in *Proceedings of the 2nd ACM international workshop on Vehicular ad hoc networks* (*VANET'05*), New York, 2005.

Douglis, F., R. Caceres, B. Marsh, F. Kaashoek, K. Li, and J. Tauber. Storage alternatives for mobile computers, in *Proceedings of the First USENIX Conference on Operating Systems Design and Implementation*, Berkeley, CA, 1994.

EPCglobal Inc. The EPCglobal architecture framework, 2009. http://www.gs1.org/gsmp/kc/epcglobal/architecture/architecture_1_4-framework-20101215.pdf

Eppler, M. and R. Burkhard. Knowledge visualization, in *Encyclopedia of Knowledge Management*, Idea Group, Hershey, PA, 2005, pp. 551–560.

Eriksson, J., L. Girod, B. Hull, R. Newton, H. Balakrishnan, and S. Madden. The pothole patrol: Using a mobile sensor network for road surface monitoring, in *the Sixth Annual International conference on Mobile Systems, Applications and Services* (*MobiSys'08*), Breckenridge, CO, 2008.

ETSI TS 102 689 V1.1.1. Machine-to-machine Communications (M2M); M2M service requirements, 2010.

ETSI TS 102 690 V0.13.3. Machine-to-machine Communications (M2M); Functional architecture, 2011.

Frenkiel, R., B. Badrinath, J. Borras, and R. D. Yates. The infostations challenge: Balancing cost and ubiquity in delivering wireless data, *IEEE Personal Communications*, 7–2 (2002): 66–71.

Ganti, R. K., N. Pham, H. Ahmadi, S. Nangia, and T. F. Abdelzaher. GreenGPS: A participatory sensing fuel-efficient maps application, in *Proceedings of the 8th International Conference on Mobile Systems, Applications, and Services* (*MobiSys'10*), New York, 2010.

Gibbons, P. B., B. Karp, Y. Ke, S. Nath, and S. Seshan. IrisNet: An architecture for a worldwide sensor web, *IEEE Pervasive Computing*, 2 (2003): 22–33.

Hull, B., V. Bychkovsky, K. Chen, M. Goraczko, A. Miu, E. Shih, Y. Zhang, H. Balakrishnan, and S. Madden. CarTel: A distributed mobile sensor computing system, in *Proceedings of the 4th International Conference on Embedded Networked Sensor Systems* (*SenSys'06*), New York, 2006.

Jiang, D., V. Taliwal, A. Meier, W. Holfelder, and R. Herrtwich. Design of 5.9 GHz DSRC-based vehicular safety communication, *IEEE Wireless Communications* 13 (2006): 36–43.

Khatib, M. G., B.-J. van der Zwaag, P. H. Hartel, and G. J. M. Smit. Interposing flash between disk and DRAM to save energy for streaming workloads, in *IEE/ACM/IFIP Workshop on Embedded Systems for Real-Time Multimedia* (*ESTIMedia*), Salzburg, Austria, October 4–5, 2007.

Koukoumidis, E., L.-S. Peh, and M. Martonosi. SignalGuru: Leveraging mobile phones for collaborative traffic signal schedule advisory, in *Proceedings of the 9th International Conference on Mobile Systems, Applications* (*MobiSys'11*), New York, 2011.

Lee, U., E. Magistretti, B. Zhou, M. Gerla, P. Bellavista, and A. Corradi. MobEyes: Smart mobs for urban monitoring with vehicular sensor networks, *IEEE Wireless Communications*, 13 (2006): 51–57.

Lee, U., S. Jung, A. Chang, D.-K. Cho, and M. Gerla. P2P content distribution to mobile bluetooth users, *IEEE Transaction on Vehicular Technology*, 59 (2010): 356–367.

Lu, H., W. Pan, N. D. Lane, T. Choudhury, and A. T. Campbell. SoundSense: Scalable sound sensing for people-centric applications on mobile phones, in *Proceedings of the 7th International Conference on Mobile Systems, Applications, and Services* (*MobiSys'09*), New York, 2009.

Mell, P. and T. Grance. The NIST definition of cloud computing, National Institute of Science and Technology, Gaithersburg, MD, 2011.

Miluzzo, M., N. D. Lane, K. Fodor, R. A. Peterson, H. Lu, M. Musolesi, S. B. Eisenman, X. Zheng, and A. T. Campbell. Sensing meets mobile social networks. The design, implementation and evaluation of the CenceMe application, in *Proceedings of the 6th ACM Conference on Embedded Network Sensor Systems (SenSys'08)*, New York, 2008.

Mohan, P., V. Padmanabhan, and R. Ramjee. Nericell: Rich monitoring of road and traffic conditions using mobile smartphones, in *Proceedings of the 6th ACM Conference on Embedded Network Sensor Systems (SenSys'08)*, New York, 2008.

Mun, M., S. Reddy, K. Shilton, N. Yau, J. Burke, D. Estrin, M. H. Hansen, E. Howard, R. West, and P. Boda. PEIR, the personal environmental impact report, as a platform for participatory sensing systems research, in *Proceedings of the 7th International Conference on Mobile Systems, Applications, and Services (MobiSys'09)*, New York, 2009.

Nath, S., J. Liu, and F. Zhao. SensorMap for wide-area sensor webs, *IEEE Computer Magazine*, 40 (2007): 90–93.

Oviatt, S. L. Advances in robust multimodal interface design. *IEEE Computer Graphics and Applications*, 23 (2003): 52–68.

Oviatt, S. L. Multimodal interactive maps: Designing for human performance, *Human–Computer Interaction*, 12 (1997): 93–129.

Quinn, A. J. and B. B. Bederson. Human computation: A survey and taxonomy of a growing field, in *Proceedings of the 2011 annual conference on Human factors in computing systems (CHI'11)*, New York, 2011.

Reddy, S., G. Chen, B. Fulkerson, S. J. Kim, U. Park, N. Yau, J. Cho, M. Hansen, and J. Heidemann. Sensor-internet share and search—Enabling collaboration of citizen scientists, in *Proceedings of the ACM Workshop on Data Sharing and Interoperability (DSI) on the World-Wide Sensor Web*, Cambridge, MA, 2007.

Sakaki, T., M. Okazaki, and Y. Matsuo. Earthquake shakes twitter users: Real-time event detection by social sensors, in *Proceedings of the 19th International Conference on World Wide Web (WWW'10)*, New York, 2010.

Santucci, G. The internet of things: Between the revolution of the internet and the metamorphosis of objects, in *Vision and Challenges for Realising the Internet of Things (CERP-IoT)*, March 2010.

SENSEI. The SENSEI real world internet architecture, Sensei Consortium, 2010.

Sheth, A., C. Henson, and S. S. Sahoo. Semantic sensor web, *IEEE Internet Computing*, 8 (2008): 78–83.

Song, E. K. and K. Lee. Understanding IEEE 1451 networked smart transducer interface standard, *IEEE Instrumentation & Measurement Magazine*, 11 (2008): 11–17.

Sung, J., L. T. Sanchez, and D. Kim. The EPC sensor network for RFID and WSN integration infrastructure, in *Proceedings of the Fifth IEEE International Conference on Pervasive Computing and Communications Workshops (PerComW'07)*, White Plains, NY, March 19–23, 2007.

Thiagarajan, A., J. Biagioni, T. Gerlich, and J. Eriksson. Cooperative transit tracking using smart phones, in *Proceedings of the 8th ACM Conference on Embedded Networked Sensor Systems (SenSys'10)*, New York, 2010.

Thiagarajan, A., L. Ravindranath, K. LaCurts, S. Toledo, J. Eriksson, S. Madden, and H. Balakrishnan. VTrack: Accurate, energy-aware traffic delay estimation using mobile phones, in *Proceedings of the 7th ACM Conference on Embedded Networked Sensor Systems (SenSys'09)*, Berkeley, CA, 2009.

Tsiatsis, V., A. Gluhak, T. Bauge, F. Montagut, J. Bernat, M. Bauer, C. Villalonga, P. Barnaghi, and S. Krco. The SENSEI real world internet architecture, in *Towards the Future Internet—Emerging Trends from European Research*, IOS Press, Amsterdam, the Netherlands, 2010, pp. 247–256.

Weiser, M. The computer for the 21st century, *ACM SIGMOBILE Mobile Computing and Communications Review*, 3 (1999): 3–11.

Woo, A. A new embedded web services approach to wireless sensor networks, in *SenSys'06 Proceedings of the Fourth International Conference on Embedded Networked Sensor Systems*, New York, 2006.

Yoon, J., B. Noble, and M. Liu. Surface street traffic estimation, in *Proceedings of the 5th international conference on Mobile systems, applications and services (MobiSys'07)*, New York, 2007.

19 Knowledge-Intensive Cloud Services

Transforming the Cloud Delivery Stack

Michael P. Papazoglou and Luis M. Vaquero

CONTENTS

19.1 INTRODUCTION

Currently enterprises are faced with a continuing demand for business expansion, and, as a result, they often need to invest in stand-alone servers or software that traditionally demand heavy capital investment but remain frequently underutilized. Information technology (IT) infrastructures have become today too complex and brittle, and, as a result, enterprise IT is unable to keep pace with their needs as almost 70% of IT investment concentrates on maintenance, leaving little time to support strategic development projects. Many enterprises are usually left with a plethora of technologies and systems, with a suboptimal level of integration, creating significant barriers to developing innovative solutions that are required to provide a competitive edge. For such enterprises in search of a cost-effective use of computing resources, cloud computing technology is rapidly becoming a commercially viable computing service delivery model.

Much like the historical client-server era, the advent of cloud computing is fundamentally changing the way we consume computing resources. Cloud computing allows enterprises to share resources, software, and information across a rapidly growing multitude of connected devices, creating new opportunities for business speed and efficiency. It reduces IT complexity by leveraging the efficient pooling of on-demand, self-managed virtual infrastructure, consumed as a service. Cloud computing eliminates the need for large capital outlays to launch new applications, moving the decision out of the investment realm into the operational. With Cloud computing practices, service providers can expand their offerings to include the entire traditional IT stack ranging from foundational hardware and platforms to application components, software services, and entire software applications.

Cloud computing is an evolving term that describes a broad movement toward the deployment of network-based applications in a highly flexible, virtualized IT environment to enable interaction between IT service providers of many types and clients. Virtualization means a form of abstraction between the user and the physical resource in a way that preserves the user's impression that he or she is actually interacting directly with the physical resource. Actually, cloud resources and software services are delivered virtually, and although they may appear to be physical (servers, disks, network segments, web or REpresentational State Transfer (REST)ful services, etc.), they are actually virtual implementations of those on an underlying physical infrastructure, which the subscriber never sees. The essence is to decouple the delivery of computing services from their underlying technology. Beyond the user interface, the technology behind the cloud is opaque to the user, abstracting away from technology in order to make cloud computing user-friendly.

With cloud computing resources and services are delivered virtually, that is, although they may appear to be physical (servers, disks, network segments, etc.), they are actually virtual implementations of those on an underlying physical infrastructure, which the subscriber never sees. This enables service providers to move toward a fully virtualized service-oriented infrastructure, eliminating silo-based complexity, acquiring and moving resources to where they are most needed, improving performance, reducing costs, and speeding up the delivery of services. The objective is to create a virtualized service-oriented infrastructure for multiple integrated services of all types. Services here include infrastructure resources (i.e., computing power, storage, and machine provisioning), software resources (i.e., middleware and platform-based development resources), and application components (which include full-fledged applications as well as stand-alone business processes).

Over the past few years, we have experienced significant technological developments that enable cloud computing. These include far more reliable Internet services, with higher throughput and resilience coupled with virtualization techniques that enable computing facilities to be replicated and reproduced easily. One consequence of this move to virtualization is that it is no longer necessary to have in-house computing infrastructures. Instead, it is possible to shift computing and storage capabilities "into the cloud" where they offer economies of scale in terms of IT support, speed, and resource sharing. At the same time, cloud computing faces a number of challenges, which include improved security and privacy models and associated regulatory compliance considerations, interoperability, latency, performance, and reliability concerns, as well as managing the very flexibility that cloud provides.

One of the greatest challenge facing longer-term adoption of cloud computing services is the ability to automatically provision services, effectively manage workload segmentation and portability (i.e., seamless movement of workloads across many platforms and clouds), and manage virtual service instances, while optimizing use of the resources and accelerating the deployment of new services. Within such a cloud environment, it is also important to be able to work with both cloud-based and enterprise-based applications using a unified approach that can function across existing applications and multiple cloud providers. This includes the ability for opportunistic and scalable provisioning of application services, consistently achieving quality of service (QoS) targets under variable workload, resource, and network conditions. The overall goal is to create a computing environment that supports dynamic expansion or contraction of capabilities (virtual machines [VMs], application services, storage, and database) for handling sudden variations in service demands. To meet these pressing demands, it is essential to transform the fabric of the current inflexible service delivery models. To transform the cloud delivery stack requires reliance on what we call *knowledge-intensive cloud services*. As we explain later in this chapter, knowledge-intensive cloud services need to rely on a framework combining definition and manipulation languages.

This chapter aims to discuss the various cloud delivery model options and provide a perspective on how to improve cloud service delivery and guarantee efficient workload segmentation and resource utilization by relying on knowledge-intensive cloud services. The chapter presents vision, challenges, and architectural elements for achieving interoperability and support of dynamic expansion or contraction of capabilities and partitioning of workloads in cloud environments. We begin our discussion by providing a brief overview of cloud computing to improve readability and understanding.

19.2 CLOUD COMPUTING OVERVIEW

In order to understand the technological direction of cloud computing and evaluate differing cloud service delivery options, it is necessary to understand its operating principles as well as the distinct dimensions of the various cloud offerings.

Defining an evolving term such as cloud computing is a hard task. There are currently several definitions for this term most comprehensive of which the one given by the National Institute of Standards and Technology (NIST). NIST defines Cloud Computing as [Mell 2009]

> A consumption and on-demand delivery computing paradigm that enables convenient network access to a shared pool of configurable and often virtualized computing resources (e.g., networks, servers, storage, middleware and applications as services) that can be rapidly provisioned and released with minimal management effort or service provider interaction.

The aim at offering every element of the IT stack as a service implies that cloud providers can also offer platforms to deliver specialized services while enabling an industrialized and pervasive deliverance of these highly specialized services. Users of IT-related services can follow an Service Oriented Architecture (SOA)-like approach that lets them focus on what the services provide them with rather than how the services are implemented or hosted. SOA is an architecture pattern while the cloud can be viewed as a target deployment platform for that architecture pattern. In particular, cloud computing provides the ability to leverage on-demand new computing resources and platforms that SOA applications require, but an organization may not own. The cloud-SOA merger offers unprecedented control in allocating resources dynamically to meet the changing needs of service-based applications, which is only effective when the service level objectives at the application level guide the cloud's infrastructure/platform management layer [Rodero-Merino 2010, Papazoglou 2012].

Cloud computing can be viewed as a holistic ecosystem of a collection of services, applications, information, and infrastructure comprised of pools of computing, network, information, and storage resources. These components can be rapidly orchestrated, provisioned, implemented, and decommissioned, and scaled up or down thereby providing for an on-demand utility-like model of allocation and consumption.

The NIST definition states the cloud computing is composed of five essential characteristics, three cloud service delivery models, and four deployment models.

The most commonly accepted characteristics of cloud computing are briefly summarized as follows [Mell 2009]:

- *On-demand self-service*: Service providers offer clients the ability to provision computing capabilities, such as cloud servers, dedicated storage, hardware load balancers, an online authorization service, some geolocation data, a testing platform, etc., on demand. This way, applications can automatically build and scale across their whole life cycle without requiring human interaction with each provider of all the services the application is using.
- *Broad network access*: The cloud provider's resources are available over the network and can be accessed through standard mechanisms.

- *Location independent resource pooling*: Provider's resources are pooled and shared among several clients (multitenancy). The client generally has no control or knowledge over the exact location of the provided resources but may be able to specify location at a higher level of abstraction (e.g., country, region, or datacenter).
- *Rapid elasticity and provisioning*: Elasticity implies the ability to scale resources as needed by applications. To the client, the cloud appears to be infinite, and the client can purchase or lease capabilities available for provisioning in any quantity at any time.
- *Pay-per-use measured service*: Clients consume resources and pay only for resources that they use. To this effect, cloud computing platforms usually employ consumption-based billing mechanisms to charge for cloud service use.

Cloud computing is typically divided into three models of service delivery (commonly referred to as cloud stack layers). These three categories are software as a service (SaaS), platform as a service (PaaS), and infrastructure as a service (IaaS):

- IaaS is at the lowest layer and is a means of delivering very basic computing capability—machines with operating systems and storage as standardized services over the network. The IaaS layer allows the infrastructure provider to abstract away infrastructure-specific details such as which exact hardware an application is using and where the application is running. IaaS incorporates the capability to abstract resources as well as deliver physical and logical connectivity to those resources and provides a set of application programming interfaces (APIs), which allow interaction with the infrastructure by clients. Currently, the most prominent example of IaaS is Amazon Web Services (AWS), whose Elastic Compute Cloud (EC2) and Simple Storage Service (S3) products offer bare-bone computing and storage services, respectively.
- Platform provided as a service or "platform as a service" (PaaS) is interposed between IaaS and SaaS and refers to an environment where developers build and run an application by using whatever prebuilt components and interfaces that particular platform provides as a service to developers over the Internet. The PaaS provider facilitates deployment of cloud applications by entirely managing the cloud software development platform (often an application hosting, middleware environment) providing all of the facilities required to support the complete life cycle of building and delivering networked services. The PaaS consumer does not manage or control the underlying cloud infrastructure including network, servers, operating systems, or storage, but has control over the deployed applications and possibly application hosting environment configurations. One serious drawback with conventional PaaS offerings is that they are constrained by the capabilities that are available through the PaaS provider and do not allow easy extensibility or customization options. Currently, there exist several examples of PaaS, the most well known being Google App Engine (GAE) and Force.com.

- SaaS is an "on-demand" software application that is owned, delivered, and managed remotely by one or more software providers. This layer builds upon the underlying IaaS and PaaS and provides clients with integrated access to software applications. The provider delivers software based on a single set of common code and data definitions that are consumed in a one-to-many model. This layer provides a self-contained operating environment used to deliver the entire user experience including the content, its presentation, the application(s), and management capabilities. SaaS provides the most integrated functionality built directly into the offering with no option for consumer extensibility. Typical examples of the SaaS level are SalesForce. com that offers Customer Relationship Management (CRM) applications accessible by subscription over the web and the Amazon Fulfillment Web Service (Amazon FWS). The most important overlap between SOA and cloud computing occurs at this level of the cloud stack.

The cloud computing layers do not simply encapsulate on-demand resources at differing operational levels; they also help define a new application development model. Each layer of abstraction provides numerous possibilities for defining services that can be offered on a pay-per-use basis. These three levels support virtualization and management of differing levels of the computing solution stack.

The cloud model gives also the opportunity to mashup services from a variety of cloud providers to create what is known as a *cloud syndication*. Cloud syndications are essentially federations of cloud providers whose services are aggregated in a single pool supporting three basic interoperability features: resource migration, resource redundancy, and combination of complementary cloud resources and services [Kurz 2011]. Cloud syndications at the SaaS level are termed "BPaaS" (or business process as a service). These comprise a distinct integrative layer above the SaaS layer that reflects the focus on enterprise-specific services. BPaaS allows creating unique end-to-end business processes that are usually syndicated with other external services (possibly provided by diverse SaaS providers). BPaaS requires effective management of cloud resources. It is important to understand that when integrating end-to-end business processes from diverse providers, the resultant ecosystem is limited by the service quality of its weakest component in the stack.

The continuous pressure of IT departments and infrastructure providers to offer computing infrastructure at the lowest possible cost have made them realize that by effectively leveraging the concepts of resource pooling, virtualization, dynamic provisioning, utility, and commodity computing, they can create cost-effective cloud solutions. This has resulted in four major cloud deployment models [Armbrust 2009]:

- The first cloud deployment model where developers leverage applications that are available from and run on public facilities is collectively referred to as the "public cloud" model.
- The second cloud deployment model where developers develop their own "on-premise" or "internal" cloud solutions is collectively referred to as the "private cloud" model.

- The third and final cloud deployment model concerns developers who integrate and federate services across both the public and private cloud. This model is collectively referred to as the "hybrid cloud" model.
- The fourth and final cloud deployment model is the "community cloud." Here, the cloud infrastructure is shared by several organizations and supports a specific community that has shared concerns (e.g., mission, security requirements, policy, and compliance considerations). It may be managed by the organizations or a third party and may exist on premise or off premise.

To ensure guarantees from cloud service providers for service delivery, businesses using cloud services typically enter into service level agreements (SLAs) with the cloud service providers. Although SLAs vary between businesses and cloud service providers, they typically include the required/agreed service level through QoS parameters, the level of service availability, the indication of the security measures adopted by the cloud service provider, and the rates of the services.

19.3   CLOUD APIs

In a cloud architecture, modules communicate with each other over APIs, usually by means of web services or REST technologies. A cloud architecture model exposes a set of software interfaces or APIs that developers and customers use to manage and interact with cloud services. Provisioning, management, orchestration, and monitoring are all performed using these interfaces. Cloud architecture offers at least three broad kinds of APIs:

- *Cloud functionality interfaces*: These are programming interfaces through which developers and consumers interact with providers to request, deploy, reconfigure, administer, and use services. These types of interfaces include managing the security-related aspects of a cloud as well as managing and modifying instances of deployed services [DMTF 2009]. In addition, they include interfaces for image and infrastructure management. These APIs control details such as firewalls, node management, VM, and network management, as well as load balancing.
- *Data functionality interfaces*: These are the channels through which data flow in and out of the cloud. These APIs encompass various forms of cloud data sources, documents, web portals, maps, etc., and are used for reading, writing, updating, and querying data. For instance, most SaaS providers offer API calls to read (and thereby "export") data records. Data functionality APIs also include simple primitives for file storage services, document storage services, and so on.
- *Application cloud interfaces*: Application cloud APIs provide methods to interface and extend applications on the web. These APIs provide functionality beyond data access. They are hallmarks of most SaaS providers, easing integration of on-premise vendor and home-grown applications. Application cloud APIs connect to applications such as CRM, Enterprise Resource Planning (ERP), financial, billing, desktop productivity, and social media/networks applications, and so on. These applications are delivered as SaaS.

19.4   CLOUD INTEROPERABILITY

Providers have their own way of implementing how a user or cloud application interacts with their cloud. This leads to cloud API propagation as rarely, if ever, do two providers implement their clouds in exactly the same way using the same components. This limits cloud choice because of vendor lock-in, portability, ability to use the cloud services provided by multiple vendors including the ability to use an organization's own existing datacenter resources seamlessly.

We are already seeing different vertical PaaS stacks that are popping up everywhere, and each is incompatible with the other. Almost every cloud has a unique infrastructure for providing network services between servers and applications and servers and storage. Differences are likely in network addressing, directory services, firewalls, routers, switches, identity services, naming services, security policies, and other resources. This exacerbates the need for interoperability between clouds so that complex business applications developed on clouds are interoperable.

Cloud interoperability refers to the situation where two distinctly separate and independent providers (who, for instance, provide PaaS services) work and orchestrate seamlessly from a customer perspective and integration. This also includes the possibility of "wiring up" the PaaS offerings with not only current and modern provider offerings but also to legacy resources and services. Another related dimension of cloud interoperability is intercloud interoperability, which leads to forming cloud federations of private and public clouds using some form of cloud orchestration services [Parameswaran 2009].

An important concept in the cloud interoperation space is that of metadata. When managing large amounts of data with differing requirements, metadata are a convenient mechanism to express those requirements in such a way that underlying data services can differentiate their treatment of the data to meet those requirements.

In the following, we provide a brief overview of the most promising cloud interoperability approaches.

19.4.1   OPEN CLOUD STANDARDS INCUBATOR

DMTF's cloud efforts are focused on standardizing interactions between cloud environments by developing specifications that deliver architectural semantics and implementation details to achieve interoperable cloud management between service providers and their consumers and developers. For this, DMTF is using the recommendations developed by its Open Cloud Standards Incubator initiative. DMTF's Open Cloud Standards Incubator aim is to aid the industry in addressing challenges that affect the interoperability, portability, and security of cloud computing environments [DMTF 2009].

The environment under consideration by the Open Cloud Standards Incubator includes all of these deployment models. The main focus of the incubator is management aspects of IaaS, with some work involving PaaS. These aspects include SLAs, QoS, workload portability, automated provisioning, and accounting and billing.

Examples of standardization areas include resource management protocols, data artifacts, packaging formats, and security mechanisms to enable interoperability.

19.4.2 Open Cloud Computing Interface

The Open Cloud Computing Interface (OCCI) comprises a set of open community-lead specifications (http://occi-wg.org/). The OCCI is a RESTful protocol and API for all kinds of management tasks. It is OCCI that provides for commonly understood semantics and syntax in the domain of customer-to-provider infrastructure management. OCCI was originally initiated to create a remote management API for IaaS model-based services, allowing for the development of interoperable tools for common tasks including deployment, autonomic scaling, and monitoring. It has since evolved into a flexible API with a strong focus on interoperability while still offering a high degree of extensibility. The current release of the OCCI is suitable to serve many other models in addition to IaaS, including PaaS and SaaS.

The OCCI specification is divided into three parts:

- OCCI core model defines a representation of instance types, which can be manipulated through an OCCI rendering implementation. It is an abstraction of real-world resources, including the means to identify, classify, associate, and extend those resources. A fundamental feature of the OCCI core model is that it can be extended in such a way that any extension will be discoverable and visible to an OCCI client at run-time. An OCCI client can connect to an OCCI implementation using an extended OCCI core model, without knowing anything in advance, and still be able to discover and understand, at run-time, the various resource and link subtypes supported by that implementation.
- OCCI infrastructure defines how an OCCI implementation can model and implement an IaaS API offering by utilizing the OCCI core model. This API allows for the creation and management of typical resources associated with an IaaS service, for example, creating a "compute" instance and "storage" instance and then linking them with StorageLink. The StorageLink type represents a link from a resource to a target storage instance. This enables a storage instance be attached to a compute instance, with all the prerequisite low-level operations handled by the OCCI implementation.
- OCCI HTTP rendering defines how to interact with the OCCI core model using the RESTful OCCI API. The document defines how the OCCI core model can be communicated and thus serialized using the HTTP protocol.

OCCI can be used in conjunction with other standards to provide for extended interoperability. It can, for instance, connect to the Storage Networking Industry Association's (SNIA) cloud storage standard, Cloud Data Management Interface (CDMI) to provide enhanced management of the cloud computing storage and data [SNIA 2010].

19.4.3 Cloud Data Management Interface

The SNIA has completed a standard aimed at making it easier for enterprises to move data around various clouds on the basis of common data storage interfaces. A data storage interface is an interface interposed between an application (or service) and the underlying set of services that enable reading and writing data, among other functions. SNIA categorizes and describes the cloud services that live behind this interface and as well as those which consume it.

SNIA's Cloud Storage Reference Model (CSRM) uses multiple types of cloud data storage interfaces that are able to support both legacy and new applications [SNIA 2010]. All of the interfaces allow storage to be provided on demand, drawn from a pool of resources. The capacity is drawn from a pool of storage capacity provided by storage services. The data services are applied to individual data elements, as determined by the data system metadata. Metadata specify the data requirements on the basis of individual data elements or on groups of data elements (containers). Storage system metadata are produced and interpreted by the cloud offering or basic storage functions, such as modification and access statistics, and for governing access control.

The CSRM suggests employing a CDMI as the functional interface that applications may use to create, retrieve, update, and delete data elements from the cloud. As part of this interface, the client will be able to discover the capabilities of the cloud storage offering and to use this interface to manage containers and the data that are placed in them. This interface may also be used by administrative and management applications to manage containers, domains, security access, and monitoring/billing information, even for storage that is functionally accessible by legacy or proprietary protocols. The capabilities of the underlying storage and data services are exposed so that clients can understand the offering. The CDMI specification uses RESTful principles in the interface design where possible.

In the SNIA model, data system metadata offer guidelines to the cloud storage system on how to provide storage data services for data managed through the CDMI interface. Such metadata contain information about the number of bytes stored in the data object, the time when a data object was created or last modified, geopolitical identifiers specifying a region where the object is permitted to or not permitted to be stored, data retention periods, desired encryption algorithms, desired maximum data rate on retrieve, and so on.

Cloud service life cycle focuses on exposing the actual cloud service functionality, which is one or more aggregated resources exposed as a single unit of management and managing the life cycle of a service in a distributed multiple-provider environment in a way that satisfies SLA with its customers.

19.5   CLOUD SERVICE LIFE CYCLE

The idea of managing applications throughout their complete life cycle is not new. The IT infrastructure library (ITIL) is a set of best practices that are widely accepted to manage IT services. These include operation and improvement of IT services [Cartlidge 2007]. Applications deployed on top of the cloud stack are also associated with their own life cycle, migrating from development to testing, staging, and production environments, with frequent rollbacks. One of the purposes of the Open Cloud Standards Incubator is to address the characteristics of the cloud service life cycle. The DMTF identified the following aspects of the life cycle of a cloud service:

- Description of the cloud service in a template
- Deployment of the cloud service into the cloud
- Offering of the service to its consumers
- Consumer entrance into contracts for the offering
- Provider operation and management of instances of the service
- Decommissioning of the service offering

The DMTF's life cycle could be seen as a subset of a more complete software life cycle, which is presented in Figure 19.1. In [Stankov 2010], the authors present an analysis on how the ITIL and the global software development and delivery (which refers to the emergence of Internet-based collaborative environments and the expansion of the global delivery of IT services) can align cloud application life cycle with the environment (platform, infrastructure, and support services) required at every phase in the life cycle.

In the DMTF's *description phase*, all information required to create a certain type of cloud service is captured in a service template. A *cloud service template* contains all knowledge that is needed to instantiate a cloud service and to manage the resulting cloud service instances. After a developer has created the components of a service, the developer begins the process of making it available to cloud consumers by creating a template that defines the content of an interface to the service.

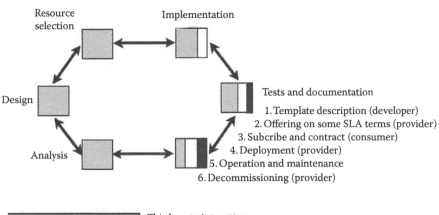

FIGURE 19.1   Mapping of a typical software life cycle to elements of the Cloud Stack*.

---

* The numbered elements in the figure correspond with the life cycle proposed by the cloud incubator in DMTF.

This cloud service template describes the service in a generic manner completely independent of how the service is exposed to potential service subscribers. For example, it does not include any pricing information or information about datacenter-specific resources required for hosting instances of the service. This template can be registered in a marketplace for other developers or providers to integrate them in their own applications or offers. This, in turn, implies the need for templates to be searched and processed so that a user could, for instance, just "plug" two of them together to get a running service or perform basic set theoretic operations on the templates such as union, intersection, and difference.

In the offering phase, a provider who creates a service *offering* for consumption by one or more consumers customizes the cloud service template. This may include aspects such as the prices associated with the respective offering and specific technical information, such as VMs, Internet protocol address ranges, and storage requirements for hosting instances of this specific offering. An offering is the unit that a consumer requests by establishing a contract with the provider for that offering.

When a cloud service offering is published in the service catalog, the *subscription* and *contracting* phase begins. From an offering catalog, a requester can select an offering and then specify the necessary parameters for the cloud service instance, such as capacity, availability, performance, and duration. The provider and customer then agree on the SLA terms that stipulate the use of this particular service offering. Subsequently, the provider provisions a service instance in the service provision phase that satisfies the constraints defined in the SLA offering, and the consumer uses the instance as defined in the contract.

The *provision phase* incorporates run-time monitoring and maintenance functionality where a provider manages a deployed service and all its resources, including monitoring resources and notifying the consumer of critical situations. After the contract is terminated, the provider decommissions the offering by reclaiming the service instance and its supporting resources.

For instance, providers of PaaS services can offer their services, which can be integrated with services from some PaaS library. This approach frees developers from the burden or reimplementing highly frequent pieces, such as authentication or data storage. The presence of a marketplace with appropriate selection tools is, therefore, essential in the cloud ecosystem.

Having PaaS online libraries integrated in a PaaS application is complemented by the presence of online tools assisting developers in this integration and in developing and debugging new pieces of code. In this implementation phase, IaaS resources can be instantiated by the developers to host online developing environments or unit testing tools. The tests can also make use of SaaS test suites or software for enabling regression tests.

State-of-the-art infrastructure management tools offer a large set of automation packages and a platform for programming workflows for higher-level provisioning operations. Indeed, a cloud workflow system can be regarded as a type of PaaS that facilitates the automation of distributed large-scale applications in the cloud. Being a PaaS, a cloud workflow system can get access to other cloud services. This way, the workflow service can control the underlying IaaS resources or control the workflow of the PaaS and SaaS components integrated in the cloud application.

19.6 ANALYSIS OF THE CLOUD DELIVERY MODEL LANDSCAPE

The three conventional cloud delivery models (SaaS, PaaS, and IaaS) are constrained by the capabilities available at their delivery level and do not allow for easy extensibility or customization options or any level of transparency on their underlying layers in order to endow developers and providers with more powerful and configurable service delivery options.

Arguably, IaaS clouds are the most explored layer in the cloud stack. IaaS allows for certain reconfiguration potential for the VMs and their (virtual) networks. Application components deployed on an IaaS cloud may exhibit complex dependencies on the configuration of multiple datacenter network, middleware (e.g., PaaS services used for the development of maintenance of the applications embedded in the VMs), and related application resources (external SaaS pieces integrated with the application). IaaS level cross-configurations of the VMs comprising a specific cloud service are currently not possible, and cross-machine configurations need to be inferred at deployment time or changed once the VMs are running, instead of remaining statically allocated as in today's IaaS solutions. Moreover, VM images, VM appliance descriptions, or even the APIs for exposing the IaaS capabilities are hardly interoperable and require a lot of adaptation work to interoperate.

PaaS clouds are supposed to ease programmers' and administrators' lives along the entire software life cycle. In general, the PaaS is supposed to assist in all these processes. One serious drawback with contemporary PaaS offerings is that they are constrained by the capabilities that are available through the PaaS provider and do not allow easy extensibility or customization options. To address some of these problems, Pandey et al. propose a PaaS service in charge of coordinating some of the tasks in the VM life cycle: selection, integration, and deployment [Pandey 2011]. They propose a "cloud market maker" service in charge of negotiation with multiple service providers to match VM requirements to VM capabilities. Once a match is found, required resources are allocated. A market directory service is needed to maintain a catalog of services. Unfortunately, this proposal leaves a few other unattended stages in the entire cloud application life cycle (e.g., operations, redeployment, development, maintenance, etc.), is statically defined (inflexible to cope with changes in long-lived applications), and heavily depends on the visibility that other service providers endow their applications with. For instance, the market maker cannot change the IaaS resources assigned to a Google Application Engine service.

In summary, current IaaS/PaaS offerings are constrained by their underlying VM capabilities and do not allow for easy extensibility, mashup, or customization options at the PaaS consumer (or developer) level. They also expose little (if any) configuration capabilities of the underlying IaaS layer, thereby lacking configurable automated elasticity. PaaS middleware and the applications built on it are not portable across different public and private clouds and cloud providers. Furthermore, many hosted middleware solutions are not integrated with any IaaS management.

Table 19.1 provides an overview of some of the most well-known IaaS and PaaS approaches along the dimensions of flexibility, life cycle management, scalability security, and data management. As shown in Table 19.1, advanced IaaS solutions include features for custom control on their automated scalability features.

TABLE 19.1

Comparison of IaaS and PaaS Approaches

Name of Approach	Cloud Level	Flexibility	App Life Cycle Mgt	Scalability	Security	Data Mgt	Technology Specification
Amazon SimpleDB, SQS, SNS, RDS, and AutoScale	IaaS—PaaS	Modular aggregation of separate services	None	Horizontal and vertical scaling plus load balancing	Key-based authentication to access the different modules	Key-value and relational DBs	Web, REST, and SOA interfaces
Azure	PaaS	Modular aggregation of separate services	App and data marketplace	Horizontal and vertical scaling plus load balancing	Federated identity and access control through rule	BLOB, table, queues, caching, and relational DB	Web, REST, and SOA interfaces
Amazon Elastic Beanstalk (Tomcat app server-based)	PaaS	Transparency into the underlying services (VM, SNS, etc.)	Tomcat-derived development support (concurrency, logging, etc.)	Horizontal and vertical scaling plus load balancing	Key-based authentication plus Tomcat features	Not native: pluggable DB modules and transaction managers	WAR file
GAE	PaaS	Some modules can be integrated by invoking GAE APIs (e-mail, memcache, user authentication, sessions, or image manipulation)	None	None	Google account—based authentication and sandboxing	High replication and master/slave datastores (easier to use: integrated with JDO/JPA or with GQL) and memcache	Servlet technology
AppScale	PaaS	Some modules can be integrated by invoking GAE APIs (user authentication, mail, etc.)	"Workflow" and some IaaS constraints	None	Google authentication	Datastore GL	Neptune: Servlet, map-reduce, MPI technology

Several modules can be activated and used, but they let little room for defining custom security mechanisms, data management options, or support for a complete cloud application life cycle.

Among the PaaS solutions, more security, life cycle, and data management capabilities are exposed (some of them directly derived from the application containers underlying most PaaS solutions). A salient example with regard to PaaS configurability and the transparency on underlying layers is Amazon's Elastic Beanstalk. Elastic Beanstalk lets users keep some visibility on the underlying IaaS resources related to the container: they can select the most appropriate Amazon VM instance type, the database, and storage options; choose the Availability Zone(s), enabling HTTPS protocol on the load balancer; and run other application components, such as a memory caching service, side-by-side in Amazon's VMs or the container scalability by using Amazon's scaling and monitoring tools for VMs (Autoscale and CloudWatch). Also, Neptune offers the option of specifying the number of nodes supporting a given service [Bunch 2011]. These are detailed in the related work section.

SaaS is not an exception to the interoperability and lack of configurability problems. The SaaS client is presented with a complete application, for example, CRM, which cannot be customized easily at the process level and cannot be simply combined with external services from other providers to offer improved functionality to the client or developer.

The cloud life cycle and the cloud-enabled marketplace are, then, far from being realized since cloud vendors expose a monolithic services with little (if any) configuration/customization options, preventing these services from diverse cloud service tiers to be linked together in a synergistic manner to address their application needs (e.g., "scale my PaaS for authentication when my IaaS-embedded application receives too many users").

There is clearly a need to mashup services from a variety of cloud providers at the different delivery levels to create a cloud ecosystem. However, the type of tailored integration of cloud services needs to fit to specific business needs, using a mixture of SaaS, PaaS, and IaaS. In addition, they need to satisfy QoS requirements. This integrated ecosystem can be realized by leveraging and weaving together all aspects of infrastructure, platform, and application services to provide a wide variety of vertically aligned business solutions and falls under the purview of BPaaS [Papazoglou 2011a]. An example on how the cloud application life cycle will influence current mobile application market is presented in the following.

In current mobile application markets, developers register, upload their applications, and publish them making them visible and available for mobile users. Mobile users connect to the marketplace, search for applications, and get them downloaded in their devices. While this model may be simple and may work very well for this segment, corporate applications are often more demanding. They are typically composed of several complex and distributed bits that need to be configured so that they can work together. These composing pieces often belong to separate vendors, present heterogeneous packaging and distribution mechanisms.

Also, these applications are often server elements (mobile market applications can be seen either as clients or as stand-alone applications), which require a powerful underlying infrastructure (IaaS) that scales with the application load under predefined and varying conditions. Mobile operating systems expose APIs that reveal some services assisting developers of client application on common tasks. These APIs are typically a service in the device itself. In a corporate market, developers may need to tie client application installation in the device with server instantiation in the available IaaS service that is closest to the user. Similarly, general applications may require interaction with "online" APIs exposing some frequently used services (e.g., GAE's data access object or authentication PaaS services) or other online services developed by third parties (SaaS).

The complex cloud ecosystem enabled by these new possibilities introduces new needs for complex interservice dependencies, huge space of solutions, and holistic cloud application solutions. These call for additional knowledge from all the actors (and supporting mechanisms exposed by the cloud platform). Here, the concepts of reuse, refinement, and dynamic (often automated) adaptation for dealing with cloud applications and their physical/platform configurations can be fused in the need for expressing, storing, querying, and manipulating some metadata (knowledge) in a formalized manner. The rising of knowledge-based cloud application engineering thus becomes a first class citizen for constructing cloud ecosystems. This topic and its relation to traditional knowledge-based engineering are dealt with in the next section.

19.7 KNOWLEDGE-INTENSIVE CLOUD SERVICES

There is indeed a need for structuring the information exchanged by any service components (across all the layers of the cloud stack) to make proper use of a cloud service during its life span. This information is essential and spans an enormous breadth of subject areas. It may originate from analysis and design tools (e.g., data from a marketplace for programmers to reuse previously developed code or integrate online APIs in their development process) all the way to the manner a given component consumes the monitoring data exposed by the IaaS provider. This range of information (knowledge) is currently widespread and compartmentalized per layer, for example, operational knowledge about a particular layer (e.g., PaaS), if it exists, stays confined strictly within this layer. It is by no means cross-correlated with knowledge originating from other layers, or not even exposed to other layer components, which could make more informed decisions and can thus increase the overall cloud application utility. It is therefore obvious that there is a pressing need for not only exposing operational knowledge at all levels of the cloud stack but also organizing and determining the visibility of the available information. This is due to the fact that many service providers require part of this information to remain private to avoid facing security or privacy vulnerabilities [Vaquero 2011].

The idea of knowledge-based cloud services finds its roots in knowledge-based engineering, and it concerns structuring metadata about cloud services. There is a very real need for interoperability and portability specifications to include all the relevant application, platform, and network infrastructure metadata necessary to move parts of an application from one cloud to another and allow invoking services from diverse clouds. The truth is that a comprehensive solution to the cloud metadata problem—apart from the practical efforts of DMFT—is currently missing.

Cloud-intensive knowledge may relate to service applications, technical capabilities, or operational level details and may serve to provide answers to such questions as follows:

- Where and how is a service implemented?
- What kind of customer service and response time does a service guarantee?
- What type of SLAs or security policies does a service come endowed with?
- What are the technical specifications that a platform supports?
- Is a platform an open-source one or a proprietary, closed one?
- Does a platform bill a fixed or variable cost for capacity?

Such knowledge can be captured, abstracted, and stored in some medium, such as DMTF's service template, and can be ultimately used to instantiate a cloud service and to manage it during its life cycle. Furthermore, this knowledge must be organized and appropriately utilized and combined with other service knowledge to fulfill the vision of a holistic and efficient cloud application life cycle management.

The need to provide a holistic and abstracted view of all cloud service and resource-specific knowledge, which possibly span disparate cloud environments, is critical in order to realize the BPaaS model, which we presented in Section 19.2. Such efforts call for defining an appropriate mechanism for breaking the cloud monolith and building a more flexible system that preserves the essential properties of the cloud: abstraction, scalability, pay as you go model, and illusion of infinite resources underneath [Papazoglou 2011a]. This topic will concern us throughout this chapter. We start first by identifying the requirements and efforts for describing, organizing, and manipulating cloud-intensive knowledge.

19.8   ANALYSIS OF CLOUD SERVICE DOMAIN LANGUAGES

Development of cloud applications that utilize knowledge-intensive cloud services requires an effective description of cloud service-related metadata, an effective organization of the required of knowledge, and an associated support environment, which supports the description and manipulation of service-related metadata. We collectively refer to the language approaches that contribute to this environment as *cloud service domain languages.*

Cloud service domain languages interlace three interrelated components:

1. A cloud service declarative definition language (CSDL), which provides the necessary abstraction constructs, called *blueprints*, to describe the operational, performance, and capacity requirements of infrastructure, platform, and application third-party services
2. A cloud service constraint language (CSCL), which specifies any explicitly stated rule or regulation that prescribes any aspect of cloud service defined in CSDL and verifies the validity of rule combinations
3. A cloud service manipulation language (CSML), which provides a set of operators for manipulating, comparing, and redefining CSDL blueprints, without compromising backward compatibility

In the following, we shall briefly review R&D activities related to cloud service domain languages. We shall commence by reviewing research work in cloud service definition languages first.

19.8.1 Overview of Cloud Service Definition Languages

As currently no definition language exists that combines IaaS, PaaS, and SaaS service elements, we shall divide work related to cloud template definition languages in three sections, each of which reflects activities in the corresponding cloud stack layer. Cloud service template languages target the definition of useful knowledge that extends from lower-level cloud interoperability challenges centered around network addressing, multicast enablement, and VM techniques to the higher-level interoperability desires of software services, for example, web or RESTful services.

19.8.1.1 IaaS Definition Languages

The description mechanisms for IaaS clouds can be categorized as "template-based" and model-driven [Papazoglou 2011a].

19.8.1.1.1 Template-Based Approaches

IaaS users who employ template-driven approaches, for example, either end users or PaaS providers, typically package the software stack (operating systems, middleware, and service components) in one or more VMs. However, each IaaS provider exposes its own interface and lets users exploit cloud capabilities and define their own services [Youseff 2008, Galán 2009]. This fact certainly complicates interoperability between clouds at the same layer of the cloud stack and, more so, across layers.

In order to solve this interoperability problem, DMTF uses templates that help manage cloud resources and support the cloud service life cycle by, for instance, describing the way in which a cloud offering is presented and consumed [DMTF 2009]. The service offering is abstracted from the specific type of cloud resources offered, and service templates are used to describe in a general form what a cloud provider can offer (see also Section 19.5). The cloud service template is then mapped into the specifics of the provider's technology and constrains.

One of the most prominent initiatives in DMTF, the open virtualization format (OVF) [OVF 2010], describes a specification of software packaging and distribution to be run in VMs. OVF is a platform-independent, efficient, extensible, and open packaging and distribution format for VMs. OVF is virtualization platform neutral, while also enabling platform-specific enhancements to be captured.

There are several extensions of OVF to address low-level interoperability problems. Galán et al. consider configuration parameters (e.g., hostnames, IP, and other application-specific parameters) for software components included in the VMs as one of the capabilities that should be exposed by an IaaS provider [Galán 2009]. Horizontal VM scalability by relying on noninfrastructure metrics is also briefly dealt with in this publication. In a continuation of this work, the authors describe a system for a more complete application life cycle management on top of an IaaS cloud [Rodero-Merino 2010]. However, all OVF-based approaches are very restrictive as they focus on specific points of the application life cycle: definition and deployment time configuration. The fact that the application horizontally scales and load balances its VMs cannot be considered beyond the configuration step.

Previous approaches focused on the computing part of the infrastructure being exposed. One of the important missing bits in most IaaS representations is the network. Bernstein et al. [Bernstein 2009] mention a series of protocols that could help to enable the exposure of communications as another IaaS service. A wealth of network description languages (e.g., Network Description Language) exist that could fit in here. However, the lack of maturity in the cloud networking arena places the analysis of network description languages beyond the scope of this chapter.

The work of Bernstein and Vij concentrates on proposing a layered set of protocols and formats (called intercloud protocols) as well as a set of common mechanisms that are needed both inside the clouds and in-between the clouds to resolve the IaaS interoperability challenge [Bernstein 2010]. The authors propose the use of intercloud directories as mediators for enabling connectivity and collaboration among disparate cloud providers. To facilitate collaboration among disparate heterogeneous cloud environments, they use the Extensible Messaging and Presence Protocol (XMPP), which forms a sound basis for asynchronous communication. This publication proposes the use of a declarative semantic model to capture both the requirements and constraints of IaaS computing resources. For this purpose, they use the resource description framework (RDF) language to specify IaaS resources and SPARQL, an SQL-like language, for querying and manipulating semantic information [Bernstein 2010].

In concluding this subsection, it is probably worth mentioning the first holistic integration approach: the Virtual private eXecution infrastructure Description Language (VXDL). VXDL allowed for an organized aggregation of heterogeneous computing and communication resources [Koslovski 2008].

19.8.1.1.2 Model-Driven Approaches

Model-driven approaches are based on considering models of the system as the main elements along all the steps of the software life cycle (design, implementation, deployment, test, and maintenance).

Model-driven approaches have been employed for automating the deployment on top of IaaS clouds. Goldsack et al. [Goldsack 2009] propose a declarative configuration framework (SmartFrog) in which component descriptions were specified as ordered hash tables. The provided description is checked against a data model representing the life cycle status of the component, and inferences are made as to what changes are required to take the component from its current status into the one that matches the hash table declaration. Configurations and descriptions may contain data that come from different stages of the processing of a description. There may be data that are known in advance provided by the writer as part of the description or data that may only be known much later, either during the processing of the description or at run-time. This enables a full delegation approach with dynamic reference resolution.

A model-based approach in which domain experts construct virtual appliances (virtual images with prebuilt configuration points) is proposed in [Konstantinou 2009]. The virtual appliance model in [Konstantinou 2009] consists of two parts: packaging and composition. Packaging includes configuration attributes of different

TABLE 19.2
Summary of Research Work in Service Definition Languages

	Operational Service Description	Performance-Oriented Service Capabilities	Resource Utilization	Policies
Rodero-Merino et al. (2010)	OVF	Ad hoc XML	Ad hoc XML	NA
Bernstein and Vij (2010)	RDF	RDF	RDF	RDF
Koslovski et al. (2008)	VXDL	VXDL	N/A	N/A
Konstantinou et al. (2009)	OVF	Ad hoc XML	N/A	Ad hoc XML
Collazo-Mojica et al. (2010)	Ad hoc XML	Ad hoc XML	N/A	Ad hoc XML
Goldsack et al. (2009)	Ordered hash tables	Ordered hash tables	N/A	N/A

products to be installed on the image, while composition determines how the appliance can be connected to other appliances. The proposed model declaratively exposes configuration points for the appliance models to be composed into solution models (using simple aggregation and cross-configuration of the appliances). This can be further customized by including infrastructure-specific requirements and may affect and drive VM placement and network configuration, which is referred to as a cloud-specific virtual solution deployment plan. In a similar manner, [Chieu 2010] describes a solution-based provisioning mechanism using composite appliances to automate the deployment of complex application services on an IaaS cloud. A key element for the composition of virtual appliances in these two publications is the virtual port. This is a typed object encapsulating all the properties required to affect a specific type of communication (it captures how images can be linked).

Virtual ports are very similar to the service endpoints presented by [Collazo-Mojica 2010]. This work exposes a visual model for service specification that resembles traditional web mashup, but relies on an ad hoc XML format.

Table 19.2 summarizes characteristics of the most important works identified relating to cloud service definition languages.

19.8.1.2 PaaS Definition Languages

The variety of services that can be exposed a la PaaS make it quite complicated to get a unified view on the models and templates employed in describing these capabilities. The reader should take into account that a workflow engine or a service marketplace can both be PaaS services that are inherently different in the data they require and present. It is also worth mentioning that, while the literature is rich in attempts on describing IaaS systems, PaaS-related work is much scarcer.

In this subsection, we shall concentrate on the approaches relevant to PaaS systems, which are compared and summarized in Table 19.3. We shall examine two commercial approaches (Azure and Beanstock) as well as an experimental system.

The Windows Azure Platform is a Microsoft cloud platform (http://www.microsoft.com/windowsazure/) is used to build, host, and scale web applications through Microsoft datacenters. Azure offers a SOA and REST imperative

TABLE 19.3

Comparison of PaaS Definition Approaches

Platform	API Type	Language/Payload
Azure	Imperative	Ad hoc XML
BeanStalk	Imperative	Ad hoc XML
CloudFormation	Imperative	Ad hoc JSON
AppScale	Imperative	Ad hoc XML, Ruby metalanguage, and key/value pairs

procedure for issuing basic CRUD operations using a series of predefined pay-loads in the HTTP requests. It defines a set of HTTP headers in its requests and responses that vary widely across the services offered by this PaaS. Information on how these requests are represented internally remains proprietary.

AWS Elastic Beanstalk (http://aws.amazon.com/elasticbeanstalk/) is a PaaS solution to quickly deploy and manage applications in the AWS cloud. Elastic Beanstalk automatically handles the deployment details of capacity provisioning, load balancing, autoscaling, and application health monitoring. Elastic Beanstalk leverages AWS services such as Amazon EC2, Amazon Simple Storage Service (S3), Amazon Simple Notification Service, Elastic Load Balancing, and Autoscaling to deliver the same highly reliable, scalable, and cost-effective infrastructure.

```
https://elasticbeanstalk.us-east-1.amazon.com/?
ApplicationName=SampleApp
&Description=Sample%20Description
&Operation=CreateApplication
&AuthParams
```

Listing 19.1 Sample BeanStalk request

```
<CreateEnvironmentResponsexmlns="https://elasticbeanstalk.
amazonaws.com/docs/2010-12-01/">
<CreateEnvironmentResult>
  <VersionLabel>Version1</VersionLabel>
  <Status>Deploying</Status>
  <ApplicationName>SampleApp</ApplicationName>
  <Health>Grey</Health>
  <EnvironmentId>e-icsgecu3wf</EnvironmentId>
  <DateUpdated>2010-11-17T03:59:33.520Z</DateUpdated>
  <SolutionStackName>32bit Amazon Linux running Tomcat
7</SolutionStackName>
```

(continued)

(continued)

```
    <Description>EnvDescrip</Description>
    <EnvironmentName>SampleApp</EnvironmentName>
    <DateCreated>2010-11-17T03:59:33.520Z</DateCreated>
    </CreateEnvironmentResult>
    <ResponseMetadata>
    <RequestId>15db925e-f1ff-11df-8a78-9f77047e0d0c</
    RequestId>
    </ResponseMetadata>
    </CreateEnvironmentResponse>
```

Source: aws.amazon.com (accessed on August 2011).

Listing 19.2 Sample BeanStalk request to create an environment for an application

We shall use representative examples based on AWS Elastic Beanstalk for illustration purposes. The code in Listing 19.1 shows a sample request to create an application on top of BeanStack. As can be observed, the user can configure it specifically and independently of all other applications. The specific behavior of the application depends on several elements underneath this top level, such as the environment or the underlying instances and how they scale and are load balanced. A sample environment is shown in Listing 19.2, while Listing 19.3 illustrates how users can "attach" S3 buckets to applications running on top of one of Amazon's exposed Tomcat application servers.

```
<CreateStorageLocationResponse xmlns="https://
elasticbeanstalk.amazonaws.com/docs/2010-12-01/">
  <CreateStorageLocationResult>
   <S3Bucket>elasticbeanstalk-us-east-1-780612358023</
   S3Bucket>
  </CreateStorageLocationResult>
  <ResponseMetadata>
   <RequestId>ef51b94a-f1d6-11df-8a78-9f77047e0d0c</
   RequestId>
  </ResponseMetadata>
</CreateStorageLocationResponse>
```

Source: aws.amazon.com (accessed on August 2011).

Listing 19.3 Sample specification of an S3 bucket to be used by the application

BeanStalk relies on Amazon's CloudFormation (http://aws.amazon.com/de/cloudformation/). CloudFormation includes the concept of templates; a JSON format, file that describes all the AWS resources needed by the application; and Stack, the set of AWS resources that are created and managed as a single unit when a template is instantiated. It also allows for provisioning of these resources in an ordered manner, letting users specify resource dependencies, deployment constraints, or application environment (e.g., use queues to access a web server).

Finally, AppScale is an open source implementation and extension to the GAE (see Table 19.3) PaaS cloud technology. AppScale has been build upon the GAE software development kit, which helps to distribute execution of GAE applications over virtualized cluster resources. This also includes IaaS cloud systems such as Amazon's AWS/EC2 and Eucalyptus. Through the framework offered by AppScale, developers can investigate the interaction between PaaS and IaaS systems. AppScale exposes a series of underlying PaaS services in a unified manner [Chohan 2009]. This means that AppScale data are handled as a combination of XML and key/value pairs.

Very recently, AppScale included Neptune, a domain-specific language for AppScale cloud configuration. Neptune automates configuration and deployment of existing high-performance computing software applications via cloud computing platforms. Neptune is a metaprogramming extension of the Ruby programming language, which allows users to specify workflows for certain languages, frameworks, and scientific applications [Bunch 2011].

Each Neptune program consists of several jobs to run in a given cloud. The name of the job and which parameters are necessary for the given jobs; these can be extended to enable each job to include any set of keywords. Neptune can specify what parameters are needed to optimize a job and what data types are required for each.

19.8.1.3 SaaS Definition Languages

There is a great deal of work related to composing services based on the concept of service mashups [Benslimane 2008, Mietzner 2008]. In this chapter, we shall restrict ourselves to cloud-specific initiatives in an attempt to distill what is new to the cloud that was not there in previous service composition models.

The work in [Hamdaqa 2011] follows model-driven approach and proposes a metamodel that allows cloud users to design applications independent of any platform and build inexpensive elastic applications. This metamodel, however, describes only the capabilities and technical interfaces of cloud application services.

The authors in [Mietzner 2008] investigate how a SOA could be extended with variability descriptors and SaaS multitenancy patterns to package and deploy multitenant aware configurable composite SaaS applications [Mietzner 2008]. As an extension of this work, an application metamodel has been proposed that enables the modeling of components of an application, component dependencies at the SaaS and at the IaaS level [Fehling 2011]. The authors in this work also define an application descriptor independent on the underlying technologies that packages all the application components.

Overall, SaaS template refinement models lack nonfunctional and constraint descriptions for arbitrary cloud services. In addition, variability metamodels at the SaaS level are inherently vague and leave actual enactment unspecified. Although additional work

TABLE 19.4

Comparison of SaaS Definition Approaches

Approach	Scope	Aim	Format	Across-Layer Transparency
Hamdaqa (2011)	IT resources	Engineering and portability	None: reference model	Some IaaS/ PaaS-transparency
Mietzner et al. (2008)	IT resources	Engineering and portability	Ad hoc XML	Some IaaS-transparency

is available for the SaaS level, for example [Cai 2009, Thrash 2010], it lacks a formal structure and detailed technical definitions thus reducing their usability.

Finally, Table 19.4 summarizes activities in the SaaS definition space.

19.8.2 OVERVIEW OF CLOUD SERVICE CONSTRAINT LANGUAGES

The cloud paradigm offers the chance for applications to implement their own policies in a dynamic manner, depending on the actual environment they are running on. In particular, cloud computing enables customers to acquire and release resources for client applications adaptively in response to load surges and other dynamic behaviors.

One pending challenge for cloud applications has to do with the means to express such control policies and how to automate dynamic cloud adaptation and, in general, take advantage of the natural elasticity of shared resources in cloud computing systems. The nature of this challenge is multipronged since it now implies delivering policy specification and adaptation across different abstraction layers in the cloud stack, which were black boxes in previous paradigms.

Existing solutions are often restricted to event-condition-action rules, where conditions are matched against metrics of the observed system, for example, incoming traffic flows. Endowing systems with dynamic adaptation and conflict resolution of policies in response to changes within the managed services has been a constant concern in policy-based frameworks (e.g., [Lupu 1999, Lymberopoulos 2003]).

In the cloud domain, we are dealing with a constraint satisfaction problem applied to cloud resources and services that need to meet specific requirements, such as QoS, or the need for a specific infrastructure. This problem can be represented by a set of variables, which represent cloud resources or services, and a set of constraints imposed on those variables in order to find an assignment of values to the variables such that all the constraints are satisfied. To improve readability, before we delve into this topic, we shall provide a brief introduction to the topic of formal representation of constraints and how to express and reason about constraints.

19.8.2.1 Formal Representation of Constraints

Policies are typically based in some sort of logic to express the relations among the objects subject to study. In general, the computational power and complexity of the constraint satisfaction problem depend on the constraint language that we are allowed to use in the instances of the constraint satisfaction problem.

Temporal logic is a brand of modal logic tailored to temporal reasoning. Temporal logic has found an important application in formal verification, where it is used to state requirements of hardware or software systems. For instance, one may wish to say that whenever a request is made, access to a resource is eventually granted, but it is never granted to two requesters simultaneously. Temporal logics are usually interpreted in modal structures where the nodes (the modal worlds) are positions in time, often called instants. These need not be just points but can be, for example, time intervals (periods). In linear time logics, all instants are linearly ordered from past to future, and there is only one possible future: future is determined. Linear temporal logic (LTL) is a modal temporal logic with modalities referring to time so that one can encode formulae about the future of paths. LTL can be shown to be equivalent to the monadic first-order logic of order a result known as Kamp's theorem.

Generally speaking, model checking is the algorithmic verification that a given logic formula holds in a given structure (the model that one checks). This is something similar to what policies do for distributed systems: checking that the system meets some conditions. Model checking deals with finite computational structures (decidable system) displaying infinite behaviors (rich set of situations that can be modeled).

Formally, given a set $AP = \{P1, P2, ...\}$ of atomic propositions, a constraint model can be defined as a tuple $S = \langle Q, R, L, I \rangle$, where

- $Q = \{s1, s2, ...\}$ is the set of configurations of the system. This set is dynamic, and new sets can appear or disappear in the lifetime of the cloud application.
- $R \subseteq Q \times Q$ is a transition relation between states.
- $L: Q \rightarrow 2AP$ is a labeling of states with propositions.
- $I \subseteq Q$ is a nonempty set of initial configurations.

Other formalizations are also possible that take into account more specific elements or resources of the model at hand. Here, we restricted ourselves to this particular formalization for the sake of keeping generality. Note that most of the works presented in the following can either be explicitly or implicitly mapped to this general type of formalization.

19.8.2.2 Overview of Candidates for a Cloud Service Constraint Language

While cloud service description languages are about describing the elements in the cloud ecosystem, a CSCL (and for computer-supported collaborative systems by implication) is more about defining boundaries for the resources and services in that system to behave as expected by clients. Moreover, the dynamic nature of cloud ecosystems, where resources can appear or disappear and the need to scale cloud capacity up or down, dictates that the rules governing the behavior of resources and services may need to be altered at different stages at any level in the service architecture (tiers, components, configurations, etc.). Therefore, any kind of CSCL needs to express such situations to achieve cloud interoperability.

Logic programming is a convenient and concise notation to exemplify the aims of a CSCL. To illustrate this, consider the snippet in Listing 19.4, which shows a set of rules for scaling service components based on business level metrics expressed as logic programming with a constraints notation. This notation can express conditions over response times, bandwidth, queue lengths, etc. In addition, it presents a set of facts that would trigger some of the rules. MonitoringValue facts are asserted in real time and trigger the necessary rules to create a VM.

```
Full(X):- MonitoringValue(X, Y), Limit(X, Z), Y>Z
CreateVM(BigDiskVM):-
        Full(QueueLength), VMCount(BigDiskVM, X),
        Limit(BigDiskVM, Y), X<Y
CreateVM(HighBandwithVM):-
        Full(ResponseTime), VMCount(HighBandwithVM, X),
        Limit(HighBandwithVM, Y), X<Y
Limit(QueueLength, 30)
Limit(ResponseTime, 210)
MonitoringValue(QueueLength, 64)
MonitoringValue(ResponseTime, 90)
```

Listing 19.4 Logic programming example for scaling a VM upon the presence of specific conditions

There is a wealth of existing languages based on logic that could potentially be used for the purposes of a CSCL.

19.8.2.2.1   Rule- and Knowledge-Based Languages

We commence our overview of formalisms and languages that could contribute to service constraint languages by considering first rule-based languages such as the rule markup language (RuleML) and the rule interchange format (RIF). RuleML is a modular markup language with Datalog at its core, which targets all aspects of web rules and their interoperation. It was developed to express both forward (bottom-up) and backward (top-down) rules in XML for deduction, rewriting, and further inferential-transformational tasks. RIF offers an abstract and XML-based syntax with a core specification and two dialects that add features to it. The production rule dialect (PRD) is used to express production rules, which specify condition and action syntax, and the basic logic dialect (BLD) adds common logic features not available on the core. These rule-based languages are imperative in nature and most of them lack features that enable inference, unless properly combined with an ontology of the cloud application, its environment and underlying infrastructure.

There are also a number of domain-specific languages for knowledge representation that could also contribute to service constraint languages. For instance, semantic web rule language (SWRL) is a semantic web rule language, combining sublanguages of the OWL web ontology language (OWL DL and Lite) with those of the RuleML. Other languages in this space include the DARPA agent mark-up language (DAML) and DAML-Service (DAML-S), which have been developed as an extension to XML and the RDF to provide a rich set of constructs. These constructs are used to describe the properties and capabilities of web resources and web services in a computer-interpretable form.

An interesting cloud constraint-based approach that uses RIF-based constraints to achieve a fine-grained level of control of a cloud service at run-time is proposed in [Morán 2011]. The authors use RIF-based rules to describe the behavior of the service that they link with its other properties, such as architecture, hardware requirements, customization, etc., which use OVF as service definition language. The extensibility mechanism for OVF is OVF sections, which are basically an XML element with a well-defined XSD syntax. This publication illustrates that RIF-PRD offers a slot-oriented, first-order logic production rule language that meets the minimum requirements to bring the power of programming logic to the service behavior specification on IaaS clouds.

19.8.2.2.2 Workflow-Based Languages

Another family of languages in this domain includes workflow languages that offer capabilities for expressing rules. Most of these approaches use the workflow language business process execution language (BPEL) as a basis for extension with rules and experimentation. BPEL is an OASIS standard executable language for specifying actions within business processes with web services. Processes in BPEL export and import information by using web service interfaces exclusively.

In this context, for instance, [Keller 2004] focused on IT system change through software upgrades, hot fixes, or hardware changes. Their system assesses the impact of the change and generates a change plan to optimize resource selection and execution as a (BPEL) workflow. This system employs deployment descriptors to annotate software packages to provide metadata about software packages and a run-time dependency model to capture dependencies that typically cross system boundaries gathered at build time about a software package.

Workflow-based languages are not specifically designed for specifying rules and complex constraints, and they usually hardcode the rules while workflow engines are highly inefficient when evaluating simple rules [El Maghraoui 2006]. In addition, the de facto standard, BPEL, is not grounded on a formal model, which implies that any BPEL specification should be transformed into a formal representation for the verification of constraint definitions against formally specified compliance rules.

19.8.2.2.3 Planning-Based Approaches

Artificial intelligence (AI) planning is an area that can contribute greatly to service constraint languages and optimize the use of cloud services. A solution to a planning problem is an ordered sequence of actions that, when carried out, will achieve the desired goals. AI planning techniques make also extensive use of constraints, and, in many cases, an AI planning problem can be reduced to solving a constraint satisfaction problem.

AI planning techniques to automatically generate workflows that bring a datacenter from its current state to another desired state are pursued in [El Maghraoui 2006]. This approach uses model-driven deployment planning on the basis of the planning domain definition language (PDDL) [Ghallab 1998] to declaratively describe the required datacenter solution, as well as the operational capabilities of existing provisioning platforms. This PDDL model is subsequently refined and translated to specific automation tools in the form of provisioning operations by means of a workflow orchestrator. For instance, this publication shows how to generate network-provisioning workflows driven by application requirements.

In [Hagen 2010], the authors propose an approach to plan and reason about state-related changes to IaaS and SaaS instances in large datacenters to dynamically scale the application layer to cope with varying loads of three-tier applications, which comprise a database, several application or web servers (WAS/Apache), and a load balancer. The authors associate a state transition system to define the configuration status of the software installed in the VM and their mutual dependencies. These dependencies can also include the status of the VM or the physical machines containing the applications. The state transition system describes a series of preconditions that, when met, trigger some reconfiguration actions on a model of the deployed service, the associated applications, VMs, and physical machines.

19.8.2.2.4   Policy- and Compliance-Based Approaches

Finally approaches that are based on policy and compliance specification and management are intrinsically related to service constraint languages.

In the policy design space, [Ramshaw 2006] present Cauldron, a policy-based design tool that uses a constraint programming approach, which employs policies, defined as configuration constraints on object-oriented models of systems, to generate configurations that meet those policies. Configuration design constraints are given as policy statements represented using the common information model (CIM) defined by DMTF. CIM is an object-oriented model that defines how managed elements—ranging from physical devices and computer systems to applications—are modeled to facilitate integration between management systems. A Cauldron policy is a class-scoped obligation policy that has to be obeyed by all instances of the class defined in the policy instance.

CSCLs are heavily related to compliance management approaches. In [Elgammal 2010], Elgammal et al. propose an LTL-based framework for compliance management. The authors focus on constraint specification at the very early stages of the business process design, such that compliance constraints are seeded into service-enabled processes. Using property specification patterns to specify compliance constraints and automatically generate formal specifications significantly facilitate the work of the compliance expert. This formal base can easily be extended to a more complete cloud service life cycle.

Finally, Table 19.5 provides a summary of the most salient features of the approaches examined earlier.

19.8.3   Overview of Cloud Service Manipulation Languages

A service manipulation language responds to the need to match, merge, compose, extract, or delete service metadata (contained in templates or blueprints as we shall see in the next section) or parts of it. Some of the required transformations are aimed at supporting the delegation and refinement process.

TABLE 19.5

Comparison of Candidates for Cloud Service Constraint Languages

Approach	Type	Across-Layer	Focus	BDL Format	Dynamic Rules
Morán et al. (2011)	Imperative	IaaS/PaaS/ SaaS	Provision and run-time reconfiguration	RIF plus ad hoc ontology	Y
Hagen and Kemper (2010)	Imperative	IaaS/SaaS	State-related adaptation at run-time	Groovy-based DSL	Y
El Maghraoui et al. (2006)	Imperative	IaaS	Provisioning automation	PDDL	N
Keller et al. (2004)	Imperative	IaaS/SaaS	Adaptation during operation and maintenance	BPEL	N
Elgammal et al. (2010)	Imperative	IaaS/PaaS/ SaaS	Design time compliance	Ad hoc LTL expressions	Y
Ramshaw et al. (2006)	Imperative	IaaS/SaaS	Software installation and configuration	CIM OO extensions	N

Most of the templates and models are either directly expressed or can easily be transformed into an XML representation. There is a wealth of work on languages that support XML management operations. For instance, XSLT (eXtensible Stylesheets Language Transformations) is a W3C standard allowing for transformations from an input XML document into an output XML document. Other useful functionality for a service manipulation language can find in its origin on activities related to XML-based query languages. Such work is summarized in the following.

W3C's XQuery defines a query language for extracting information that matches specific criteria from one or more XML documents. An Xquery processor reads the template, resolving the queries to generate the final document. Xquery uses Xpath in order to address specific parts of the XML document, along with FLOWR (For, Let, Order By, Where, and Return) clauses to combine and restructure information (similar to relational SELECT-FROM-WHERE joins in SQL). FLOWR can use if/then/else sentences to implement conditional results. Xquery can also use built-in and user-defined functions in Xpath predicates or to process query results. Conceived as a query language similar to classical query languages for relational databases, it is also possible to embed Xquery expressions in XML templates to produce one output XML document from the input queried ones.

The QVT language (queries/views/transformations) is OMG's standard to query (expressions are evaluated over a model to produce an output result), view (models that are completely derived from another base model to which a direct connection exists) and transformations. The QVT standard defines three model transformation languages all of which operate on models, which conform to MOF. The QVT/MOF/ OMG family of languages is well suited to the conversion of logical schema expressed in one physical encoding into an alternative physical encoding. A transformation in any of the three QVT languages can itself be regarded as a model, conforming to one of the metamodels specified in the standard. These are as follows:

- *QVT-Operational*: an imperative language designed for writing unidirectional transformations
- *QVT-Relations*: a declarative language designed to permit both unidirectional and bidirectional model transformations to be written
- *QVT-Core*: a declarative language designed to be simple and to act as the target of translation from QVT-Relations

The QVT standard integrates the OCL 2.0 standard and also extends it with imperative features.

Manipulations of database schemas and matching operations and mechanisms have been extensively studied as expected in the database literature. Several schema-matching operators can be used for constructing similar operators for manipulating cloud services. In this publication, the authors propose schema-level, instance-level, structure-level, and language-based and constraint-based matching operations for integrating diverse database schemas. Matching algorithms are also proposed.

The activities mentioned earlier are summarized in Table 19.5. This table shows that by specifying appropriate sets of rules and operations that rely on XML standards, query extensions, or ontological specifications, a cloud service can, for instance, compose, compare, or match services from diverse PaaS platforms.

19.9   NEED FOR A MORE FLEXIBLE CLOUD DELIVERY MODEL

The analysis of the cloud delivery models conducted in Section 19.6 and the literature analysis on cloud service domain languages in Section 19.8 have revealed several shortcomings. There have been several attempts to provide more flexibility and bestow cloud-deployed services with integration and customization possibilities in an increased number of phases in the cloud service life cycle exposed by cloud providers [Konstantinou 2009, Hagen 2010, and Morán 2011]. In particular, increased emphasis was given to cloud service integration possibilities. Unfortunately, as will be explained in the following, these efforts were not holistic. They only tackle a limited set of aspects of the cloud application life cycle and their incorporation within the cloud stack.

19.9.1   INFLEXIBLE MONOLITHIC CLOUD SOLUTIONS

Current cloud solutions are fraught with problems. Current solutions hinder interoperability by introducing the following:

1. A monolithic SaaS/PaaS/IaaS stack architecture in which a one-size-fits-all mentality and vendor lock-in prevails. They do not let developers mix and match services from diverse cloud service tiers and configure them dynamically to address client and application needs.

2. Rigid service orchestration practices tied to a specific resource/infrastructure configuration for cloud services at the application level. These issues hamper the (re)configuration and customization of cloud applications on demand to reflect evolving interorganizational collaborations.

Delivering flexible cloud applications that brake vendor lock-in and promote interoperability requires configuring the right mixture of SaaS, PaaS, and IaaS components, which are managed in a consistent manner throughout all the phases of the cloud service life cycle. Clearly, developers must be able to mashup services from a variety of cloud providers to create a cloud ecosystem.

19.9.2 HIGH DIFFUSION OF CLOUD DOMAIN LANGUAGES

The literature analysis on cloud service domain languages in the preceding section has revealed several disturbing phenomena. It has revealed serious gaps and discontinuities between the three families of cloud service domain languages that we examined. This perpetuates the diffusion of cloud development efforts and leads to noninteroperable cloud services that quickly become isolated.

As explained in the preceding sections, most of the work in service domain languages concentrates on IaaS cloud services. Several noteworthy attempts have been made at the IaaS level to introduce IaaS-centered domain languages to promote interoperability either using template- or model-driven approaches. OVF is gaining weight as standard for defining virtual appliances, although many features (e.g., normal network setups and standard inclusion of CSCLs) are still missing.

PaaS cloud service domain languages are scarcer and are mainly based on custom XML formats, which hinder interoperability efforts. An application using a PaaS will likely be locked-in, since its APIs are totally dependent on the format specified by the underlying platform. In addition, some of the PaaS domain languages are based on predefined templates. Templates capture static information while it is exceedingly hard to combine them with other compatible templates to render composite offerings.

The situation is more acute at the SaaS level. SaaS applications introduce a vertical isolation instigated by the nature of the application supported. The very limited approaches in the area are confined to extensions with variability points at the SaaS level to facilitate configurable SaaS applications. These approaches lack nonfunctional and operational constraint descriptions for arbitrary cloud services and focus mostly on integrated application-type services with little consideration for technical cloud service capabilities at the PaaS/IaaS layers that could offer better options for application deployment.

If we now look at the review of CDLs, we can conclude that most of them contain a set of basic information. They describe identifiers, main resources involved, and configuration points across resources (typically linking components at the same level of the cloud stack, for example, configuring Apache web server with the IP of a previously deployed database). CSDLs usually include a description of the functional features of the composing blocks and how they can be connected together to include some interdependencies. Very few approaches, for example, AppScale or Beanstalk, are offering configuration points to express dependencies across the different layers of the cloud stack.

The diffusion between cloud domain languages is evident when it comes to express service dependencies and service behavior rules across languages at the CSCL level. The dependencies are typically expressed as extensions or inclusions of a CSCL into an existent CSDL. First-order logic or LTL-based policies implementing a procedural expression of the rules composing the policy seem to be the predominant choice for CSCLs. Typically, service dependencies are confined among component belonging to the same layer of the cloud stack and do not span across cloud stack layers. The analysis of CSCLs has also revealed that there is not a fully comprehensive proposal covering service dependencies throughout the full cloud application life cycle. Indeed, most of the CSCL efforts remain highly focused on software installation and configuration.

At the CSML level, only essential operations are supported and only basic configuration options can be applied. CSML efforts are usually decoupled from existing CSDL and CSCL approaches. For instance, the algorithms to perform the merger of two CSDL descriptions or two associated CSCL rules detecting and resolving conflicts are not fully supported by most approaches.

19.10   SCAFFOLDING THE CLOUD MONOLITH

The previous section revealed a clear need for emphasizing the development of syndicated cloud applications that provide the ability to provision multiple in-the-cloud services. These syndicated services are implemented by tying together diverse virtual and physical resources across platforms, providers, and geographies on demand.

In cloud syndications, resources have to be dynamically rearranged on the basis of customers' demands and a cost-effective usage of resources to satisfy dynamically changing conditions in accordance with SLAs [Wu 2011].

Cloud syndications allow enterprises to move workloads seamlessly across internal and external clouds according to business and application requirements. This permits BPaaS level applications to dynamically expand or resize their provisioning capability based on sudden spikes in workload demands by acquiring available computational and storage capabilities from diverse cloud service providers. To provide effective solutions, cloud syndications should be formed on demand at any level of the cloud stack that could potentially involve a mixture of SaaS/PaaS/IaaS providers. This requirement can be addressed best by breaking up the current monolithic approach. In fact, breaking the current "cloud monolith" is becoming a priority target to actually deliver the full potential of the cloud [Papazoglou 2011a].

Breaking up the cloud monolith into a number of intermixable SaaS, PaaS, and IaaS components introduces a *syndicated multichannel delivery model*. This model enables a causal correlation of application-level operations to configurable platforms and infrastructure services. It also promotes upstream or downstream vertical aggregation of cloud services from diverse providers who furnish out services at different levels of the cloud stack. Such service aggregations adhere to the principles of separation of service processing concerns to minimize dependencies. This addresses *vertical cloud stack interoperability* concerns.

The objective of the syndicated multichannel delivery model is to deliver reliable, cost-effective, and QoS aware services based on virtualization techniques while ensuring high performance standards. Syndicated multichannel delivery allows any service at any layer to be appropriately combined with a service at the same (horizontal) level of the cloud stack or swapped in or out without having to stop and modify other cloud components elsewhere. This principle addresses *horizontal cloud stack interoperability* concerns.

The syndicated multichannel delivery model encapsulates an interoperability continuum that spans both horizontal and vertical cloud stack interoperability. It therefore enables multiple options of composed resource/infrastructure or implementation options for a given service at any level of the cloud stack. This undertaking is assisted by a knowledge-centered approach.

In this section, we shall describe an attempt to structure and organize cloud-related knowledge required for syndicated multichannel delivery. In particular, we shall concentrate on a knowledge-based framework developed on the basis of what we call *cloud (service) blueprinting* [Papazoglou 2011a]. This approach makes heavy use of knowledge-intensive facilities to support the development of interoperable and portable cloud applications.

Figure 19.2 illustrates how cloud blueprinting helps transform enforce one-way vertical deployment "channels," which are espoused by monolithic cloud stack solutions that permeate the cloud today, into a flexible cloud delivery model. Cloud blueprinting is the missing link, providing a structure that bridges disparate

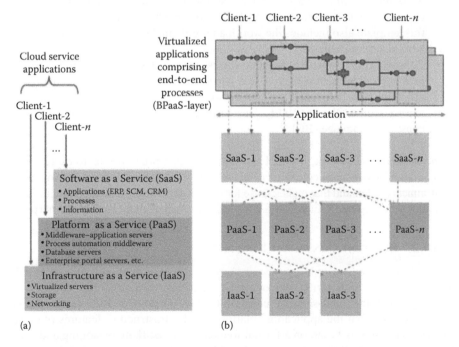

FIGURE 19.2 Transforming a (a) monolithic cloud stack to a (b) syndicated multichannel cloud delivery model.

cloud environments that can become as seamless as it needs to be in a highly controlled fashion. Cloud blueprinting relies on a convergence of cloud service domain definition, constraint, and manipulation languages, which are fused within a unifying framework. In the following, we shall provide an overview of language support in the blueprint framework [Papazoglou 2011a, Nguyen 2011].

19.10.1 BLUEPRINT DEFINITION LANGUAGE

Service providers publish their offerings using a blueprint definition language (BDL) in a cloud service marketplace. For instance, they could describe communication APIs exposed by a Telecom or online billing tools integrated with the contents generated by some TV producers, using a *source blueprint model* in a marketplace. Providers can then choose offerings, compose, extend, and customize this blueprint model to develop full-featured service-based applications.

A blueprint is based on a clear separation of service processing concerns and can be composed of the following number of interrelated templates [Papazoglou 2011a]:

- *Operational service description*: This template focuses on the description of functional characteristics of service such as service types, messages, interfaces, third party included services and operations.
- *Performance-oriented service capabilities*: This template includes key performance indicators (KPIs) associated with cloud services.
- *Resource utilization*: This template describes the resources that are required to run a particular service described in the blueprint model. In general, this template also includes the workload profile including average and peak workload requirements.
- *Policies*: This template prescribes, limits, or specifies any aspect of a business agreement necessary to use a particular service and includes among other things security, privacy, and compliance requirements. This template can be seen as the "hook" to plug in policies expressed in the blueprint constraint language (BCL).

In the blueprint framework, the BDL provides the appropriate means to express visibility constraints, manage compliance, and configure cloud resources and a computing environment on demand that makes sense for a particular workload and computing performance needed. The BCL, which is a natural extension of BDL, fulfills this purpose.

19.10.2 BLUEPRINT CONSTRAINT LANGUAGE

The BCL is the constraint language used to formally express diverse types of BDL policies, such as SLA terms, deployment constraints, data residency constraints, auditing constraints, and security constraints. The purpose of BCL is to achieve consistency between the application, platform, and infrastructure features of cloud services. The BCL is based on a formal foundation to facilitate reasoning and verification by ensuring that services comply with regulations and rules that demarcate their operational behavior.

The developer specifies configurability points (or hooks) in the performance part of the BDL template to capture IaaS or PaaS provided metrics and KPIs that control the behavior of cloud services. The BDL configurability points extend the BDL template with BCL-specific business constraints. BCL can express, for instance, that an online library increases its provided servers and databases to accommodate the surge of workload that occurs yearly during the summer and, more critically, Christmas periods. The provider can specify a set of BCL rules that govern service scalability, portability, and load balancing preferences that rely on resource utilization patterns supplied by a developer in BDL template. In this way, the cross-correlation of appropriate information from the policy section in the BCL template with resource utilization patterns in the BDL template will therefore result in claiming resources in advance to accommodate such high-traffic spikes.

19.10.3 BLUEPRINT MANIPULATION LANGUAGE

An important requirement for the blueprint model is to expose its information in a manner that facilitates comparison and simple composition of blueprints. In this way, developers can customize offerings from various providers. This is achieved by the blueprint manipulation language (BML). BML is a metalanguage that operates on top of BDL- and BCL-defined blueprints on the basis of a set of model-management algebraic operators, such as match, merge, compose, extract, delete, etc. BML is a closed algebra: its operators accept source blueprint templates and return a single modified or aggregated blueprint as a result. The language can also allow blueprints to evolve on the basis of the notion of service version compatibility [Papazoglou 2011b]. Such capabilities allow for changing the blueprint itself to accommodate future (or very specific/customized) demands (Table 19.6).

Essentially, BML can support two different ways of dealing with the blueprint. It can specify a model where:

1. The user has adequate knowledge across layers and decides to specify the behavior of syndicated cloud components that may cut across different layers of the cloud stack using BML operators (see Figure 19.2b).
2. The user fully specifies important blueprint fragments (no matter the layer that they belong to) while leaving other fragments opaque. These opaque fragments are delegated and specified by a chosen provider at run-time. Opaque blueprint fragmentation can be nested: the provider can either meet the request himself or herself or delegate the request further by invoking the services offered by yet another provider.

Using blueprint delegation is an attractive solution as a cloud provider does not need to bind statically to a specific service delivery option but keeps some degree of freedom to pursue a service provision option dynamically to minimize costs, while respecting the original request. For example, consider the case where a client decides to deploy a service from an online service marketplace and chooses to rely on PaaS services exposed by a specific PaaS provider. The PaaS blueprint offerings in the

TABLE 19.6

Operators Support BML Functionality

Specification	Translate	Match	Merge	Compose	Extract	Delete
XSLT	Y	Y	Y	Y	N	N
XQuery	Y	Y	N	N	N	N
QVT	Y	Y	N	N	N	N
Database schema matching ops	Y	Y	Y	Y	Y	Y

marketplace library are described in BDL (with QoS constraints specified in BCL). This situation is illustrated in Figure 19.3a. Now, the PaaS provider may determine the actual service enactment model. For instance, it may decide to deploy the services locally over an Azure cluster in its local datacenter and provide an inexpensive computing platform in the cloud to avoid paying for external services, for example, AWS. Alternatively, it may decide to deploy AppScale on top of Amazon's AWS. This is perfectly acceptable solution as long as the consumer of the deployed service accepts the QoS guarantees stipulated in the BDL/BCL template. In turn, an IaaS provider associated with this PaaS service may place VMs in the same racks in a datacenter (for all the deployed services, PaaS or SaaS). In this way, providers

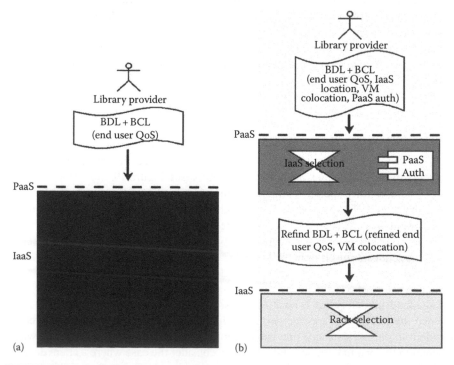

FIGURE 19.3   (a) Delegation and (b) refinement of a user request in DML as it traverses the cloud stack.

downstream can optimize their own resources in an opaque manner, that is, they can outsource some services to third parties, while fully preserving the overall QoS.

Figure 19.3b illustrates a situation that involves a hybrid request. In this figure, a provider in the marketplace library is shown to request a specific authentication services from a PaaS provider (rather than deploying it itself). The library provider may also specify some geographical and deployment constraints. For instance, that all VMs must reside in the same location and must be placed in the same rack. In Figure 19.3b, the provider in the marketplace library specifies that the geographical and deployment options associated with the request, for example, select a specific PaaS container for the authentication services, select among different IaaS providers for the best rack to host the VM, and so on, can be delegated to a third party (e.g., another PaaS provider) in the marketplace.

19.10.4 LIFE CYCLE SUPPORT

As we already explained in our earlier example, cloud services in the blueprinting approach are available via a cloud marketplace. This comes in a form of an online platform that manages service distribution and bid management. In a marketplace, providers store their offerings, and clients can discover and deploy third-party cloud applications that they can integrate with their own. The concept is similar to that of the Google Apps Marketplace. To understand how the cloud blueprint model facilitates this process, consider Figure 19.4.

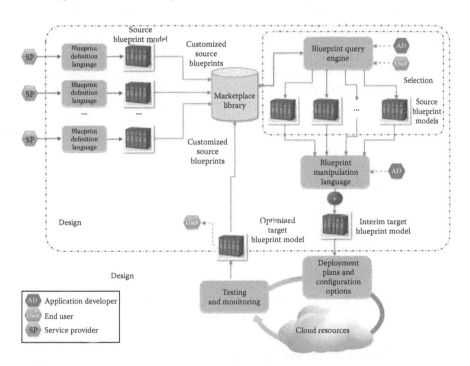

FIGURE 19.4 Blueprint support for the cloud service life cycle.

In Figure 19.4, we illustrate how the elements of the blueprint model (the BDL, BCL, and BML), the marketplace library, and blueprint templates are connected and work in tandem to address application needs. This figure also shows how the various constructs of the blueprint framework interact to support the phases of cloud service life cycle.

After a provider (or developer) has created all the components of a discrete service, it uses BDL to describe all its relevant aspects in a structure called a source blueprint, which it stores in the marketplace library. The provider or developer also customizes the source blueprint templates to create a service offering for different types of consumers. Each such offering is accompanied by an SLA to delimit the range of service instances defined in the offering. This happens during the blueprint model's design phase.

Now consider a virtual service operator who wishes to bundle together several interactive telecommunications services to produce a turnkey application. This developer must first discover these discrete services' source blueprint models. This happens during the selection phase as Figure 19.4 shows. Subsequently, the developer creates an *interim target blueprint* model by combining the set of source blueprint models it selected. For this purpose, it uses the BML operators. At this point, the interim blueprint model generates a deployment plan, which might drive PaaS resources and determine VM placement and network configuration. The developer could, for instance, choose to deploy different PaaS options to address customization requirements. The application might also deliver the same content over diverse devices, such as TVs, PCs, and smart phones, and provide discounts to certain customers during low network-utilization periods. PaaS services for this type of application could involve workflow facilities, event-processing functionality, deployment, and hosting. As usual, not all these services must come from the same third-party PaaS provider.

The deployment plan matches the service-based application's demands, addresses scalability estimates, and, in general, tries to optimize the service assembly's performance according to QoS requirements specified in the interim target blueprint model. This helps to provision resources and adjust the workload and traffic during the deployment phase. It also provides upstream and downstream alternatives to automate the dynamic configuration and deployment of application instances onto available cloud resources.

Finally, during the testing and monitoring phase, the developer gradually refines the abstract information contained in the interim target blueprint to reach the level of rigor and concreteness required for a production-ready cloud application. This results in an *optimized target blueprint* model describing the integrated application. The developer then publishes this blueprint model in the marketplace library for potential clients to discover and use it.

19.11 SAMPLE CLOUD APPLICATION USING BLUEPRINTING

To exemplify the blueprint approach, we introduce in this section a fairly simple cloud application that involves interacting telecommunications services, such as a mobile video game downloaded from a telecommunications marketplace (e.g., Android's) into a user's device. This mobile video game application uses

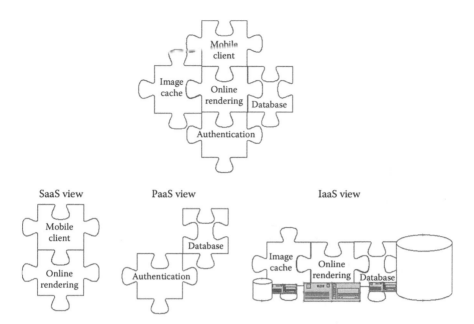

FIGURE 19.5   Views of a sample gaming application formed by integrating different IaaS/PaaS/SaaS cloud elements.

rendering techniques to generate images from scene files. A scene file contains objects that contain geometry, viewpoint, texture, lighting, and shading information as a description of the virtual scene. The top view in Figure 19.5 shows a complete perspective of all the integrated components required to create a fully functional mobile video game application.

The perspective of the application developer who creates a mobile video game is given by the SaaS view in the bottom part of Figure 19.5. The SaaS view provides an overall view of the different SaaS elements of the application, such the mobile client and the rendering service. The other parts in the bottom portion of Figure 19.5 represent the PaaS and IaaS layers in the cloud stack where different components, such as authentication, databases, image caches, etc., reside.

In our example, we assume that the application developer decides to use available services in the cloud, rather than reimplementing them. To create a complete mobile video game application, a developer configures the application by appropriately combining distinct software services from diverse cloud service tiers, such as the ones exposed by disparate providers and shown as untied jigsaw pieces at the bottom part of Figure 19.5. These providers describe their respective service using an appropriate combination of BDL and BCL source blueprints. As usual, the blueprints are stored in a service marketplace for service developers to discover and integrate them into a composite application as needed.

Figure 19.6 shows a set of blueprints required to deploy and operate a part of the mobile video game application. The figure shows diverse configuration options: a dark gray shade means complete lack of visibility and no configuration capabilities, while a white shade means complete visibility and diverse configuration options.

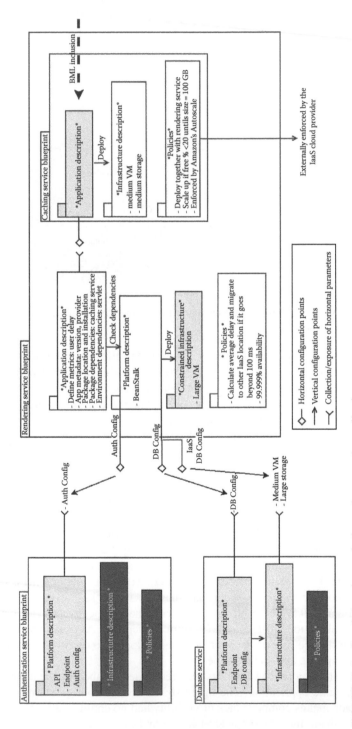

FIGURE 19.6   Set of blueprints to deploy and operate the application.

In the latter case, nonmandatory parameters can be left unspecified for an arbitrary provider. For instance, unspecified parameters at the SaaS level imply that an arbitrary provider at a downstream level of the cloud stack (e.g., PaaS or IaaS) can fulfill them. As usual, blueprint descriptions in Figure 19.6 are in BDL, while policies are expressed in BCL.

A developer may use the merge operation to aggregate two different source blueprints and specify several cross blueprint dependencies at various levels of the cloud stack. The merge operation is used when the blueprints are fully compatible. For instance, to improve performance, the game service provider may decide to deploy a rendering server in close proximity to a given number of clients. This rendering service, in turn, may be tied to an image cache that stores the most used rendered images (these components always need to be colocated). The cache presents a variety of clients (players) with the same set of images. This means that players at a given stage of the game will have a common view of the scenario they are engaged in. Performance levels are denoted by a series of application-level KPIs. In our scenario, such KPIs could include rendering time (including network delays), which is calculated and sent by each client application to the rendering service. These constraints are application-related and aim to maximize the quality of experience of the online player.

In Figure 19.6, we assume that the authentication service is offered by a third-party online. Consequently, there is no need for the developer to package it and deploy it inside a previously deployed physical or VM. As shown in Figure 19.6, specifying appropriate horizontal configuration hooks can achieve this outcome. In this figure, we assume that the PaaS provider who supplies the "authentication service" delivers a service that does not reveal any detail of the underlying platform or infrastructure for others who use it to configure their services. Figure 19.6 signifies this fact by means of the two dark gray boxes in the bottom part of the "authentication service blueprint." This implies that a hook with appropriate configuration points, such as a list of potential users and permissions, and groups, is the only visible item exposed from the authentication service.

In addition to the aforementioned, we assume that the developer chooses an online database service to avoid dealing with replication or backups for the game images. In this case, the database PaaS provider offers some degree of transparency for its users: they could decide about the size of the VM and the storage attached to that VM. This is again shown in Figure 19.6. In this way, the developer can chose larger chunks of storage for a mid-sized VM instance to improve performance. Both the authentication and the database service and its associated images are deployed only once and are shared among all users of the game. The developer itself is not burdened by configuring, maintaining, backing-up, or replicating the database service. The PaaS provider accomplishes all these tasks on behalf of the developer. The rendering service is, however, required to configure and populate the database.

The developer can now focus on creating a rendering sever. The rendering service blueprint definition indicates that reusing code from an existing Servlet is the preferred option. This means that the developer needs to specify the required container for the Servlet to run. This can be denoted as a vertical configuration point (check dependencies) in Figure 19.6. We assume that in order to maintain the code unchanged as much as possible, the developer decides to use a PaaS container with reasonable Servlet compatibility, for example, BeanStalk. The developer also specifies in the rendering service BDL that it needs a large VM for hosting the PaaS container. The light gray box, which is part of the "Rendering Service Blueprint" in Figure 19.6, signifies this fact.

An interesting case is when a developer needs to ascertain that two arbitrary services are compatible before it merges them. Figure 19.6 illustrates this situation again. It shows that as part of the rendering service blueprint, the developer has placed a requirement for using a rendered image cache to reduce the response time perceived by a player. Here, we note that the caching service may itself include a server embedded in a Tomcat-like container (PaaS-SaaS dependency) that keeps images in a mid-sized repository. The BDL description for this software package is imported from the marketplace into the caching service blueprint (the dashed arrow at the right hand side of Figure 19.6 signifies the BML include operator). This container service needs to be ultimately aggregated with the rendering service. Here, we assume that these two services are partially compatible. To discover how these two services can be transformed to create a cached rendering service, the developer needs to use the BML match operator. This operator will indicate which mappings need to be performed between the rendered image cache service and the rendering source blueprints to eventually create a cached rendering service target blueprint. In this example, we assume that the BCL of the caching service specifies to scale the storage capacity of the cache up to a maximum level when new replicas are triggered. This could be accomplished by specifying additional BCL configuration points (e.g., scaling rules based on current storage size in Amazon's AutoScale), which are introduced by the cloud service provider.

In addition to defining and configuring individual services, the game application may include KPIs and associated rules that cannot be directly mapped into a single component, but belong to the application as a whole. In our example, user-perceived delay and the geocoordinates of the players are such application-wide KPIs. Several options exist:

1. The application provider includes a series of predefined probes monitoring the relevant KPI (e.g., CPU probes offered by the infrastructure provider, such as Amazon's CloudWatch) and a prepackaged "QoS Manager" provided by a third party that collects the metrics and checks whether the required conditions are met to trigger subsequent user-specified actions.
2. The application provider specifies an application component in order to deal with these application-wide elements.

If we consider the second option, the target BCL template of the aggregated cached rendering service must accommodate these KPIs, rules, and actions. For instance, the final implementation of the rendering service may include a specific thread to trigger the repositioning of the rendering service on top of an IaaS provider that delivers less delay to players.

19.12   SUMMARY

One of the greatest challenges facing longer-term adoption of cloud computing is the ability to automatically provision services, effectively manage workload segmentation and portability (i.e., the seamless movement of workloads across many platforms and clouds), and manage virtual service instances, all while optimizing the use of cloud resources and accelerating the deployment of new services. Within such a cloud environment, it is also important to equip developers with a unified approach that lets them develop cloud applications on top of existing applications at any layer of the cloud stack from multiple cloud providers. This includes the ability for opportunistic and scalable provisioning of application services, consistently achieving QoS targets under variable workload, resource, and network conditions.

The cloud blueprinting approach that we presented in this chapter greatly helps cloud application developers deal with such challenges. It greatly improves cloud service interoperability while alleviating the pains of vendor lock-in. The overarching goal of cloud blueprinting is to break the cloud stack monolith by creating a configurable cloud computing environment that syndicates services on demand across all layers of the cloud stack no matter whether they are supplied by diverse providers. Cloud blueprinting supports dynamic expansion or contraction of capabilities (VMs, application services, storage, and databases) for efficient workload segmentation and handling of variations in cloud service demands. This approach transforms the fabric of the current inflexible service delivery models by making heavy use of knowledge-intensive techniques. Such knowledge-intensive techniques rely on the use of cloud service definition, constraint specification, and manipulation languages.

REFERENCES

[Armbrust 2009] M. Armbrust et al., Above the clouds: A Berkeley view of cloud computing, University of California at Berkeley, Technical Report, February 2009, available from: http://berkeleyclouds.blogspot.com/2009/02/above-clouds-released.html (accessed on August 2011).

[Benslimane 2008] D. Benslimane, S. Dustdar, A. Sheth, Services mashups: The new generation of web applications, *IEEE Internet Computing*, 12(5) (September–October 2008), 13–15.

[Bernstein 2009] D. Bernstein et al., Blueprint for the inter-cloud: Protocols and formats for cloud computing interoperability, *International Conference on Internet and Web Applications and Services ICIW 2009*, Venice, Italy, May 2009.

[Bernstein 2010] D. Bernstein, D. Vij, Intercloud directory and exchange protocol detail using XMPP and RDF, *Proceedings of the 2010 6th World Congress on Services (SERVICES'10)*. IEEE Computer Society, Washington, DC, pp. 431–438.

[Bunch 2011] C. Bunch et al., Neptune: A domain specific language for deploying HPC software on cloud platforms, *Proceedings of the 2nd Workshop on Scientific Cloud Computing (ACM ScienceCloud)*, Chicago, IL, June 2011.

[Cai 2009] H. Cai et al., Customer centric cloud service model and a case study on commerce as a service, *Proceedings of the 1st IEEE International Conference on Cloud Computing*, Bangalore, India, September 2009.

[Cartlidge 2007] A. Cartlidge et al., An introductory overview of ITIL, The IT Service Management Forum, 2007. http://www.itsmfi.org/ (accessed on August 2011).

[Chieu 2010] T.C. Chieu et al., Solution-based deployment of complex application services on a cloud, *Proceedings of International Conference on Service Operations and Logistics and Informatics*, Qingdao, Shandong, China, IEEE CS, August 2010.

[Chohan 2009] N. Chohan et al., AppScale: Scalable and open AppEngine application development and deployment, *ICST International Conference on Cloud Computing*, Bangalore, India, September 2009.

[Collazo-Mojica 2010] X. Collazo-Mojica et al., Virtual environments: Easy modeling of interdependent virtual appliances in the cloud, *Proceedings of SPLASH 2010 Workshop on Flexible Modeling Tools*, Reno, NV, October 2010.

[DMTF 2009] Distributed Management Task Force, Interoperable clouds, White Paper CIM Version 1.0, document DSP-IS010, November 2009, available from: http://www.dmtf.org (accessed on August 2011).

[Elgammal 2010] A. Elgammal et al., Root-cause analysis of design-time compliance violations on the basis of property patterns, *8th International Conference on Service-Oriented Computing (ICSOC: 2010)*, San Francisco, CA, December 2010.

[El Maghraoui 2006] K. El Maghraoui et al., Model driven provisioning: Bridging the gap between declarative object models and procedural provisioning tools, *Proceedings of the ACM/IFIP/USENIX 2006 International Conference on Middleware (Middleware'06)*, Melbourne, Victoria, Australia, December 2006.

[Fehling 2011] C. Fehling, R. Mietzner, Composite as a service: Cloud application structures, provisioning, and management, *IT—Information Technology*, 53(4), 2011, pp. 188–194.

[Galán 2009] F. Galán et al., Service specification in cloud environments based on extensions to open standards, *4th International Conference on Communication System Software and Middleware*, Dublin, Ireland, June 2009.

[Ghallab 1998] M. Ghallab et al., PDDL: The planning domain definition language, AIPS Planning Competition Committee, Yale Center for Computational Vision and Control, Technical Report TR-98-03, October 1998.

[Goldsack 2009] P. Goldsack et al., The SmartFrog configuration management framework, *SIGOPS Operating Systems Review*, 43(1), January 2009, 16–25.

[Hagen 2010] S. Hagen, A. Kemper, Model-based planning for state-related changes to infrastructure and software as a service instances in large data centers, *3rd International Conference on Cloud Computing (CLOUD)*, Miami, FL, July 2010.

[Hamdaqa 2011] M. Hamdaqa, T. Livogiannis, L. Tahvildari, A reference model for developing cloud applications, *Proceedings of 1st International Conference on Cloud Computing and Services Science (CLOSER'11)*, Noordwijkerhout, the Netherlands, May, 2011.

[Keller 2004] A. Keller et al., The champs system: Change management with planning and scheduling, *IEEE/IFIP Network Operations and Management Symposium (NOMS 2004)*, Seoul, Korea, April 2004.

[Konstantinou 2009] A.V. Konstantinou et al., An architecture for virtual solution composition and deployment in infrastructure clouds, *3rd International Workshop on Virtualization Technologies in Distributed Computing*, Barcelona, Spain, June 2009.

[Koslovski 2008] G. Koslovski et al., VXDL Virtual resources and interconnection networks description language, *Proceedings of 2nd International Conference on Networks for Grid Applications (GridNets 2008)*, Beijing, China, October 2008.

[Kurz 2011] T. Kurz et al., Cloud federation, *Proceedings of the 2nd International Conference on Cloud Computing, GRIDs, and Virtualization, CLOUD COMPUTING 2011*, Rome, Italy, September 2011.

[Lupu 1999] E. Lupu, M. Sloman, Conflicts in policy-based distributed systems management, *IEEE Transactions on Software Engineering*, 25(6), November/December 1999, 852–869.

[Lymberopoulos 2003] L. Lymberopoulos, E. Lupu, M. Sloman, An adaptive policy-based framework for network services management, *Journal of Network and Systems Management*, 11(3), 2003, 277–303.

[Mell 2009] P. Mell, T. Grance, The NIST definition of cloud computing, *National Institute of Standards and Technology*, 53(6), 2009, 50, available from http://csrc.nist.gov/groups/SNS/cloud-computing/cloud-def-v15 doc (accessed on August 2011).

[Mietzner 2008] R. Mietzner, F. Leymann, M. Papazoglou, Defining composite configurable SaaS application packages using SCA, variability descriptors and multi-tenancy patterns, *3rd International Conference on Internet and Web Applications and Services*, Athens, Greece, June 2008.

[Morán 2011] D. Moran, L.M. Vaquero, F. Galan, Elastically ruling the cloud: Specifying application's behavior in federated clouds, *4th IEEE International Conference on Cloud Computing (CLOUD 2011)*, Washington, DC, July 2011.

[Nguyen 2011] D.K. Nguyen et al., Blueprint template support for cloud-based service engineering, *Proceedings of the 4th European Conference ServiceWave 2011*, Poznan, Poland, Springer-Verlag LNCS 6994, October 2011.

[OVF 2010] Virtualization management (VMAN) initiative, distributed management task force, Inc., available at: http://www.dmtf.org/standards/mgmt/vman/ (accessed on August 2011).

[Pandey 2011] S. Pandey, D. Karunamoorthy, R. Buyya, Workflow engine for clouds, in *Cloud Computing: Principles and Paradigms*, R. Buyya, J. Broberg, A. Goscinski (eds.), Wiley Press, New York, February 2011.

[Papazoglou 2011a] M.P. Papazoglou, W.J. van den Heuvel, Blueprinting the cloud, *IEEE Internet Computing*, 15(6), November 2011, 74–79.

[Papazoglou 2011b] M.P. Papazoglou, V. Andikopoulos, S. Bernbernou Managing Evolving Services, *IEEE Software*, 28(3), May/June 2011, 49–55.

[Papazoglou 2012] *Web Services and SOA: Principles and Technology*, 2nd edn., Prentice-Hall, Upper Saddle River, NJ, January 2012.

[Parameswaran 2009] A.V. Parameswaran, A. Chaddha, Cloud interoperability and standardization, *SETLabs Briefings*, 7(9), 2009, 19–27.

[Ramshaw 2006] L. Ramshaw et al., Cauldron: A policy- based design tool, *Proceedings of 7th IEEE International Workshop on Policies for Distributed Systems and Networks (POLICY 2006)*, Ottawa, Ontario, Canada, June 2006, IEEE Computer Society.

[Rodero-Merino 2010] L. Rodero-Merino et al., SOA and cloud technologies, *CEPIS Upgrade*, 11(4), August 2010, available from: http://www.cepis.org/upgrade/media/rodero-merino.IV.20101.pdf (accessed on August 2011).

[SNIA 2010] Storage Networking Industry Association, Cloud data management interface, SNIA Technical Position, April 12, 2010, available from: http://www.snia.org/cdmi (accessed on August 2011).

[Stankov 2010] I. Stankov, R. Datsenka, Platform-as-a-service as an enabler for global software development and delivery, *Proceedings of the Multikonferenz Wirtschaftsinformatik— (MKWI Software-Industrie)*, Göttingen, Germany, February 23–25, 2010.

[Thrash 2010] R. Thrash, Building a cloud computing specification: Fundamental engineering for optimizing cloud computing initiatives, Computer Science Corporation (CSC) Whitepaper, August 2010, available from: http://assets1.csc.com/lef/downloads/CSC_Papers_2010_Building_a_Cloud_Computing_Specification.pdf (accessed on August 2011).

[Vaquero 2011] L.M. Vaquero, L. Rodero-Merin, D. Morán, Locking the sky: A survey on IaaS cloud security *Computing*, Springer, 91(1), January 2011, 93–118.

[Wu 2011] L. Wu, et al., SLA-based resource allocation for a software as a service provider in cloud computing environments, *Proceedings of the 11th IEEE/ACM International Symposium on Cluster, Cloud and Grid Computing*, Los Angeles, CA, May 2011, IEEE Computer Society Press.

[Youseff 2008] L. Youseff, M. Butrico, D. Da Silva, Towards a unified ontology of cloud computing, *Grid Computing Environments Workshop (GCE'08)*, Austin, TX, November 2008.

20 Complex Service Systems

Knowledge-Based User-Centered Systems Engineering for Performance Improvement

Tareq Z. Ahram and Waldemar Karwowski

CONTENTS

20.1 INTRODUCTION

Businesses thrive to differentiate themselves under pressures of rapid technological change, competitors, and regulatory challenges. These pressures force companies to continuously search for new technologies that help improve their internal and external processes to reduce the time required to market new or improved products and services. The most prominent issue is the lack of automation when trying to integrate new business requirements into existing services and systems, adding new components, and adapting existing processes as priorities and perspectives change. Increased complexity of service systems and the underlying knowledge required for those systems has influenced industrial and economic growth in many nations. These developments have reinforced the need for emphasizing the role of information and knowledge in service systems. Service systems sparked a revolution whereby the traditional industrial society that emerged over the last two centuries is being rapidly overtaken by the new information society (Hauknes, 1996; Ahram et al., 2010c). The design of complex service systems considers qualitative attributes between human–human and human–machine interactions. These considerations encompass service employees and associates (i.e., those who provide the service) as well as service users or stakeholders (i.e., those who receive and use the service). The service system design process also includes the necessary objects and/or components that constitute successful business decisions and therefore competitive service offering.

User-centered design (UCD) is a proactive approach for making informed and appropriate design decisions. Executives may feel pressured to make quick, unfounded, and aggressive business decisions. Such impulsive choices can be disastrous in detailed and far-reaching design activities. Business process modeling ontology (BPMO) defines the link between processes and organizations in an effort to enhance the decision-making activities. Business process modeling through UCD provides adequate data to make rational and substantiated decisions regarding business process management actions, user experience choices, and usability of services and products. A reactive approach to feedback is no longer adequate. Key decisions must ensure that the root causes of problems are addressed and that a user-centric methodology be employed. According to Nadel and Piazza (2005), two of the many justifications for employing a user-centric method are as follows:

1. Users of the service often have unspoken needs that are hard to articulate. Through systematic data gathering, a skilled analyst can uncover how users conceptually think about a task. Only then is it possible to design an intuitive navigational scheme that reflects this mental model.
2. Usability problems are often symptomatic of deeper business issues.

A major consumer electronics company found that its e-commerce site was drastically underperforming that of the competition's site. It turned out that the site was confusing customers due to its structure. The site was organized around the company's separate business units for internal political reasons. Each division was so territorial that they couldn't agree on a unified site structure—each wanted their own homepage. As a result, customers could not compare and order a product from the same section of the site. If they found a television matching their preference, they had to return to the main corporate homepage and then navigate to another area to actually buy it!

UCD requires stakeholder analysis to determine which business needs are being addressed and what the organization really wants to achieve. A design that attempts to achieve business objectives but ignores the actual users is bound to fail.

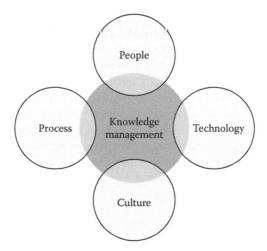

FIGURE 20.1 Four pillars of knowledge management. (Modified from the original by Bourdreau, A. and Couillard, G., *Inform. Syst. Manage.*, 16, 4, 2006.)

The Center for Business Knowledge has identified four pillars of knowledge management, as shown in Figure 20.1: people, process, culture, and technology. With process and user knowledge as a foundation, this will enable conducting end-user data gathering. The insight this provides about user behavior lets us design an interface/architecture that is intuitive and logical. Success must ultimately be measured against the following parameters: Does the site/application meet business needs and can users efficiently accomplish what they want to do?

Management of complex service systems is characterized into four main dimensions (Fähnrich and Meiren, 2007; Karwowski et al., 2009):

1. *Structure*—human, material, information, communication, technology, resources, and operating facilities
2. *Process*—process model and service provision
3. *Outcome*—product model, service content, consequences, quality, performance, and standards
4. *Market*—requirement model, market requirements, and user needs

Although the field of UCD is well suited for evaluation and analysis, many service system designers stumble when using user-centered approaches as tools for iterating process design solutions. Human factors engineering (HFE) and usability evaluation (UE) often occur too late in the design process to effect meaningful change, or changes cannot be made due to cost or schedule. However, HFE and usability require some semblance of a design to analyze, so evaluations cannot be implemented from the very start either. To mediate this, heuristic general HFE design practices should be employed at conceptual stages, with appropriate HFE and UE practices occurring at service maturity levels. Such early intervention and ongoing iterative evaluation would greatly aid in the design and execution of existing business process service prototypes.

This chapter defines complex service system design factors from a user-centered prospective by adopting concepts from Metters et al. (2003), who defines 13 distinct service system design components: facility location, facility layout, service product, scheduling, worker skills, quality control, demand and capacity planning, technology level, sales opportunity, standardization or customization, user contact time, personnel interaction, and user participation. These components share features with the user-centered approach in service system design and modeling. Service system design extends basic design concepts to include experiences that clients have with products and services. It also applies to the processes, strategies, and systems that are behind the experiences. Previous research in service engineering indicated the importance of the following user-centered service system design components:

- Users or service users (goals, needs, behavior, demographics, and psychographics)
- Contexts (political, legislation, economic, social, technological, and competition)
- Service employees or providers (resources, constraints, processes and systems, and delivery languages)
- Relationships (opportunities, and other providers)

20.2 HUMAN FACTORS AND BUSINESS PROCESS MODELING FOR SERVICE INDUSTRY

According to Durkin (1994), knowledge is represented by the understanding of the field of study, including concepts, facts, and relations among different variables and mechanisms for how to combine them to solve problems. Artificial intelligence (AI) attempts to approximate human reasoning by organizing and manipulating factual and heuristic knowledge. AI tries to understand the nature of human knowledge and mental skills. Since computers do not have as efficient a mechanism for collecting, representing, and sharing knowledge as that of their human counterparts, AI has emulated a number of representations for various types of human knowledge, each associated with a structured method to encode knowledge in a system and a specific related data structure. Key issues in AI are to study human thinking in terms of representational mind structures and computational procedures that operate on those structures (Hofstadter, 1994). Davis et al. (1993) and Gašević et al. (2006) listed five crucial characteristics for efficient human knowledge representations:

1. Knowledge representation serves as a substitute for concepts in the real world.
2. Ontological representation operates by expressing and highlighting only relevant parts of information about the world important to logic and ignoring the rest; the essential part is not the form but the content.
3. Every representation reflects at least a partial aspect about how people reason intelligently and what it means to reason intelligently.

4. Every representation, by its nature, provides some guidance (i.e., medium for efficient computation) about how to organize knowledge in order for computation with it to be most efficient.
5. Knowledge representation is a medium for human expression such that humans can create and communicate representations to machines and to each other.

Performance requires, in addition to technical skills, a raft of decision making and process skills to achieve success. The skills development process must be derived from the strategic business plans. It should be geared toward building the competence required to achieve the service-specific goals and objectives in terms of the delivery of services or products (Hattingh, 2009). The most important success factor is to integrate system skills development into the broader skills planning process of the service provider.

Skills development is the structured process of analyzing the skill needs of employees in the organization to determine what skills must be developed and then planning and implementing interventions that will build the capacity of individuals to be able to do things and perform service tasks while adhering to relevant standards and regulations. Skills development and training should not only be focused on developing the skills of underperformers. It is equally important to nurture the top performers, because they are generally the organization's most valuable human capital. Their value could be increased through more in-depth training so that they can be even more valuable to the organization. Other options would be to promote them so that they can apply their skills in more senior positions where they will have a greater impact, or use them to mentor new employees or for on-the-job training. Neglecting the development of already skilled employees often results in the loss of these skills as employees retire or resign to move to organizations where they feel that their competence is valued. Service system skills and functions planning includes all the processes and steps taken to ensure that smarter service system providers have the necessary skills to perform the functions they are required to perform. This includes processes such as recruitment, succession planning, mentoring, and coaching, as well as planning for skills development (Ahram et al., 2011). This would cover the short- and long-term plans relating to the acquisition of skilled people, nurturing the skills that are in the organization, as well as training people to develop new skills.

20.2.1 HUMAN PERFORMANCE FOR SERVICE SYSTEMS

The human performance improvement (HPI) approach in service systems recognizes that any development effort implemented in isolation will produce only limited results. Implementing new service functions or procedure without appropriate training or coaching will not produce improvement. Similarly, any training program implemented without tools and manager support will also have limited impact. According to Wilson Learning (2011), HPI is a process and creative approach to training that acknowledges the critical role of tools, measurement, and management support that will enhance and extend the impact of learning. There are three elements to service system training and development effort that are critical to creating maximum results:

1. *Establishing a business case*—establishing a business case for performance improvement by linking specific strategic drivers to skill requirements
2. *Understanding the challenge*—making effective decisions about what skills to focus on, selecting delivery methods, and determining how to integrate services into the organization and how to align all key stakeholders to support this approach to performance improvement
3. *Creating integrated solutions*—creating the performance improvement elements necessary to address the challenge and accomplish the strategy, this includes
   a. Developing learning components to deliver the knowledge, skills, and abilities
   b. Developing work tools and process to support the use of the learning
   c. Providing organizations with the ability to track the impact of the learning on performance (e.g., business performances analytics)
   d. Ensuring service planners and manager support and that they are prepared to coach the application of the skills

Quality skills development that makes a measurable impact in the organization requires proper planning and preparation. This involves a structured process of analysis, design, development, and implementation. Implementation must be accompanied by the evaluation of the quality and relevance of all learning programs. Figure 20.2 shows where the skills training needs analysis fits into the overall process of ensuring that service system providers have the skills required to fulfill their core processes.

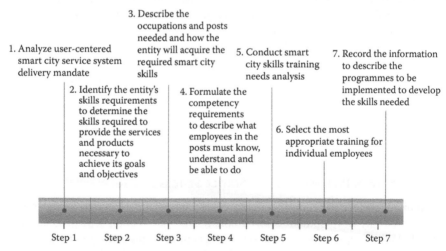

FIGURE 20.2   Service system skills and training needs analysis in service development process. (Adapted from the framework by Ahram, T.Z. et al., *Electronic Globalized Business and Sustainable Development through IT Management: Strategies and Perspective*, IGI Global, ISBN-13: 978-1615206230, 2010c, and Skills Development Facilitation (SDF), *Guide for the Public Sector*, Chapter 3, p. 12, 2006, http://www.pseta.gov.za/)

This is one of the most important parts of the skills development process. A proper needs analysis will ensure the relevance of the training that is delivered (Hattingh, 2009). The purpose of the training needs analysis is to identify the skill gaps and training needs of employees. This information will then be used to determine which employees should be trained in which areas. To determine an employee's training needs, planners start with a description of the competence employees should have. They need to know what employees in particular posts should know and understand to be able to perform competently in their jobs (see Figure 20.2). This will be used as a basis for determining what their current competence is. Once this information is collected, management will be able to determine the training needed to fill the gap between what their competence should be and their current competence level.

Development of a conceptual skills and performances model determines whether the ideas generated about how the systems should look and behave will be perceived clearly by the end user in the manner intended. The framework provided by Norman (1988, 1993) (see Figure 20.3) illustrates the relationship between the design of a conceptual model and the end user understanding of product usage. In Figure 20.3, there are three interacting components: the designer, the end user, and the system. Preece et al. (2002) indicated that behind each of these are three interlinking conceptual models:

- *The design model*—designer's impression of how system should work
- *The system component*—how the system actually works
- *The user's model*—the end user's understanding of how the system works

Preece et al. (2002) also indicated that in an ideal world, all three conceptual models should map onto each other. End users should be able to carry out their tasks in the way intended by the designer through interaction with the system. However, if the system

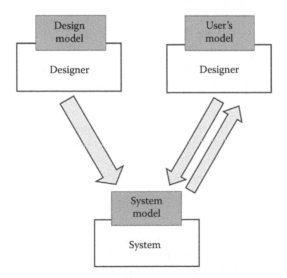

FIGURE 20.3 Conceptual model development. (From Preece, J. et al., *Interaction Design: Beyond Human-Computer Interaction*, John Wiley & Sons, New York, 2002.)

image does not make the design model clear to the end users, it is likely that they will end up with an incorrect understanding of system operation and, as a result, make errors. A sequence of learning about the technology, science, engineering, and math (STEM) concepts, as well as the processes of research, design, manufacture, construction, testing, and communication, needs to be created. Novice and expert system architects and designers need the background knowledge and skills before they are able to successfully contribute to a project or design task. These engineering skills can be summarized into the following general categories:

- An understanding of basic science principles
- Application of the principles of science to the solution of engineering problems
- Applying computer application skills
- Recognizing the system components and processes of a technological system
- Solving problems using engineering tools and resources
- Developing and following a plan for the solution of a design problem
- Understanding the properties of different materials
- Selecting the materials and processes necessary for developing a solution to engineering problems
- Carrying out prototype development and testing
- Applying critical and logical thinking skills
- Explaining the impact of business on engineering

The fundamental principle of a competitive service design strategy is to give novice designers a design challenge and let them develop conceptual design solutions and critical thinking skills by applying collaborative design and problem solving (Karwowski et al., 2009). If the design challenge results in something tangible, for example, a manufactured or constructed object, this dramatically increases the motivation of teammates to push themselves. The challenge could take the form of a competition, which is an exhilarating experience for novice designers and researchers.

20.3   PRINCIPLES OF USER-CENTERED SKILLS AND PERFORMANCE IMPROVEMENT FOR SERVICE SYSTEMS

Service system design is characterized by relationships between knowledge and technology. This includes the human and business process knowledge required to deliver the service, whether it is invested in the technology or core service applications. Knowledge requirements in service system design and modeling have been categorized into three main categories: *knowledge based, knowledge embedded, and knowledge separated* (McDermott et al., 2001). The knowledge-based service system, such as teaching, depends on user knowledge to deliver the service. This knowledge may become embedded in a product or process that makes the services accessible to more people. An example of this is logistics providers, where the technology of package delivery is embedded in service system computers that schedule and route the delivery of packages. It is important to note, however, that delivery personnel contribute to the critical components of both delivery and pickup. Their process knowledge is crucial to satisfying users and providing quality services.

This approach contributes to the business process development rather than replacing them. The following key principles of the user-centered business process service systems have been identified (Karwowski et al., 2009):

- *Clear understanding of user and task requirements*—Key strengths of user-centered service system design are the spontaneous and active involvement of service users and the understanding of their task requirements. Involving end users will improve service system acceptance and increase the commitment to the success of the new service.
- *Consistent allocation of functions between users and service system*—Allocation of functions should be based on full understanding of user capabilities, limitations, and task demands.
- *Iterative service system design approach*—Iterative service system design solutions include processing responses and feedback from service users after their use of proposed design solutions. Design solutions could range from simple paper prototypes to high-fidelity service system mock-ups.
- *Multidisciplinary design teams*—User-centered service system design is a multitasking collaborative process that involves multidisciplinary design teams. It is crucial that the service system design team comprise professionals and experts with suitable skills and interests in the proposed service system design. Such a team might include end users, service handlers (front-stage service system designers), managers, usability specialists, software engineers (backstage service system designers), interaction designers, user experience architects, and training support professionals.

Users of the service develop knowledge in order to use the service. In knowledge-separated service systems, the service may be accessible to users without the need to interact with another human in the service loop. An example of this is the electronic ticketing systems and ticketing kiosks at airports or cinema theaters, which replace the airline representative or sales representative. The knowledge of the sales representative is now fully embedded in the ticketing kiosk logic, and a user must only have the knowledge to operate the system. All these components are incorporated and organized in a scheme originating from a generalized definition of a system (Checkland, 1981). For example, the goals of the electronic ticketing system are

- To provide anytime anyplace service to the customer
- To provide services that are not available in the existing system
- To minimize the number of staff
- To reduce waiting times and increase service efficiency (i.e., higher profits)
- For the collection of statistical information

The class diagram shown in Figure 20.4 shows the service system attributes, main functions, and the relationship between classes when buying a ticket from a kiosk or the online service. The electronic ticketing system supports data concurrence so that time allocations will be recognized when two or more customers request the same seat at the same time. Also the system is compatible with existing systems and services.

Class diagram of buying ticket

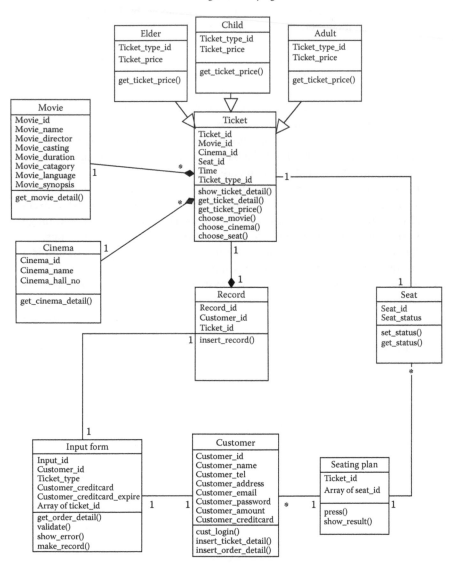

FIGURE 20.4 Example SysML class diagram of buying e-ticket.

User-centered service system design for business process engineering involves three main components: service problem structuring, idea generation, and idea evaluation and selection. This approach helps smarter service system designers to make new connections between various service elements, recognize key processes and elements in the system and recombine them in different ways, identify elements of purpose, and focus on goals. The primary mechanism of user value creation is divided between users and their knowledge, on the one hand, and machines, human, and process technological knowledge, on the other. User-centered service system design can be conceptualized as a sequence composed of nine interlinked classes of objects, where each class interacts with the other class in a dynamic relation to provide service benefits to users. Taxonomies have been used in systems engineering (SE) to classify and organize large bodies of information during business process modeling service system design. For example, Gershenson and Stauffer (1999) defined a taxonomy for extracting service design requirements from end users, while Hauge and Stauffer (1993) used taxonomies as a technique for eliciting business knowledge from end users. White and Edwards (1995) incorporated taxonomies to specify requirements for complex systems. Taxonomies can also serve as a basis for human factors (HF) and business process knowledge management. For example, Gershenson and Stauffer (1999) propose unique business process service system taxonomies for four design requirement types: *end user, corporate, technical, and regulatory requirements.*

Karwowski et al. (2009) and Kaner and Karni (2007) described models for service system design that facilitate the construction of a taxonomy for a design concept, enabling the categorization of the features and attributes contributing to a total service system, while incorporating both requirements and specifications. The taxonomy model introduced by Kaner and Karni (2007) breaks down the nine categories "major classes" of system elements (Table 20.1) to 75 features or objects within these categories ("main classes"). For a complete taxonomy structure, see Karwowski et al. (2009).

Each feature or service object is then further associated with a set of attributes where a set of possible (qualitative) values is coupled with each attribute. This taxonomy defines a category of objects that constitutes the conceptualized user-centered service system. The parametric nature of user-centered service design is reflected in the last level of the taxonomy, which provides the "quantitative" value assigned to an attribute when a specific service design concept is formulated.

The multilevel business processes and service system taxonomy facilitates a multilevel user-centered approach to the service system design and modeling. At the top

TABLE 20.1
HF Business Process Modeling Taxonomy

Classes of Objects	Description
Users	Those benefiting from the system or otherwise affected by it
Goals	Aims, purposes, or central meaning of the system for the organization and its users
Inputs	Physical, human, financial, or informational entities to be processed by the system
Outputs	Physical, informational, or human entities after processing by the system
Processes	Transformations for obtaining outputs from inputs
Employees (human enablers)	Human resources owning and/or operating the system
Physical enablers	Physical and technological resources that aid in operating the system
Information enablers	Information and knowledge resources supporting the system
Environment	Physical, economic, technological, social, ecological, or legal factors and contexts influencing the system

TABLE 20.2
HF Business Process Model for User-Centered Service Organizations

Classes of Objects	Examples
Users	User organization, user features, user association, user attitudes, and user preferences
Goals	Strategic goals (general), strategic goals (service related), service goals, user goals, economic goals, and enterprise culture
Inputs	Physical factors, HF, demand factors, utilization factors, user factors, constraint factors, financial factors, payment factors, and informational factors
Outputs	Physical factors (service product), HF (service product), informational factors (service product), organizational factors, financial factors, and efficiency factors
Processes	Service configuration, service variability, service initiation, service provision, service operations, service delivery (physical), service payment, service recovery, user contact, user relationships, service support, planning and control, waiting line management, and call center (management)
Human resources	Owner organization (enterprise), service providers, support providers, employee management, employee culture, and employee competence
Physical resources	Service facilities, amenities (hospitality), equipment, furnishing, service vehicles, geography (location), access, call center (physical), and information technology
Informational resources	Service information, promotion, official references, service configuration, product (physical), component reparability, procedures and processes, decision support, performance measures, prices and charges, and information sources
Organizational environment	Market factors, geographic factors, economic factors, technological factors, social factors, ecological factors, and legal factors

level of the taxonomy, the service system general concepts are defined by available system categorizations. The service system interaction designer must decide what features and attribute values can achieve the desired service output (i.e., unguided design). In the middle level, the service system interaction designer conceptualizes the set of necessary features, a subset of the capstone model shown in Table 20.2, and assigns attribute values to accomplish each feature's goal or purpose. At the bottom level, the interaction designer assigns values to attributes provided (i.e., guided design).

The multilevel user-centered service system taxonomy requires more input from the designer at top levels, providing better opportunities for service system design and creativity based on user-centered approaches. The hierarchal architecture of the HF business process model for user-centered service organizations enables the business process managers and service designers to define the service system at several levels while seamlessly moving from one level to another more detailed one.

Service system design is about modeling service engineering systems, starting with the most important subsystems and gradually reaching the level of detail necessary to build a given service system. Capturing the business process context of use is important for helping to specify user requirements as well as for evaluation and testing. Best practices indicate that effective service systems strongly promote service usability, end-user health and safety, and proper understanding of the service context of use. Service system context of use can be gathered using established structured methods for eliciting detailed information. This information will help facilitate the user requirements specification and service system evaluation. Service system context of use information provides details about the user's profile and characteristics, as well as task and environment service usage.

UCD methods and strategies are concerned with incorporating the user's perspective into the systems development process to achieve usable systems and services or to improve existing ones. This section adopts the framework of ISO 13407, where each step in the UCD cycle is evaluated with supporting usability methods. Service usability and ease of use are now widely recognized as one of the critical success factors of an interactive service system or product development process (Fowler, 1991; Nielsen, 1994a,b; ISO, 1997b).

Unfortunately, poorly designed and unusable service systems exist, which end users find difficult and frustrating to use. Poor service provisions are costly for an organization and negatively affect the reputation of the service vendor. Dissatisfied users may find and prefer a substitute vendor with a more desirable service system. UCD processes and methods help design better service systems and increase service quality to meet user expectations. The benefits of following UCD principles in business process service systems have been summarized by Maguire (2001):

- *Reduced training and support*—UCD and usability principles help reduce service provider training time and the need for user support. This is of special importance to novel service systems, since newly introduced service systems most often require dedicated training and support.
- *Reduced errors*—Poorly designed service systems significantly increase human error due to inconsistencies, ambiguities, or other interface design faults.
- *Increased productivity*—A service system designed employing UCD and usability principles will enable users to concentrate on the task rather than the interface in order to operate effectively.
- *Improved acceptance*—Most users would be more likely to trust a service system that provides well-presented service information that is easily accessed, UCD end-user acceptance and enhancing user satisfaction.
- *Enhanced reputation*—A well-designed service system will enhance the service vendor's reputation in the marketplace.

20.3.1 User-Centered Service System Development Cycle

The ISO 13407 human-centered design framework is considered the cornerstone for incorporating different business process design techniques, all of which can be merged to support a UCD process. According to the ISO 13407 standard (ISO, 1999), appropriate UCD processes are composed of five iterative steps (see Figure 20.5), which will guarantee the fulfillment of all requirements into the business process design as follows:

1. Planning service system design processes
2. Service system context of use
3. Requirements specification

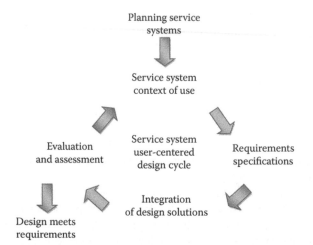

Planning service systems

Service system context of use

Evaluation and assessment

Service system user-centered design cycle

Requirements specifications

Design meets requirements

Integration of design solutions

FIGURE 20.5 Service system UCD cycle. (Modified from the original framework ISO 13407, *Human-Centered Design Processes for Interactive Systems*, International Standards Organization, Geneva, Switzerland. Also available from the British Standards Institute, London, U.K., 1999.)

4. Integration of design solutions
5. Service system evaluation and assessment
6. Planning service system design processes

20.3.1.1 Planning Service System Design Processes

User-centered service system planning plays a crucial role in projects by reducing the risks of system failure, thereby preventing additional costs and disruptions in service. Service system planning will guarantee meeting the process requirements and information flow between service users and the project team. Service system planning helps maintain the maximum integration of the business activities as part of the service system strategy. The first step in planning a service system is to discuss the business needs with stakeholders and users to reach an agreement on how UCD techniques can contribute to the service system objectives. In addition, the planning process prioritizes business requirements and highlights the potential benefit gained from including business process activities within the service system development process.

20.3.1.2 Service System Context of Use

Service system context of use defines all aspects of the business process service system's intended usage as well as the user population characteristics (i.e., user profile). Developed business process service systems will be used within a certain set of tasks by users with defined results and goals by performing certain activities. The service system will also be used within a known context of physical, environmental, and organizational conditions. Capturing the business process context of use is important for helping to specify user requirements as well as for evaluation and testing. Best practices indicate that effective business process service systems strongly promote service usability, end-user health and safety, and the proper understanding of the business context of use.

20.3.1.3   Requirements Specification

Requirements specification is one of the most crucial activities of business process system design The two most common causes of system failure are that there is insufficient effort to identify user requirements and the lack of end-user involvement in the design process. ISO 13407 design framework (ISO, 1999) provides guidance on specifying end-user requirements and objectives; the framework states that the following elements should be covered in the specification:

* Identification of users and other personnel in the business process design (i.e., users, employees, associates, designers, and support)
* Clear statement of the business process service's design goals
* Inclusion of appropriate priorities for the different requirements
* Establishment of measurable benchmarks for which the business process design can be tested
* Acceptance of design requirements by end users and stakeholders
* Acknowledgment of mandatory or legislative requirements
* Documentation of requirements change management as the service system development progress

20.3.1.4   Integration of Design Solutions

Business process solutions start with innovative and creative ideas through the iterative development process. Low-fidelity prototypes of business process components are necessary inclusions to the design life cycle. Simple design prototypes can be produced by HF professionals and the design team. Major problems can be identified before system development proceeds too far along, as it is always cheaper and easier to make changes sooner rather than later in the system design life cycle (SDLC).

20.3.1.5   Service System Evaluation and Assessment

The service system should be evaluated during all design and development stages. Evaluation helps confirm that business process objectives have been met and provide further information for refining the design. Evaluation starts with low-fidelity prototypes, followed by more sophisticated high-fidelity prototypes. Evaluation and assessment helps improve the business processes as part of the iterative development process and assures the business service can be used successfully by intended users. Business process evaluation and assessment can highlight problems by either user- or expert-based methods. Expert-based methods can help find service weaknesses that may not be revealed by a small number of users. User-based testing is required to find out whether intended users can interact with the service successfully. When running user testing, the emphasis may be on identifying business process problems and addressing them in the design process.

20.4   SYSTEMS ENGINEERING

The contemporary SE process is an iterative, hierarchical, top-down decomposition of system requirements (Hitchins, 2007). The hierarchical decomposition includes functional analysis, allocation, and synthesis. The iterative process begins with a system-level decomposition of systems and services and then proceeds through the functional subsystem level, all the way to the assembly and program level. The activities of functional analysis, requirements allocation, and synthesis will be completed before proceeding to the next lower level in the system development. Table 20.3 summarizes the *SE process activity hierarchy*, while Table 20.4 illustrates a *SE management framework*. Key tasks in modeling systems are performed using Systems Modeling Language (SysML). SysML is a general-purpose visual modeling language for specifying, analyzing, designing, and verifying complex systems, which may include hardware, software, information, personnel, procedures, and facilities (OMG SysML: http://www.omgsysml.org). SysML provides visual semantic representations for modeling system requirements, behavior, structure, and parametrics, which is used to integrate with other engineering analysis models (Friedenthal et al., 2008). Model-based interactive HSI approaches for design and modeling of smarter systems and services differentiate between human performance and effectiveness criteria. These criteria determine a total system mission performance level and acceptability that is directly attributable to specific actions allocated to human performance metrics. These are indicators that measure which performance effectiveness criteria are met (Ahram et al., 2009; Karwowski and Ahram, 2009). The smarter products human systems integration (HSI) paradigm can be used *"to develop a system where the human and machine synergistically and interactively cooperate to conduct the mission,"* and the *"low-hanging fruit"* of performance improvement lies in the human–machine interface block.

SE teams along with service system planners and designers are responsible for verifying that developed services and systems meet all the requirements defined in the system specification documents. The following procedures outline the relevant SE process steps (*DAU Guidebook*, 2004):

1. *Requirements analysis*—review and analyze the impacts of operational characteristics, environmental factors, and functional requirements and develops measures suitable for ranking alternative system designs in a consistent, objective manner. Each requirement should be reexamined for consistency, desirability, applicability, and potential for improved return on investment (Ahram and Karwowski, 2009). This analysis verifies that the requirements are appropriate or develops new requirements for the smart product operation.
2. *Functional analysis*—systems engineers and service designers use the input of performance requirements to identify and analyze system functions in order to create alternatives to meet system requirements. SE then establishes performance requirements for each function and subfunction identified.
3. *Performance and functionality*—SE allocates service design requirements and performance to each system function. These requirements are stated in appropriate detail to permit allocation to software, system components, or personnel. Performance and functionality allocation process identifies any special personnel skills or design requirements.

TABLE 20.3
SE Process Activity Hierarchy

	SE Process Activity	Subactivity	Description
Planning and preparation phase	1. City services needs	State the problem	The problem (i.e., why do we need this service or system?)
		Define needs	Through interviewing the users and other stakeholders of the service, the user needs can be conducted
		Analysis	Preliminary analysis has to indicate if the project is feasible and beneficial. System planners and project manager decide whether or not to continue with the project
	2. Planning	Plan services and systems	Draft service plan that will be updated and revised through whole the technical processes phase
		Assess project	How is the project going? Are there any changes?
Assessment phase	3. Assessment and control	Control risks Control data Control configuration	System control includes technical management activities required to measure progress, evaluate and select alternatives, and document data and decisions. Control activities include: Risk management Configuration management Data management
Technical smart product design phase	4. Technical phase	Requirements definition and analysis	Analyze the customer's needs to identify functional and systems requirements (i.e., performance and design constraint requirements)
		Identify functional requirements	Functional requirements define quantify (how many), qualify (how good), coverage (how far), time lines (when and how long), and availability (how often)
		Define performance and design constraint requirements	Design constraints define those factors that limit design flexibility
		Model the system	The list of requirements presented in the functional architecture of the design. This architecture can help to identify bottlenecks. It can be either a written document or a graphical representation of the functions of the system

(continued)

TABLE 20.3 (continued)
SE Process Activity Hierarchy

SE Process Activity	Subactivity	Description
	Evaluate alternatives	Evaluate and investigate alternative designs and look which requirement is met by each design
	Integrate the smart product	Subsystems or functionalities integration
	Smart product implementation	The creation of the product is completed and can be implemented and delivered to the customer
Evaluation phase 5. Evaluation	Verification Validation	It is important to verify and validate if the user needs are met and if the product fulfills the intended use
		Several evaluation methods are available for this purpose. The outcomes are described in an evaluation report. Also during the technical processes, evaluation will constantly take place, especially the evaluation of requirements
	Maintenance	Together with possible complaints, the evaluation report is the basis for the maintaining process (This is a closed concept and out of the scope of SE. If adjustments to the product have been made, the product should be evaluated again)

Source: Modified from the original by *DAU Guidebook*, 2004.

4. *Design synthesis*—system designers and other appropriate engineering specialties develop a system architecture design to specify the performance and design requirements that are allocated in the detailed design. The design of the system architecture is performed simultaneously with the allocation of requirements and analysis of system functions. The design is supported with block and flow diagrams. Such diagrams support the following:
   a. Identifying the internal and external interfaces
   b. Permitting traceability to source requirements
   c. Portraying the allocation of items that make up the design
   d. Identifying system elements along with techniques for its test and operation
   e. Providing a means for comprehensive change control management
5. *Documentation*—is the primary source for developing, updating, and completing the system and subsystem specifications. Smart service requirements and design drawings should be established and maintained.

TABLE 20.4
Systems Engineering Process Management Framework

System Engineering Process Concept	Definition
Top level requirements	Identification of the user needs. (Talking to customers and suppliers is very useful for this matter). This will result in a general overview and initial requirements of the system
Problem definition	Description of the intended use of the system. The problem statement starts with a reason for change followed by vision and mission statements for the company (Bahill and Dean, 2005)
Return on investment (feasibility report)	This contains benefits and costs of the software product that is to be made
Project management	Project management is the planning, organizing, directing, and controlling of company resources to meet specific goals and objectives within time, within cost, and at the desired performance level (Bahill and Dean, 2005)
Risk management	Identification of risks and opportunities for improvement
Document management	Control of all documentation and activities of the SE process
Configuration management	Configuration management ensures that any changes in requirements, design, or implementation are controlled. Configuration management involves configuration identification, control, status accounting, audit, and planning
Requirement list	Customer needs and requirements are translated into a set of requirements that define what the system must do and how well it must perform Requirements analysis must clarify and define functional requirements and design constraints
Requirement	A requirement consists of a number, type, description, and evaluation method that can be used to test the requirement
Functional architecture	Functions are analyzed by decomposing higher-level functions identified through requirements
Product	Definition of SE according to INCOSE: *"Systems Engineering is an engineering discipline whose responsibility is creating and executing an interdisciplinary process to ensure that the customer's and stakeholder's needs are satisfied throughout a system's entire life cycle"* (INCOSE SE handbook; Vasquez, 2003). The outcome of the SE process is a product that satisfies the needs of the customer and stakeholders
Evaluation	Each requirement must be verified and validated to determine whether the software product fulfills the requirements and if the final as-built smart product fulfills its specific intended use

Source: Modified from the original by *DAU Guidebook*, 2004.

6. *Specifications*—to transfer information from the system requirements analysis, system architecture design, and system design tasks. The specifications should assure that the requirements are testable and are stated at the appropriate specification level.
7. *Specialty engineering functions*—participate in the SE process in all phases. They are responsible for the systems maintainability, testability, producibility, HSI, HF, safety, design-to-cost, and performance analysis to assure the design requirements are met.
8. *Requirements verification*—SE and test engineering verify the completed system design to assure that all the requirements contained in the requirements specifications have been met.

20.5 HUMAN FACTORS ENGINEERING SERVICE SYSTEM ONTOLOGIES

The word "ontology" comes from the Greek word *ontos,* for "being," and logos, for "word." It refers to the study of the categories of things that exist in a domain of knowledge (Sowa, 2000). The idea of ontologies emerged as a means for sharing knowledge. Ontologies rely on well-defined and semantically powerful concepts in AI, such as logical statements, reasoning, and rule-based systems. Ontologies enable access to a huge network of machine-understandable and easily machine-processed human knowledge (Gruber, 1991). As an example of the representation of ontologies, consider the concept of an operations manager. Assume that the concepts used to describe essential knowledge about the concept of operations that are the manager; operations; some products of his or her work, namely, the company operations the manager was responsible for and departments in which he or she has direct responsibility to manage; and employees, who maintain the company performance. Also, assume that the variety of relationships among these concepts that can be considered may be reduced to just a few of the most essential ones, such as the fact that each operations manager manages some departments; that when giving business decisions, he or she checks each department's performance; and that he or she also monitors overall business performance. We deliberately avoid in this simple example the numerous kind-of and part-of relationships to other concepts associated with operations management and their detailed responsibilities. These natural-language statements represent the conceptualization of the operations management ontology (Ahram and Karwowski, 2011).

The literature describes explorations of potential of software engineering tools and methodologies for ontology development. Pioneering research proposed the unified modeling language (UML), a well-known software modeling language, to be used for ontology development (Cranefield, 2001). Gaševic et al. (2006) explored further the similarities, differences, and equivalences between UML and ontology languages, as well as the potential of the most recent software engineering initiative called the model-driven architecture (MDA).

There are a number of knowledge management representational languages. Web ontology language or OWL was adopted by the World Wide Web Consortium (W3C) (http://www.w3c.org); OWL is currently the most popular ontology representation language. OWL facilitates greater machine interpretability of web content than that supported by XML, RDF, and RDF Schema (RDF-S) by providing additional vocab-ulary along with a formal semantics (Smith et al., 2004). An important feature of the OWL vocabulary is its extreme richness for describing relationships among classes, properties, and individuals. The OWL is designed for use by applications that need to process the content of information or "knowledge" instead of just presenting information to humans. OWL has three increasingly expressive sublanguages: OWL Lite, OWL DL, and OWL Full. Ontologies differ from taxonomies; an ontology is a full specification of a domain. According to Gaševic et al. (2006), a taxonomy is a hierarchical classification of entities within a domain or field of study, whereas an ontology is the reflection of knowledge and relations between entities within a domain of knowledge and between different domains. Ontologies do not merely serve as a tool for knowledge sharing and knowledge reuse. Ontologies support the concepts and relationships that can exist in a domain and that can be shared and reused. Ontologies reflect the most important part contained, which is knowledge and logical reasoning underpinning such hierarchies. However, every ontology also details a taxonomy in a machine-readable format. Ontologies facilitate knowledge sharing and reuse across various domains of knowledge. Fikes (1998) listed the importance of ontologies in four key application areas: collaboration, interoperation, education, and modeling.

- *Collaboration*—Ontologies provide a unified knowledge architecture that can be used as a common, shared reference for further development and participation. Researchers and scientists from different domains and backgrounds can talk more easily to each other when they have such a stable, consensual knowledge base to rely on.
- *Interoperation*—Ontologies allow integration of databases from different sources. Distributed applications may need to access different knowledge sources in different formats and in different levels of detail in order to obtain relevant information. Recognizing the same ontology, data conversion and information integration will be easier to allocate.
- *Education*—Domain experts can use ontologies for education and referencing to share their understanding of the conceptualization and structure of the domain.
- *Modeling*—Ontologies support the development of knowledge-based applications by providing important reusable building blocks.

The standard set of tools for ontology development includes graphical development environments and representation languages. The W3C supports ontology representation on the semantic web, also called "semantic web languages." In all ontology representation languages, there is a graphical ontology editor to help organize the conceptual structure of the ontology that integrates concepts, properties, relationships, constraints, and logical and semantic inconsistencies among the various attributes of the ontology. Corcho et al. (2002) defined a graphical ontology development environment (ODE) as the one that integrates an ontology editor with other tools and usually supports multiple ontology representation languages. ODEs aim at providing support for the entire ontology development process and for the subsequent use of the ontology (Gaševic et al., 2006).

Protégé (http://protege.stanford.edu/) is the leading ontology development editor and ODE. Protégé was developed at Stanford University (Protégé, 2010). Protégé's plug-in-based extensible architecture allows integration with a number of other tools, applications, knowledge bases, and storage formats (e.g., OMG SysML, UML, Visual OWL, XML, and XMI for storing knowledge bases) and facilitates concept (class) definition, taxonomies, properties, and restrictions, as well as class instances or actual data in the knowledge base. Protégé has a standard ontology editor GUI with a tab menu for knowledge acquisition and knowledge collection into a knowledge base conforming to the ontology. The knowledge base can then be used with a problem-solving method to perform various inference or decision-making tasks and logical statements and tasks.

HF knowledge management classifications involve two major categories, physical and cognitive aspects of interaction. The cognitive attributes of HF in knowledge management concentrate on conscious and subconscious mental activities involved in business process decision making (Carey, 1988). The knowledge gained in both aspects of cognitive processes is used to create systems and work environments, which help to make people more productive and more satisfied with their work life. A number of different taxonomies exist that break HF into multiple categories of research.

The fundamental research approach is to represent both the business perspective and the HFE perspective of enterprise business processes using a set of ontologies and to apply machine reasoning for carrying out or supporting the translation tasks between the two representations. Ontologies enable access to a huge network of machine-understandable human knowledge (Gašević et al., 2006).

Once the essential knowledge of a certain domain has been represented in the form of interconnecting ontologies, it creates a solid basis for further development of intelligent applications in that domain because it alleviates the problem of knowledge acquisition. More specifically, ontologies play multiple roles in the architecture of the semantic web knowledge bases (Gašević et al., 2006; Ahram and Karwowski, 2011):

- Enable web-based knowledge processing, sharing, and reuse between applications, by the sharing of common concepts and the specialization of the concepts and vocabulary for reuse across multiple applications
- Establish levels of interoperability in terms of mappings between terms within the data, which requires content analysis
- Enable intelligent services (information brokers, search agents, HSI connectors, filters, smart information integration, human knowledge management systems)

A comprehensive HF and BPMO framework facilitates the design of agile business process services and systems by capturing the core business knowledge into reusable modules and components (Janusch, 2007). In Figure 20.6, HF ontology is connected to business maps and strategy ontologies to aid in decision making and process improvement. This process is supported by the understanding of business roles through a business functions ontology that establishes process frameworks and architecture. These components can be seen as a complement to the traditional agile design engineering process (see Figure 20.7).

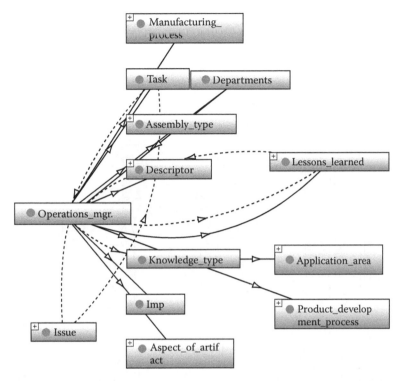

FIGURE 20.6 Model visualization of the business operations management decisions OWL ontology.

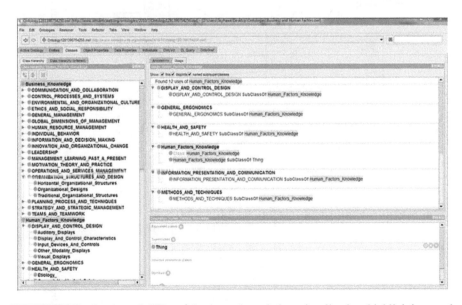

FIGURE 20.7 Integrated HF and business knowledge visualization highlighting work design and organization.

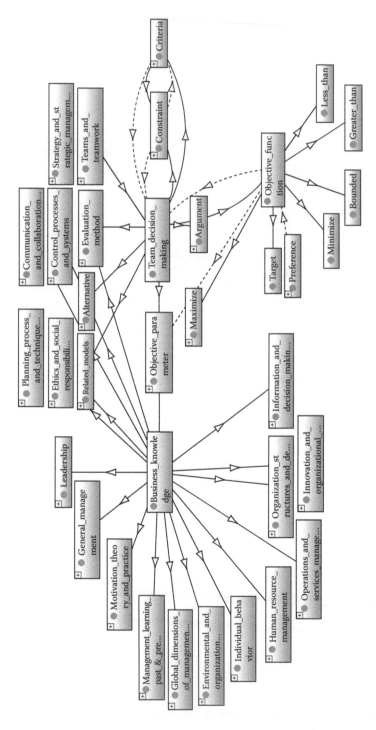

FIGURE 20.8   Example decision support and business knowledge ontology that illustrates team decision making.

Business knowledge is independent of the technology that implements it. Diverse business process knowledge sources characterize and formulate the processes of an organization as (1) business documents and (2) tacit knowledge (i.e., knowledge that is difficult to transfer to another person), this knowledge is maintained by business process decision makers and business owners. This knowledge is can only be accessed and reused if it is extracted and stored as information in a knowledge repository (Janusch, 2007). Ontologies provide appropriate means to formally describe and encode the concepts and relationships between HF and business process modeling. Making the business knowledge explicit in terms of an ontology would enable systems to manipulate meanings within program code. Following this notion, Figure 20.8 demonstrates a formal visual ontological model for describing business processes, which integrates different workflow perspectives with HF knowledge and allows for expressive querying and reasoning on business process models. Business Process Discoverer is part of the SUPER project (http://www.ip-super.org). SUPER refers to the *semantics utilized for process management within and between enterprises.* SUPER framework identifies process improvement options and serves as a basis for allocating executable business processes. The SUPER business process modeling architecture supports the development of business process models created by business users to (Janusch, 2007)

- Capture business requirements
- Enable a better understanding of business processes
- Facilitate two-way communication between business analysts and software experts

20.6 ROADMAP FOR DESIGNING SERVICE SYSTEMS OF THE FUTURE

The concept of smartness of systems and services has been investigated by several authors. This section presents a synthesis and summary of the most innovative work that influenced research in this field. Allmendinger and Lombreglia (2005) highlighted smartness in a system from a business perspective. They regard *"smartness"* as the system's capability to predict business errors and faults, thus *"removing unpleasant surprises from the users' lives."* Ambient Intelligence (AMI) group describes a vision where distributed services, mobile computing, or embedded devices in almost any type of environment (e.g., homes, offices, and cars) all integrate seamlessly with one another using information and intelligence to enhance user experiences (Weiser, 1991; Ahola, 2001; Arts and de Ruyter, 2009). Rapid technological advancements and agile manufacturing created what is called today smart environments.

Definitions of smart environments may be taken into account as a first reference point since smart systems have to be considered in the context of their environment. For example, Das and Cook (2006) define a smart environment as the one that is able to acquire and apply knowledge about an environment and adapt to its inhabitants in order to improve their experience in that environment.

Mühlhäuser (2008) refers to smart system characteristics that are attributed to future smart environments, that is, *"integrated interwoven sensors and computational systems seamlessly embedded in everyday systems and tools of our lives, connected through a continuous network."* In this respect, smarter systems can be viewed as the collective systems that facilitate daily tasks and augment everyday objects. It is noticed that the knowledge aspect has been recognized as a key issue in this definition. In 2007, AMI identified two motivating goals for building smart systems (Sabou et al., 2009):

1. Increased need for simplicity in using everyday systems as their functionalities become ever more complex. Simplicity in using smart systems facilities is desirable during the entire life cycle of design and development and to support, repair, or use.
2. Increased number, sophistication, and diversity of system components (e.g., in the aerospace industry), as well as the tendency of the suppliers and manufacturers to become increasingly independent of each other, which requires a considerable level of openness on the system side.

According to Mühlhäuser (2008), system simplicity can be achieved with an improved system to user interaction (*p2u*). Furthermore, openness of a system requires an optimal system to system interaction (*p2p*). Ahram et al. (2010b) and Mühlhäuser (2008) observed that these system characteristics can now be developed due to recent advances in information technology as well as ubiquitous computing that provide real-world awareness in these systems through the use of sensors, smart labels, and wearable, embedded computers.

Knowledge-intensive techniques enable better p2p interaction through self-organization within a system or a group of systems. Indeed, recent research on semantic web service description, discovery, and composition may enable self-organization within a group of systems and, therefore, reduce the need for top-down constructed smart environments (Chandrasekharan, 2004) (Ahram et al., 2010a,b). Smart systems also require some level of internal organization by making use of planning and diagnosis algorithms (Mühlhäuser, 2008):

> A Smart product is an entity (tangible object, software, or service) designed and made for self-organized embedding into different (smart) environments in the course of its lifecycle, providing improved simplicity and openness through improved p2u and p2p interaction by means of context-awareness, semantic self-description, proactive behavior, multimodal natural interfaces, AI planning, and machine learning.

Major characteristics of smart systems are illustrated by comparing their essential features. For example, Maass and Varshney (2008) define six major characteristics for smart systems illustrated in Table 20.5. In addition to the aforementioned characteristics, smart systems should support their entire life cycle and special care should be devoted to offering multimodal interaction with the potential users in order to increase the simplicity characteristics of the systems. The conceptual SysML integrated services/systems architecture shown in Figure 20.9 is composed of 12

TABLE 20.5
Smart System Characteristics

Characteristic	Description
Personalization	Customization of systems according to buyer's and consumer's needs
Business awareness	Consideration of business and legal constraints
Situatedness	Recognition of situational and community contexts
Adaptiveness	Change system behavior according to buyer's and consumer's responses to tasks
Network ability	Ability to communicate and bundle with other smart systems and new infrastructure
Proactivity	Anticipation of user's plans and intentions

Source: Modified from the original by Maass, W. and Varshney, U., *Electron. Markets*, 18(3), 211, 2008.

subsystems. For example, SysML block definition diagram (BDD) allocates system resources to the following subsystems:

- Residents or user needs (i.e., population)
- Efficient transportation system
- Public safety system
- Health care system
- Public service system
- Communications system
- Support system
- Education system
- Maintenance system
- Utilities and support
- Information technology center for data collection, processing, and analytics
- Research and development (technology innovation systems)

20.7 INTEGRATION OF USER-CENTERED DESIGN PRINCIPLES INTO SERVICES AND SYSTEMS

Knowledge management for service systems is challenging, especially for large organizations with complex projects involving multiple disciplines. Despite the increasing ability to communicate and share knowledge, it seems that many designers and engineering groups do not share their findings outside of their own group (Stasser and Stewart, 1992). An often encountered phenomenon is that of a *"tribal knowledge,"* where a certain group or individual acquires a skill or trade and keeps it, employing it when called upon. Such groups rarely leave a legacy or ability to transfer this knowledge to their replacements, forcing the organization to relearn, and HFE groups and SE practitioners have realized the limitations and coined the term to recreate that which it already knew (Ahram et al., 2010a). Obviously, this is counterproductive to a fast-paced design world. In order to track significant findings found during the design process, these findings are then categorized and organized

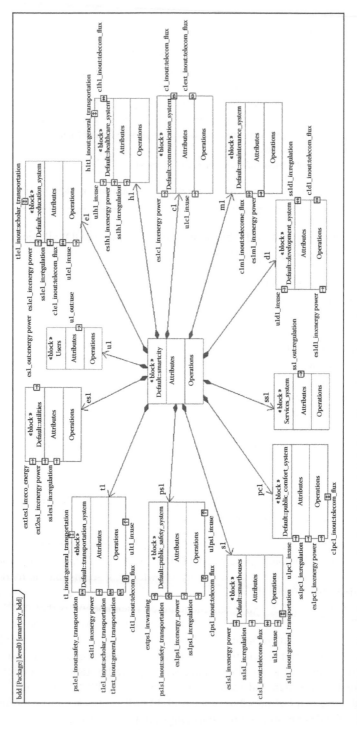

FIGURE 20.9 Example smarter city SysML block definition diagram showing major system subsystems.

logically into a database. This database can take many forms, from the commercial off-the-shelf (COTS) synthetic environment database solutions to highly specified custom-coded internal database. Newly hired systems engineers and practitioners transferred to the design program are encouraged to review the database to avoid the dreaded "duplication of effort."

The commercial market has already realized the importance of SE and HF knowledge management, and, thus, a few software systems such as the *IBM Rational Focal Point™* have been created to assist organizations and governments managing complex projects involved in designing and optimizing systems and infrastructure, for example, a portfolio for smarter service system projects management console developed in *IBM Rational Focal Point™*. *Focal Point™* supports SE, human trade-off analysis, and data integration to optimize energy usage and maximize distribution efficiency. Successful practices for smart systems (software based on these practices) require the following features (Karwowski and Ahram, 2009):

- *Service ease of use*—It is unlikely that something awkward or difficult to use will continue to be used in the long term.
- *Varied format input*—Aside from scribbling on notebook paper, the software should accept many document and file formats.
- *Traceability*—Inputs should be traced to their owner.
- *Security*—All users may not require access to all elements of the knowledge database; proprietary, secret, and competitiveness concerns must be addressed.
- *Routine*—Inputs should be encouraged while the idea, solution, or process is still fresh in the creator's mind
- *Organization*—Without a structured method of entry or effective indexing method, acquired knowledge is meaningless; it should be viewable top-down or searchable and browsable on varied related topics.

One of the applications that support UCD within an SE process is DOORS™ by IBM Rational. DOORS™ stands for *"dynamic object oriented requirements system"* and specifically tracks requirements for smarter systems or software design. Since the requirements process has many shared elements to knowledge management, DOORS™ facilitates HF performance requirements entry, organization into hierarchies, and user-centered service system requirements allocation (Karwowski et al., 2009) (Karwowski, 2006). DOORS™ users make changes and link any system or service HF requirement to subrequirements and related requirements. DOORS™ requires individual users to have accounts. Each account can be restricted to elements of the human performance database and given read-only or administrative-level rights to HF practitioners (Ahram et al., 2010a,b). Changes made are tracked by user, allowing managers to trace changes down to the individual user level. CATIA™ allows smart system designers and HF practitioners to simulate all the industrial design processes, from the preproject phase through detailed design and systems optimization, analysis, simulation, assembly, and maintenance. CATIA™ has been used by *Boeing* to design massive airplanes and by *NASA* to help design the space shuttle. These are just a few systems out in the market employing knowledge management practices and suitable for smart system design and modeling. Ultimately, organizations responsible for employing a process or a set of processes for HF knowledge creation, implementation, and utilization will be able to minimize or eliminate duplication of efforts and streamline the system design cycle.

20.8 CONCLUSIONS

As an introductory contribution to the application of knowledge management UCD principles and SE process for the design and development of service systems, this chapter provides a motivation and quest for an integrated UCD approach to SE principles and development life cycle. While a large number of disciplines and research fields must be integrated toward the development of intelligent and smarter services, considerable advancements achieved in these fields in recent years indicate that the adaptation of these results can lead to highly intelligent and sophisticated yet widely usable systems and services. It is believed that SE and UCD approach to design and development should prove useful in supporting and facilitating the development and applications of smarter services in the near future.

ACKNOWLEDGMENT

The authors acknowledge Mr. Anthony Cauvin, Mr. Benjamin Baron, and Mr. Fabrice Faure from the Department of Systems Engineering at the Institute National des Sciences Appliquées, Toulouse, France, for their valuable software modeling assistance while working on the smarter service system project during their internship at the University of Central Florida.

REFERENCES

Ahram, T. Z. and Karwowski, W. (2009). Measuring human systems integration return on investment. *The International Council on Systems Engineering—INCOSE Spring 09 Conference: Virginia Modeling, Analysis and Simulation Center (VMASC)*, Suffolk, VA.

Ahram, T. Z. and Karwowski, W. (2011). Developing human social, cultural, behaviour (HSCB) ontologies: Visualizing & modeling complex human interactions, Presented at the Office of Secretary of Defense Human Social, Culture Behavior Modeling (HSCB Focus 2011), February 8–10, Chantilly, VA.

Ahram, T. Z., Karwowski, W., and Amaba, B. (2010a). User-centered systems engineering & knowledge management framework for design & modeling of future smart cities, *54th Annual Meeting of the Human Factors and Ergonomics Society (HFES 2010)*, San Francisco, CA.

Ahram, T. Z., Karwowski, W., and Amaba, B. (2010b). User-centered systems engineering approach to design and modeling of smarter systems, *5th IEEE International Conference on System of Systems Engineering (SoSE)*, Loughborough University, Loughborough, U.K., June, 2010.

Ahram, T. Z., Karwowski, W., Amaba, B., and Obeid, P. (2009). Human systems integration: Development based on SysML and the rational systems platform, *Proceedings of the 2009 Industrial Engineering Research Conference*, Miami, FL.

Ahram, T. Z., Karwowski, W., and Andrzejczak, C. (2010c). Interactive management of human factors knowledge for human systems integration, in: Patricia Ordonez De Pablos (ed.). *Electronic Globalized Business and Sustainable Development through IT Management: Strategies and Perspective*. IGI Global, ISBN-13: 978-1615206230.

Ahram, T. Z., Karwowski, W., and Soares M. (2011). Smarter products user-centered systems engineering. In: W. Karwowski, M. Soares and N. Stanton (eds.) *Handbook of Human Factors and Ergonomics in Consumer Product Design. Methods and Techniques*. Taylor & Francis, Boca Raton, FL.

Allmendinger, G. and Lombreglia, R. (2005). Four strategies for the age of smart services. *Harvard Business Review*, 83(10):131–145.

Arts, E. and de Ruyter, B. (2009). New research perspectives on ambient intelligence. *Journal of Ambient Intelligence and Smart Environments*, 1:5–14.

Bahill, T. and Dean, F. (1999). Discovering system requirements. In: A.P. Sage and W.B. Rouse (eds.) *Handbook of Systems Engineering and Management*. John Wiley & Sons, New York, pp. 175–220.

Bahill, T. and Dean, F. (2005). What is systems engineering? A consensus of senior systems engineers. http://www.sie.arizona.edu/sysengr/whatis/whatis.html (accessed on January 15, 2009).

Bourdreau, A. and Couillard, Guy (2006). Systems integration and knowledge management. *Information Systems Management*, 16(4).

Carey, J. (1988). *Human Factors in Management Information Systems*. Ablex Publishing Corporation, Greenwich, CT, pp.7–24.

Chandrasekharan, S. (2004). The semantic web: Knowledge representation and affordance.

Checkland, P. B. (1981). *Systems Thinking Systems Practice*. Wiley, New York.

Corcho, O., Fernández-López, M., and Gómez-Pérez, A. (2002). Methodologies, tools and languages for building ontologies. Where is their meeting point? *Data and Knowledge Engineering*, 46(1): 41–64.

Cranefield, S. (2001). Networked knowledge representation and exchange using UML and RDF. *Journal of Digital Information*, 1(8) article no. 44, 2001-02-15.

Das, S. and Cook, D. (2006). Designing smart environments: A paradigm based on learning and prediction. In: R. Shorey, A. Ananda, M.C. Chan, and W.T. Ooi (eds.) *Mobile, Wireless, and Sensor Networks: Technology, Applications, and Future Directions*. Wiley, Chichester, U.K., pp. 337–358.

Davis, R., Shrobe, H., and Szolovits, P. (1993). What is a knowledge representation? *AI Magazine*, 14(1): 17–33.

Defense Acquisition University (DAU) Guidebook (2004). Systems engineering, Chapter 4.

Durkin, J. (1994). *Expert Systems: Design and Development*. Macmillan, New York.

Fähnrich, K.-P. and Meiren, T. (2007). Service engineering: State of the art and future trends. In: D. Spath and K.-P. Fähnrich (eds.) *Advances in Service Innovations*. Berlin: Springer, pp. 3–16.

Fikes, R. (1998). Multi-use Ontologies, Stanford University [Online]. Available: http://www.ksl.stanford.edu/people/fikes/cs222/1998/Ontologies/sld001.htm (accessed 2010-08-07).

Fowler, C. (1991). Usability evaluation-usability in the product lifecycle. Usability Now! Newsletter, Issue 3, Spring, 6–7. HUSAT Research Institute, The Elms, Elms Grove, Loughborough, Leicestershire, U.K.

Friedenthal, S., Moore, A., and Steiner, R. (2008). *A Practical Guide to SysML: The Systems Modeling Language*. Morgan Kaufmann, Elsevier Science, Amsterdam, the Netherlands.

Gašević, D., Djuric, D., and Devedžic, V. (2006). *Model Driven Architecture and Ontology Development*. Springer, Berlin, Germany.

Gershenson, J. K. and Stauffer, L. A. (1999). A taxonomy for design requirements from corporate users. *Journal of Research in Engineering Design*, 11(2):103–115.

Gruber, T.R. (1991). The role of common ontology in achieving sharable, reusable knowledge bases. In: J.A. Allen, R. Fikes, and E. Sandewall (eds.) *Principles of Knowledge Representation and Reasoning: Proceedings of the Second International Conference*, Cambridge, MA, pp. 601–602, Morgan Kaufmann.

Hattingh, S. (2009). Skills development for performance improvement: Optimum utilization of the WSP to enhance performance and service delivery, retrieved from: http://www.learningroadmap.co.za/ (accessed on June 2011).

Hauge, P. and Stauffer, L. A. (1993). ELK: A method for eliciting knowledge from users. *Proceedings of the Fifth ASME Design Theory and Methodology Conference*, Albuquerque, NM, September 13–16, 1993.

Hauknes, J. (1996). Innovation in the service economy, STEP group Storgt. 1 N-0155 Oslo, ISSN 0804-8185.

Hitchins, D. K. (2007). *Systems Engineering: A 21st Century Systems Methodology*. John Wiley & Sons, Chichester, U.K.

Hofstadter, D. (1994). *Fluid Concepts and Creative Analysis—Computer Models of the Fundamental Mechanisms of Thought*. Basic Books/Harper Collins, New York.

ISO 9241-11 (1997). *Ergonomic Requirements for Office Work with Visual Display Terminals (VDTs). Part 11: Guidelines for Specifying and Measuring Usability*. International Standards Organization, Geneva, Switzerland. Also available from the British Standards Institute, London, U.K.

ISO 13407 (1999). Human-centered design processes for interactive systems. International Standards Organization, Geneva. Also available from the British Standards Institute, London, U.K.

Janusch, W. (2007). Business process discoverer: Semantics utilised for process management within and between enterprises (SUPER). Website: http://www.ip-super.org (accessed on June 2011).

Kaner, M. and Karni, R. (2007). Engineering design of a service system: An empirical study. *Information Knowledge Systems Management*, 6:235–263, IOS Press.

Karwowski, W. (2006). *International Encyclopedia of Ergonomics and Human Factors*, 2nd edn. CRC Press, Inc., Boca Raton, FL.

Karwowski, W. and Ahram, T. Z. (2009). Interactive management of human factors knowledge for human systems integration using systems modeling language. Special issue for information systems management. *Journal of Information Systems Management*, 26(3):262–274.

Karwowski, W., Salvendy, G., and Ahram, T. Z. (2009). Customer-centered design of service organizations. In: G. Salvendy and W. Karwowski (eds.) *Introduction to Service Engineering*, Chapter 9. John Wiley & Sons, Hoboken, NJ, pp. 179–206, ISBN-10: 0470382414.

Maass, W. and Varshney, U. (2008). Preface to the focus theme section: 'Smart Systems.' *Electronic Markets*, 18(3):211–215.

Maguire, M. (2001). Methods to support human-centered design. *International Journal of Human–Computer Studies*, 55:587–634.

McDermott, C. M., Kang, H., and Walsh, S. (2001). A framework for technology management in services. *IEEE Transactions in Engineering Management*, 48(3):333–341.

Metters, R., King-Metters, K., and Pullman, M. (2003). *Successful Service Operations Management*. South-Western, Mason, OH.

Mühlhäuser, M. (2008). Smart systems: An introduction. *Constructing Ambient Intelligence—AmI 2007 Workshop*, pp. 154–164.

Nadel, J. and Piazza, G. (2005). *Managing the Knowledge Behind Business Decisions through User-Centered Design*. Center for Business Knowledge.

Nielsen, J. (1994a). Special issue: Usability laboratories. *Behavior and Information Technology*, 13:3–197.

Nielsen, J. (1994b). *Usability Engineering*. Morgan-Kauffman, San Francisco, CA.

Norman, D. (1988). *The Design of Everyday Things*, Basic Books, New York.

Norman, D. (1993) *Things That Make Us Smart*. Addison-Wesley, Reading, MA.

Object management group (OMG) systems modeling language (SysML): http://www.omgsysml.org

Preece, J., Rogers, Y., and Sharp, H. (2002). *Interaction Design: Beyond Human-Computer Interaction*. John Wiley & Sons, New York.

Protégé Ontologies Library (2010). *ProtegeOntologiesLibrary* [Online]. Available: http://protegewiki.stanford.edu/wiki/Protege_Ontology_Library (accessed: 2010-08-08).

Sabou, M., Kantorovitch, J., Nikolov, A., Tokmakoff, A., Zhou, X., and Motta, E. (2009). Position paper on realizing smart systems: Challenges for semantic web technologies, Report by Knowledge Media Institute, Milton Keynes, U.K.

Skills Development Facilitation (SDF) (2006). *Guide for the Public Sector, 2006* Chapter 3, p. 12. Website: http://www.pseta.gov.za/

Smith, M. K., Welty, C., and McGuinness D. (2004). *OWL Web Ontology Language Guide*, W3C Recommendation, http://www.w3.org/TR/2004/REC-owl-guide-20040210/. Latest version available at http://www.w3.org/TR/owl-guide/ (accessed on June 2011).

Sowa, J.F. (2000). *Knowledge Representation: Logical, Philosophical, and Computational Foundations*. Brooks Cole, Pacific Grove, CA.

Stasser, G. and Stewart, D. D. (1992). Discovery of hidden profiles by decision-making groups: Solving a problem versus making a judgment. *Journal of Personality and Social Psychology*, 63(3):426–434.

Vasquez, J. (2003). *Guide to the Systems Engineering Body of Knowledge—G2SEBoK*. International Council on Systems Engineering, San Diego, CA.

Weiser, M. (1991). The computer of the 21st century. *Scientific American*, 265(3):66–75.

Wilson Learning Worldwide Website. *Approach to Learning: How We Enhance Human Performance Through Learning*, http://wilsonlearning.com/research_insights/position_papers/ (accessed on June 2011).

White, S. and Edwards, M. (1995). A requirements taxonomy for specifying complex systems. *Proceedings of the First IEEE International Conference on Engineering of Complex Computer Systems (IECCS'95)*, Southern Florida, pp. 373–376.

21 Strategic Knowledge Services

Hannu Vanharanta, Camilla Magnusson,
Kari Ingman, Annika H. Holmbom,
and Jussi I. Kantola

CONTENTS

21.1 INTRODUCTION

With the development of the knowledge society, we have entered the postindustrial era of the service economy. Knowledge is strategic in nature and can lead to significant competitive advantage and added value if it is understood and used correctly in business situations. Explicit knowledge comes from data and information. Applicable explicit and tacit knowledge, in turn, is created through people and novel tools. The question is how we can manage all these system inputs—including data warehouses, information flows, knowledge creation and perception, and situation-specific life and business information—and whether it is even possible to create more knowledge using the capabilities of the organization and modern technology. The service economy uses technology to meet these challenges and requirements, as well as to reach clients, organizations, and groups to offer them value-added services. Service, in broad terms, refers to the action of helping or doing work for someone else (cf. Oxford English Dictionary 2011); the concept is apt for the knowledge society and integral in knowledge services.

In this chapter, we examine the strategic knowledge services that the company can use to understand the recommended response and action in different business situations in order to continuously create added value for its customers. We first introduce the continuous strategy ontology, which shows the main dimensions, constructs, and variables of strategic knowledge services in different organizations. We then present three case studies: the first is a case study in customer segmentation with the self-organizing map (SOM) (from data to knowledge service), the second case is a collocation network-based knowledge service (from information to knowledge service), and the third case is a collective human knowledge-based service using an ontology (from knowledge to knowledge service).

21.2 CONTEXT OF STRATEGY MAKING AND STRATEGIC KNOWLEDGE SERVICES

Strategy makers try to predict and influence the future state of the organization. They often work together with their executives and the decision makers, as the work involves linking the organization's goals with the external environment. Success in strategy-making activity requires that all the people involved arrive at a shared vision as the basis for progression. Often though, the process of strategy making is fraught with misunderstanding and conflict, especially at its outset. Participants may lack a shared understanding of the future path of the organization; they may not have a common view of the operating environment or may lack a shared appreciation of the overlying world structure. On top of this, different participants perceive managerial issues and organizational characteristics in particular in different ways, and, thus, the strategy-making process becomes even more complex and problematic. This is true of any organization, public or private, because frequently the decision makers themselves do not understand the organization's structure and variables or information and knowledge well enough.

Faced with these problems, strategy makers often express the need for comprehensive, reliable, and commonly assimilable information that they can use to analyze and synthesize the current performance of their organization and to estimate its future potential. The software industry has tried to offer knowledge and information storage systems for this purpose for many years, yet the computer-based executive support systems (ESS) and decision support systems (DSS) developed so far have only provided partial solutions. Such systems support either specific activities or specific processes, but they do not provide executives with the kind of support that would enable a collective, holistic understanding of the organizational situation and also of the relationships and interrelationships that must be mastered in strategic management, and especially in sustainable development nowadays.

Prior-art ESS and DSS lack a statement of direction that would serve as a general framework for strategic management and a content-specific construct as a generic ontology for support systems. They also lack a basic strategic construct for guiding the organization toward purposeful progress and integrating its internal and external worlds. However, rapid advances in research and technology have propelled us into a new era where it is possible to incorporate also "soft" and "unstructured" abstract concepts, like those encountered in strategic management generally, into computer-based working environments. We can therefore demand more now that strategy making and planning can benefit from the new generation of computerized applications. Of course, strategy making this way requires new ways of thinking, as well as novel technological software and approaches.

These new ontology and content-oriented approaches to support systems development will take pride of place over prior-art activity and process-oriented approaches. New theories and methodologies, as well as technology, will facilitate the integration of the organization's internal world with its external world. This will definitely and radically influence ESS and DSS design and thinking to the benefit of strategic management and leadership. Computers will then become integrative tools of communication within members of the organization, and they will become more collective in nature than specific. This, in turn, will advance the use of computer applications in strategic decision-making work; those applications require more from management. The tools will enhance the use of data acquisition and knowledge creation in strategic management in a two-way or three-way fashion, that is, top-down, bottom-up, and middle-top-down, as previously referred to in strategic management literature.

To reach these goals and objectives in practice, software technology must be developed so that the executive support and decision support structures are coherent with the real structure of the world. The system architecture must emulate the reality of the organizational and individual behavior and allow the strategy makers to gain a holistic view of the organization's activities, its operating environment, and its external structure. Metaphors, as tools, can assist us in achieving these aims, and ontologies help us to understand more through different perceptions of the same content.

For several years, we have worked to develop ESS and DSS that will enhance actual decision making through visual perception, as well as through textual meanings and meta-knowledge formation. These new support systems are based on the ontologies of various organizational constructs and on the overall conceptual framework the continuous strategy. The basic principles of the continuous strategy ontology are described in the following chapter. With this ontology, we have good opportunities to fix the case research and applications to the dimensions, contents, and concepts of it. The resulting data can help executives in their daily tasks and turn their activities into strategic knowledge services. Through case studies, we want to show how important and meaningful it is for decision makers to process data, information, and knowledge from many viewpoints. Presentation of the data, constructs, and variables in a visual form offers new possibilities to improve decision making in strategic management.

21.3   CONTINUOUS STRATEGY ONTOLOGY

The continuous strategy ontology is a framework for strategic planning processes. The framework is built on metaphorical insights into the "company"—an "organism" seen as part of the living system. The continuous strategy ontology is supported by a chain of construction metaphors: the external world metaphor, the business world metaphor, the company world metaphor, the product world metaphor, and the buyer world metaphor. These represent conceptual models, and they are used to construct a coherent picture of the real world that exists in and around the company.

The metaphors described here are designed to facilitate the holistic understanding of management issues, business interrelationships, and company characteristics, so that they may be better recognized and understood in strategic management. The metaphors enhance the visual perception of decision makers and help them to arrive at a shared strategic vision for the company. As a chain, the metaphors depict a world structure that is coherent with the basic goals of the company, and they describe this structure in terms of three basic components, or cornerstones, of the business world and the companies within it. These cornerstones are capital, work, and people, and they are arranged in each metaphor on three perpendicular axes to form a multidimensional space. The projections formed by the axes vary according to the scale and the content of the metaphor (cf. Vanharanta 1995).

21.3.1   EXTERNAL WORLD METAPHOR

In the largest scale metaphor, which depicts the external environment as an overlying world structure, the dimensions are environment (capital), organized world (work), and people (their organizational role). The projections of these dimensions are ecosystem, infrastructure, and organized knowledge. See Figure 21.1.

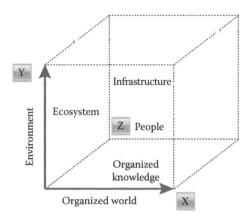

FIGURE 21.1   The external world metaphor.

21.3.2 BUSINESS WORLD METAPHOR

Similar metaphors are used to depict the general framework of the business world. The business world metaphor represents the company's external commerciopolitical environment to which it is exposed. It has the same components, that is, capital, work, and people, as the external world metaphor. See Figure 21.2.

All business activity originates from the combination and interplay of capital, work, and people. These components are very close to those described by Archibald (1973) (resources, labor force, and technical knowledge), Galbraith (1963) (capital, production, and technostructure), and Ohmae's Japanese view (1982) (kane, mono, and hito). The axes form three projections: business assets, business structure, and business knowledge.

21.3.3 COMPANY WORLD METAPHOR

The company world metaphor determines the company's characteristics in the same three dimensions and can be extended to cover the corresponding activities, that is,

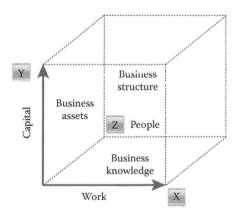

FIGURE 21.2   The business world metaphor.

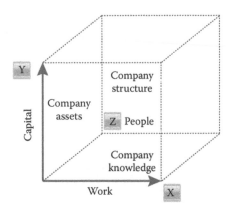

FIGURE 21.3 The company world metaphor.

financing, operations, and management. The three axes form the projections company assets, company structure, and company knowledge. The combined vector of these projections is akin to company performance. The metaphor holistically depicts the living company, its main characteristics, and its performance. See Figure 21.3.

The company world metaphor can be used to create a wide variety of active computer templates for incorporation into human–computer interfaces of company-specific ESS and DSS.

21.3.4 PRODUCT WORLD METAPHOR

The fourth business metaphor is the product world metaphor. The metaphor is designed according to the concept of supply and depicts the framework of supply as a subclimate within the company environment. Supply is crucial to company survival. The product world metaphor is constructed according to the same principles and the same sequential assembly process as the company world metaphor; see Figure 21.4.

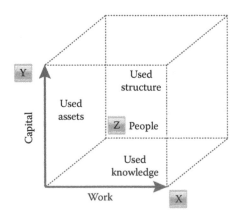

FIGURE 21.4 The product world metaphor.

The three planes of projection characterize the creation of a product or service, a process that consumes a finite quantity of assets (assets used), requires specific activities inside and outside the company (structure used), and is created using a finite amount of knowledge (knowledge used). The metaphor defines the characteristics and performance of a product or service in the market. It is also possible to think through the added-value concept of each dimension. The metaphor is directly convertible into a graphical template for incorporation into human–computer interface of ESS and DSS.

21.3.5 BUYER WORLD METAPHOR

The fifth metaphor is the buyer world metaphor. It is a metaphor for the concept of demand, showing it as a subclimate in the company environment. Demand is just as crucial for company survival as supply. The buyer world metaphor, shown in Figure 21.5, defines buyer characteristics. Again, there are three planes of projection, this time characterizing buyers. The metaphor emphasizes the relationships and interrelationships between the producer, product, and buyer.

Buyers have finite assets and related financial potential (buyer assets), they consume specific products and belong to various buyer groups (buyer structure), and they use a finite amount of knowledge to make their purchasing decisions (buyer knowledge). A template can be developed for incorporation into human–computer interfaces of ESS and DSS from this metaphor.

21.3.6 COMPANY CONTINUUM METAPHOR

These five metaphors can be combined through a sixth metaphor, that is, the company continuum metaphor. In this metaphor, each of the previous metaphors has a specific place and a specific content, which gives us a dynamic picture of the living company and its relationships and interrelationships. The company continuum metaphor is both a static and dynamic representation of a continuous company strategy, in which the company is part of the living system and is represented by capital,

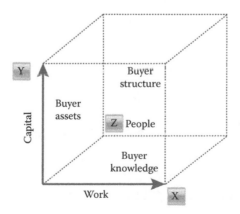

FIGURE 21.5 The buyer world metaphor.

work, and people. Company activities and company characteristics are formed in the company world, in terms of financing (capital), operations (work), and management (people). The formation of the supply concept in the product world is through assets used (capital), structure used (work), and knowledge used (people). The formation of the demand concept in the buyer world is through buyer assets (capital), buyer structure (work), and buyer knowledge (people). The formation of the main components of the business world is through business assets (capital), business structure (work), and business knowledge (people). In the external world, projections are formed through the ecosystem (capital), infrastructure (work), and organized knowledge (people).

The visual and cognitive perceptions, both static and dynamic, show the mutual interdependence of these five metaphors, but they are not easily perceivable holistically and do not show the dynamic dimension of the metaphorical content. Thus, it was necessary to create the company continuum metaphor as a template. See Figure 21.6.

The company continuum template is an integrated working method for use in management. It is a concept map for navigating and combining the dynamic content of the metaphors in the chain, that is, data, information, and knowledge. The template illustrates the interdependence of the five metaphors from macrocosm to microcosm. It shows that all company activities start and are maintained within the freedom and constraints of the supporting frameworks provided by the five-metaphor chain, that is, by the continuous strategy ontology. The template itself is self-explanatory and guides users onward, for example, in the computer context toward other templates and applications.

Managers are responsible for, and should be committed to, development of the company. If they are active users of the metaphor chain and its computer applications, their responsibility and commitment also become an integral part of the supporting framework. In other words, responsibility and commitment are conditional on the freedom and constraints of the metaphor framework.

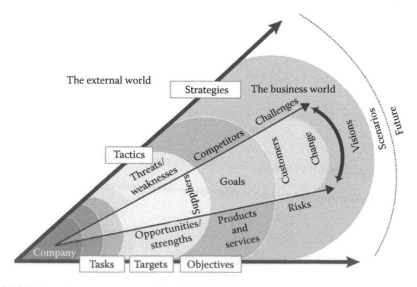

FIGURE 21.6   The company continuum metaphor.

Integration of the top-down overview with the bottom-up market-driven view results in a holistic perception of the company, that is, its business issues, interrelationships, characteristics, its present position, and its progress in the competitive environment. If this visual perception can, on a continuous basis, be instilled into decision makers as a shared vision, it will give them improved competence in strategic management and will diminish the misunderstanding, isolation, and conflict that may otherwise arise. This will result in better strategies and business activities for the survival, progress, and continuity of the company. The following case examples show how knowledge services help managers to create possibilities within strategic decision making.

21.4 FROM DATA TO KNOWLEDGE: A CASE STUDY IN CUSTOMER SEGMENTATION WITH THE SELF-ORGANIZING MAP

In this case study, customer segmentation (Lingras et al. 2005), one of the most important tasks in customer relationship management (CRM) (Datta 1996; Heinrich 2005; Chalmeta 2006), is used to identify and group customers with similar profiles or requirements. The grouping is based on customer information readily available through ERPs, corporate data warehouses, and the Internet (Rygielski et al. 2002; Buttle 2004). Problems arising from the vast amounts and varying quality of the data are solved using data mining methods (Famili et al. 1997; Berson et al. 2000; Shaw et al. 2001; Rygielski et al. 2002; Berry and Linoff 2004).

The aim is to get an overview of the customer base, its demographic attributes, and consumer behavior, with focus on converting large amounts of customer data into actual knowledge of the customer base (cf. Product World Metaphor and Buyer World Metaphor). More specifically, questions such as who buys, how much, what products, how often, and how recently need to be answered. The data are analyzed with the SOM (Kohonen 2001), and a data-driven exploratory customer segmentation (Berson et al. 2000; Holmbom et al. 2011) is carried out. Customers are grouped according to their shopping behavior, and an analysis is conducted based on the demographic information (Frank et al. 1972; Wedel and Kamakura 1999; Tsai and Chiu 2004).

21.4.1 DATA

The study focused on the customers of four major department stores and spanned a period of 2 years, from August 2007 to July 2009. The data consisted of the following:

- Demographic data, such as customer loyalty point class, service level, customer duration, gender, child decile, estimate of income, and age
- Product data, consisting of eight product categories, labeled A to H
- RFM (recency, frequency, monetary, and RFM score) information, calculated with IBM SPSS Modeler

The data were preprocessed in order to give better data mining results. Some limitations were made, for example, customers that made purchases for less than

EUR 100.00 (monetary > EUR 100) or less than two times (frequency > 2) during the 2-year period were excluded. This eliminated about 30% of the customers of the data set.

21.4.2 TRAINING THE MODEL

The data were obtained in the form of text files consisting of transaction data and customer demographic data. IBM SPSS Modeler (http://www.ibm.com/) was used to preprocess and translate the data into the right format, and Viscovery SOMine 5.0 (http://www.viscovery.net/) was used to create a map based on the demographic data. The results of the demographic segmentation were then matched with the sales information for each product category.

The resulting data contained significant outliers, and even though the SOM is fairly tolerant toward noisy or missing data (Bigus 1996; Smith and Gupta 2002), sigmoid (or logistic) transformation of the data (Bishop 1995) was used. Sigmoid transformation reduces the influence of extreme input values and emphasizes center input values (Bishop 1995; Larose 2005), while variance scaling makes the variables comparable. Other input parameters used when training the map were a map size of 1000 nodes, a tension of 0.5, and Gaussian neighborhood function. The clusters on the final map, displayed in Figure 21.7, were identified with the help of Ward's hierarchical clustering method and resulted in seven clusters of various sizes, labeled C1–C7. The color of the cluster in the figure signifies cluster membership and does not imply value.

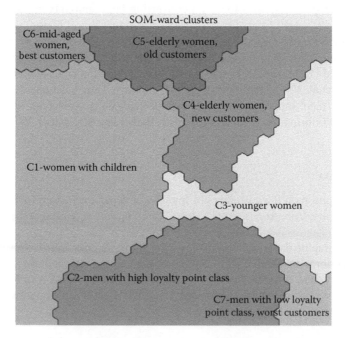

FIGURE 21.7 A clustered SOM model of the segmentation. The segments are labeled according to their characteristic features.

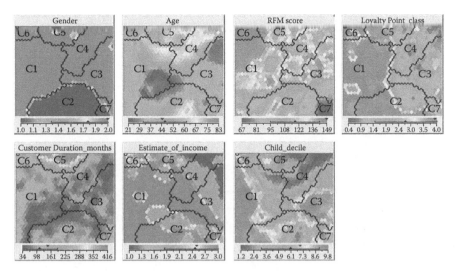

FIGURE 21.8 The feature planes of the main variables (gender, age, RFM score, loyalty point class, customer duration in months, estimate of income, and child decile) that contributed to the map, showing the values by each component.

The feature planes, displayed in Figure 21.8, were used to interpret the characteristics of each cluster. They show the distribution of values across the map according to one variable at a time. The values for each variable are displayed by the color of the neuron, where "warm" colors (red, orange, and yellow) illustrate high values and "cool" colors (blue) illustrate low values. The approximate values are indicated by the scale under each component plane. The map is interpreted by reading the feature planes for each cluster. For example, Cluster C5 displays medium to high values in recency, customer duration, gender, and age and low values for all the product categories. We can conclude that these customers were mainly women of a higher age, who have been customers for a longer time and have visited the department store recently, but that they do not buy much in terms of Euros.

21.4.3 DEMOGRAPHIC SEGMENTATION

Demographic segmentation aimed to find out who buys, what products, how much, and how often, based on customers' demographic information (cf. External World Metaphor). Therefore, the demographic and RFM variables were given equal weight in the training process, but the variables describing the product categories, along with the RFM score, were not given any weight. Thus, the model is entirely based on the demographic attributes of the customers, and purchasing behavior is associated with the demographic segmentation.

Similar results were found for all four department stores. As Figures 21.7 and 21.8 show, segmentation was mainly based on gender, age, and the purchase amount. Figure 21.9 shows closer analysis of the clusters.

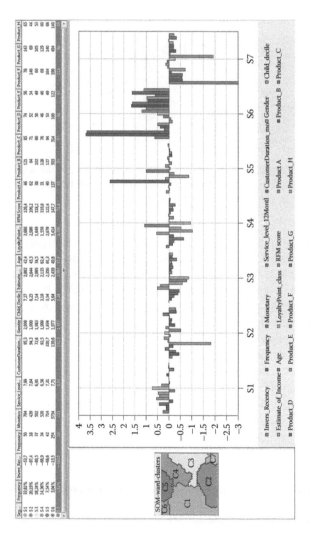

FIGURE 21.9 The statistical analysis of the created clusters. The clusters are displayed separately at the top of the figure, including their size (%) and an average of each of the variables. The bars in the diagram show the deviation from the average (in standard deviation) for each of the variables. The explanation for each bar (from left to right) is found underneath the diagram, showing first the RFM variables, then the demographic variables, and last the products.

With the help of the clustering, the feature planes, and statistical analysis of the created clusters (Figures 21.7 through 21.9), we obtain an overview of the customer base and are able to answer the strategic questions:

- *Who buys?* The purchases are mainly made by customers from two of the clusters: elderly women (average 60 years) in Cluster 5 (7.54%), with a low loyalty point class, service level, and income and who have been customers for a longer time, and middle-aged women (average 45–50 years) in Cluster 6 (3.04%), with a high loyalty point class, service level, and income. They visit the store often and make a significant amount of the total purchases.
- *How much?* The middle-aged women in Cluster 6 buy the most. They spent an average EUR 3750 within the 2 year period. Men in Cluster 7 (2.82%), who do not visit the store often and have a low loyalty point class and service level, are the ones who buy the least—they spent an average EUR 220. Customers from the other clusters purchased between EUR 440 and EUR 800 in the period.
- *Which products?* Women with children in Cluster 1 (33.8%), who have a high loyalty point class, service level, and estimate of income, buy products mainly in Category D. Men with a high loyalty point class, service level, and income in Cluster 2 (20.1%) buy products mainly in Categories A and F. Elderly women with a low loyalty point class, service level, and income in Cluster 4 (14.3%) have not been customers for long. They largely purchase products in Category B. The elderly women in Cluster 5 buy products mainly in Categories B, C, and H. The middle-aged women in Cluster 6 buy from every product category. The younger women (average 30 years) in Cluster 3 (18.2%), with a low income, loyalty point class, and service level, and men in Cluster 7 do not buy from any specific product category.
- *How recently and how often?* The middle-aged women in Cluster 6 have visited the store recently, and they visit the store often. The men with a low loyalty point class in Cluster 7 do not visit the store often and have not done so recently. Customers from the other clusters fall in between these two extremes.

21.4.4 CASE DISCUSSION

Customer data from a chain of department stores were successfully analyzed with the SOM to obtain a demographic and behavioral overview of the customer base. Customers were grouped according to their shopping behavior through data-driven exploratory customer segmentation and analyzed based on the demographic information, resulting in a map consisting of seven clusters.

21.5 FROM INFORMATION TO KNOWLEDGE: A CASE STUDY OF COLLOCATIONAL NETWORKS

This case study demonstrates how DSS can assist in creating strategic knowledge from information. Telecommunications companies' annual reports were visualized using the text visualization method collocational topic networks and then shown to interviewees who follow the industry closely. The interviews show that the visualizations were able to capture what the interviewees considered to be actual developments in the telecommunications industry. Furthermore, the visualizations were found to provide new insights into this market.

21.5.1 Background on Competitive Intelligence

Competitive intelligence, sometimes also known as business intelligence or market intelligence, has been defined by the Society of Competitive Intelligence Professionals (SCIP) as "a systematic and ethical programme for gathering, analyzing, and managing any combination of data, information, and knowledge concerning the business environment in which a company operates that, when acted upon, will confer a significant competitive advantage or enable sound decisions to be made" (SCIP, 2011). Competitive intelligence practice owes much to Porter's (1980) work on competitive strategy, which introduces five forces that shape a company's business environment: current competitors, the threat of substitute products, the threat of possible new entrants to the market, the bargaining power of suppliers, and the bargaining power of customers. It is essential that companies know these forces and how they are changing as best as they can; however, some issues make this very difficult.

First, there is information overload. Companies are overwhelmed by the amount of data that is publicly available about their competitors, customers, and suppliers. They need tools to be able to distinguish the relevant from the irrelevant.

Second, competitive analysis in strategic management often relies heavily on quantitative financial data. Fleisher and Bensoussan (2003) call this phenomenon "ratio blinders." According to them, many organizations make the mistake of relying too much on the financial data of their business environment. This can lead to a situation where companies see a financial gap between their organization and a competitor but cannot see the reasons behind it, and thus do not have the means to close it. This issue relates to a commonly made mistake—the confusion of operational data with strategic data, as mentioned by Zahra and Chaples (1993). To deepen analysis based on operational financial data, there is a demand for methods that allow competitive intelligence practitioners to systematically include qualitative industry data in their analysis. Such data are available, for example, in public texts produced by competitors or other companies in the industry.

In order to incorporate qualitative data effectively into the analysis, a systematic methodology for analyzing large quantities of text is needed. In today's complex business environment, with a high number of companies that need to be tracked and a constantly increasing amount of texts being produced, a manual scan is ineffective. Text mining and text visualization are possible solutions to this issue, and they will be discussed in more detail next.

21.5.2 Text Mining and Text Visualization

Text mining or text data mining, defined as the discovery of trends and patterns within textual data (Hearst, 1999), promises to alleviate the task of processing texts produced in a company's business environment.

Fan et al. (2006) provide a brief overview of text mining methods intended for business users that are available as commercial software packages. However, there is still a long way for these methods to become everyday tools in companies. Krier and Zacca (2002) point out that although various text mining methods have been available for some time within the competitive intelligence community, their use by intelligence practitioners has been limited because they are perceived as "black box" methods, that is, the workings of these methods have been difficult to understand by users who do not have a background in linguistics.

This case study addresses these issues by presenting a text mining method that produces a visual network consisting of words as its outcome. Text visualization is something that allows users to interpret the results in a more intuitive way. However, when working with any intelligence tools, the final interpretation of the results and their significance is always a qualitative process that cannot be automated. Qualitative interviews were therefore conducted in order to evaluate the collocational topic networks as a text visualization method.

21.5.3 Evaluation of Collocational Topic Networks as a Text Visualization Method

Temporal text mining is a term that covers the methods used to discover temporal patterns in texts collected over time (Mei and Zhai, 2005). Temporal text mining suits the purposes of strategic competitive analysis well, because an essential part of competitive intelligence analysis lies in understanding not only competitors' or partners' current strategic decisions, but also their development over time. This creates a starting point for predicting their future decisions.

In this case study, various public companies' annual reports were used as visualization material. Annual reports are ideal material for temporal competitive analysis, as they reflect changes in a company's strategy and competitive situation over time. They are also particularly suitable for the task from a text visualization point of view, as the document structure is usually similar (passages, length) from 1 year to the next.

For the purposes of this case study, six visualizations from 2003 to 2008 were created from the annual reports of seven large telecommunications companies (see Table 21.1).

TABLE 21.1
Texts Included in the Visualization

Company	Country of Origin	Type of Document	Years Included
AT&T	United States	Annual reports	2003–2008
BT	United Kingdom	Annual reports	2004–2009
France Télécom	France	Annual reports	2003–2008
Telecom Italia	Italy	Annual reports	2003–2008
Telenor	Norway	Annual reports	2003–2008
TeliaSonera	Sweden	Annual reports	2003–2008
Verizon	United States	Annual reports	2003–2008

A text file was created for each year included in the visualization by merging the seven annual reports published by the seven companies that year. A similar visualization could be made out of the reports of a single company, but it was decided that for the purposes of this evaluation, a general view of the industry (cf. Business World Metaphor) would be produced based on all company reports, as this would be something that all the interviewees would have an opinion on and would be easier to discuss than developments at a single company.

21.5.3.1   Visualization Method

The visualization method applied in this study, *collocational topic networks*, has its roots in lexicographical research (Williams, 1998) and, more generally, text linguistics (Phillips, 1985). A collocational topic network is a network containing a user-defined topic word as its central node and a user-defined number of links to words that occur in a statistically significant way in the vicinity of the topic word in the text that is visualized. The method is based on the assumption that changes in the networks will reflect changes in how the topic is discussed in the underlying texts.

A collocation is defined as "the occurrence of two or more words within a short space of each other in a text" by Sinclair (1991). Significant collocation takes place when two or more words occur together more frequently than would be expected by coincidence. The significance of collocation is measured using the mutual information or MI score, which compares the frequency of two words with the frequency of their occurrence independent of each other; it is widely used in linguistics when dealing with text masses.

The initial stage of producing the networks consisted of calculating the MI score for all words occurring within a span of four words. A maximum span of this size has been recommended by Sinclair (1991) for studying collocations in English. So-called stop words with little semantic content were left out, such as prepositions, articles, conjunctions, and words referring to figures and currency. Because the case company provided value-added services to telecommunications companies, the word "service" was selected to be the central node in the networks as the context in which the companies discussed the concept of service that was deemed to be of interest to the interviewees. This process produced the six topic networks depicted in Figure 21.10.

Theoretically, the size of the networks is limited only by the contents of the text. It would be possible to make an extensive network where each appearing word is linked further to its own most significant collocations. However, for the purposes of this case study, simple networks containing only the six main collocates of the word "service" were produced. No further links from these six words were drawn. The networks were then shown to interviewees. The eight interviewees represented the company's management team, product management team, and the sales department. All of the interviewees had been working within the technology industry for over 10 years. The interviews were semistructured, allowing for a conversation between interviewer and interviewee and for the introduction of new topics by the interviewee.

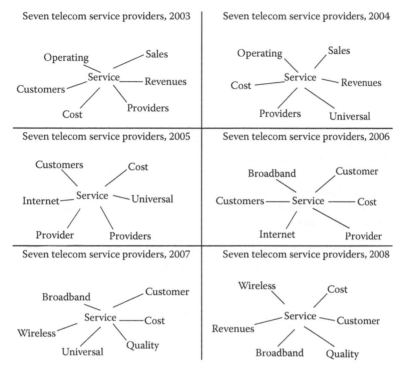

FIGURE 21.10 The topic networks for the word "service" in seven telecommunications companies' annual reports 2003–2008.

21.5.3.2 Interpretation of Visualizations

The interviewees were shown the list of companies included in the study (Table 21.1) and the collocational topic networks in Figure 21.10. The basic workings of the visualization were explained to them in the following way: "These networks are based on annual reports by all seven telecommunications service providers, year by year. The six words around the word 'service' occur in these reports in a statistically significant way together with 'service.'" The interviewees where then asked to look at the networks and discuss any thoughts that emerged. Further clarification of the method was provided if requested.

After studying the networks, most interviewees mentioned at least two of the following observations:

- The word "cost" appears in connection with service in every network. Interviewees believed this reflected telecommunications companies' struggles with cost reduction throughout the period.
- The word "revenues" appears in the first 2 years then disappears and reappears in 2008. Several interviewees noticed this, but did not quite know how to interpret it. During the period, revenues from new services had been rising, while traditional service revenues had declined.

- The earliest networks do not mention any specific service types, but in 2005, "Internet" occurs for the first time. In 2006, the more specific "broadband" appears, and in 2007, "wireless" appears. The interviewees found this interesting and believed it showed that the companies have become more elaborate about the different types of service they provide and now discuss the services in more detail.
- The word "universal" is present in 2004, 2005, and 2007. The interviewees believed this was because most of the included companies are incumbents in their home country and so they may be required by law to provide a universal service, that is, a service that reaches all residents of that country.
- The word "quality" appears in 2007 and remains in the network in 2008. The interviewees saw this as an indication that fierce price competition has driven telecommunications companies to emphasize service quality in their external communication.

Generally, the interviewees considered the networks to reflect actual developments in the telecommunications service industry during 2003–2008. It should be noted that this method allows users to go to the original source texts, if needed, to look for explanations for the occurrence of unusual, ambiguous, or otherwise particularly interesting words in the networks. In that sense, this is not a "black box" method, and user interpretations of changes in the networks can usually find support (or be disconfirmed) through the original texts.

21.5.3.3   Interview Themes on Text Visualization

During the interviews, three recurring themes on the subject of text visualization began to emerge from the conversations. The themes were as follows:

1. Visualization of qualitative versus quantitative data. All interviewees said that they work with both qualitative and quantitative data. They considered qualitative data to include both written and verbal nonnumeric data. The interviewees said that they were familiar with, and frequently worked with, quantitative data models such as pie charts or bar charts. In contrast to this, the use of text visualization and text mining methods in a corporate context is still rare. None of the interviewees had encountered such methods during their working career.
2. Uses of text visualization. The interviewees were asked to spontaneously present situations where they would consider using such visualizations. All interviewees quickly responded with various scenarios where they could see the topic networks to be useful. Such scenarios include an analysis of competitors' marketing materials, including differences between competitors and changes within the industry over time.
3. Visualization for analysis versus visualization for presentation. The interviewees' responses to the networks and their suggestions for usage scenarios show that they see the visualizations as accurate representations of the underlying texts. Furthermore, they see them as tools that could be used to make users aware of issues that would otherwise go undetected. In some cases, they could also be used in presentations as a basis for discussion.

21.5.4 Case Discussion

In conclusion, all interviewees considered the collocational topic networks that they were shown to reflect actual developments in the telecommunications service industry during 2003–2008. Furthermore, all interviewees found the visualizations to provide interesting insights and thought of several possible uses for the method within the domain of competitive intelligence.

The collocational topic network method presented here could easily be used as a tool for assisting competitive intelligence analysis in situations where textual data from an organization's competitive environment are needed to balance financial data. The tool could produce networks out of several user-specified topics and highlight changes according to the user's wishes. Building on Porter's (1980) theory of five basic competitive forces, the tool could be used not only for competitor analysis but also to keep an eye on strategic changes in potential entrants to the market or vendors of substitute products. It would also be useful for following developments at suppliers and customers, particularly in a business-to-business context.

Topic networks could, in some cases, also be shown to an audience, internal or external, but the need for making this type of presentation has to be carefully considered; the unfamiliarity of the visualization method to most viewers means that the visualization itself might get more attention than the argument that the presenter is trying to make using it.

21.6 FROM KNOWLEDGE TO KNOWLEDGE: FROM INNER PERCEPTIONS TO STRONG SALES CULTURE USING ONTOLOGY

The medical technology company in this case study has experienced a strong expansion of its sales force outside the domestic markets. Sales personnel sell the same products in different market areas, but the cultural background of the customer base varies greatly. Single salespersons have remote locations in different continents without close connection to the company's technical staff, and, therefore, multicultural understanding and cooperation is needed at the headquarters. Cultural understanding is also required to understand customers' needs in the correct way based on the globally incoming information.

The objectives in this case were, first, to define the culture in the industrial environment and, second, to build an ontology of sales culture and its relation to organizational culture in order to understand underlying processes within sales culture. The third objective was to use the new ontology to study and analyze the export sales force in the company to find out which areas of the sales culture needed to be developed. The fourth objective was to see how well the results match the literature on the most important organizational competencies in creating a strong sales culture. Finally, after bridging the results and the literature, a prioritized list of actions was suggested for strategic decision making based on the individuals' collective inner perceptions.

21.6.1 SALES CULTURE ONTOLOGY

As sales culture is one aspect of the company culture, each culture affects the other. The company culture comprises the company's vision, company values, and company strategy. Though the sales culture has its own strategy, it naturally must be in concordance with the company strategy. Company culture is affected by the environment, stakeholders, and customer feedback. Sales culture adapts the relevant elements of these factors via vision, values, and strategy from the company culture. Sales culture consists of five main components: sales culture support, sales force, sales system, sales process, and sales culture assumptions.

Sales culture support consists of three concepts. The first contains suitable elements of company strategy, vision, and values. The second is the ethics of sales, and this probably has the most important role in the company because sales processes vary so much depending on the customers and market areas. Also, the opportunity to act in an unethical way is biggest during sales and marketing activities, for example, selling unsuitable products or selling for personal benefit. The third and the most important subconcept is leadership, which is to build, change, and strengthen the culture. Leadership can further be divided into leadership, management, and time management, but to keep this ontology lean, the term represents all three aspects in this study.

The *sales force* concept establishes the paths through which a single sales activity takes form. These paths are represented by dotted lines in Figure 21.11. Some of the elements can be regarded as tools of leadership and are affected by leaders, but some will be developed over a longer time period and cannot be controlled directly. Elements along a path constitute the logical chain that is behind a single action, for example, training and development improves salespersons' skills and thus improves their understanding of selected actions. Following the elements on a path, leaders and managers can focus their development efforts on the correct issues.

Sales system defines the objects of sales activities and the market areas in which these activities are performed. The general sales strategy is set within the sales system subconcept and defines the company's strategic position in the marketplace. When leaders/top managers communicate the company strategy to the whole organization, the sales system is the detailed projection of the strategy from the sales point of view. Sales system also includes the prescription of functions and resources outside the sales organization that are needed to support sales processes. For example, the more technically advanced the products the company is selling, the more technical expertise is needed from other functions. The attitude toward sales and how well sales are integrated in the other functions are essential to the organization's success. If sales remain an isolated function, the company may lose many opportunities to improve its customer service; in customer meetings, it might be hard to convince a customer without the presence of experts, but, on the other hand, if salespersons are ignored, much valuable information from the marketplace might not reach the technical staff at all.

The *sales process* concept comprises the theoretical context of the sales procedure and describes all the processes that are needed to realize sales efforts in the company. The quantity of processes varies depending on the company's business.

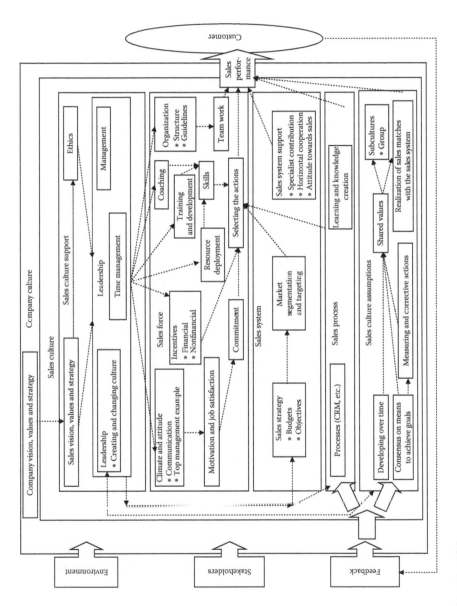

FIGURE 21.11 The SCO.

The simplest sales process is for agencies, which have only a few major customers in the domestic market. The most complicated processes are for multinational companies, which have their own sales offices and agencies serving customers worldwide. A vital part of the sales process is continuous learning and knowledge creation. If training—which is included in the sales force concept—is more focused on individual salespersons, improving their personal sales competencies, learning, and knowledge creation includes processing sales-related innovations and processes, where tacit knowledge is converted to explicit knowledge. Continuous learning and knowledge creation should be a controlled process to ensure proper results.

Sales culture assumptions reflect the underlying assumptions that guide salespersons' professional decisions and daily actions. Most of these assumptions are not conscious, but are behind decisions. These assumptions develop over time, and if no assumptions can be found, it could be that there is no culture or the culture is weak. Furthermore, in the case of young companies, certain elements of the culture may have not yet evolved. Leaders have a very strong effect on the progression of the sales culture. Typically, the organization follows the example of the leader (often the leader is also the founder): what the leader pays attention to, how the leader reacts to certain situations, etc. By repeating certain patterns of behavior, leaders build the underlying assumptions for the organization. When the organization as a whole can agree on the objectives and the applicable methods to reach those objectives, the culture starts to form and strengthen. One sign of a more advanced culture is the consensus on measuring results and especially on the actions needed if the objectives are not reached. Though the leaders can develop and change the culture, the culture exists among the individuals in the organization. Therefore, the culture affects how well the sales strategy is realized in operations, the sales systems, and sales processes on a daily basis. The difference between defined and actual behavior reflects the level of culture. It is critical that the group shares values that are important to it, and from the company point of view, those values should be in conjunction with the company values. When the organization grows larger, it may have several groups with different value bases. This is acceptable as long as the most important values are the same for each group and a single group's values do not contradict those of the company. The sales culture can be seen through customers' eyes as the performance of sales force. A theoretical "best way" of sales processes exists, but how the underlying assumptions are affecting in practice, that is, whether the written method is actually followed or not, is at the core of the sales culture. It is easy to have an ideal way of handling a sales case and other customer management actions on paper, but in reality, how well each salesperson operates according to this way is crucial. As shown in Figure 21.11, several elements influence sales performance, as illustrated by the dotted lines leading up to sales performance. The customer's positive or negative feedback on the sales performance is mainly handled in the sales process concept. Customer feedback also affects sales culture assumptions because there is a two-way relationship between sales culture assumptions and leadership. Customer feedback is also handled by leaders because they are able to change the culture, and feedback-based changes can often be tracked to leadership. In the next section, we explain how the sales culture ontology (SCO) was used to analyze the case company's sales force.

21.6.2 EVOLUTE SYSTEM

The evolute system supports the use of fuzzy logic (Zadeh 1965, 1973) applications on the Internet (Kantola et al. 2006; Kantola 2009). The evolute system "hosts" domain ontologies and presents the ontologies online to target groups through semantic entities, such as statements. Different classes (concepts) in the SCO are described in familiar language in the statements. A few indicative statements relate to each class in the SCO. One statement can relate to more than one concept, and each statement has a weight. Statements are compared to linguistic labels on a fuzzy scale, which means that the meaning of the statement in the individual's mind can be captured and a conversion from the meaning to the crisp numerical value during the evaluation process is eliminated. When the statements are evaluated, inputs are converted into fuzzy sets (fuzzification). The inference engine evaluates the fuzzified inputs using rules in the rule base. This results in one fuzzy set for each SCO class (inferencing). Fuzzy sets are then converted into crisp class values and, furthermore, into visual reports. The evolute system works as a generic fuzzy rule base system, as described in Figure 21.12.

Participants evaluate the current reality and the future vision of their surroundings according to the statements in the evolute system. The difference between the self-perceived current reality and the vision is called creative tension (Senge 1994). In the case of the SCO (external object), the difference between the perceived current reality and future vision can be called the proactive vision. Evolute provides SCO-based "answers" or instances (cf. Kantola 2009). The collection of instances forms the instance matrix (cf. Kantola 2009):

$$\text{ONTOLOGY}_{\text{SCO}} \ (\text{individuals}_{1-n}, \ \text{instance})$$

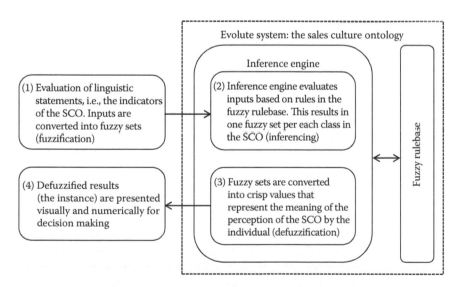

FIGURE 21.12 The SCO is evaluated through indirect statements.

The instance matrix describes the perceived state of the object under scrutiny in the organization. The instance matrix, as a function of time, can be stated as follows:

$$ONTOLOGY_{SCO} \text{ (individuals}_{1-n}, \text{ instance}_{1-k})$$

The instance matrix represents the collective perception/collective mind of stakeholders regarding the SCO in the company. In this case, the goal is to capture a true bottom-up view of the current reality and envisioned future of the features and practices of the company's sales culture.

The SCO and its propositions in the evolute knowledge base can be fine-tuned by adjusting the fuzzy sets and fuzzy rules as more is learned about the sales culture domain. The content of the SCO will develop over time as the domain naturally evolves and as researchers learn more about it (Gomez-Perez 2004).

21.6.3 Dataset

From total 19 inquiries, a completed evaluation was received from 14 participants, so the answering ratio was 74%. After all the employees completed the inquiry, the results were summarized so that no individual's answers could be seen. The results, therefore, show the collective status in the whole company. The results are shown in graphical format.

21.6.4 Results

In Figure 21.13, the results are sorted top-down according to the biggest proactive vision (vision–current). The bigger the proactive vision, the more employees want to improve the concept. Development efforts focusing on the concepts with the biggest proactive vision can be expected to give the best results due to the high internal demand of employees. The biggest proactive vision in this case company was for "selecting the actions," the second was "development over time," and the third "strategy." At the other end, there was practically no proactive vision for "ethics," which means that employees do not see any need/place for improvements in ethics-related issues. Therefore, efforts to improve ethics in the company would probably not be successful. We concentrate on the three concepts with the biggest proactive vision and more closely examine the indicative statements behind the results.

21.6.4.1 Selecting the Actions

There are two statements behind this concept: (1) *In our company, the rewarding of certain behavior is logical*, and (2) *there are enough sales meetings to give guidelines for sales operations*. Employees answered on average that there is the biggest improvement potential according to these two indicators, in other words that rewarding behavior is not logical and there are not enough sales meetings.

Evolute/Kappa/Kappa_ALL/Features/Creative tension

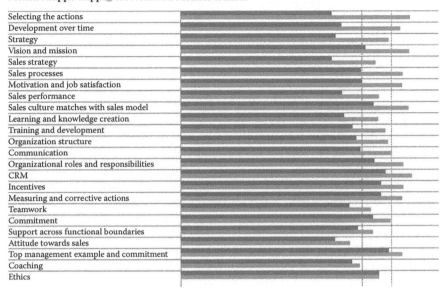

FIGURE 21.13 The concepts in the SCO sorted by biggest proactive vision.

21.6.4.2 Development over Time

The statements behind this concept are as follows: (1) *new members in sales team adapt to the team habits*, (2) *I know how my colleagues/team members will act when the customer gives negative feedback*. Statements in this class describe the level of culture in the purest form, that is, how well a group has learned how to react in challenging situations. If there is large proactive vision in this concept, it means that the culture has not been developed. There are mainly two reasons for this: either the organization is so young that the culture has not yet been developed or the leaders have not been able to establish the culture, for example, through own example and being logical in operations.

21.6.4.3 Strategy

In this concept, the indicative statements are the following: (1) *our company offerings are based on global market demand*, (2) *our company offerings are based on local market demand*, (3) our company is customer/product oriented, (4) our company is service /technology oriented, (5) *our company recognizes changes in the environment*, and (6) *the relationship between the marketing strategy and sales strategy is defined in the company strategy*. Interpretation of these results shows how well the case company has been able to develop its offerings to meet customers' requirements. According to sales personnel, there is clearly room for improvement. Also, the sales personnel consider the attitude to be a typical engineering attitude, that is, technology is the driver instead of the customer and service. However, this is an area that will probably never be—from the sales personnel's point of view—at a satisfactory level.

The company's capability to recognize changes in the environment is vital in order to adapt to changes in the strategy and meet customer requirements in the future. If no changes are recognized, the company keeps its old methods and offerings, and inevitable disaster follows. The last statement refers to the roles of sales and marketing and especially to the relationship between them, which is often poorly defined. The literature frequently shows sales under marketing, even though the sales personnel can be significantly larger. However, the distinction in roles between these two functions should be clear in order to avoid overlapping tasks or even neglecting them.

To find out the level of the case company's sales culture, one option is to compare the results to the literature regarding the creation of a strong culture. The literature gives the following as the main components when creating a strong culture: (1) the leader's role: leaders show their own example and operate logically (Oedewald and Reiman 2003; Benson and Rutigliano (2004)); (2) the organization acts as defined in process descriptions (Oedewald and Reiman 2003; Holland and Laine 2008); (3) personnel is committed to work and company (Oedewald and Reiman 2003); (4) the culture develops over time, that is, groups have had enough common experiences (Schein 1992; Leppänen 2006); and (5) the sales strategy is clear to everyone (Oedewald and Reiman 2003). Elements found from literature are not exactly matching with the concepts in the SCO, but still there are clear similarities.

21.6.5 RECOMMENDED ACTIONS

For culture development, it is important to have repeated patterns of behavior for certain situations. Typically, these patterns evolve over time through the selection of methods that will solve problems. If top management does not support the evolution taking place within a group by operating logically, it weakens the culture.

The consensus on illogical and even unfair rewarding is obvious among the sales force. Recommended action for the management is to clarify what has been the basis for rewarding, analyze it, and set new guidelines for rewarding. The final step is to inform the whole sales force of these guidelines and have top management commit to them. The literature supports this; the leader's role is very important. Rewarding must be seen on a wide base; it does not only include incentives but also promotions, getting certain sales territories, getting customers or customer groups, being allowed to operate with larger authorizations, or public rewarding. For example, in a big sales case, a top management takeover would lead to an unsatisfactory situation from the salesperson's point of view—the top management's role should be more supportive in order to improve the salesperson's knowledge.

Culture development takes place in groups and therefore needs forums and face-to-face meetings. For this purpose, sales meetings are a good and natural option. Also information can be distributed in this kind of meetings. Although it would be easy to propose more sales meetings for the whole sales force, there would be limitations due to expenses and time usage. Rather the way to improve this issue is to develop the quality of the meetings, use modern negotiation media (net meetings) instead of traveling, and increase the amount of local meetings. Support from the literature can be found here, too: culture develops over time, that is, when groups have had enough common experiences.

Culturally, sales meetings foster the development of the subculture. When the company grows and operations become global, meetings that the whole company can join are impossible to arrange. Subcultures are acceptable as long as they share the main elements of the company culture.

21.7 DISCUSSION AND CONCLUSIONS

In strategic management, a lot of data, information, and knowledge are needed to make a sound strategy for the organization. This strategy work seems to require knowledge services, but the supply side still tries to find out new ways and better tools to serve the demanding requirements coming from turbulent business environments. Strategic management also has its problems, because many strategy frameworks presented in the literature are close to each other, but the contents inside these frameworks vary and are not simple enough for daily management.

In this chapter, we show that by using an ontology-based strategic management framework, we can reach a model that can easily be connected to the modern requirements of strategic management. Novel tools and methods can offer new types of knowledge to help managers make strategic choices when they implement their overall company strategies in turbulent business markets. We have shown in detailed case examples how it is possible to perform customer segmentation based on business data and carry out in-depth analysis with feature planes concerning the main variables of the segmentation, as well as how we can create new, relevant knowledge of competitors with text visualization methods. In the third case example, the ontology-based method and fuzzy logic tools were used to visualize current demand to improve the sales culture and show this kind of internal, bottom-up strategic information to top management.

All the knowledge from the case studies can easily be adapted to the general strategy framework presented at the beginning of this chapter. The connection to the continuous strategy ontology gives managers a strong basis to demand more situation-dependent knowledge to help their strategic decision making. We have shown that visual images contain a lot of information and are easy to understand and perceive. We believe these kinds of new knowledge services are strongly needed by strategic management and will change future ideas of strategy making and implementation.

ACKNOWLEDGMENTS

The author thanks the case organization for its participation in the study. The author also gratefully acknowledges the financial support of the National Agency of Technology (Titan, grant. no 33/31/08).

REFERENCES

Archibald, G. C. (1973). *The Theory of the Firm*. Great Britain, U.K.: Penguin Education; Richard Clay Ltd., pp. 9–10.
Benson, S. and Rutigliano, T. (2004). Discover your sales strengths, *Gallup Series*, Random House, 256p.

Berry, M. J. A. and Linoff, G. S. (2004). *Data Mining Techniques: For Marketing, Sales, and Customer Relationship Management*. Indianapolis, IN: Wiley Publishing Inc.

Berson, A., Smith, S., and Thearling, K. (2000). *Building Data Mining Applications for CRM*. New York: McGraw-Hill Companies Inc.

Bigus, J. P. (1996). *Data Mining with Neural Networks: Solving Business Problems from Application Development to Decision Support*. New York: The McGraw-Hill Companies Inc.

Bishop, C. M. (1995). *Neural Networks for Pattern Recognition*. Avon, Oxford, U.K.: Oxford University Press.

Buttle, F. (2004). *Customer Relationship Management Concepts and Tools*. Oxford, U.K.: Butterworth-Heinemann.

Chalmeta, R. (2006). Methodology for customer relationship management, *The Journal of Systems and Software* (79:7), 1015–1024.

Datta, Y. (1996). Market segmentation: An integrated framework, *Long Range Planning* (29:6), 797–811.

Famili, A., Shen, W.-M., Weber, R., and Simoudis, E. (1997). Data preprocessing and intelligent data analysis, *Intelligent Data Analysis* (1:1), 3–23.

Fan, W., Wallace, L., Rich, S., and Zhang, Z. (2006). Tapping the power of text mining, *Communications of the ACM* (49:9), 76–82.

Fleisher, C. S. and Bensoussan, B. (2003). Why is analysis performed so poorly and what can be done to improve it? In *Controversies in Competitive Intelligence: The Enduring Issues* (Fleisher, C.S. and Blenkhorn, D.L., Eds.). Westport, CT: Praeger Publishers, pp. 110–122.

Frank, R. E., Massy, W. F., and Wind, Y. (1972). *Market Segmentation*. Englewood Cliffs, NJ: Prentice-hall Inc.

Galbraith, J. K. (1963). *The Industrial State*. New York: The American Library Inc., 418pp.

Gomez-Perez, A. (2004). Ontology evaluation. In *Handbook on Ontologies* (S. Staab and R. Studer, Eds.). Springer, Berlin, pp. 251–273.

Hearst, M. (1999). Untangling text data mining, In *Proceedings of ACL'99: the 37th Annual Meeting of the Association for Computational Linguistics*, University of Maryland, June 20–26, 1999 (invited paper).

Heinrich, B. (2005). Transforming strategic goals of CRM into process goals and activities, *Business Process Management Journal* (11:6), 709–723.

Holland, J. and Laine, P. (2008). *Improvisointi Kuriin Myynnin Johtaja* (Stop Improvising, Sales Manager), Fakta. Talentum, Helsinki, Finland, December 2008.

Holmbom, A. H., Eklund, T., and Back, B. (2011). Customer portfolio analysis using the SOM, *International Journal of Business Information Systems* (8:4), 396–412.

Kantola, J. (2009). Ontology-based resource management, *Human Factors and Ergonomics in Manufacturing & Service Industries* (19:6), 515–527.

Kantola, J., Vanharanta, H., and Karwowski, W. (2006).The evolute system: A co-evolutionary human resource development methodology, *International Encyclopedia of Ergonomics and Human Factors*, 2nd edn. Boca Raton, FL: CRC.

Kohonen, T. (2001). *Self-Organizing Maps*. Berlin, Germany: Springer-Verlag.

Krier, M. and Zacca, F. (2002). Automatic categorisation applications at the European patent office. *World Patent Information* (24:3), 187–196.

Larose, D. T. (2005). *Discovering Knowledge in Data. An Introduction to Data Mining*. Hoboken, NJ: John Wiley & Sons Inc.

Leppänen, J. (2006). *Yritysturvallisuus Käytännössä*. (Company Security in Practice). Talentum, Helsinki, Finland, 403p.

Lingras, P., Hogo, M., Snorek, M., and West, C. (2005). Temporal analysis of clusters of supermarket customers: Conventional versus interval set approach, *Information Sciences* (172:1–2), 215–240.

Mei, Q. and Zhai, C. (2005). Discovering evolutionary theme patterns from text: An exploration of temporal text mining. In *Proceedings of the eleventh ACM SIGKDD international conference on knowledge discovery in data mining*. ACM, p. 198.

Oedewald, P. and Reiman, T. (2003). Core task modeling in cultural assessment: A case study
in nuclear power plant maintenance. *Cognition, Technology and Work* (5:4), 283–293.

Ohmae, K. (1982). *The Mind of the Strategist*. New York: McGraw-Hill, Inc., 283p.

Oxford English Dictionary. (2011). www.oed.com, Oxford University Press.

Phillips, M. (1985). *Aspects of Text Structure: An Investigation of the Lexical Organisation of
Text*. Amsterdam, the Netherlands: North-Holland.

Porter, M. E. (1980). *Competitive Strategy: Techniques for Analyzing Industries and
Competitors*. New York: The Free Press.

Rygielski, C., Wang, J.-C., and Yen, D. C. (2002). Data mining techniques for customer relationship management, *Technology in Society* (24:4), 483–502.

Schein, E. H. (1992). *Organizational Culture and Leadership*, 2nd edn. San Francisco, CA:
Jossey-Bass management series.

SCIP (2011). Glossary of terms used in competitive intelligence and knowledge management.
http://scip.cms-plus.com/files/Resources/Prior%20Intelligence%20Glossary
%2009Oct.pdf (accessed 11/08/2011).

Senge, P., The Fifth Discipline: The Art & Practice of the Learning Organization, Doubleday
Business; 1st edition (October 1, 1994), ISBN-10: 0385260954, ISBN-13: 978-
0385260954, 424p.

Shaw, M. J., Subramaniam, C., Tan, G. W., and Welge, M. E. (2001). Knowledge management
and data mining for marketing, *Decision Support Systems* (31:1), 127–137.

Sinclair, J. (1991). *Corpus Concordance and Collocation*. Oxford, U.K.: Oxford University Press.

Smith, K. and Gupta, J. (2002). *Neural Networks in Business*. Hershey, PA: IDEA Group
Publishing.

Tsai, C.-Y. and Chiu, C.-C. (2004). A purchase-based market segmentation methodology,
Expert Systems with Applications (27:2), August, 265–276.

Vanharanta, H. (1995). Hyperknowledge and continuous strategy in executive support systems. *Acta Academiae Aboensis*. PhD thesis, Ser. B, Vol. 55, No. 1. Turku, Finland.

Wedel, M. and Kamakura, W. (1999). *Market Segmentation Conceptual and Methodological
Foundations*. Boston, MA: Kluwer Academic Publishers.

Williams, G. (1998). Collocational networks: Interlocking patterns of lexis in a corpus of plant
biology research articles. *International Journal of Corpus Linguistics* (3:1), 151–171.

Zadeh, L. (1965). Fuzzy sets, *Information and Control* (8:3), 338–353.

Zadeh, L. (1973). Outline of a new approach to the analysis of complex systems and decision
processes, *IEEE Transactions on Systems, Man, and Cybernetics* (1:1), 28–44.

Zahra, S. A. and Chaples, S. S. (1993). Blind spots in competitive analysis. *The Academy of
Management Executive* (7:2), 7–28.

Index